H. T. Fortune

MATHEMATICAL AND THEORETICAL PHYSICS

Volume I

PAINTING BY VILLE AARSETH OSLO 1967, NORWAY

MATHEMATICAL AND THEORETICAL PHYSICS
Volume I

EGIL A. HYLLERAAS

WILEY-INTERSCIENCE
a Division of John Wiley & Sons
New York · London · Sydney · Toronto

Copyright © 1970, by John Wiley & Sons, Inc.

All rights reserved. No part of this book may be reproduced by any means, nor transmitted, nor translated into a machine language without the written permission of the publisher.

10 9 8 7 6 5 4 3 2 1

Library of Congress Catalogue Card Number: 76-91157

SBN 471 42601 6

Printed in the United States of America

Acknowledgment

When the wish for an English edition of Egil A. Hylleraas' treatise on *Mathematical and Theoretical Physics* was first expressed to his family, we immediately understood that without the aid of skilled and able people, there was nothing we could do to prepare such an edition.

Therefore I want to express my warmest thanks to the two persons at the Physics Department of the University of Oslo, Norway, who have made this English edition possible, namely, to my father's friend and colleague *försteamanuensis* John Midtdal, who, with his thorough knowledge of my father's works, has given invaluable help and advice in the preparation of this edition, and to *cand. real.* Andreas Quale, who at present holds a University Fellowship here in Oslo and, with his combined knowledge of physics and of the English language, willingly and fearlessly undertook the great work of translating that half of the treatise which my father had not translated himself, as well as to do the author's proofreading of the entire work. For all this I am very grateful.

And I also wish to extend my thanks to all those people who in different ways have been involved in the making of this book, and who with their initiative, their conscientiousness, and their workmanship have contributed to make this English edition what my father hoped it would be.

Oslo, Norway INGER HYLLERAAS BÖ
July 1969

Foreword

Egil A Hylleraas was a very forthright, humble man with a very keen analytical mind. In common with such physicists as Einstein and Feynman, he was gifted with exceptional physical intuition. Most of the techniques that we now use in molecular quantum mechanics were conceived by Hylleraas. To him, mathematics was a tool for solving physical problems. However, he spent long hours in trying to find the most direct mathematical approach. When he found a "cute" mathematical trick that simplified the treatment, he was as delighted as a child with a new toy.

In this book, he explains in his own simple but elegant fashion the development of theoretical physics. At each step he derives the mathematics most appropriate for the solution of the physical problem under consideration. And with his sly sense of humor, he occasionally adds a historical tidbit which makes the treatment more interesting. Many of the derivations are quite novel, others were standard fifty years ago but have long since been forgotten. All of his derivations are neat, simple, and self contained; yet they are completely rigorous. In reading this book, the reader cannot help being carried away with Hylleraas' joy in discovering physics. Primarily the treatment is directed at beginning graduate students in physics. However, because of the clever methods that Hylleraas employs and his down-to-earth approach to difficult problems, this book is of interest to physicists of all degrees of experience and background.

The first edition of *Mathematical and Theoretical Physics* was published in Norwegian and served as the basis for a sequence of graduate courses that Professor Hylleraas taught at the University of Oslo, and was regarded as the "Bible" by a whole generation of Norwegian physicists. Hylleraas gave me a copy when he came to spend the year 1962–1963 at the University of Wisconsin. Although I can barely read Norwegian, I discovered in this text a myriad of simple derivations and important theorems that are difficult to find anywhere else. I then persuaded Hylleraas to bring this work up to date and to translate it into English. At the time of his death, he had completed about half of the manuscript. Professor John Midtdal, a former student and for many years a close colleague of Professor Hylleraas at the University of Oslo, brought the remainder of the text up to date by making use of Hylleraas' lecture notes. University Fellow Andreas Quale of the University of Oslo translated this material. Midtdal and Quale, as well as Hylleraas' other students who participated in this project, are to be complimented on the excellence of this new English edition!

Because of its wealth of simple appropriate problems illustrating each point, *Mathematical and Theoretical Physics* is an excellent text. It could be used, as in Norway, to teach a whole sequence of graduate courses covering most aspects of theoretical physics. Or, it could be used to teach individual courses in each of the following areas: The Mathematical Preparations for Theoretical Physics; Classical Mechanics; Thermodynamics, Kinetic Theory of Gases, and Statistical Mechanics; Electricity and Magnetism; and Elementary and Advanced Quantum Mechanics

1. *Mathematical Preparations for Theoretical Physics.* The subject material for this section of the book is very carefully chosen so as to be useful in many different branches of physics. I suppose that this part should be compared with Morse and Feshbach's treatise on the *Methods of Theoretical Physics.* However, Hylleraas has condensed his treatment to 208 pages, whereas Morse and Feshbach require two volumes of 1975 pages. This section begins with a thorough treatment of vectors and vector fields which includes a good discussion of various types of discontinuous vector fields. Then follow vector transformations and the properties of tensors, orthogonal and nonorthogonal coordinate systems, determinants and their evaluations, matrix algebra, quadratic forms, and their eigenvalues and invariants. The vector space is then generalized to infinite dimensions, which of course leads to expansions of functions in orthogonal polynomials and the concept of completeness. The logical sequence continues with a good discussion of the calculus of variations (with lots of examples), Euler's differential equations of a variational integral, eigenvalue problems and how they are solved, the distribution of eigenvalues, and the completeness of solutions to an eigenvalue equation. The remainder of this section of the book is concerned with the theory of functions: analytic functions, convergence and analytic continuation of power series, contour integrals, the gamma and delta functions and solution of differential equations by power series (stressing the effects of various types of singular points). Finally Hylleraas considers the hypergeometric and Bessel functions, giving the generating functions, integral representations, special cases, and so forth.

2. *Classical Mechanics.* This section begins with a discussion of trajectory and oscillation problems. The treatment of Lagrangian and Hamiltonian

mechanics is truly excellent and contains applications to relativistic electromagnetic problems. The chapter on the motion of solid bodies has a particularly good section on the coefficients of elasticity. The chapter on hydrodynamics provides a very complete treatment of ideal fluids, including conformal mappings of a two-dimensional flow, gravity waves, and the theory of sound. The motion of a sphere through a viscous fluid, and the flow of a viscous fluid through a pipe are also considered.

3. *Thermodynamics, Kinetic Theory of Gases, and Statistical Mechanics.* The treatment of thermodynamics is simple, concise, and appears rather standard. The kinetic theory of gases is presented from the Jeans or Maxwell point of view. Classical and quantum statistical mechanics are explained in a simple fashion. This section ends with a very interesting set of applications of quantum statistics to gas kinetics: the photon gas and the law of radiation, the virial theorem in quantum statistics, sums of states in quantum statistics, ideal gases with weak degeneracy and strong degeneracy of both a Bose-Einstein and a Fermi gas.

4. *Electricity and Magnetism.* The treatment of electrostatics is very complete with many examples, including multipoles, dipole layers, charged spheres, and disks. The dielectric constant is explained in the usual way in terms of a spherical cavity. The Maxwell stress tensor, the field energy in dielectrics, and Kelvin's theorem of minimum energy are given. The section on electrostatics is concluded with a discussion of various types of condensers, the electric balance, and the quadrant electrometer. Magnetostatics is treated in very much the same way as electrostatics. Measurements of magnetic moments and field strengths are discussed. The vector potential is introduced, and the equivalence of the line vortex to the magnetic double layer is noted. After examining the basic laws of electromagnetism, Hylleraas considers the reactive forces, the forces on moving charges, and the mutual potential energy of current-carrying circuits. This leads to a discussion of magnetic field energy, Maxwell's displacement current, and energy transport by radiation. This in turn leads to the mechanical properties of radiation. With respect to the units used in electromagnetism, Hylleraas explains with his sense of humor and knowledge of history exactly how such a sad situation could have arisen. This part of the book concludes with a brilliant treatment of electromagnetic radiation. He starts with a very thorough discussion of Hertzian dipoles and multipole radiation. He shows how to solve the inhomogeneous wave equations, making use of the spectral distribution of the wave function and Green's function in four dimensions. This leads to the Lienard-Wiechert potentials, and finally to a discussion of Cherenkov radiation.

5. *Elementary and Advanced Quantum Mechanics.* Since Hylleraas is one of the fathers of quantum mechanics as we know it today, this part of the book is especially interesting and full of historical reminiscences. He starts with a thorough discussion of blackbody radiation, heat capacity of solids, the photoelectric effect and the Compton effects, and shows how the experimental facts molded the theoretical concepts. His treatment of the Bohr atomic theory is the most thorough and complete which I have ever seen. It makes one wonder how much of this old work might be made use of in connection with modern quantum mechanics. I had not realized before that Morse potentials had been used in the Bohr theory to explain the vibrational energy levels of diatomic molecules and the transition probabilities between different states. Hylleraas gives a very interesting discussion of the origin of wave mechanics. Harmonic oscillators, free rotators, and the hydrogen atom are considered in great detail, including transition coefficients. I particularly like the treatment of the Dirac equation and the relativistic hydrogen atom. The derivations and explanations are simple and convincing. Returning to the Schrödinger equation, Hylleraas has an excellent chapter on the perturbation of both degenerate and nondegenerate discrete states, with lots of problems and examples to illustrate the procedures. Of course, he uses the Hylleraas variational principle to solve for the first-order perturbation wave function! Among these examples are the perturbed simple harmonic oscillator, the $1/Z$ expansion of atoms (or ions) with a given number of electrons, and the Stark and Zeeman effects in atomic hydrogen. Periodic perturbations are considered from the standpoint of time-dependent forces. This leads to a good discussion of the absorption and emission of an atom in a radiation field (Einstein's A's and B's), refraction and dispersion, as well as the scattering of long-wave electromagnetic radiation. The next chapter is devoted to the helium atom considering various methods of approximation, reductions to three independent variables for S-states, and a discussion of the coordinate systems which Hylleraas has found useful in considering two-electron atoms. The remainder of the book is devoted to scattering problems: coherent and incoherent scattering of short-wave radiation, atomic form factors, sum rule for total scattering, scattering of electrons in a potential field, Rutherford's formula, the Born approximation, the perturbation series. The method of partial waves is then considered with considerable detail and illustrated by an excellent set of problems. The effect of electron exchange is considered for the scattering of fast electrons in a real atom. Electron scattering is then viewed in terms of a time-dependent perturbation. Next, the electromagnetic field corresponding to radiation is quantized. The interaction of this quantized field with atoms is considered. Finally, the quantum-electrodynamic method is used to develop the formula for the scattering of coherent and incoherent short-wave radiation.

This is truly a great book written by one of the greatest physicists of the twentieth century. Nowhere else will one find all of theoretical physics, presented in such a simple, elegant, rigorous fashion!

University of Wisconsin
Theoretical Chemistry Institute
Madison, Wisconsin
December 1969

JOSEPH O. HIRSCHFELDER

Contents

PART I The Mathematical Foundation of Physics

CHAPTER 1 *Vectors and Vector Fields*

1. DEFINITIONS AND NOTATIONS — 1
   Scalars – Vectors – Tensors – Notations
2. ADDITION OF VECTORS. DECOMPOSITION BY MEANS OF UNIT VECTORS — 7
   Addition and subtraction – Multiplication by a scalar – Decomposition by orthogonal system of unit vectors
3. SCALAR AND VECTOR MULTIPLICATION — 9
   The scalar product – Commutative and distributive law – Scalar products expressed by orthogonal components of the vectors – The vector product – Positive normal of a surface area and positive direction on the limiting curve – Distributive law for vector products – Analytical expression of vector products
4. MULTIPOLE PRODUCTS OF VECTORS — 13
   The triple scalar product – Expression of the volume by the vector components – Polar and axial vectors – The triple vector product $\mathbf{A} \times (\mathbf{B} \times \mathbf{C})$ – Product of 4 vectors
5. VECTORS AS DEPENDENT VARIABLES — 16
   Differentiation of vectors – Curvature and torsion of a curve
6. SCALAR AND VECTOR FIELDS — 19
   Definitions – The gradient of a scalar field – The divergence of a vector – The curl of a vector – The curl as a true vector – The vectorial differential operator
7. INTEGRAL THEOREMS BY GAUSS, GREEN, AND STOKES — 26
   Elementary curve and surface integrals – The Gauss and the Green theorem – Equivalence of the Gauss and the Green theorems in a plane
8. LAMELLAR AND SOLENOIDAL VECTOR FIELD. DETERMINATION OF A FIELD FROM ITS DIVERGENCE AND CURL — 29
   Potential and gradient – Vector potential and curl

9. DETERMINATION OF VECTOR FIELDS FROM THEIR DIVERGENCE AND CURL   32
   Separation into an irrotational and a solenoidal field – Solution of the Poisson equation – Solution for a finite region – Uniqueness of the solution – Laplacian fields
10. DISCONTINUOUS FIELDS   35
    The source of a field. Point and surface divergence – Point and surface double sources – Discontinuities of solenoidal fields – Identity of fields from a double layer and a bordering vortex line

CHAPTER 2   *Tensors and Tensor Fields*

11. LINEAR TRANSFORMATIONS OF VECTORS   41
    Linear vector functions – Interchange of factors and transposition of tensors
12. THE CHARACTERISTICS OF VECTOR TRANSFORMATIONS   44
    Addition and subtraction of tensors – Decomposition in symmetric and antisymmetric tensors – Tension and shears – Antisymmetric tensors and vector products – Main axis of a symmetric tensor – The invariants of the tensor
13. TENSOR FIELDS AND THEIR DERIVATIVES   48
    Divergence of a tensor – Gradient of a vector – Displacement and deformation
14. RECIPROCAL VECTOR SYSTEMS   50
    Vector components in nonorthogonal coordinate systems – Reciprocal basic vectors – Solution of linear equations
15. CURVILINEAR COORDINATES   53
    Three-dimensional general coordinates – Divergence and the Laplacian in curvilinear coordinates – The Laplacian in curvilinear orthogonal coordinates – Spherical coordinates – Prolate and oblate rotational elliptic coordinates – Parabolic coordinates

CHAPTER 3   *Linear Transformation. Extremum and Variational Problems*

16. THE GENERALIZED VECTOR SPACE. ROTATION OF COORDINATE SYSTEMS   61
    Vectors in the n-dimensional space – The scalar products of n-dimensional vectors – Unit vectors and orthonormal systems
17. DETERMINANTS AND LINEAR EQUATIONS   64
    The volume in n-dimensional space – Development of determinants in terms of minors. Reciprocal vectors – Solution of inhomogeneous linear equations – Homogeneous linear

equations – Inhomogeneous linear equations with vanishing determinant – Numerical evaluation of determinants

18. MATRIX ALGEBRA 69
    Linear transformation and matrices – Composite transformations. Multiplication of matrices – Product determinants – Inverse transformations and reciprocal matrices

19. QUADRATIC FORMS AND THEIR STATIONARY VALUES 73
    Quadratic forms with symmetric coefficients – The eigenvalues of a quadratic form – Orthogonality of eigenvectors – Transformation on the principal axis – Simultaneous transformations – Invariants of a quadratic form – Solution of inhomogeneous eigenvalue equations

20. VECTORS IN A SPACE OF INFINITE DIMENSIONS. ORTHOGONAL SYSTEMS OF POLYNOMIALS AND FUNCTIONS 79
    Vibration of a continuum as a limiting case of vibrating mass points – Orthogonal polynomials

21. EXPANSION OF FUNCTIONS BY ORTHONORMAL SYSTEMS 88
    The components of a function on the axis of an orthonormal system by least square deviation. – Completeness of a functional system

22. ELEMENTS OF CALCULUS OF VARIATION 91
    Statement of problems. Functions of functions – Fermat's principle – The isoperimetric problem – The area of a surface of rotation and the brachystochrone – The chain line – Maupertuis's principle or the principle of least action

23. THE EULER DIFFERENTIAL EQUATIONS OF A VARIATIONAL INTEGRAL WITH BOUNDARY CONDITIONS 98
    The variation of a function – The Euler differential equation – Boundary conditions – Several dependent variables. Simultaneous Euler equations – Several independent variables. Partial differential equations – Homogeneous and linear Euler equations and linear homogeneous boundary conditions

24. VARIATIONAL PROBLEMS WITH ADDITIONAL CONDITIONS. EIGENVALUE PROBLEMS 103
    Variational problems leading to linear differential equations – Normalization as an additional condition

25. EIGENVALUE PROBLEMS 106
    Eigenvalues and eigenfunctions – Coincidence of eigenvalues or degeneration – Direct methods of solution. The Ritz method – Distribution of approximate eigenvalues by the Ritz method – The minimum properties of eigenvalues

26. NUMBER AND DISTRIBUTION OF EIGENVALUES 112
    Comparison with simpler eigenvalue equations – The total number of eigenvalues – Asymptotic behavior of eigenvalues
27. COMPLETENESS PROPERTIES FOR SOLUTIONS OF EIGENVALUE EQUATIONS 115
    Proof of the completeness of a system – Completeness of systems in an infinite region – Noncomplete systems of eigenfunctions – Independent proof of the completeness relation

CHAPTER 4 *Theory of Functions*

28. COMPLEX NUMBERS 121
    Complex numbers as vectors – Operations with complex numbers – Complex numbers in polar coordinates – The numbers zero and infinity. The unit circle and the number sphere
29. ANALYTIC FUNCTIONS 125
    Definition of an analytic function – Construction of analytic functions – Complex integrals and Cauchy's theorem – Deformation of path of integration – The Cauchy integral theorem – Connection with vector analysis
30. POWER SERIES FOR ANALYTIC FUNCTIONS 134
    Expansion at a regular point – Some examples of power series
31. CONVERGENCE OF INFINITE SERIES 140
    Absolute and conditional convergence – Interchange of terms in conditionally convergent series – Comparison series – Uniform convergence
32. ANALYTIC CONTINUATION OF POWER SERIES 145
    Different power series of an analytic function – Many-valued functions
33. COMPLEX INTEGRATION AND CONTOUR INTEGRALS 150
    Evaluation of definite integrals by complex integration – The theorem of residues – Examples of contour integrals
34. THE FACTORIAL OR GAMMA FUNCTION 154
    Definition of the gamma function – Alternative definition of the gamma function – The gamma function expressed by infinite products – The Euler integral of the first kind – The logarithmic derivative of the gamma function – The duplication and multiplication formulas
35. THE DELTA FUNCTION AND THE COMPLETENESS OF SYSTEMS OF FUNCTIONS 161
    The concept of the delta function – Orthogonality of functions with a continuous parameter – The δ-function of Fourier series

CHAPTER 5 *Differential Equations and Particular Functions*

36. SOLUTION OF DIFFERENTIAL EQUATIONS BY POWER SERIES — 167
    Classification of differential equations – The Wronskian of second order equations – Solution by power series. The homogeneous equation in a regular point – Power series at singular points – Logarithmic solutions – Irregular singular points – Asymptotic solutions

37. ARRANGEMENT OF DIFFERENTIAL EQUATIONS ACCORDING TO THEIR SINGULAR POINTS — 176
    Differential equations with not more than two singular points. Confluence of singular points – Differential equations with three or more singular points – The hypergeometric equation. Legendre's equation – The confluent hypergeometric equation

38. Hypergeometric functions — 180
    The hypergeometric series – Integral representation of the hypergeometric series – Legendre functions of the first kind – Integral representation of $P_n(z)$ – Generating function for Legendre polynomials – Gegenbauer's polynomials – The delta function of Gegenbauer polynomials – The completeness of Legendre polynomial systems – Legendre functions of the second kind – Associate Legendre functions of the second kind

39. CONFLUENT HYPERGEOMETRIC FUNCTIONS — 192
    Laguerre polynomials and orthogonal functions – Integral representations and generating functions of Laguerre polynomials – Hydrogen atomic functions – Hermite's polynomials and functions

40. BESSEL FUNCTIONS — 201
    First solution of the Bessel equation – Bessel functions of the second kind. The Neumann function $N(z)$ – Bessel functions of the third kind or Hankel functions – Asymptotic values and asymptotic expansions – Bessel functions with imaginary arguments – Integral representations

PART II Mechanics and Statistics

CHAPTER 1 *Elementary Mechanical Problems*

1. HISTORICAL BACKGROUND — 213
2. FUNDAMENTAL LAWS OF MECHANICS. THE NEWTONIAN AXIOMS — 215
    The law of inertia – The fundamental of mechanics – Newton's

third law – Force and work – The parallelogram theorem – Coordinate systems – "Absolute" space

3. FALLING AND OTHER ONE-DIMENSIONAL MOTIONS   220
General conditions for integration of equations of motion for one-dimensional problems – Some elementary examples

4. FREE AND FORCED ELASTIC OSCILLATIONS   225
Proper oscillations and proper frequency – Forced oscillations – Resonance – Electrical oscillatory circuit with self-induction and capacitance.

5. DAMPED OSCILLATIONS   228
Variation of frequency and amplitude – Periodic and aperiodic oscillations – Amplitude variation and phase shift for forced damped oscillations – Forced damped electrical oscillations – Work of external force and energy of oscillator for damped oscillations – Forced damped oscillations

6. OTHER OSCILLATION PROBLEMS   234
The mathematical pendulum – The cycloid pendulum – Anharmonic oscillators – Molecular vibrations and planetary motion

7. HIGHER-DIMENSIONAL PROBLEMS   241
Kinetic and potential energy for conservative forces – Integration of the equations of motion – Motion of a free particle

8. LAWS OF LINEAR AND ANGULAR MOMENTUM FOR PARTICLE SYSTEMS OR SOLID BODIES   247
Definitions of center of mass, angular momentum, and moment of inertia – The center of mass law – The law of areas – Work and kinetic energy referred to the center of mass – Application of the law of areas to the determination of planetary orbits

9. RELATIVE MOTION   253
Coordinate systems with nonuniform motions – Relative velocity – Relative acceleration – Equations of motion in relative coordinates – Free fall relative to the rotating earth – The Foucault pendulum

CHAPTER 2 *Lagrangian and Hamiltonian Mechanics*

10. THE PRINCIPLE OF VIRTUAL WORK AND D'ALEMBERT'S PRINCIPLE   261
Degrees of freedom for a system of mass points – Virtual displacements and virtual work – Generalized coordinates – D'Alembert's principle

11. LAGRANGE'S EQUATIONS IN GENERALIZED COORDINATES   268
Inertial forces expressed in terms of kinetic energy

12. THE PRINCIPLES OF FERMAT, MAUPERTUIS, AND HAMILTON 269
    Fermat's principle of shortest light-time or shortest "optical" light-path – Maupertuis's principle of least action – Hamilton's principle
13. OTHER DEDUCTIONS AND APPLICATIONS OF LAGRANGE'S EQUATIONS 275
    Deduction of Lagrange's equations from Hamilton's principle – Direct deduction from the equations of motion – Formulation of Lagrange's equations for nonconservative forces – Lagrangian formulation of the relativistic equation
14. PARTIAL INTEGRATION OF LAGRANGE'S EQUATIONS 280
    Generalized momenta – Lagrange's equations and the energy function – Kinetic energy in relativistic mechanics. Kinetic energy for electromagnetic forces
15. HAMILTON'S EQUATIONS 283
    Energy as a function of coordinates and momenta – Deduction of Hamilton's equations from Lagrange's equations – The connection between the canonical equations and Hamilton's principle
16. GENERAL SOLUTION METHODS FOR THE CANONICAL EQUATIONS 287
    Cyclic variables and constants of motion – The Hamilton-Jacobi differential equation and the action function – Integration of the equations of motion
17. CANONICAL TRANSFORMATIONS 291
    Coordinate-momentum space – The canonical transformation function
18. SOLUTION OF THE PROBLEM OF MOTION BY SEPARATION OF THE VARIABLES 294
    Orbits and orbital motion

CHAPTER 3 *Dynamics of Continuously Extended Bodies*

19. MOTION OF SOLID BODIES 299
    Moment of inertia and products of inertia for a body – Equations of motion for rigid bodies – Eulerian angles for rotational motion
20. MOTION OF A BODY AROUND A FIXED POINT OR AROUND THE CENTER OF MASS 303
    The gyroscope – Precession of the equinoxes – The gyrocompass
21. INTERNAL MOTION IN SOLID BODIES 311
    Displacement and strain – The strain tensor – Examples of strain. One-, two-, and three-dimensional continua

22. HAMILTON'S PRINCIPLE FOR CONTINUOUS ELASTIC MEDIA 316
    Equations of motion and potential energy for a vibrating string – Potential energy and stress for a vibrating membrane – Hamilton's principle for three-dimensional elastic bodies
23. STRESSES AND COEFFICIENTS OF ELASTICITY 322
    Surface forces or stresses – Number of coefficients of elasticity for a solid body – Reduction of the number of independent coefficients of elasticity for higher lattice symmetries
24. WAVES IN ISOTROPIC ELASTIC MEDIA 327
    Deduction of the elastic force density – Transversal and longitudinal waves – Longitudinal waves in fluid media

CHAPTER 4 *Hydrodynamics*

25. KINEMATICS OF FLUIDS 331
    Instantaneous coordinates and identification coordinates – Velocity field – Acceleration and total derivative in hydrodynamics – Equation of continuity for compressible and incompressible fluids – Velocity potential and Laplace's equation for incompressible fluids – Kinematic boundary conditions
26. THE HYDRODYNAMIC EQUATIONS OF MOTION 337
    Equations of motion for frictionless fluids – Equations of motion for fluids with internal friction
27. HYDRODYNAMICS OF IDEAL FLUIDS 340
    Definition of ideal fluids – Stationary current in ideal fluids – Torricelli's law – Vena contracta
28. ROTATIONAL FLOW IN IDEAL FLUIDS 343
    Kelvin's circulation theorem – Vortices and vortex filaments – Helmholtz' law of conservation of vortex strength – Deduction of the velocity field when the vorticity is known – The rectilinear homogeneous vortex filament
29. FLOW IN AN INCOMPRESSIBLE IDEAL FLUID 349
    Two-dimensional flow and conformal mappings – Flow around a sphere
30. WAVES IN IDEAL FLUIDS 354
    Gravity waves – Waves due to molecular forces – Pressure and sound waves – Energy and energy transport in a pressure wave
31. FLUIDS WITH INTERNAL FRICTION 360
    General considerations – A fluid between two rotating cylinders – Poiseuille's law – Flow of a fluid through a pipe – Motion of a sphere through a viscous fluid

CHAPTER 5 Thermodynamics

32. TEMPERATURE AND TEMPERATURE SCALES — 370
Measurement and definition of temperatures – Absolute zero and absolute temperature

33. THE EQUATION OF STATE — 372
States – State variables – Connection between pressure, temperature, and volume – State diagrams

34. THE FIRST LAW OF THERMODYNAMICS — 375
Heat and energy. The law of energy conservation – Internal energy of a gas – Ideal gases – Entropy of an ideal gas

35. THE SECOND LAW OF THERMODYNAMICS — 379
Reversible and irreversible processes – Perpetual motion machines of the second kind – Efficiency of a heat engine – Efficiency of reversible processes – Comparison of the efficiencies of two heat engines – Entropy as a state variable – The second law and entropy – Application of the entropy law to evaporation and melting

36. THERMODYNAMIC POTENTIALS — 388
Entropy and free energy. Gibbs' potential and enthalpy – Maxwell's thermodynamic relations – The Joule-Kelvin experiment – Determination of absolute temperatures by means of nonideal gases

37. EQUATIONS OF STATE FOR NONIDEAL GASES — 392
The van der Waals equation of state – Critical constants of a van der Waals gas

CHAPTER 6 Kinetic Theory of Gases

PRELIMINARY REMARKS — 397

38. CONNECTION BETWEEN KINETIC ENERGY AND PRESSURE IN A GAS — 398
Momentum and pressure for molecular collisions – Equation of state for a molecular gas – Partial pressures in mixtures of gases

39. VELOCITY DISTRIBUTION FOR DISORDERED HEAT MOTION — 400
Molecular collisions – Energy exchanges between molecules – The equipartition theorem

40. MAXWELL'S VELOCITY DISTRIBUTION — 407
Energy exchanges and the distribution of velocities – A more rigorous deduction of Maxwell's distribution law

41. MEAN VALUES FOR COLLISION PROCESSES — 413
Mean speeds and mean square speeds – Mean relative speeds – Collision number and mean free path of molecules in a gas

42. TRANSPORT OF MASS, ENERGY AND MOMENTUM IN A GAS ... 417
    Smoothing-out of the properties of an inhomogeneous gas – Diffusion – The diffusion law – Disordered migration of molecules – Heat conduction – Viscosity
43. INTERMOLECULAR FORCES AND THE EQUATION OF STATE ... 424
    The virial theorem of Clausius – Gas pressure and virial – Form of the virial for central forces – The equation of state for molecules of finite size – The equation of state for cohesive forces

CHAPTER 7 Statistical Mechanics

44. PHASE SPACE AND PHASE CELLS ... 431
    Phase space – Invariance of volume under canonical transformations – Liouville's theorem
45. THERMODYNAMIC PROBABILITY AND ENTROPY ... 434
    Statistical description of complex systems – The *a priori* probability of the phase cells – Thermodynamic probability and thermodynamic equilibrium – Entropy
46. THE BOLTZMANN DISTRIBUTION. CONNECTION WITH THERMODYNAMICS ... 437
    Stirling's formula – Condition for maximum entropy. The Boltzmann law – Temperature, pressure, and free energy – Deduction of Stirling's formula
47. APPLICATIONS OF STATISTICAL MECHANICS TO GAS KINETICS ... 441
    Mean kinetic energy and the equipartition theorem – Equation of state for an ideal gas – Mean free energy of an oscillator
48. GIBBS' FORMULATION OF STATISTICAL MECHANICS ... 445
    Macrosystems and microsystems – Statistical ensembles and canonical ensembles – Thermodynamic quantities in the Gibbs statistics – Equation of state for a nonideal gas
49. QUANTUM STATISTICS ... 449
    Elementary particles and individuality – A new formulation of the Boltzmann statistics – The Fermi-Dirac statistics – The Bose-Einstein statistics – Entropy and Nernst's theorem
50. APPLICATIONS OF THE NEW STATISTICS TO GAS KINETICS ... 457
    The photon gas and the law of radiation – The virial theorem in quantum statistics – Sums of state in quantum statistics – Ideal gases with weak degeneration – Strong degeneration of a Bose-Einstein gas – Zero-point energy of a Fermi gas – Strong degeneration of a Fermi gas

AUTHOR INDEX ... 467

SUBJECT INDEX ... 469

MATHEMATICAL AND THEORETICAL PHYSICS

Volume I

PART I

The Mathematical Foundation of Physics

Introduction

Because there is no lack of textbooks in theoretical physics nowadays, it may seem daring, in such a small country as ours, to publish a large book about a subject that has relatively few admirers. There are reasons for this step, however.

In our century, especially during the last 25 years, theoretical physics has developed more rapidly than any other science. It follows that many text and reference books, however modern they may have been in their time, have become more or less old fashioned when judged by modern standards. Therefore the choice of books useful for education in modern theoretical physics is very limited. Supplements of special works covering modern subjects have the disadvantage of easily becoming too difficult for students with weak and varied backgrounds.

Any selection of material for a completely modern textbook can be criticized, since different demands and different needs lead to different approaches. Each textbook has its roots in its environment, on the one hand, and in the author's personal scientific accomplishments and ideas, on the other. Each book, in other words, is both traditional and original. Different surrounding, with a very different basis of education and perhaps with other types of natural research projects, will naturally produce modifications. In the scope and thoroughness of the representation as well, several different levels are possible.

With these points in mind, I have concluded that a new publication can be useful, at least to some people. I have tried to build the description in this book on my experiences both in active research and in educating students from all parts of our country.

This book is intended as a textbook for students—in a first course, when they should be instructed as to what material can be skipped without harm, and in advanced courses, when the material here should be supplemented with specialized works. The book is also intended to prepare for individual scientific research, and I have therefore included much of that which has been useful to me or in which I have taken a special interest. Not every reader will have the same aim, of course. Especially the first purely mathematical part of the book may seem too extensive to some readers.

For these reasons, I did not wish to call this book a textbook on theoretical physics; instead, I have chosen the title *Mathematical and Theoretical Physics* in order to mark a certain amount of freedom. I have also tried to write the book so that it would give a fairly complete guidance in classical as well as modern theoretical physics.

I express my gratitude to the printers and to the publishing house Grøndahl & Son for the tremendous work done to bring out the typographical layout in the most desirable form. Above all, I wish to thank my many young collaborators who eagerly and with interest helped in correcting the material. Werner Romberg, Aadne Ore, Marius Kolsrud, Tor B. Staver, Vidar Risberg, Gustav Mathisen, and Helene-Marie Voldner all helped with different parts of the work.

June 1950
Blindern, Oslo

Egil Hylleraas

CHAPTER 1

Vectors and Vector Fields

1. DEFINITIONS AND NOTATIONS

Scalars

In the mathematical description of physical processes the values of a great many quantities can be specified by a single real number. For instance, length and temperature are such quantities; their values can be arranged on a single scale. These quantities are called *scalars*.

To obtain such a scale we must first choose a measuring unit. For the sake of consistency we cannot always choose the unit arbitrarily. What we can choose are a few so-called fundamental units. Other units are derived from these units and thus are uniquely determined by a set of fundamental units. For instance, if we choose a given unit of length, such as cm, m, or km, the units of area and volume are already given.

The number of necessary fundamental units for physical quantities is bound to be small, although to some extent this is a question of convention. It is usually agreed upon that a number smaller than three is of no practical interest. In the cgs system the fundamental units are the units of length, mass, and time: the centimeter, the gram, and the second. Even electrical and magnetic units are derived from this system. In the mks system the units are the meter, the kilogram, and the second. Usually the ampere is taken to be a fundamental electrical unit; this turns the mks system into the mksa or the Giorgi system.

Scalars may be essentially positive, as mass and volume, or they may be both positive and negative, such as the density of electric charge. Any physical quantity, whether scalar or not, has what is called a given "dimension" as defined by the measuring unit. The ratio between two quantities of the same kind, however, is "dimensionless" or a pure number.

The calculus used for pure numbers is valid for scalars. However, only scalars of the same kind and of the same dimension can be added or subtracted. By multiplication and division we get quantities of different dimensions that must be expressed in other units.

Vectors

Not all geometrical and physical quantities can be measured by a single number. Many of them belong to a class that can be specified by three numbers and are called *vectors*. A quantity that can be described uniquely by a displacement in a three-dimensional space is a vector. More precisely, it is a three-dimensional vector. A vector that can be visualized in a plane is a two-dimensional vector, but this may be considered a special kind of restricted three-dimensional vector. One may also imagine vectors in a space of higher, say n, dimensions, but by vectors we usually mean ordinary three-dimensional ones.

Physical quantities of the vector type are known in abundance. Displacements and velocities of particles, forces, torques, and so on, are quantities of this type. They can all be characterized by a certain positive scalar value, namely, their absolute value. But this is not all. We must also mention their direction in space. Vectors may have the same absolute value and still differ widely; for instance, they may point exactly in opposite directions or they may be orthogonal to each other.

Tensors

Physical quantities may be of still higher complexity than scalars and vectors, needing more than one or even three numbers for their specification. Usually we shall mean by tensors those of the so-called second rank, scalars and vectors being tensors of the zeroth and first rank. A tensor of the second rank can generally be specified by three vectors, or nine scalars. For instance, applying a vector force to a body held in space by elastic springs, the displacement of the body can be separated into three vectors, each of which may be considered to result from either of the three orthogonal components of the force. A tensor can usually be said to define the dependence of a vector upon another vector. If the elastic binding of a body is isotropic, the tensor defining the elastic binding degenerates into a scalar or, more precisely, into a scalar times the so-called unit tensor.

Notations

For scalars we shall use all sorts of ordinary letters: m, M, and μ for mass, ρ for density of mass or electricity, and so forth. In mathematical equations we shall as a rule use italic type. For vectors a particular notation is needed. From a logical point of view $\vec{A}$, $\vec{b}$, and so on, is a good one, but sometimes in handwriting $\underline{A}$, $\underline{b}$ or some other system may be preferred. In printed text boldface letters are commonly used (except in German literature where Gothic letters are frequently in use). For tensors of the second rank further specialization might easily lead to over-complications. Hence we shall mostly return to ordinary letters.

The absolute value or "length" of a vector **A** is usually denoted by |**A**| or, in handwriting, simply A.

2. ADDITION OF VECTORS. DECOMPOSITION BY MEANS OF UNIT VECTORS

Addition and Subtraction

Since a vector is given by three scalar numbers—positive or negative—it can be visualized by an arrow in a three-dimensional space starting from the origin of a Cartesian coordinate system and ending at the point in space given by the three scalar numbers. The sum of two vectors is obtained by taking the vectors in succession as shown in Fig. 1-1 and drawing the sum vector from the origin of the first vector to the end point of the second. As is intuitively seen, the sum vector is independent of the order of succession. Hence we write

$$\mathbf{A} + \mathbf{B} = \mathbf{B} + \mathbf{A}, \qquad (2.1)$$

the *commutative law*. From Fig. 1-2 we easily see that the *distributive law* also must hold; that is,

$$\mathbf{A} + (\mathbf{B} + \mathbf{C}) = (\mathbf{A} + \mathbf{B}) + \mathbf{C} = \mathbf{A} + \mathbf{B} + \mathbf{C}. \qquad (2.2)$$

Hence an arbitrary number of vectors can be added to give the same sum vector irrespective of the order of succession of the vectors. With respect to addition (and subtraction) there is no difference between scalars and vectors.

The subtraction of a vector **B** from a vector **A** being defined as the inverse operation of addition

$$(\mathbf{A} - \mathbf{B}) + \mathbf{B} = \mathbf{A} \qquad (2.3)$$

can also be expressed as

$$\mathbf{A} - \mathbf{B} = \mathbf{A} + (-\mathbf{B}), \qquad (2.4)$$

that is, as addition of the reversed vector $(-\mathbf{B})$. Fig. 1-3.

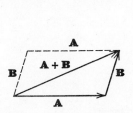

Fig. 1-1. Addition of vectors.

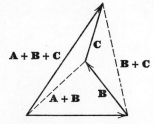

Fig. 1-2. Addition of several vectors.

8 / The Mathematical Foundation of Physics

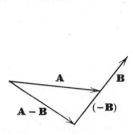

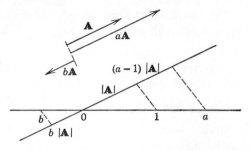

Fig. 1-3. Subtraction.   Fig. 1-4. Multiplication of a vector by a scalar.

Multiplication by a Scalar

Multiplication by a scalar is a generalization of repeated addition, which we write $A + A = 2A$, $A + A + A = 3A$, and so forth, and consider as a multiplication by positive integers. Obviously, multiplication by positive nonintegral numbers can be defined as follows. The absolute value of a vector $\mathbf{A}$ is multiplied by the scalar a and the vector $a\mathbf{A}$ is a vector of this magnitude pointing in the direction of $\mathbf{A}$ itself. If a is negative, we multiply by $-a$ and the product vector $a\mathbf{A}$ is a vector of length $(-a)\mathbf{A}$ pointing in the opposite direction as shown in Fig. 1-4.

Decomposition by Orthogonal System of Unit Vectors

A vector can be added up in a number of ways from quite different series of vectors. A useful procedure is to use just three mutually orthogonal vectors. By given directions the procedure is unique and the three vectors are called Cartesian components of the vector. If unit vectors $\mathbf{i}$, $\mathbf{j}$, and $\mathbf{k}$ in the same three directions are introduced, the components can be written $A_x\mathbf{i}$, $A_y\mathbf{j}$, $A_z\mathbf{k}$, where the three scalars A_x, A_y, A_z are also frequently called the Cartesian components of $\mathbf{A}$. Hence we can write

$$\mathbf{A} = A_x\mathbf{i} + A_y\mathbf{j} + A_z\mathbf{k}. \qquad (2.5)$$

The length of $\mathbf{A}$ is given by

$$A = |\mathbf{A}| = \sqrt{A_x^2 + A_y^2 + A_z^2} \qquad (2.6)$$

as shown in Fig. 1-5.

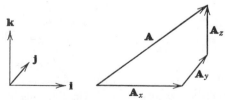

Fig. 1-5. Orthogonal unit vectors. Decomposition of a vector.

3. SCALAR AND VECTOR MULTIPLICATION

The Scalar Product

We have seen that the multiplication of a vector by a scalar produces a new vector. We can also form products of two vectors, but in this case there is more than one possibility. The product that appears most frequently is the *scalar product*. It is a scalar quantity which may be defined as the product of the magnitudes of the vectors times the cosine of the angle between the vectors. The scalar product of two vectors **A** and **B** with magnitudes A and B is $AB \cos \varphi$, as illustrated in Fig. 1-6. The product may be positive or negative. If the angle φ between the vectors is larger than 90°, the product is negative. If the vectors are strictly orthogonal to each other, the product is 0.

There are several notations for the scalar product. Gibbs used the notation **A . B** which many authors have modified into **A · B**, a notation that appears to be the most common one used now in English and American literature. The notation **A ˙ B** has also been proposed but unfortunately this notation has been used by others for the so-called *dyadic products* which are tensors (compare Joos, *Lehrbuch der theoretischen Physik*).

In German literature there is the old-established form **(AB)** in which the parentheses play the role of operator. Deleting the parentheses, we arrive at the very simplest notation **AB** which, for instance, was used by Joos. In this book we adopt this notation rather than the more common notation **A · B**.

Commutative and Distributive Law

It is obvious from the definition of the scalar product that the order of succession of the factors **A** and **B** is immaterial. Hence we have the commutative law

$$\mathbf{AB} = \mathbf{BA}. \tag{3.1}$$

The scalar product **AB** may also be interpreted as the length of **A** times the projection $B \cos \varphi$ of the vector **B** on **A**. Considering another vector **C**, **AC** is A times $C \cos \psi$ (Fig. 1-6). Now the projection of **B + C** on **A** is just

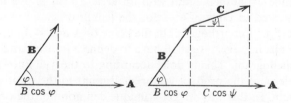

Fig. 1-6. Scalar multiplication. Distributive law.

$B \cos \varphi + C \cos \psi$. Hence

$$A(B + C) = AB + AC, \qquad (3.2)$$

an equation that expresses the *distributive law* for scalar products.

Scalar Products Expressed by Orthogonal Components of the Vectors

It is obvious from our definitions that $(aA)(bB) = ab(AB)$, where a and b are scalars. Aided by this result and by the commutative and distributive law, we can easily obtain an expression for the scalar product based on the vector decomposition formula (2.5). For the mutually orthogonal unit vectors **i**, **j**, **k** we get the obvious results

$$\mathbf{i}^2 = \mathbf{j}^2 = \mathbf{k}^2 = 1, \qquad \mathbf{jk} = \mathbf{kj} = \mathbf{ij} = 0, \qquad (3.3)$$

writing $\mathbf{i}^2$ for $\mathbf{ii}$, and so forth. Hence, multiplying term by term, we obtain

$$\mathbf{AB} = (A_x\mathbf{i} + A_y\mathbf{j} + A_z\mathbf{k})(B_x\mathbf{i} + B_y\mathbf{j} + B_z\mathbf{k}) = A_xB_x + A_yB_y + A_zB_z. \qquad (3.4)$$

The scalar product thus is equal to the sum product of the Cartesian components of the two vectors. In particular $A^2 = A^2 = A_x^2 + A_y^2 + A_z^2$, corresponding to the obvious definition

$$A = \sqrt{A_x^2 + A_y^2 + A_z^2} \qquad (3.5)$$

of the vector length.

The components of a vector in a given Cartesian system can also be expressed by scalar products

$$A_x = \mathbf{i}A, \qquad A_y = \mathbf{j}A, \qquad A_z = \mathbf{k}A \qquad (3.6)$$

as easily seen from the expression (2.5).

The Vector Product

From two vectors **A** and **B** it is also possible to form a new vector which may naturally be called a product of the two vectors. The two vectors together define a parallellogram with an area $AB \sin \varphi$. We now take this area as expressed in units of area for the length of the product vector and draw it in a direction orthogonal to the plane or **A** and **B**. In order to obtain a unique result, however, we shall have to define a positive and negative side of the parallellogram. This is done according to the well-known, right-hand or screw rule, and the product vector shall point to the positive side of the parallellogram. In this way we get a unique definition of the product, which is simply called the vector product and is denoted by **A** × **B**.

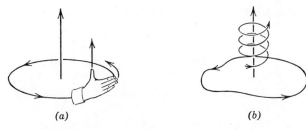

Fig. 1-7. (*a*) The right-hand rule; (*b*) the screw rule.

Positive Normal of a Surface Area. Positive Direction on the Limiting Curve

In analytical geometry two different coordinate systems were formerly in use, the right-hand system and the left-hand system. Today only the right-hand system appears to have survived. In conformity with this system we can define a positive direction on the circumference curve of a surface area when the positive side of the surface has been chosen. Conversely, a given positive direction of the circumference curve uniquely defines the positive side of the surface, as illustrated in Fig. 1-7.

Now the vectors in the given succession **A**—**B** uniquely defines the positive side of the parallellogram, as shown in Fig. 1-8. Hence the vector product **A** × **B** is uniquely defined. The product is not commutative with respect to the vectors. On the contrary, it is subject to the anticommutative law

$$\mathbf{A} \times \mathbf{B} = -(\mathbf{B} \times \mathbf{A}). \tag{3.7}$$

Distributive Law for Vector Products

We shall now show that the distributive law

$$\mathbf{A} \times (\mathbf{B} + \mathbf{C}) = \mathbf{A} \times \mathbf{B} + \mathbf{A} \times \mathbf{C} \tag{3.8}$$

is valid for vector products. When this law is established, we can easily find the vector product **A** × **B** expressed by the Cartesian components of the two vectors.

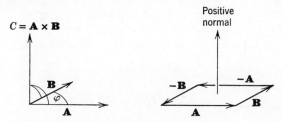

Fig. 1-8. Vector multiplication. $C = |\mathbf{C}| = |\mathbf{A} \times \mathbf{B}| = AB \sin\varphi$.

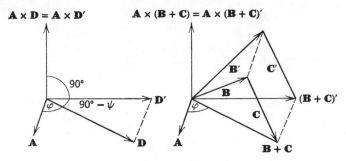

Fig. 1-9. To the left: construction of **A** × **D** by **A** × **D**′. To the right: **B**′ + **C**′ = (**B** + **C**)′.

Consider a vector **A** and two vectors, **D** and its projection **D**′ in a plane orthogonal to **A**. As shown in Fig. 1-9, **A** × **D** and **A** × **D**′ have the same direction and the same magnitude; hence

$$\mathbf{A} \times \mathbf{D} = \mathbf{A} \times \mathbf{D}'. \tag{3.9}$$

Writing D quite arbitrarily as a vector sum **B** + **C**, we find that the projection vector **D**′ is unchanged, that is,

$$\mathbf{D}' = \mathbf{B}' + \mathbf{C}'. \tag{3.10}$$

Now the absolute values of **A** × **D**′, **A** × **B**′, and **A** × **C**′ are found by multiplication of **D**′, **B**′, and **C**′ by **A**. The directions for the vector products are found by turning the vectors 90° about **A** as shown at the right in Fig. 1-8. Hence the distribution law (3.8) is true.

Analytical Expression of Vector Products

For a set of mutually orthogonal unit vectors (for brevity, orthonormal vectors), the following vector products are easily found

$$\mathbf{i} \times \mathbf{i} = \mathbf{j} \times \mathbf{j} = \mathbf{k} \times \mathbf{k} = 0,$$

$$\mathbf{j} \times \mathbf{k} = \mathbf{i}, \quad \mathbf{k} \times \mathbf{i} = \mathbf{j}, \quad \mathbf{i} \times \mathbf{j} = \mathbf{k}, \tag{3.11}$$

$$\mathbf{k} \times \mathbf{j} = -\mathbf{i}, \quad \mathbf{i} \times \mathbf{k} = -\mathbf{j}, \quad \mathbf{j} \times \mathbf{i} = -\mathbf{k}.$$

Hence we find

$$\mathbf{A} \times \mathbf{B} = (A_x \mathbf{i} + A_y \mathbf{j} + A_z \mathbf{k}) \times (B_x \mathbf{i} + B_y \mathbf{j} + B_z \mathbf{k})$$
$$= (A_y B_z - A_z B_y)\mathbf{i} + (A_z B_x - A_x B_z)\mathbf{j} + (A_x B_y - A_y B_x)\mathbf{k}. \tag{3.12}$$

This result can be expressed formally in a very convenient way by the determinant

$$\mathbf{A} \times \mathbf{B} = \begin{vmatrix} \mathbf{i} & \mathbf{j} & \mathbf{k} \\ A_x & A_y & A_z \\ B_x & B_y & B_z \end{vmatrix}. \tag{3.13}$$

In conformity with the anticommutative law (3.7), the interchange of **A** and **B** components reverses the sign of the determinant.

4. MULTIPLE PRODUCTS OF VECTORS

The Triple Scalar Product

From the vectors **A**, **B**, and **C** we can form a triple scalar product, for instance (**A** × **B**)**C**, as illustrated in Fig. 1-10. The product is uniquely defined by the two operations of vector and scalar multiplication; in that sense it needs no further explanation. Our main interest in this product is in its geometrical interpretation as the volume of a parallelepiped formed by the three vectors. Hence the three vectors may be interchanged without changing the volume product. This, however, is true only by cyclic permutation. By the other cyclic succession of the vectors, the sign of the product is reversed:

$$\mathbf{A}(\mathbf{B} \times \mathbf{C}) = \mathbf{B}(\mathbf{C} \times \mathbf{A}) = \mathbf{C}(\mathbf{A} \times \mathbf{B}) = |ABC|$$
$$= -\mathbf{A}(\mathbf{C} \times \mathbf{B}) = -\mathbf{B}(\mathbf{A} \times \mathbf{C}) = -\mathbf{C}(\mathbf{B} \times \mathbf{A}) = -|ACB| \quad (4.1)$$

According to the definition of a vector product, **A** × **B** is a vector (as shown in Fig. 1-10) that is perpendicular to **A** and **B** and points to the positive side, forming together with **A** and **B** a right-hand system. Its magnitude is $AB \sin \varphi$. If the angle between **A** × **B** and **C** is ψ, as in Fig. 1-10, the product is

$$(\mathbf{A} \times \mathbf{B})\mathbf{C} = ABC \sin \varphi \cos \psi \tag{4.2}$$

which is just the volume of the parallelepiped ABC. If $\psi > \pi/2$, and the vector **A** × **B** thus pointing downward, the product is negative but still equal in size to the volume. In this case the succession of the vectors in a

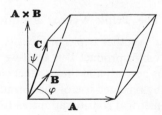

Fig. 1-10. (**A** × **B**)**C**, the volume of a parallelepiped.

right-hand system is **B—A—C**. It is natural, therefore, to generalize the concept of volume to include negative values by adopting a definition of it by the triple scalar product. This product is sometimes called a *pseudoscalar*.

Expression of the Volume by Vector Components

With the aid of the formal expression (3.13) the dependence of the triple product on its vector components is easily found. In the product **A × B**, we have only to replace the vectors **i**, **j**, **k** by C_x, C_y, C_z. Hence, the triple scalar product is equal to the ordinary determinant

$$(\mathbf{A} \times \mathbf{B})\mathbf{C} = \mathbf{A}(\mathbf{B} \times \mathbf{C}) = \begin{vmatrix} A_x & A_y & A_z \\ B_x & B_y & B_z \\ C_x & C_y & C_z \end{vmatrix}. \tag{4.3}$$

Polar and Axial Vectors

We have seen that the vector product of two vectors is a vector of a type somewhat different from the vectors from which the product is built up. In the definition of an ordinary vector we thought of the vector as representing some sort of displacement, in the sense that the vector was characterized by a definite magnitude and a definitite direction. Such vectors are called *polar vectors*.

The vector product is connected with a parallellogram, that is, with an orientated surface area of given sign of rotation or peripheral direction. Such vectors are called *axial vectors*. The description of the process underlying the definition of such a vector by an ordinary vector is only artificial. It stresses, however, the fact that in the case of the vector product we shall not distinguish between parallellograms of different shape, only between those of different size and orientation.

A vector that is undoubtedly of the axial type is the rotational velocity ω of a body. In other cases it may sometimes be doubtful whether vectorial physical quantities can be characterized with certainty as polar or axial vectors. What can definitely be said is that the vector products of two polar or two axial vectors are axial vectors; whereas the vector product of a polar and an axial vector is a polar vector.

An example of the latter case is provided by Fig. 1-11.

The Triple Vector Product A × (B × C)

Another triple product is the vector product of a vector **A** and another vector product **B × C**. If we use the triple scalar product, no new definition is needed, and the product can easily be expressed by the vector components by repeated use of the formula (3.13). It can, however, be evaluated by geometrical considerations as follows.

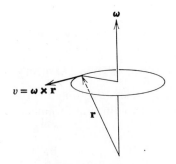 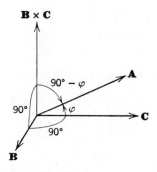

Fig. 1-11. Rotational (axial) and linear (polar) velocity.

Fig. 1-12. Triple vector product when B is orthogonal to A and C.

Since **B** × **C** is orthogonal to either of **B** and **C**, it is evident that the triple product is a vector in the plane of **B** and **C**,

$$\mathbf{A} \times (\mathbf{B} \times \mathbf{C}) = b\mathbf{B} - c\mathbf{C}. \tag{4.4}$$

It is also orthogonal to **A**; hence by scalar multiplication with **A**

$$b(\mathbf{AB}) - c(\mathbf{AC}) = 0 \quad \text{or} \quad b:c = \mathbf{AC}:\mathbf{AB}, \tag{4.5}$$

or

$$\mathbf{A} \times (\mathbf{B} \times \mathbf{C}) = \gamma[(\mathbf{AC})\mathbf{B} - (\mathbf{AB})\mathbf{C}] \tag{4.6}$$

where, obviously, γ is a pure number, since the product is linear in either of the components of the three vectors. It suffices then to determine γ in a particular case.

We may, for instant, take **B** orthogonal with respect to both **A** and **C**. Denoting the angle between **A** and **C** by φ as in Fig. 1-12, we have $\mathbf{AB} = 0$ and

$$\mathbf{A} \times (\mathbf{B} \times \mathbf{C}) = \gamma(\mathbf{AC})\mathbf{B} = \gamma AC \cos \varphi \mathbf{B}. \tag{4.7}$$

Now **B** × **C** is a vector which is also orthogonal to **B** and makes an angle $(90° - \varphi)$ with **A**. Its vector product with **A** as a prefactor therefore has the magnitude $ABC \sin (90° - \varphi) = ABC \cos \varphi$ and points in the direction of **B**. The net result is that in this case

$$\mathbf{A} \times (\mathbf{B} \times \mathbf{C}) = AC \cos \varphi \, \mathbf{B}, \tag{4.8}$$

which by (4.7) proves that $\gamma = 1$.

By Eq. (4.7) the general result, then is

$$\mathbf{A} \times (\mathbf{B} \times \mathbf{C}) = (\mathbf{AC})\mathbf{B} - (\mathbf{AB})\mathbf{C}. \tag{4.9}$$

Product of Vectors

The scalar product of two vector products is easily obtained by the combination of formulas for the scalar and vector triple products:

$$(A \times B)(C \times D) = A[B \times (C \times D)] = (AC)(BD) - (AD)(BC). \quad (4.10)$$

For the vector product of $A \times B$ and $C \times D$, we need only the formula for the triple vector product. We get, however, two different formulas according to the application of the triple formula

$$(A \times B) \times (C \times D) = [D(A \times B)]C - [C(A \times B)]D$$
$$= [B(D \times C)]A - [A(D \times C)]B. \quad (4.11)$$

This conforms with the fact that the product is a vector pointing in the direction of the line that is common to either of the planes through A and B and through C and D.

5. VECTORS AS DEPENDENT VARIABLES

Differentiation of Vectors

A vector may be a function of one or more independent variables. For instance, in the dynamics of mass points the space vector $r(t)$, velocity $v(t)$, and acceleration $a(t)$ are functions of a single variable, the time t. We denote this by writing $A(t)$ for the vector itself and $A_x(t)$ for its components. If it is a function of several variables, as for instance space coordinates x, y, z, and time t, we write $A(x, y, z, t)$ or sometimes simply A, it being understood that the number and kind of independent variables have been clearly stated.

The derivative of a vector $A(t)$ of a single variable can be defined in the very same way as the derivative of a single variable or a scalar

$$\frac{dA}{dt} = \lim_{\Delta t \to \infty} \frac{A(t + \Delta t) - A(t)}{\Delta t} = \lim_{\Delta t \to 0} \frac{\Delta A}{\Delta t} \quad (5.1)$$

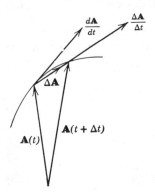

Fig. 1-13. dA/dt obtained as the limit of $\Delta A/\Delta t$.

This means that the derivative of a vector is obtained by simply differentiating its components. The process of differentiating is shown in Fig. 1-13.

It is easily proved that the differentiation of various products of vectors or vectors and scalars obeys the same rules as the differentiation of products of ordinary functions and scalars; for instance,

$$\frac{d(ab)}{dt} = \frac{da}{dt}b + a\frac{db}{dt},$$

$$\frac{d(a\mathbf{A})}{dt} = a\frac{d\mathbf{A}}{dt} + \frac{da}{dt}\mathbf{A},$$

$$\frac{d(\mathbf{AB})}{dt} = \frac{d\mathbf{A}}{dt}\mathbf{B} + \mathbf{A}\frac{d\mathbf{B}}{dt}, \qquad (5.2)$$

$$\frac{d(\mathbf{A}\times\mathbf{B})}{dt} = \frac{d\mathbf{A}}{dt}\times\mathbf{B} + \mathbf{A}\times\frac{d\mathbf{B}}{dt},$$

$$\frac{d}{dt}\mathbf{A}(\mathbf{B}\times\mathbf{C}) = \frac{d\mathbf{A}}{dt}(\mathbf{B}\times\mathbf{C}) + \mathbf{A}\left(\frac{d\mathbf{B}}{dt}\times\mathbf{C}\right) + \mathbf{A}\left(\mathbf{B}\times\frac{d\mathbf{C}}{dt}\right).$$

Curvature and Torsion of a Curve

If a curve in space is defined by the parameter representation

$$x = x(t), \qquad y = y(t), \qquad z = z(t) \qquad \text{or} \qquad \mathbf{r} = \mathbf{r}(t), \qquad (5.3)$$

it is possible by the equation

$$\left(\frac{ds}{dt}\right)^2 = \dot{x}^2 + \dot{y}^2 + \dot{z}^2 \qquad (5.4)$$

to express the arc length $s = s(t)$ of the curve as a function of t, since in (5.4), $\dot{x} = \dot{x}(t)$, and so forth, are known functions of t. Substituting for $t = t(s)$ in (5.3), we arrive at the representation

$$\mathbf{r} = \mathbf{r}(s) \qquad (5.5)$$

as a function of the arc length s. Since, by (5.4), $(d\mathbf{r})^2 = ds^2$, it follows that

$$\mathbf{t} = \frac{d\mathbf{r}}{ds} \qquad (5.6)$$

is a unit vector, which according to Fig. 1-14a points tangentially to the curve. Again $d\mathbf{t}/ds$ is a vector orthogonal to $\mathbf{t}$. From Fig. 1-14b it is seen that approximately $\rho \Delta\varphi = \Delta s$, ρ meaning the radius of curvature and $\Delta\varphi$ the angle between the curve normals in the point s and $s + \Delta s$. It is also seen from the figure that approximately $|\Delta \mathbf{t}| = \Delta\varphi$. Hence

$$\frac{d\mathbf{t}}{ds} = \frac{\mathbf{n}}{\rho}, \qquad (5.7)$$

18 / The Mathematical Foundation of Physics

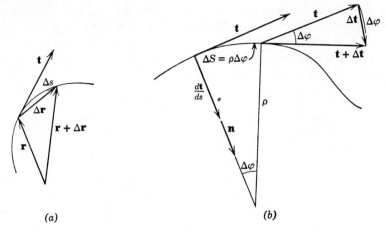

Fig. 1-14. (a) The derivation of **r**(s) with respect to arc length s. (b) Derivation of the tangent vector.

n meaning the unit normal vector of the curve which is situated in the so-called osculating plane and points toward the center of curvature. The unit vector **n** is called the *principal normal:* the unit vector

$$\mathbf{b} = \mathbf{t} \times \mathbf{n} \tag{5.8}$$

is called the *secondary normal*.

If **b** is constant in direction, the curve is a plane curve. If **b** changes, the curve is not plane, and the deviation is measured by $d\mathbf{b}/dt$. This again is a vector orthogonal to the unit vector **b**. Next, because of (5.7),

$$\frac{d\mathbf{b}}{dt} = \mathbf{t} \times \frac{d\mathbf{n}}{dt}, \tag{5.9}$$

and the vector is orthogonal to **t**. Hence

$$\frac{d\mathbf{b}}{dt} = -\tau\mathbf{n} \tag{5.10}$$

τ being a scalar factor named *torsion*.

Consider, for instance, the curve illustrated in Fig. 1-15b whose parameter representation by the independent variable φ, the equatorial angle, is given by

$$\mathbf{r} = a(\mathbf{i} \sin \varphi + \mathbf{j} \sin \varphi) + \mathbf{k}h\varphi. \tag{5.11}$$

It can be proved by the above equations that

$$\rho = a + \frac{h^2}{a} \quad \text{and} \quad \frac{1}{\tau} = \frac{a^2}{h} + h. \tag{5.12}$$

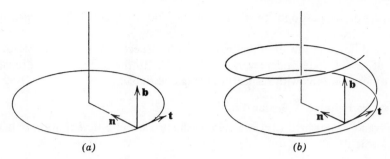

Fig. 1-15. (a) Plane curve of constant curvature. (b) Spiral curve with constant curvature and torsion.

Both curvature $1/\rho$ and torsion τ tend to zero in either of the cases $a \to \infty$ or $h \to \infty$.

6. SCALAR AND VECTOR FIELDS

Definitions

A physical or geometrical quantity may be of a kind that has a definite value for any point in space. We may consider it a single quantity that is a function of the space coordinates x, y, z, or, for brevity, a function of the space vector $\mathbf{r} = \mathbf{i}x + \mathbf{j}y + \mathbf{k}z$. The quantity may be a scalar or a vector $\varphi = \varphi(x, y, z) = \varphi(\mathbf{r})$, $\mathbf{A} = \mathbf{A}(x, y, z) = \mathbf{A}(\mathbf{r})$ and is said to form a *scalar* or a *vector field*. The field may also depend on other variables; for instance, it may vary with the time, $\mathbf{A} = \mathbf{A}(\mathbf{r}, t)$. In certain cases, as when we are considering fields acting on a particle that moves through space, $\mathbf{r} = \mathbf{r}(t)$, the quantity may again become a function only of time, $\mathbf{A} = \mathbf{A}(\mathbf{r}(t), t)$.

A well-known example of a vector field is provided by the instantaneous velocity distribution $\mathbf{v} = \mathbf{v}(\mathbf{r}, t)$ in a fluid or, eventually, a stationary distribution $\mathbf{v} = \mathbf{v}(r)$. Such a field is frequently used for illustration purposes, leading to phrases like vector *flow* or *flux* (through a surface element or a closed surface).

The flow of a fluid cannot, however, be quite an arbitrary one. In a highly compressible fluid the flow is less restricted than in an incompressible fluid, but in any fluid the possibilities are limited. By an artificial procedure, however, even the flow of an incompressible fluid can be made arbitrary, namely, by the conception of artificial production or destruction of the fluid in the various parts in space. For instance, if through a closed surface in space there is a total positive outward flux, it is necessary to have a production of an equally large quantity of fluid within the surface in order to prevent formation of empty regions in the fluid. This production of fluid (with negative sign for

20 / The Mathematical Foundation of Physics

destruction) corresponds to what in vector analysis is called the *divergence* of a vector field.

The differentiation of a vector field with respect to one coordinate or to one of several independent variables is subject to the same rules as are valid for differentiations with respect to a single variable.

The Gradient of a Scalar Field

Given a scalar field $\varphi(\mathbf{r})$, we may by the process of differentiation obtain a vector field

$$\operatorname{grad} \varphi = \mathbf{i}\frac{\partial \varphi}{\partial x} + \mathbf{j}\frac{\partial \varphi}{\partial y} + \mathbf{k}\frac{\partial \varphi}{\partial z}, \tag{6.1}$$

which is called the *gradient* of the scalar φ. The gradient is a true vector in the sense that its definition is independent of the particular coordinate system in which the differentiation is expressed. This can be shown as follows.

Consider a point in space and a curve $\mathbf{r} = \mathbf{r}(s)$ through the point, s being the arc length of the curve. The quantity $d\varphi/ds$ obviously is the same in any coordinate system. On the other hand,

$$\frac{d\varphi}{ds} = \frac{\partial \varphi}{\partial x}\frac{\partial x}{\partial s} + \frac{\partial \varphi}{\partial y}\frac{\partial y}{\partial s} + \frac{\partial \varphi}{\partial z}\frac{\partial z}{\partial s} = \mathbf{t} \operatorname{grad} \varphi, \tag{6.2}$$

which means that the projection of the gradient vector on the unit tangent vector $\mathbf{t}$, hence on any given unit vector in space, is the same in any coordinate system.

A scalar field or, for brevity, a potential can conveniently be illustrated by means of equipotential surfaces defined by equations

$$\varphi(\mathbf{r}) = c + nd, \quad n = \cdots -2, -1, 0, 1, 2, \cdots. \tag{6.3}$$

Since along such a surface the potential is constant, it follows from

$$d\varphi = \frac{\partial \varphi}{\partial x} dx + \frac{\partial \varphi}{\partial y} dy + \frac{\partial \varphi}{\partial z} dz = d\mathbf{r} \operatorname{grad} \varphi = 0 \tag{6.4}$$

that the gradient vector is everywhere orthogonal to the equipotential surface of the scalar.

If, in (6.2), the vector $\mathbf{t}$ is taken in the direction of $\operatorname{grad} \varphi$, we have

$$\frac{d\varphi}{ds} = |\operatorname{grad} \varphi|. \tag{6.5}$$

On the other hand, in the direction of the gradient there are m equipotential surfaces or levels per unit length. We have approximately

$$\frac{d\varphi}{ds} = m. \tag{6.6}$$

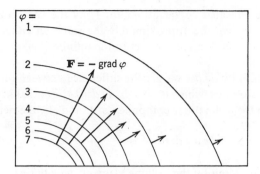

Fig. 1-16. Equipotential surfaces and gradient.

Hence $|\text{grad } \varphi| = m$, and the gradient is proportional to the number of levels per unit length.

In Fig. 1-16 a scalar field and its negative gradient field are illustrated by means of equipotential surfaces.

The Divergence of a Vector

From a vector field $A(\mathbf{r})$ we can obtain a scalar field called the divergence of the vector field

$$\text{div } \mathbf{A} = \frac{\partial A_x}{\partial x} + \frac{\partial A_y}{\partial y} + \frac{\partial A_z}{\partial z}. \tag{6.7}$$

As already mentioned in discussing the vector field as the velocity field of a flowing fluid, the divergence of the field may be thought of as the flux of fluid per unit time from a volume element as referred to unit volume.

By the aid of Fig. 1-17 the flux is easily calculated. Consider the infinitesimal volume element $dx\,dy\,dz$. Let us calculate the flux through the sides $dy\,dz$ due to the x-component of the field. From x to $x + dx$, there will be a general difference in A_x of the first order in dx

$$[A_x(x + dx, y, z) - A_x(x, y, z)] = dx \frac{\partial A_x}{\partial x}. \tag{6.8}$$

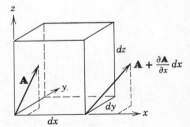

Fig. 1-17. Calculation of the flux from a volume element.

22 / The Mathematical Foundation of Physics

Of course, there will also be variations in A_x by the variations of y and z, but these variations will be approximately the same on either side and can only lead to quantities of higher orders in the infinitesimal quantities dx, dy, and dz.

Multiplying (6.7) by $dy\, dz$, we get the difference between outward and inward flux at the two sides, which is now $dx\, dy\, dz(\partial A_x/\partial x)$. Taking into account the contribution from the other components A_y and A_z, the net flux becomes

$$dx\, dy\, dz\left(\frac{\partial A_x}{\partial x} + \frac{\partial A_y}{\partial y} + \frac{\partial A_z}{\partial z}\right), \tag{6.9}$$

from which, on dividing by the volume element, we obtain

$$\text{div } \mathbf{A} = \frac{\partial A_x}{\partial x} + \frac{\partial A_y}{\partial y} + \frac{\partial A_z}{\partial z}. \tag{6.10}$$

Assuming this expression to be invariant with respect to the orientation of the coordinate system, we may omit a special proof of the Gauss theorem as that usually given in textbooks. The result may become even more convincing, however, if it is deduced by another, quite different, procedure.

Consider a closed surface of quite a different shape that does not alter its shape by rotation of the coordinate system, namely, a sphere of a small radius ρ. Let ξ, η, ζ be the coordinates on its surface in a system with its origin at the center. We introduce some mean values of coordinates and coordinate products, the mean values being referred to equal weight of equal areas on the sphere. Thus

$$\bar{\xi} = \bar{\eta} = \bar{\zeta} = 0, \qquad \overline{\xi\eta} = 0, \text{ etc.}, \qquad \overline{\xi^2} = \overline{\eta^2} = \overline{\zeta^2} = \tfrac{1}{3}\rho^2. \tag{6.11}$$

Again we first calculate the mean flux through the surface in the x-direction by expanding $A_x(x + \xi, y + \eta, z + \zeta)$ in powers of ξ, η, ζ, the center of the sphere having the coordinates x, y, z. See Fig. 1-18. Since we end with the limiting case $\rho \to 0$, only the first powers of the expression need be considered.

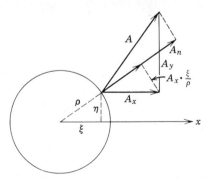

Fig. 1-18. Calculation of the divergence by means of the flux from a sphere.

It is easily seen that the mean flux per surface area is found by taking the mean value of

$$\left[A_x(x, y, z) + \frac{\partial A_x}{\partial x} \xi + \frac{\partial A_x}{\partial y} \eta + \frac{\partial A_x}{\partial z} \right] \frac{\xi}{\zeta}$$

over the surface. By this process all terms except the second vanish, the latter yielding a mean value $\frac{1}{3}\rho(\partial A_x/\partial x)$. Multiplying by the surface area $4\pi\rho^2$ and adding the flux in the y- and z-directions, we find a total flux

$$\frac{4\pi}{3} \rho^3 \left(\frac{\partial A_x}{\partial x} + \frac{\partial A_y}{\partial y} + \frac{\partial A_z}{\partial z} \right) = dV \text{ div } \mathbf{A}. \qquad (6.12)$$

Making $\rho \to 0$ in order to increase the accuracy of the expansion and dividing by $dV = (4\pi/3)\rho^3$, we get the same expression as before for the divergence of the field.

The Curl of a Vector

From a given vector field $\mathbf{A}(\mathbf{r})$, another vector field can be derived, called the curl of the field, which we denote by curl $\mathbf{A}$. Analytically, it may be written as

$$\text{curl } \mathbf{A} = \begin{vmatrix} \mathbf{i} & \mathbf{j} & \mathbf{k} \\ \dfrac{\partial}{\partial x} & \dfrac{\partial}{\partial y} & \dfrac{\partial}{\partial z} \\ A_x & A_y & A_z \end{vmatrix} \qquad (6.13)$$

or, for example,

$$\text{curl }_z \mathbf{A} = \frac{\partial A_y}{\partial x} - \frac{\partial A_x}{\partial y}. \qquad (6.13a)$$

It is possible to interpret either of the components of this vector as the *circulation* of the vector $\mathbf{A}$ along a closed curve in the plane perpendicular to the component divided by the area of the enclosed surface.

Consider the simple infinitesimal surface element $dx\, dy$ orthogonal to A_z. By the circulation we mean what is later more precisely defined as a curve integral along the periphery $dx, dy, -dx, -dy$, which is in the first approximation $A_x\, dx + A_y\, dy - A_y\, dy - A_x\, dx = 0$. In the next approximation we have to take into account the difference in the component A_y for the coordinates x and $x + dx$, which is in the first approximation $dx\, \partial A_y/\partial x$, see Fig. 1-19. Similarly, the difference in A_x from $y + dy$ to y is $-dy\, \partial A_x/\partial y$. Hence in this approximation the circulation is

$$\left(\frac{\partial A_y}{\partial x} - \frac{\partial A_x}{\partial y} \right) dx\, dy = \text{curl }_z \mathbf{A}\, dx\, dy. \qquad (6.14)$$

24 / The Mathematical Foundation of Physics

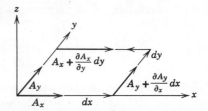

Fig. 1-19. Calculation of curl component.

Using the above interpretation as a definition, we obtain the expression (6.13a). In addition, it is obvious from the definition that the integral $\int \text{curl}\,_z A\, dx\, dy$, that is, the surface integral of curl $_z A$ taken over any finite area in the xy-plane, equals the circulation along the periphery of the area in the positive sense. This means that—except for some additional considerations—the Stokes theorem of the next section needs no further particular proof.

As in the case of divergence, it may also be more convincing here to have an alternative and independent calculation of the curl components as defined above. Consider, for instance, a circular curve in the xy-plane, which has the property of being unchanged by rotation of the coordinate system around the z-axis. Let ρ be the radius and ξ and η the coordinates of a point on the circle as measured from the coordinates x and y of its center. The mean values of coordinates and coordinate products are now

$$\bar{\xi} = \bar{\eta} = 0, \qquad \overline{\xi\eta} = 0,$$
$$\overline{\xi^2} = \overline{\eta^2} = \tfrac{1}{2}\rho^2. \tag{6.15}$$

We now consider the x- and y-components which in the first approximation of a power expansion are

$$A_x + \frac{\partial A_x}{\partial x}\xi + \frac{\partial A_x}{\partial y}\eta, \qquad A_y + \frac{\partial A_y}{\partial x}\xi + \frac{\partial A_y}{\partial x}\eta.$$

The tangential component is obtained by multiplying by $-\eta/\rho$ and ξ/ρ and adding the two products,

$$A_s = \frac{1}{\rho}\left[A_y\xi - A_x\eta + \left(\frac{\partial A_y}{\partial y} - \frac{\partial A_x}{\partial x}\right)\xi\eta + \frac{\partial A_y}{\partial x}\xi^2 - \frac{\partial A_x}{\partial y}\eta^2\right] \tag{6.16}$$

yielding the mean value

$$\bar{A}_s = \tfrac{1}{2}\rho\left(\frac{\partial A_y}{\partial x} - \frac{\partial A_x}{\partial y}\right). \tag{6.17}$$

Upon multiplying by the circumference $2\pi\rho$ and dividing by the area $\pi\rho^2$ of the circle, the result (6.13a) is again obtained.

In the representation of the vector field by a flowing fluid the curl is twice the rotational velocity vector. This may be seen by considering the same circle as above and by multiplying the linear velocity $\omega_z \rho$ by circumference,

$$2\pi\rho \cdot \rho\omega_z = \pi\rho^2 \cdot 2\omega_z. \tag{6.18}$$

Dividing by the area $\pi\rho^2$, we have

$$\text{curl}\,_z\mathbf{A} = 2\omega_z \quad \text{or} \quad \text{curl}\,\mathbf{A} = 2\boldsymbol{\omega}. \tag{6.19}$$

The Curl as a True Vector

The present considerations of the vector curl so far are not as complete as those of the scalar divergence. To remove this lack consider the component of curl $\mathbf{A}$ in an arbitrary direction as defined above and let $\mathbf{n}$ be the unit vector in that direction.

We now choose an arbitrarily orientated Cartesian system whose axes are cut by the normal plane of $\mathbf{n}$ in the points L, M, N. Consider the triangle LMN, together with the triangles MON, and so forth, cut from the yz-plane. As shown in Fig. 1-20 the sum of the circulations of the vector field $\mathbf{A}$ around the latter three triangles obviously equals the circulation around LMN. The areas of the three triangles are, respectively, $n_x f, n_y f, n_z f$, f being the area of LMN. Hence

$$f \,\text{curl}\,_n\mathbf{A} = f(n_c \,\text{curl}\,_x\mathbf{A} + n_y \,\text{curl}\,_y\mathbf{A} + n_z \,\text{curl}\,_z\mathbf{A}). \tag{6.20}$$

Cancelling the f, we obtain an equation that proves the above defined components of the curl to be in reality the components of a vector.

The Vectorial Differential Operator

By use of the symbolic vector

$$\boldsymbol{\nabla} = \mathbf{i}\frac{\partial}{\partial x} + \mathbf{j}\frac{\partial}{\partial y} + \mathbf{k}\frac{\partial}{\partial z}, \tag{6.21}$$

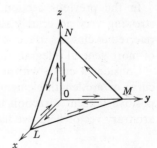

Fig. 1-20.   The curl as a true vector.

which is in reality a vectorial differential operator, some of the above results can be expressed considerably simpler. With reference to formulas from vector algebra many differential operations can be even more easily performed. We may write, for instance,

$$\operatorname{grad} \varphi = \left(\mathbf{i}\frac{\partial}{\partial x} + \mathbf{j}\frac{\partial}{\partial y} + \mathbf{k}\frac{\partial}{\partial z}\right)\varphi = \nabla \varphi$$

$$\operatorname{div} \mathbf{A} = \left(\mathbf{i}\frac{\partial}{\partial x} + \mathbf{j}\frac{\partial}{\partial y} + \mathbf{k}\frac{\partial}{\partial z}\right)\mathbf{A} = \nabla \mathbf{A} \qquad (6.22)$$

$$\operatorname{curl} \mathbf{A} = \left(\mathbf{i}\frac{\partial}{\partial x} + \mathbf{j}\frac{\partial}{\partial y} + \mathbf{k}\frac{\partial}{\partial z}\right) \times \mathbf{A} = \nabla \times \mathbf{A}.$$

By repeated use of the operators ∇ and $\nabla \times$, a number of new scalar and vector fields or scalar and vector equations can be obtained, such as

$$\operatorname{curl} \operatorname{grad} \varphi = \nabla \times \nabla \varphi = \begin{vmatrix} \mathbf{i} & \mathbf{j} & \mathbf{k} \\ \frac{\partial}{\partial x} & \frac{\partial}{\partial y} & \frac{\partial}{\partial z} \\ \frac{\partial \varphi}{\partial x} & \frac{\partial \varphi}{\partial y} & \frac{\partial \varphi}{\partial z} \end{vmatrix} = 0, \qquad (6.23)$$

$$\operatorname{div} \operatorname{curl} \mathbf{A} = \nabla(\nabla \times \mathbf{A}) = (\nabla \times \nabla)\mathbf{A} = 0, \qquad (6.24)$$

$$\operatorname{div} \operatorname{grad} \varphi = \nabla(\nabla \varphi) = \nabla^2 \varphi = \left(\frac{\partial^2}{\partial x^2} + \frac{\partial^2}{\partial y^2} + \frac{\partial^2}{\partial z^2}\right)\varphi \qquad (6.25)$$

$$\operatorname{curl} \operatorname{curl} \mathbf{A} = \nabla \times (\nabla \times \mathbf{A}) = \nabla(\nabla \mathbf{A}) - \nabla^2 \mathbf{A}$$
$$= \nabla \operatorname{div} \mathbf{A} - \nabla^2 \mathbf{A}. \qquad (6.26)$$

7. INTEGRAL THEOREMS BY GAUSS, GREEN, AND STOKES

Elementary Curve and Surface Integrals

In the previous section we used surface and curve integrals without stressing very accurately the definition of the integrals. Such integrals as referred to closed surfaces or closed curves are of course only particular cases of more general integrals for open finite surfaces and curves.

Consider a curve with a given positive direction (Fig. 1-21a). If ds is a curve or arc length element, a vector $d\mathbf{s}$ of the same length pointing in the positive tangent direction is called the vectorial curve element. If $\mathbf{G}$ is an arbitrary vector field, we have

$$\mathbf{G}\,d\mathbf{s} = G_s\,ds, \qquad (7.1)$$

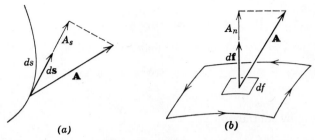

Fig. 1-21. (*a*) Vectorial line element $d\mathbf{s}$. (*b*) Vectorial surface element $d\mathbf{f}$.

G_s being the tangential component of G. This expression, if integrated between two points of the curve, is called a *curve* integral

$$\int_a^b G_s \, ds = \int_a^b \mathbf{G} \, d\mathbf{s}. \tag{7.2}$$

In Fig. 1-21*b* we have a surface bounded by an oriented closed curve. A scalar surface element df may be used for definition of a corresponding vectorial element

$$d\mathbf{f} = \mathbf{n} \, df \quad (\text{or } d\mathbf{f} = \mathbf{n} \, d\mathbf{f}), \tag{7.3}$$

$\mathbf{n}$ being the positive normal vector of the surface element as defined in accordance with the positive direction on the peripheral curve.

For an arbitrary vector field $\mathbf{F}$ we now have

$$\mathbf{F} \, d\mathbf{f} = F_n \, df \tag{7.4}$$

and a surface integral for a bound region as in Fig. 1-21*a* is defined as

$$\int_C F_n \, df = \int_C \mathbf{F} \, d\mathbf{f}. \tag{7.5}$$

For closed curves we usually write $\oint \mathbf{G} \, d\mathbf{s}$ for a curve integral; for surface integrals we very often write simply $\int \mathbf{F} \, d\mathbf{f}$, whether referring to open or to closed surfaces.

The Gauss and the Green Theorems

The *Gauss theorem* relating to a closed surface and the volume integral of its interior may be written

$$\int \mathbf{F} \, d\mathbf{f} = \int \operatorname{div} \mathbf{F} \, dv. \tag{7.6}$$

It will here be considered to be valid on the basis of the definition of the divergence. The corresponding *Green theorem* is virtually a particular form of the Gauss theorem which is obtained as follows. By partial

derivation
$$\text{div}(\varphi \mathbf{A}) = \varphi \text{ div } \mathbf{A} + \mathbf{A} \nabla \varphi \qquad (7.7)$$
or, putting $\mathbf{A} = \nabla \psi$ and next interchanging φ and ψ
$$\begin{aligned}\text{div}(\varphi \nabla \psi) &= \varphi \nabla^2 \psi + \nabla \varphi \nabla \psi, \\ \text{div}(\psi \nabla \varphi) &= \psi \nabla^2 \varphi + \nabla \psi \nabla \varphi,\end{aligned} \qquad (7.8)$$
and finally by subtraction
$$\text{div}(\varphi \nabla \psi - \psi \nabla \varphi) = \varphi \nabla^2 \psi - \psi \nabla^2 \varphi. \qquad (7.9)$$

Integrating the equation for a region in space within a closed surface, we obtain the Green theorem

$$\int (\varphi \nabla^2 \psi - \psi \nabla^2 \varphi) \, dv = \int \left(\varphi \frac{\partial \psi}{\partial n} - \psi \frac{\partial \varphi}{\partial n} \right) df, \quad \frac{\partial \psi}{\partial n} = \mathbf{n} \nabla \psi \quad (7.10)$$

which is often used for integration purposes.

The Stokes Theorem

The Stokes theorem relates to a closed curve and any surface bounded by this curve. The curve and the surface are orientated by positive tangent direction and positive normal vector according to the right-hand rule. It reads

$$\oint \mathbf{G} \, d\mathbf{s} = \int \text{curl } \mathbf{G} \, d\mathbf{f} = \int \text{curl }_n \mathbf{G} \, df. \qquad (7.11)$$

As in the former case of the Gauss theorem, the Stokes theorem is considered valid on the basis of definitions already used. The definition of a curl component, however, was related to the circulation or line integral along a closed plane curve. Consider therefore a nonplane surface bounded by a curve. As shown in Fig. 1-22 we divide the surface into smaller regions with common boundaries of neighboring regions. Along these boundaries all circulations or line integrals appear twice and in either direction. Summing up all smaller line integrals, we find that the net result is the line integral of the boundary curve. On the other hand, the smaller regions can be made plane to any desired approximation. The Stokes theorem, being valid for any of the smaller

Fig. 1-22. Sum of smaller contour integral equals that of the bordering curve.

plane regions of the surface is therefore valid for the whole surface and its boundary curve.

The form of the surface thus is immaterial. The surface integral depends solely on the boundary curve. This result can be confirmed by means of the Gauss theorem as applied to two different surfaces with the same boundary curve and to the space region enclosed by the two surfaces. The difference of the two surface integrals obviously may be considered a single integral over a closed surface when reversing the sign of the normal vector for one of them, as shown in Fig. 1-24. By the Gauss theorem this integral equals the inside volume integral of the divergence.

If now the vector appearing in these integrals is, say, curl **G**, the volume integral vanishes because of the general result

$$\text{div curl } \mathbf{G} = 0, \tag{7.12}$$

and the Stokes theorem is true whether referred to the one or the other of equally large surface integrals.

Equivalence of the Gauss and the Green Theorem in a Plane

The Gauss theorem is easily generalized to a space of higher or lower dimensions. In a two-dimensional plane it obviously reads

$$\oint F_n \, ds = \int \left(\frac{\partial F_x}{\partial x} + \frac{\partial F_y}{\partial y} \right) dx \, dy, \tag{7.13}$$

where n indicates the curve normal. Now consider another vector **G** defined by $G_x = -F_y$, $G_y = F_x$, then applying for a moment the vector algebra of three-dimensional space

$$\mathbf{k} \times \mathbf{F} = \mathbf{k} \times (\mathbf{i}F_x + \mathbf{j}F_y) = -\mathbf{i}F_y + \mathbf{j}F_x = \mathbf{G}. \tag{7.14}$$

This means that the vector **G** is obtained by turning **F** 90° around the plane normal **k**; that is, $G_s = F_n$. Rewriting (7.13) in terms of **G**, we find

$$\oint G_s \, ds = n \int \left(\frac{\partial G_y}{\partial x} - \frac{\partial G_x}{\partial y} \right) dx \, dy, \tag{7.15}$$

which proves that in a plane the two theorems have virtually degenerated into a single one.

8. LAMELLAR AND SOLENOIDAL VECTOR FIELD.

Potential and Gradient

Since curl $\nabla \varphi = 0$, it is to be expected that an *irrotational* field F with vanishing curl, curl $\mathbf{F} = 0$, can be represented by a gradient. According to the

30 / The Mathematical Foundation of Physics

Stokes theorem any contour integral of the vector will be zero. This means that any line integral between a and b in Fig. 1-23 has the same value irrespective of the path of integration,

$$\int_a^b \mathbf{F} \, d\mathbf{s} = \Phi(\mathbf{r}_a, \mathbf{r}_b) = \varphi(r_b) - \varphi(r_a). \tag{8.1}$$

The latter form of the right-hand side is a consequence of the equation being valid for both terminals. The equation can be true only if $\mathbf{F}\,d\mathbf{s}$ is a total differential of a scalar function of the coordinates x, y, z,

$$\mathbf{F}\,d\mathbf{s} = \frac{\partial \varphi}{\partial x} dx + \frac{\partial \varphi}{\partial y} dy + \frac{\partial \varphi}{\partial z} dz = \operatorname{grad} \varphi \, d\mathbf{s}. \tag{8.2}$$

Hence

$$\mathbf{F} = \operatorname{grad} \varphi. \tag{8.3}$$

The scalar is called the potential of the gradient. Hence an irrotational or gradient field can be represented by a set of equipotential surfaces dividing the space into lamellas of variable thickness, for which reason the gradient field may also be named a *lamellar* field.

Vector Potential and Curl

From the equation div curl $\mathbf{A} = 0$, it follows from the Gauss theorem that any closed surface integral of curl $\mathbf{A}$ is zero. By considering a surface consisting of a cylinder along the vector lines, together with surfaces normal to the lines at both ends, it follows that the flux of curl $\mathbf{A}$ through either of the latter surfaces is the same. If we imagine the density of the vector line to be proportional to the field strength, then the number of vector lines in the field curl $\mathbf{A}$ cutting either of the end surfaces is the same. A tube of this kind is sometimes called a solenoid, for which reason the field curl $\mathbf{A}$ may be named *solenoidal*. (See Fig. 1-24.)

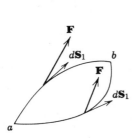

Fig. 1-23. $\oint \mathbf{F}\,d\mathbf{s} = \int_a^b \mathbf{F}\,d\mathbf{s}_1 - \int_a^b \mathbf{F}\,d\mathbf{s}_2 = 0.$

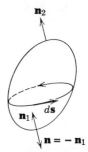

Fig. 1-24.

Since any nondivergent field **G**, div **G** = 0, is solenoidal for the same reasons as above, it is expressible as the curl of another vector **A**. If we write

$$\mathbf{G} = \operatorname{curl} \mathbf{A}, \tag{8.4}$$

the vector **A** is said to be the *vector potential* of **G**.
Because of the equation

$$\operatorname{div} \mathbf{G} = 0, \tag{8.5}$$

the three components of **G** are not mutually independent. The same is true for the components of **A** which need not be presumed to be entirely independent. We therefore try the sometimes very useful restriction

$$\operatorname{div} \mathbf{A} = 0. \tag{8.6}$$

It may be appropriate at this stage to show by direct calculations that for an arbitrary field **G**, which is restricted only by (8.5), the equations (8.4) and (8.6) can be satisfied.

If, for instance, G_x and G_y are arbitrarily chosen, (8.5) is satisfied by

$$G_z = -\int_{z_0}^{z} \left(\frac{\partial G_x}{\partial x} + \frac{\partial G_y}{\partial y} \right) dz. \tag{8.7}$$

On the other hand, if for the moment A_z is an arbitrary function of the coordinates, the expressions

$$A_x = \int_{z_0}^{z} \left(G_y + \frac{\partial A_z}{\partial x} \right) dz, \quad A_y = \int_{z_0}^{z} \left(-G_x + \frac{\partial A_z}{\partial y} \right) dz, \tag{8.8}$$

are obtained from the first two component equations (8.4). Next we obtain from (8.8)

$$\frac{\partial A_y}{\partial x} - \frac{\partial A_x}{\partial y} = -\int_{z_0}^{z} \left(\frac{\partial G_x}{\partial x} + \frac{\partial G_y}{\partial y} \right) dz = G_z, \tag{8.9}$$

so that the third component equation of (8.4) is satisfied.

To satisfy an additional condition as div $A = 0$ it is now sufficient to put a restriction on the so-far arbitrary function A_z. According to (8.8)

$$\operatorname{div} \mathbf{A} = \int_{z_0}^{z} \left(\frac{\partial G_y}{\partial x} - \frac{\partial G_x}{\partial y} + \frac{\partial^2 A_z}{\partial x^2} + \frac{\partial^2 A_z}{\partial y^2} + \frac{\partial^2 A_z}{\partial z^2} \right) dz. \tag{8.10}$$

Hence to obtain div $\mathbf{A} = 0$ we require

$$\nabla^2 A_z + \operatorname{curl}_z \mathbf{G} = 0. \tag{8.11}$$

It is interesting to note that corresponding equations for A_y and A_x can be obtained directly from (8.4) by taking the curl of the equation and assuming div $\mathbf{A} = 0$,

$$\operatorname{curl} \mathbf{G} = \operatorname{curl} \operatorname{curl} \mathbf{A} = \operatorname{div} \mathbf{A} - \nabla^2 \mathbf{A} = -\nabla^2 \mathbf{A}. \tag{8.12}$$

9. DETERMINATION OF VECTOR FIELDS FROM THEIR DIVERGENCE AND CURL

Separation into an Irrotational and a Solenoidal Field

Consider a field **D** defined by the equations

$$\text{div } \mathbf{D} = 4\pi\rho, \qquad \text{curl } \mathbf{D} = 4\pi\mathbf{u}, \tag{9.1}$$

$\rho(x, y, z)$ and $\mathbf{u}(x, y, z)$ being known functions of space. We shall try to separate the field into an irrotational and a solenoidal part, **F** and **G**, by writing

$$\mathbf{D} = \mathbf{F} + \mathbf{G},$$
$$\text{curl } \mathbf{F} = 0, \qquad \text{div } \mathbf{F} = 4\pi\rho, \tag{9.2}$$
$$\text{div } \mathbf{G} = 0, \qquad \text{curl } \mathbf{G} = 4\pi\mathbf{u}.$$

Apart from additional fields **F'** and **G'** which are both irrotational and solenoidal, the fields **F** and **G** are uniquely determined by the above equations. Putting

$$\mathbf{F} = -\text{grad } \Phi, \qquad \mathbf{G} = \text{curl } \mathbf{A}, \qquad \text{div } \mathbf{A} = 0, \tag{9.3}$$

we are led to the equations

$$\nabla^2 \Phi = -4\pi\rho, \qquad \nabla^2 \mathbf{A} = -4\pi\mathbf{u} \tag{9.4}$$

of the Poisson type for the potential Φ and either of the components of **A**.

Solution of the Poisson Equation

The Poisson equation $\nabla^2 \Phi = -4\pi\rho$ plays a predominant role in electrostatics. It can be solved by the so-called *Green method* which makes use of an auxiliary function

$$\psi = \frac{1}{r} \tag{9.5}$$

that is a particular solution of the corresponding *Laplace equation* $\nabla^2 \psi = 0$.

We now apply the Green theorem (7.10) to a region in space bounded by an inner and an outer closed surface, 1 and 2. The inner surface may be taken to be a sphere around a point ξ, η, ζ, where we want to know the potential Φ, and

$$r = ((x - \xi)^2 + (y - \eta)^2 + (z - \zeta)^2)^{1/2} \tag{9.6}$$

is the distance of this point from the integration point x, y, z. Using the Poisson and the Laplace equations for Φ we obtain

$$\int \frac{\rho}{r} dv = \frac{1}{4\pi} \int \left(\frac{\Phi}{r^2} + \frac{1}{r} \frac{\partial \Phi}{\partial r} \right) df_1 + \frac{1}{4\pi} \int \left(\Phi \frac{\partial \psi}{\partial n} - \psi \frac{\partial \Phi}{\partial n} \right) df_2 \tag{9.7}$$

where $\partial/\partial n = \mathbf{n}_2 \nabla$ and $\partial/\partial r = -\mathbf{n}_1 \nabla$, since the surface normal $\mathbf{n}_1$ is pointing inward to the center of sphere 1.

If the outer surface now withdraws toward infinity, and Φ decreases at least as r^{-1}, the second surface integral will vanish.

For a sufficiently small sphere around ξ, η, ζ the second term in the first integral can be made arbitrarily small. In the first term Φ is approximately constant and may be replaced by $\Phi(\xi, \eta, \zeta)$ and taken outside, whereby the remaining integral becomes 1. Hence the result for Φ and correspondingly for $\mathbf{A}$ is

$$\Phi(\xi, \eta, \zeta) = \int \frac{\rho(x, y, z)}{r} \, dv, \qquad \mathbf{A}(\xi, \eta, \zeta) = \int \frac{u(x, y, z)}{r} \, dv. \qquad (9.8)$$

Solution for a Finite Region

If the outer surface remains at finite distances, we have the modified solution

$$\Phi(\xi, \eta, \zeta) = \int \frac{\rho}{r} \, dv + \frac{1}{4\pi} \int \left(\Phi \frac{\partial}{\partial n} \frac{1}{r} - \frac{1}{r} \frac{\partial \Phi}{\partial n} \right) df. \qquad (9.9)$$

In this equation the sign of the surface normal of the outer boundary, and consequently all the signs of the integral, have been reversed. In Part 3, Chapter 1, on electrostatics, the solution is discussed in more detail. We shall be satisfied to state here that the solution is determined by the divergence of the field $\mathbf{F} = -\nabla \Phi$ in the bounded region, together with the values of Φ and $\partial \Phi/\partial n$ on the outer surface.

Uniqueness of the Solution. Laplacian Fields

Apart from the possible additional fields $\mathbf{F}'$ and $\mathbf{G}'$, as mentioned before, the solutions discussed above are unique. It remains to prove that $\mathbf{F}'$ or $\mathbf{G}'$ are zero fields. Both the divergence and curl of these vectors are zero; hence the two fields can be united into one field called $\mathbf{F}'$. Being irrotational, it can be taken as a gradient field

$$\mathbf{F}' = -\nabla \Phi', \qquad (9.10)$$

and from div $\mathbf{F}' = 0$ it follows that

$$\nabla^2 \Phi' = 0. \qquad (9.11)$$

The field $\mathbf{F}'$ itself obeys the same equation and is called a Laplacian field.

Consider first an infinite region and let us look for a solution of (9.11) that is finite and continuous to the first-order derivatives everywhere in space. In addition, it shall be zero to the order r^{-1} at infinite distances from the origin of our coordinate system.

Continuous functions must be expressible by power series in x, y, z; for instance, $\Phi' = 1$ and $\Phi' = x$ or y or z are solutions of (9.11) but neither of them tend to zero as $r \to \infty$.

A multitude of solutions of the form

$$\Phi' = \sum_{i+j+k=n} c_{ijk} x^i y^j z^k \tag{9.12}$$

can be found, for example, the polynomials $2z^2 - x^2 - y^2$ and xy of the second degree. However, all the polynomials (9.12) become infinitely large as $r \to \infty$ instead of decreasing as r^{-1}.

Writing the Laplace equation in spherical coordinates,

$$\left[\frac{\partial^2}{\partial r^2} + \frac{2}{r}\frac{\partial}{\partial r} + \frac{1}{r^2}\left(\frac{\partial^2}{\partial r^2} + \frac{\cos\vartheta}{\sin\vartheta}\frac{\partial}{\partial\vartheta} + \frac{1}{\sin^2\vartheta}\frac{\partial^2}{\partial\varphi^2}\right)\right]\Phi' = 0, \tag{9.13}$$

the general solution has been found to be

$$\Phi' = \sum_{n=0}^{a}\sum_{m=-n}^{n} a_{nm} r^n + b_{nm} r^{-n-1} P_n^m(\cos\vartheta) e^{im\varphi}, \tag{9.14}$$

$P_n^m(\cos\vartheta)e^{im\varphi}$ being spherical harmonics (to be discussed later). To avoid irregularities at $r = 0$, the last terms with coefficients b_{nm} must be removed, leaving the first terms which can be identified step by step with the solutions of (9.12), as expressed in powers of the Cartesian coordinates.

Hence we see that no Laplacian field can exist which conforms with the restrictions put upon the field F' or its potential function Φ'.

For a finite region a more convenient proof is as follows: For the additional field potential Φ' the boundary conditions are $\Phi' = 0 = \partial\Phi'/\partial n$. Since $\nabla^2\Phi' = 0$,

$$\text{div } \Phi' \nabla\Phi' = (\nabla\Phi')^2 + \Phi' \nabla^2\Phi' = (\nabla\Phi')^2. \tag{9.15}$$

Applying the Gauss theorem and observing that because of the boundary conditions

$$\int \Phi' \frac{\partial\Phi'}{\partial n} dS = 0 \tag{9.16}$$

on the surface of the region, we obtain

$$\int (\nabla\Phi')^2 dv = 0. \tag{9.17}$$

The result applies also to an infinite region. Hence $\nabla\Phi'$ is zero everywhere and the solution (9.9) is unique.

10. DISCONTINUOUS FIELDS

The Sources of a Field. Point and Surface Divergence

Picturing an arbitrary vector field as the velocity field of an incompressible fluid, we are forced to imagine the outward flow of fluid from a given volume element to be maintained by an equal amount of fluid which is in some way brought into, or created within, the volume element. In this way we are led to an artificial conception of sources and sinks of the field, or sources of positive and negative sign. Sources may be distributed continuously to produce continuous fields of the gradient type. They may as well, however, be concentrated into point or surface sources, which cause field discontinuities of characteristic types. We could also have line divergences as shown in an example in Part 3 on electrostatics.

A very simple discontinuous field is that of a single discontinuity, say, in the origin $r = 0$,

$$\Phi = \frac{q}{r}, \mathbf{F} = -\text{grad } \Phi = \frac{q}{r^2}\frac{\mathbf{r}}{r}, \qquad (10.1)$$

where q may be termed the intensity of the point source. The rate of its production of fluid or its point divergence is $4\pi q$. This is found by evaluating the surface integral

$$\int \mathbf{F} \, d\mathbf{f} = 4\pi q, \qquad (10.2)$$

whose value, if taken on a sphere around the source, is easily found. Relying upon the Gauss theorem, it follows from the equation div $\mathbf{F} = 0$ outside the source point that the result is true for any surface around the point. This can also be seen directly from Fig. 1-25, where a point source, its field $\mathbf{F}$, the surface vectorial element $d\mathbf{f}$, and its projection df' normal to $\mathbf{F}$ have been drawn. If $d\omega$ is a solid angle subtended by df as seen from q, we have

$$\frac{q}{r^2}\frac{\mathbf{r}}{r} \, d\mathbf{f} = q\frac{df'}{r^2} = q \, d\omega. \qquad (10.3)$$

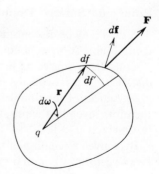

Fig. 1-25. Surface integral of a point source.

36 / The Mathematical Foundation of Physics

This proves the proposition (10.2), since for the whole surface the solid angle is 4π.

The effect of a surface source or surface divergence is illustrated in Fig. 1-26. Let 0 be the source intensity per unit area of the surface. According to (10.2) the outward flow through a closed surface cutting a unit area of the discontinuity will be $4\pi\sigma$. If the height of the pillbox-shaped surface is sufficiently small, the flow is simply the difference between the two normal components of the field on either side of the discontinuity. Taking the normal **n** in the direction from region 1 to region 2,

$$F_{1n} - F_{2n} = 4\pi\sigma. \tag{10.4}$$

Hence a surface source produces a discontinuity in the field.

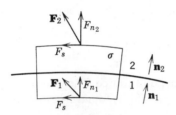

Fig 1-26. Discontinuity of field due to surface sources.

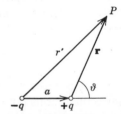

Fig. 1-27. Point doublets of intensity qa.

If the source is thought of as formed by the concentration of continuous volume sources, writing $\mathbf{F} = -\nabla\Phi$, we find by a simple limiting procedure that

$$\Phi = \int \frac{\sigma}{r} dr. \tag{10.5}$$

Hence the potential Φ is the same on either side. Therefore the potential is continuous, as is also the tangential component of $\mathbf{F} = -\nabla\Phi$.

Point and Surface Double Sources

A field may be discontinuous without any visible sources. Such fields result from double sources of equal magnitude and opposite sign.

Consider a source as in Fig. 1-27 and a sink of intensity q and $-q$ at a distance a. This is called a point doublet. In the ideal case $a \to 0$, and by finite q, any effect of the doublet would disappear. The strength of the doublet is therefore given by the product $p = qa$. To keep p constant, q and a must vary inversely as $a \to 0$.

By a given strength p of the doublet, or, for the sake of completeness, by a given vectorial strength $\mathbf{p} = q\mathbf{a}$, the potential of the field in a point P may be calculated as follows:

$$\Phi = \frac{q}{r} - \frac{q}{r'} + a\frac{r'-r}{rr'}, \tag{10.6}$$

where the various quantities are explained by Fig. 1-27. As $a/r \to 0$ by constant qa, we may replace $r' - r$ with $a \cos \vartheta$ and drop the distinction between r' and r to obtain

$$\Phi = \frac{qa \cos \vartheta}{r^2} = \frac{p \cos \vartheta}{r^2} = \frac{\mathbf{pr}}{r^3}, \tag{10.7}$$

r denoting the vector from the source to point P.

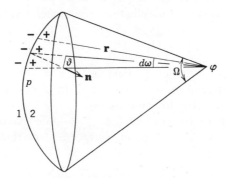

Fig. 1-28. Potential of doublet source layer $\varphi = \int p \, d\omega$.

The field of the doublet or the *dipole field*, as it is usually called with reference to electrostatics, is

$$\mathbf{F} = \operatorname{grad} \Phi = \frac{3\mathbf{r}(\mathbf{rp})}{r^5} - \frac{\mathbf{p}}{r^3}. \tag{10.8}$$

Consider now a continuous surface distribution of double sources consisting of two adjacent surfaces at a distance a and with source strength σ and $-\sigma$ per unit area. This means that the strength of the double layer is $p = \sigma a$ which, for convenience, we take to be a constant. Let **n** be the normal unit vector of the layer as pointing from the negative to the positive surface sources. Then, obviously, the potential of the layer is

$$\Phi = \int \frac{p}{r^2}\left(\mathbf{n}\frac{\mathbf{r}}{r}\right) df. \tag{10.9}$$

It is possible to convert this formula into a more compact one by observing that the bracket in (10.9) is $\cos \vartheta$ as given in Fig. 1-28. Therefore $d\omega = df \cos \vartheta / r^2$ is simply the differential solid angle subtended by df as viewed from the field point. Hence

$$\Phi = \int p \, d\omega = p\Omega, \tag{10.10}$$

Ω being the corresponding total solid angle subtended by the positive side of the layer or, for brevity, its boundary curve.

The vector field is only apparently continuous. Since $\sigma \to \infty$, as $a \to 0$, the field between the surface sources is reversed and of infinite strength as $a \to 0$. To this discontinuity of the field there corresponds a less serious discontinuity in the potential. On the positive side of the layer the solid angle Ω is positive and of the order 2π, the accurate value being determined by the position of the boundary curve. At a neighboring point on the other side it is negative and of the order -2π. Whatever the accurate values may be, their difference is exactly

$$\Omega_2 - \Omega_1 = 4\pi. \tag{10.11}$$

Therefore the curve integral of the field vector $\mathbf{F} = -\nabla\Phi$ along any closed curve from the positive to the negative side of the layer, hence encircling the boundary curve in positive direction, is $4\pi p$.

The potential of an irrotational field $\mathbf{F} = -\nabla\varphi$ having a continuous div $\mathbf{F} = \rho$, together with given point, surface and double surface sources, can be written

$$\varphi = \int \frac{\rho}{r} \, dv + \int \frac{\sigma}{r} \, df + \int \frac{p}{r^2} \left(\mathbf{n} \frac{\mathbf{r}}{r}\right) df + \sum_k \frac{q_k}{r_k}. \tag{10.12}$$

Discontinuities of Solenoidal Fields

In a nondivergent or solenoidal field $\mathbf{G} = \operatorname{curl} \mathbf{A}$, the field curl $\mathbf{G}$ may be thought of as heavily concentrated near given surfaces or curves, hence by a limiting process, it produces *surface* or *line vortices*. In studying irrotational fields particular attention was paid to point and surface discontinuities. In the case of solenoidal fields, line discontinuities shall be considered the most important.

It should be noted that the vector lines representing curl $\mathbf{G}$ can never end somewhere in the field. They always continue toward infinity if they do not return to an arbitrarily chosen starting point and thus form closed curves. This is a consequence of Gauss's theorem combined with the fact that div curl $\mathbf{G} = 0$.

From Stokes's theorem we learn that the curl of a vector field can be measured by a line integral arond its vortex lines,

$$\int \mathbf{G} \, d\mathbf{r} = \int \operatorname{curl} \mathbf{G} \, df. \tag{10.13}$$

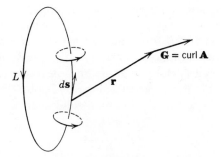

Fig. 1-29. The field of a vortex line.

If the curl is concentrate on single curve, which for the sake of simplicity we take to be a finite closed curve as in Fig. 1-29, the strength of this line vortex can be measured by a quantity I as defined by the equation

$$\oint \mathbf{G} \, d\mathbf{r} = 4\pi I. \tag{10.14}$$

To obtain a suitable representation of the field we put

$$\mathbf{G} = \operatorname{curl} \mathbf{A} \tag{10.15}$$

as in (9.3) and in analogy to (9.8) we find

$$\mathbf{A} = I \oint \frac{d\mathbf{s}}{r}, \tag{10.16}$$

the volume integral (9.8) degenerating into the above curve integral. The field itself is obtained by differentiation and is given by the curve integral

$$\mathbf{G} = \operatorname{curl} \mathbf{A} = I \oint \frac{d\mathbf{s} \times \mathbf{r}}{r^3}, \tag{10.17}$$

$\mathbf{r}$ being the space vector pointing from the vortex to the field point.

Identity of Fields from a Double Layer and a Bordering Vortex Line

Since curl $\mathbf{G} = 0$ outside the vortex line, the field $\mathbf{G}$ can also be expressed as a gradient field

$$\mathbf{G} = -\operatorname{grad} \Phi \tag{10.18}$$

Multiplying this equation by the displacement vector $d\mathbf{r}$, we obtain

$$d\Phi = -\mathbf{G} \, d\mathbf{r} = I \oint \frac{\mathbf{r}(-d\mathbf{r} \times d\mathbf{s})}{r^3}. \tag{10.19}$$

This integral can be given a geometrical interpretation by means of the solid angle subtended by the curve as viewed from the field point. In calculating the variation of this solid angle we may replace the displacement $d\mathbf{r}$ of the field point by a translation $-d\mathbf{r}$ of the whole curve. Since $d\mathbf{r} \times d\mathbf{s}$ is the vectorial area of the parallelogram $d\mathbf{r}, d\mathbf{s}$, the integrand in (10.9) is the solid angle subtended by this area as seen from the field point. Integrating over the curve, we obtain the increase of the total solid angle Ω subtended by the curve; that is,

$$d\Phi = I\, d\Omega. \tag{10.20}$$

The potential of the gradient field (10.18) is thus

$$\Phi = I\Omega. \tag{10.21}$$

This means that the potential is the same as that of a double layer with strength $p = I$.

We thus arrive at the interesting result that the field of a vortex line can be replaced by a field of any double layer of corresponding strength having the vortex line as its boundary curve. This applies, however, only to the field outside the layer. Whereas in the case of the double layer the inside field must be thought of as reversed and of infinite strength, the field of the vortex line is continuous. Correspondingly, the potential of the double layer increases from about $-2\pi p$ to approximately 2π on passing the layer, whereas the potential of the vortex line decreases continually at any point of a curve encircling the vortex line in the positive sense.

CHAPTER 2

Tensors and Tensor Fields

11. LINEAR TRANSFORMATIONS OF VECTORS

Linear Vector Functions

A great many physical laws can be expressed by very simple vector equations such as, say,

$$\mathbf{F} = e\mathbf{E} \tag{11.1}$$

which states the proportionality of the electric intensity $\mathbf{E}$ and the mechanical force $\mathbf{F}$ on a charged particle e. Another example is provided by Newton's second law:

$$\mathbf{K} = m\ddot{\mathbf{r}} = m(\ddot{x}\mathbf{i} + \ddot{y}\mathbf{j} + \ddot{z}\mathbf{k}) \tag{11.2}$$

which relates the acceleration of a particle to the applied force.

In general, however, we should not require a physical law to do more than, say, relate one vectorial physical quantity to another vector in a unique way, that is, to express one of the vectors as a function of the other. Very often, however, we will find that the relation is linear: that the components of either vector are linear functions of the components of the other. In this case we speak of *linear vector functions*, and the equations connecting the two vectors are called a linear vector transformation.

To illustrate quantities that are not simply proportional consider a body kept at a definite equilibrium by some special device: symbolized, say, by three sets of spring directed perpendicularly to each other. (See Fig. 2-1.) By smaller displacements of the body the equilibrium equations connecting the displacements and elastic forces are

$$K_x = -a_1 x, \quad K_y = -a_2 y, \quad K_z = -a_3 z. \tag{11.3}$$

These equations obviously are different from (11.1) and (11.2) except in the case of $a_1 = a_2 = a_3$. Quite formally, however, we may write

$$\mathbf{K} = -a\mathbf{r} \tag{11.4}$$

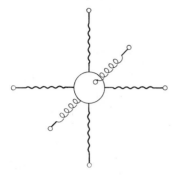

Fig. 2-1.

if the right-hand side of this equation is interpreted as the scalar product of the displacement vector **r** and a quantity a called a *tensor* and formally written as

$$a = a_1 \mathbf{i} \cdot \mathbf{i} + a_2 \mathbf{j} \cdot \mathbf{j} + a_3 \mathbf{k} \cdot \mathbf{k}. \tag{11.5}$$

The products $\mathbf{i} \cdot \mathbf{i}$, and so forth, in the above expression are themselves tensors or *dyadics* as presented in a special form of so-called *dyadic products* of vectors. The meaning of the dyadics in the expression (11.5) is that their postfactors should be used as in ordinary scalar multiplication, the prefactors being retained as vector elements. Another form of (11.4) is, therefore,

$$\mathbf{K} = -(\mathbf{i}a_1 x + \mathbf{j}a_2 y + \mathbf{k}a_3 z). \tag{11.6}$$

We now proceed to a formal generalization of (11.4), (11.5), and (11.6) by introducing

$$\mathbf{a}_1 = \mathbf{i}a_1, \qquad \mathbf{a}_2 = \mathbf{j}a_2, \qquad \mathbf{a}_3 = \mathbf{k}a_3, \tag{11.7}$$

and by writing

$$\mathbf{K} = -(\mathbf{a}_1 x + \mathbf{a}_2 y + \mathbf{a}_3 z). \tag{11.8}$$

So far, the vector transformation is defined by three independent vectors $\mathbf{a}_1$, $\mathbf{a}_2$, and $\mathbf{a}_3$, which according to (11.7) are restricted to be mutually orthogonal. Dropping this restriction and making the three vectors entirely arbitrary, we have by

$$a = \mathbf{i} \cdot \mathbf{a}_1 + \mathbf{j} \cdot \mathbf{a}_2 + \mathbf{k} \cdot \mathbf{a}_3 \tag{11.9}$$

the most general definition of a tensor. Applied to any vector **F** as a postfactor, we have

$$\mathbf{G} = a\mathbf{F} = \mathbf{i}(\mathbf{a}_1 \mathbf{F}) + \mathbf{j}(\mathbf{a}_2 \mathbf{F}) + \mathbf{k}(\mathbf{a}_3 \mathbf{F}). \tag{11.10}$$

This is obviously the most general form of a vector **G** which is a linear function of another vector **F**. Moreover, the equation yields directly the Cartesian component of the transformed vector **G**.

Interchange of Factors and Transposition of Tensors

Equations (11.9) and (11.10) can be used as the general definition of a tensor or a vector transformation. It may, however, be convenient to have alternative and equivalent definition. For instance, we might as well take the expression

$$\mathbf{H} = \mathbf{F}a = F_x\mathbf{a}_1 + F_y\mathbf{a}_2 + F_z\mathbf{a}_3 \tag{11.11}$$

which is in fact also a linear vector transformation defining the vector $\mathbf{H}$ as a linear function of $\mathbf{F}$ and in some respect of an even simpler form than (11.10).

Returning for a moment to (11.3) and (11.5), we find that in this case there is no difference between the products (11.10) and (11.11). We shall learn very soon the reason for this, namely, that the tensor (11.5) is of a particular kind called a *symmetric* tensor. In general, the products $a\mathbf{F}$ and $\mathbf{F}a$ are different.

Expressing the three fundamental vectors of the transformation by their components,

$$\mathbf{a}_1 = a_{11}\mathbf{i} + a_{12}\mathbf{j} + a_{13}\mathbf{k}, \quad \text{etc.} \tag{11.12}$$

we arrive at the same expression

$$\begin{aligned} a = {} & a_{11}\mathbf{i}\cdot\mathbf{i} + a_{12}\mathbf{i}\cdot\mathbf{j} + a_{13}\mathbf{i}\cdot\mathbf{k} \\ & + a_{21}\mathbf{j}\cdot\mathbf{i} + a_{22}\mathbf{j}\cdot\mathbf{j} + a_{23}\mathbf{j}\cdot\mathbf{k} \\ & + a_{31}\mathbf{k}\cdot\mathbf{i} + a_{32}\mathbf{k}\cdot\mathbf{j} + a_{33}\mathbf{k}\cdot\mathbf{k} \end{aligned} \tag{11.13}$$

as composed by nine dyadic products with nine coefficients which are called the coefficients of the tensor.

Introducing

$$\tilde{\mathbf{a}}_1 = \mathbf{i}a_{11} + \mathbf{j}a_{21} + \mathbf{k}a_{31}, \quad \text{etc.,} \tag{11.14}$$

and collecting the terms differently from (11.9), we obtain the alternative form

$$a = \tilde{\mathbf{a}}_1 \cdot \mathbf{i} + \tilde{\mathbf{a}}_2 \cdot \mathbf{j} + \tilde{\mathbf{a}}_3 \cdot \mathbf{k}. \tag{11.15}$$

Applying this form to the vector $\mathbf{F}$ as a postfactor,

$$a\mathbf{F} = \tilde{\mathbf{a}}_1 F_x + \tilde{\mathbf{a}}_2 F_y + \tilde{\mathbf{a}}_3 F_z. \tag{11.16}$$

and comparing with (11.11), we see that the effect of interchanging tensor and vector as pre- and postfactors is the same as that of interchanging $\mathbf{a}_1$, $\mathbf{a}_2$, $\mathbf{a}_3$ and $\tilde{\mathbf{a}}_1$, $\tilde{\mathbf{a}}_2$, $\tilde{\mathbf{a}}_3$. The latter procedure, resulting in the tensor

$$\begin{aligned} \tilde{a} &= \mathbf{i}\cdot\tilde{\mathbf{a}}_1 + \mathbf{j}\cdot\tilde{\mathbf{a}}_2 + \mathbf{k}\cdot\tilde{\mathbf{a}}_3 \\ &= \mathbf{a}_1\cdot\mathbf{i} + \mathbf{a}_2\cdot\mathbf{j} + \mathbf{a}_3\cdot\mathbf{k} \end{aligned} \tag{11.17}$$

is called a transposition of the tensor a. Quite generally, we have

$$\tilde{a}\mathbf{F} = \mathbf{F}a. \tag{11.18}$$

44 / The Mathematical Foundation of Physics

The equation
$$aF = Fa \qquad (11.19)$$
is valid only in the case that a tensor is unchanged by transposition, that is, if the tensors have symmetrical coefficients $a_{mn} = a_{nm}$ or formally $\tilde{a} = a$.

12. THE CHARACTERISTICS OF VECTOR TRANSFORMATIONS

Addition and Subtraction of Tensors

Given the tensors a and b, we find that the meaning of the sum tensor
$$a = b + c \qquad (12.1)$$
is that
$$aF = bF + cF \qquad (12.2)$$
for any vector F. This simply means that one has to add the tensor components,
$$a_{mn} = b_{mn} + c_{mn}. \qquad (12.3)$$
For the subtraction of tensors we get analogous results.

Decomposition in Symmetric and Antisymmetric Tensors

Any tensor a can be written as
$$a = b + c, \quad b = \tfrac{1}{2}(a + \tilde{a}), \quad c = \tfrac{1}{2}(a - \tilde{a}), \qquad (12.4)$$
$\tilde{a}$ being the transposed tensor $\tilde{a}_{mn} = a_{nm}$. Here b is called a *symmetric* and c an *antisymmetric* tensor. It is easily found by $aF = F\tilde{a}$
$$bF = \tfrac{1}{2}(a + \tilde{a})F = \tfrac{1}{2}F(\tilde{a} + a) = Fb, \qquad (12.5)$$
$$cF = \tfrac{1}{2}(a - \tilde{a})F = \tfrac{1}{2}F(\tilde{a} - a) = -Fc, \qquad (12.6)$$
since the interchange of the tensor and the vector in a product has the same effect as a transposition of the tensor, we may write this as
$$\tilde{b} = b, \quad \tilde{c} = -c \qquad (12.7)$$
or
$$\tilde{b}_{mn} = b_{nm}, \quad \tilde{c}_{mn} = -c_{nm}. \qquad (12.8)$$
We may take (12.5) and (12.6) or (12.7) and (12.8) as the characteristics of symmetric and antisymmetric tensors, respectively; from (12.4) we see that there is only one way of decomposing an arbitrary tensor into a sum of a symmetric and an antisymmetric tensor.

Tension and Shears

To visualize the effect of a given tensor we apply it to a simple field as that of the space vector $\mathbf{r} = x\mathbf{i} + y\mathbf{j} + z\mathbf{k}$, comparing $\rho = a\mathbf{r} = \xi\mathbf{i} + \eta\mathbf{j} + \zeta\mathbf{k}$

Tensors and Tensor Fields | 45

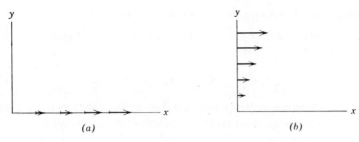

Fig. 2-2.

or rather $\rho - r = ar - r$ with r. This means that to obtain ρ we shall have to add $\rho - r$ to r. Studying the components separately, we begin with the diagonal element $a_{11}i \cdot i$ which yields a contribution

$$(a_{11} - 1)x\mathbf{i} \tag{12.9}$$

to the displacement of the vector. As is easily seen from Fig. 2-2a, this may be called a *tension* of the r space in the x-direction and similarly for other diagonal elements.

Next consider the nondiagonal element $a_{12}i \cdot j$. Its contribution to the displacement field $\rho - r$ is

$$a_{12}y\mathbf{i}, \tag{12.10}$$

as illustrated in Fig. 2-2b. This displacement is called a *shear*.

In the case of a symmetric tensor $a_{21} = a_{12}$, two corresponding shears result in tension diagonally to the axis as shown in Fig. 2.3a. If the tensor is antisymmetric, $a_{21} = -a_{12}$, we have a displacement similar to a rotation as shown in Fig. 2-3b, or, more precisely, an infinitesimal rotation for infinitesimal tensor elements.

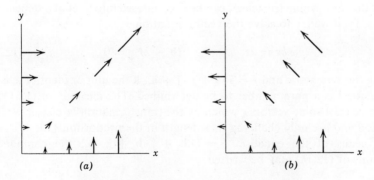

Fig. 2-3. (*a*) Diagonal dilatation. (*b*) Rotation.

Antisymmetric Tensors and Vector Products

We shall study a little more closely the effect of an antisymmetric tensor $c = -\tilde{c}$. For brevity we write

$$c_{23} = \omega_x, \qquad c_{31} = \omega_y, \qquad c_{12} = \omega_z. \tag{12.11}$$

The above three components can be taken as the component of a vector **ω** which is obviously an axial vector. We can write c in the symbolic form

$$c = \begin{Bmatrix} \mathbf{i} \cdot (\omega_z \mathbf{j} - \omega_y \mathbf{k}) \\ + \mathbf{j} \cdot (-\omega_z \mathbf{i} + \omega_x \mathbf{k}) \\ \mathbf{k} \cdot (\omega_y \mathbf{i} - \omega_x \mathbf{j}) \end{Bmatrix} = \begin{vmatrix} \mathbf{i} & \mathbf{j} & \mathbf{k} \\ \mathbf{i} & \mathbf{j} & \mathbf{k} \\ x & y & z \end{vmatrix}, \tag{12.12}$$

it being understood that the vector products are dyadics. Applying the tensor to a vector **r**, we obtain

$$c\mathbf{r} = \begin{vmatrix} \mathbf{i} & \mathbf{j} & \mathbf{k} \\ x & y & z \\ \omega_x & \omega_y & \omega_z \end{vmatrix} = \mathbf{r} \times \boldsymbol{\omega} \tag{12.13}$$

The vector on the right-hand side is illustrated in (12.11) with the only difference that in that equation ω means an angular velocity, whereas in (12.13) **ω** is a rotation vector or an angle.

Again, as discussed above, for infinitesimal **ω** the product $c\mathbf{r}$ represents a rotation, whereas for finite **ω** there is also tension in any direction orthogonal to **ω**. This tension has the magnitude $\sqrt{1 + \omega^2}$.

Main Axis of a Symmetric Tensor

Since the effect of two symmetric shears is to produce tension, we can expect that any symmetric tensor can be decomposed into a set of independent tensors. First we try to solve the vector equation

$$b\mathbf{r} = \lambda \mathbf{r} \qquad \text{or} \qquad (b - \lambda I)\mathbf{r} = 0, \tag{12.14}$$

b being the symmetric and $I = \mathbf{i} \cdot \mathbf{i} + \mathbf{j} \cdot \mathbf{j} + \mathbf{k} \cdot \mathbf{k}$ the unit or identity tensor. The factor λ is a pure number to be determined. The meaning of (12.14) is that there exist some vectors **r** which by the transformation $b\mathbf{r}$ conserve their direction and are only changing their length in the proportion $1:\lambda$.

By means of the equation $b = \mathbf{i} \cdot \mathbf{b}_1 + \mathbf{j} \cdot \mathbf{b}_2 + \mathbf{k} \cdot \mathbf{b}_3$, the component equations of (12.14) may be written

$$(\mathbf{b}_1 - \lambda \mathbf{i})\mathbf{r} = 0, \qquad (\mathbf{b}_2 - \lambda \mathbf{j})\mathbf{r} = 0, \qquad (\mathbf{b}_3 - \lambda \mathbf{k})\mathbf{r} = 0. \tag{12.15}$$

Hence r is orthogonal to three different vectors and this is not possible unless the vectors be coplanar, and their volume product zero,

$$\mathbf{b}_1 - \lambda\mathbf{i}, \quad \mathbf{b}_2 - \lambda\mathbf{j}, \quad \mathbf{b}_3 - \lambda\mathbf{k} = 0 \tag{12.16}$$

or

$$\begin{vmatrix} b_{11} - \lambda & b_{12} & b_{13} \\ b_{21} & b_{22} - \lambda & b_{23} \\ b_{31} & b_{32} & b_{33} - \lambda \end{vmatrix} = 0. \tag{12.17}$$

This algebraic equation of the third degree has three roots $\lambda = \lambda_1, \lambda_2, \lambda_3$, which lead to three independent solutions $\mathbf{r} = \mathbf{r}_1, \mathbf{r}_2, \mathbf{r}_3$. It is easily seen that all roots are real and that the three solutions are orthogonal.

Consider the equations

$$\begin{aligned} b\mathbf{r}_1 &= \lambda_1 \mathbf{r}_1, \\ b\mathbf{r}_2 &= \lambda_2 \mathbf{r}_2, \end{aligned} \tag{12.18}$$

as obtained from (12.15), and for the sake of simplicity assume $\lambda_1 \neq \lambda_2$. Multiplying the equations by $\mathbf{r}_2$ and $\mathbf{r}_1$, respectively, and keeping in mind that

$$\mathbf{r}_1(b\mathbf{r}_2) = (\mathbf{r}_2 b)\mathbf{r}_1 = \mathbf{r}_2(b\mathbf{r}_1) \tag{12.19}$$

when b is symmetric, we obtain by subtraction

$$(\lambda_1 - \lambda_2)\mathbf{r}_1\mathbf{r}_2 = 0. \tag{12.20}$$

Hence $\mathbf{r}_1\mathbf{r}_2 = 0$ and the two vectors are mutually orthogonal.

It can be seen from the same equation that the roots must be real. Assume λ to be a complex root with the corresponding complex solution $\mathbf{r}$. Then λ^* would also be a root of (12.17) with a corresponding solution $\mathbf{r}^*$. But $\mathbf{rr}^*$ is a positive real number different from zero; hence $\lambda - \lambda^* = 0$ and the root is real.

For many purposes it is convenient to use normalized solutions determined by $\mathbf{r}^2 = 1$; for instance, $\mathbf{r}_1 = \mathbf{i}'$, $\mathbf{r}_2 = \mathbf{j}'$, $\mathbf{r}_2 = \mathbf{k}'$, where $\mathbf{i}', \mathbf{j}', \mathbf{k}'$ are a set of orthonormal vectors called the main axis of the tensor b. Since $b\mathbf{i}' = \lambda_1 \mathbf{i}'$, and so forth, the tensor can be written

$$b = \lambda_1 \mathbf{i}' \cdot \mathbf{i}' + \lambda_2 \mathbf{j}' \cdot \mathbf{j}' + \lambda_3 \mathbf{k}' \cdot \mathbf{k}! \tag{12.21}$$

The Invariants of the Tensor

The three roots $\lambda_1, \lambda_2, \lambda_3$, of (12.17) will be the same, irrespective of the orientation of the coordinate axis. Equation 12.17 will be unchanged by the rotation of the coordinate system and may be written

$$(\lambda_1 - \lambda)(\lambda_2 - \lambda)(\lambda_3 - \lambda) = c_0 - c_1\lambda + c_2\lambda^2 - \lambda^3 = 0. \tag{12.22}$$

48 / The Mathematical Foundation of Physics

Hence there are, in all, three independent so-called invariants of the tensor b with respect to the rotation of the coordinate system, namely, λ_1, λ_2, λ_3 or c_0, c_1, c_2.

It is easily found from (12.16) or (12.17) that

$$c_0 = \lambda_1 \lambda_2 \lambda_3 = \mathbf{b}_1 \cdot (\mathbf{b}_2 \times \mathbf{b}_3) = \begin{vmatrix} b_{11} & b_{12} & b_{13} \\ b_{21} & b_{22} & b_{23} \\ b_{31} & b_{32} & b_{33} \end{vmatrix}. \quad (12.23)$$

This determinant of the tensor is the invariant most frequently used. The invariant next to it in importance is c_2 or the "spur,"

$$c_1 = \lambda_1 + \lambda_2 + \lambda_3 = b_{11} + b_{22} + b_{33} = \text{sp}(b). \quad (12.24)$$

The third invariant is easily found to be the sum of three minor determinants

$$c_2 = \lambda_2 \lambda_3 + \lambda_3 \lambda_1 + \lambda_1 \lambda_2 = \begin{vmatrix} b_{22} & b_{23} \\ b_{32} & b_{22} \end{vmatrix} + \begin{vmatrix} b_{33} & b_{31} \\ b_{13} & b_{11} \end{vmatrix} + \begin{vmatrix} b_{11} & b_{12} \\ b_{21} & b_{22} \end{vmatrix}. \quad (12.25)$$

13. TENSOR FIELDS AND THEIR DERIVATIVES

Divergence of a Tensor

If the nine components of a tensor are functions of the space coordinates x, y, z, we shall have different tensors for each point in space which together constitute what is called a *tensor field*. Applying the operators div, grad, and curl (∇, $\nabla \cdot$, and $\nabla \times$) to scalar, vector, and tensor fields, we obtain new fields which may in general be characterized as tensor fields of different rank. The operator $\nabla \times$ leaves the rank unchanged, whereas the rank is lowered or raised one unit by the operators div and grad, respectively. For instance, the divergence of a tensor is a vector.

We put

$$T = \mathbf{i} \cdot \mathbf{T}_1 + \mathbf{j} \cdot \mathbf{T}_2 + \mathbf{k} \cdot \mathbf{T}_3 = \tilde{\mathbf{T}}_1 \cdot \mathbf{i} + \tilde{\mathbf{T}}_2 \cdot \mathbf{j} + \tilde{\mathbf{T}}_3 \cdot \mathbf{k}, \quad (13.1)$$

$$\mathbf{T}_1 = T_{11}\mathbf{i} + T_{12}\mathbf{j} + T_{13}\mathbf{k}, \qquad \tilde{\mathbf{T}}_1 = \mathbf{i}T_{11} + \mathbf{j}T_{21} + \mathbf{k}T_{31}, \qquad \text{etc.}$$

The divergence may be written in two different ways as

$$\text{div } T = \nabla T = \frac{\partial \mathbf{T}_1}{\partial x} + \frac{\partial \mathbf{T}_2}{\partial y} + \frac{\partial \mathbf{T}_3}{\partial z} \quad (13.2)$$

or

$$\text{div } T = \text{div } \tilde{\mathbf{T}}_1 \mathbf{i} + \text{div } \tilde{\mathbf{T}}_2 \mathbf{j} + \text{div } \tilde{\mathbf{T}}_3 \mathbf{k}.$$

Gradient of a Vector

On the other hand, consider the operator grad $= \nabla \cdot$ applied to the vector field **F**,

$$\text{grad } \mathbf{F} = \nabla \cdot \mathbf{F} = \mathbf{i} \cdot \frac{\partial \mathbf{F}}{\partial x} + \mathbf{j} \cdot \frac{\partial \mathbf{F}}{\partial y} + \mathbf{k} \cdot \frac{\partial \mathbf{F}}{\partial z}$$

$$= \text{grad } F_x \cdot \mathbf{i} + \text{grad } F_y \cdot \mathbf{j} + \text{grad } F_z \cdot \mathbf{k}. \quad (13.3)$$

Again we can apply the operator div $= \nabla$ or the scalar multiplication of a vector **A** to this tensor field. It is easily seen without going into detail that the result is the same as that obtained by combining the two successive operators into one:

$$\text{div grad } \mathbf{F} = \nabla^2 \mathbf{F},$$

$$\mathbf{A} \text{ grad } \mathbf{F} = (\mathbf{A} \text{ grad}) \mathbf{F}, \quad (13.4)$$

$$\mathbf{A} \text{ grad} = \mathbf{A} \nabla = A_x \frac{\partial}{\partial x} + A_y \frac{\partial}{\partial y} + A_z \frac{\partial}{\partial z}. \quad (13.5)$$

In such cases, where the result of two successive operators is a single scalar operator leaving the rank of a tensor (in the general sense) unchanged, we should preferably use the product operator.

Displacement and Deformation

In kinematics and dynamics of elastic solids we shall have to use the displacement vector $\mathbf{u}(x, y, z)$ of points with equilibrium positions x, y, z to describe a permanent or a time-dependent deformation of the body. More important than the displacement itself is the relative displacement of, say, the point $\mathbf{r} + d\mathbf{r}$ with respect to $\mathbf{r}$,

$$d\mathbf{u} = dx \frac{\partial \mathbf{u}}{\partial x} + dy \frac{\partial \mathbf{u}}{\partial y} + dz \frac{\partial \mathbf{u}}{\partial z} = (d\mathbf{r} \text{ grad}) \mathbf{u}. \quad (13.6)$$

In this case we prefer to interpret the right-hand side as $d\mathbf{r}(\text{grad } \mathbf{u})$, since to analyze the problem fully we shall have to study the tensor grad **u**. It can be split into a symmetric and an antisymmetric tensor U and U', respectively,

$$\text{grad } \mathbf{u} = \nabla \cdot \mathbf{u} = U + U' \quad (13.7)$$

$$U_{11} = \frac{\partial u_x}{\partial x}, \quad U_{12} = \frac{1}{2}\left(\frac{\partial u_y}{\partial x} + \frac{\partial u_x}{\partial y}\right), \quad \text{etc.} \quad (13.8)$$

$$U'_{11} = 0, \quad U'_{12} = \frac{1}{2}\left(\frac{\partial u_y}{\partial x} - \frac{\partial u_x}{\partial y}\right), = \frac{1}{2} \text{curl}_z \mathbf{u} = \omega_z, \quad \text{etc.} \quad (13.9)$$

In analogy with (12.12 and 12.13) we have

$$d\mathbf{r}U' = d\mathbf{r}\begin{vmatrix} \mathbf{i} & \mathbf{j} & \mathbf{k} \\ \mathbf{i} & \mathbf{j} & \mathbf{k} \\ \omega_x & \omega_y & \omega_z \end{vmatrix} = \begin{vmatrix} dx & dy & dz \\ \mathbf{i} & \mathbf{j} & \mathbf{k} \\ \omega_x & \omega_y & \omega_z \end{vmatrix} = \boldsymbol{\omega} \times d\mathbf{r}. \quad (13.10)$$

The antisymmetric part of the relative displacement tensor grad **u**, therefore, merely describes an infinitesimal rotation of a volume element that is not a deformation in the ordinary sense. Disregarding this contribution to the tensor $\nabla \mathbf{u}$, we are left with only the symmetric part U, which is called the *deformation* or *strain tensor*.

Again, for any given point in space the strain tensor can be transformed to its main axis $\mathbf{i}'$, $\mathbf{j}'$, $\mathbf{k}'$,

$$U = U_1 \mathbf{i}' \cdot \mathbf{i}' + U_2 \mathbf{j}' \cdot \mathbf{j}' + U_3 \mathbf{k}' \cdot \mathbf{k}', \quad (13.11)$$

U_1, U_2, and U_3 being the relative increases of length in direction of the main axis. Here again, the spur is invariant with respect to rotation of the coordinate system

$$sp(U) = U_1 + U_2 + U_3 = \frac{\partial u_x}{\partial x} + \frac{\partial u_y}{\partial y} + \frac{\partial u_z}{\partial z} = \text{div } \mathbf{u}. \quad (13.12)$$

It represents the relative increase of volume of the body at a given point x, y, z.

14. RECIPROCAL VECTOR SYSTEMS

Vector Components in Nonorthogonal Coordinate Systems

Vectors can be represented in a great variety of ways by using a non-orthogonal system of basic vectors $\mathbf{a}_1$, $\mathbf{a}_2$, $\mathbf{a}_3$, corresponding to what is called an *affine* coordinate system. For the sake of generality, the basic vectors may be thought of as having arbitrary directions and magnitude, the only restriction to put upon them being that they shall not be coplanar, that is, the parallelepiped volume must not vanish,

$$D = \mathbf{a}_1(\mathbf{a}_2 \times \mathbf{a}_3) = |\mathbf{a}_1, \mathbf{a}_2, \mathbf{a}_3| \neq 0. \quad (14.1)$$

An arbitrary vector **F** can be written as

$$\mathbf{F} = F^1 \mathbf{a}_1 + F^2 \mathbf{a}_2 + F^3 \mathbf{a}_3. \quad (14.2)$$

The reason for introducing upper indices is that we need a distinction between two sets of vector components called *covariant* and *contravariant* components. The above components obviously are inversely proportional to the length of the corresponding basic vectors. They are therefore called *contravariant components*.

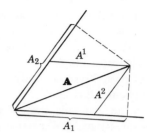

Fig. 2-4. Covariant and contravariant components of a vector A in a plane with unit basic vectors.

In analogy with the representation by orthonormal basic vectors **i, j, k**, we may also study here the scalar products

$$F_1 = \mathbf{a}_1 \mathbf{F}, \qquad F_2 = \mathbf{a}_2 \mathbf{F}, \qquad F_3 = \mathbf{a}_3 \mathbf{F}. \tag{14.3}$$

These quantities are called the *covariant components* of **F** with respect to the basic vector $\mathbf{a}_1, \mathbf{a}_2, \mathbf{a}_3$. The difference between the two kinds of components is shown in Fig. 2-4. Apart from multiplication factors, the covariant components are found by projecting the vector on the coordinate axis in the usual way, whereas in the case of contravariant vectors we have to project the vector by means of planes parallel to the coordinate planes.

Reciprocal Basic Vectors

To give a definite meaning to the covariant component (14.3) we write

$$\mathbf{F} = F_1 \mathbf{b}^1 + F_2 \mathbf{b}^2 + F_3 \mathbf{b}^3. \tag{14.4}$$

This equation is valid only if

$$\mathbf{a}_i \mathbf{b}^k = \delta_{ik} = \begin{cases} 1, & i = k \\ 0, & i \neq k \end{cases} \quad i, k = 1, 2, 3. \tag{14.5}$$

As is easily seen, (14.5) is fulfilled by

$$D\mathbf{b}^1 = \mathbf{a}_2 \times \mathbf{a}_3, \qquad D\mathbf{b}^2 = \mathbf{a}_3 \times \mathbf{a}_1, \qquad D\mathbf{b}^3 = \mathbf{a}_1 \times \mathbf{a}_2. \tag{14.6}$$

Multiplying (14.2) by the vectors **b**, we find

$$F^1 = \mathbf{b}^1 \mathbf{F}, \qquad F^2 = \mathbf{b}^2 \mathbf{F}, \qquad F^3 = \mathbf{b}^3 \mathbf{F}, \tag{14.7}$$

and from this and (14.4) we see that the roles of the covariant and contravariant components will be interchanged if we replace the basic vector systems $\mathbf{a}_1, \mathbf{a}_2, \mathbf{a}_3$ by $\mathbf{b}^1, \mathbf{b}^2, \mathbf{b}^3$. By well-known operations we find

$$D(\mathbf{b}^2 \times \mathbf{b}^3) = \frac{1}{D}(\mathbf{a}_3 \times \mathbf{a}_1)(\mathbf{a}_1 \times \mathbf{a}_2) = \mathbf{a}_1, \qquad \text{etc.} \tag{14.8}$$

and

$$\mathbf{b}^1(\mathbf{b}^2 \times \mathbf{b}^3) = \frac{1}{D}. \tag{14.9}$$

52 / The Mathematical Foundation of Physics

Hence the system $\mathbf{a}_1, \mathbf{a}_2, \mathbf{a}_3$ is obtainable in just the same way from $\mathbf{b}^1, \mathbf{b}^2, \mathbf{b}^3$ as the system $\mathbf{b}^i$ from $\mathbf{a}_i$. The two systems are reciprocal in every respect.

Reciprocal vector systems play an important role in mathematical physics. The most familiar use is in crystallography in the macroscopic description of crystal systems as well as in X-ray analysis of the distribution of atoms in crystal lattices.

The generalization of products of vectors leads to somewhat different expressions. The scalar product of two vectors $\mathbf{F}$ and $\mathbf{G}$ writes

$$\mathbf{FG} = F^1 G_1 + F^2 G_2 + F^3 G_3 = F_1 G^1 + F_2 G^2 + F_3 G^3. \tag{14.10}$$

It is the product sum of a covariant and a contravariant set of components. The vector product, of course, can be expressed differently according to whether we use the system $\mathbf{a}_i$ or $\mathbf{b}^i$ as basic vectors

$$\mathbf{F} \times \mathbf{G} = D \begin{vmatrix} \mathbf{b}^1 & \mathbf{b}^2 & \mathbf{b}^3 \\ F^1 & F^2 & F^3 \\ G^1 & G^2 & G^3 \end{vmatrix} = \frac{1}{D} \begin{vmatrix} \mathbf{a}_1 & \mathbf{a}_2 & \mathbf{a}_3 \\ F_1 & F_2 & F_3 \\ G_1 & G_2 & G_3 \end{vmatrix}. \tag{14.11}$$

Solution of Linear Equations

A most useful application of the reciprocal vector system is that of solving linear equations in three variables. Consider the system

$$\begin{aligned} a_{11}x_1 + a_{12}x_2 + a_{13}x_3 &= y_1, \\ a_{21}x_1 + a_{22}x_2 + a_{23}x_3 &= y_2, \\ a_{31}x_1 + a_{32}x_2 + a_{33}x_3 &= y_3, \end{aligned} \tag{14.12}$$

which can obviously be written as

$$\mathbf{a}_1 \mathbf{x} = y_1, \quad \mathbf{a}_2 \mathbf{x} = y_2, \quad \mathbf{a}_3 \mathbf{x} = y_3. \tag{14.13}$$

This means that y_1, y_2, y_3 are the covariant components of the unknown vector $\mathbf{x}$ in the system of basic vector $\mathbf{a}_1, \mathbf{a}_2, \mathbf{a}_3$, whose components are given in the three equations (14.12). Hence by the aid of the reciprocal vector system,

$$x = y_1 \mathbf{b}^1 + y_2 \mathbf{b}^2 + y_3 \mathbf{b}^3. \tag{14.14}$$

It is interesting to note that the transposed vector systems $\tilde{\mathbf{a}}_i, \tilde{\mathbf{b}}^i$ are also mutually reciprocal. (See Fig. 2-5.) Multiplying (14.12) by ordinary orthonormal vectors $\mathbf{i}, \mathbf{j}, \mathbf{k}$ and adding, we obtain

$$\mathbf{y} = x_1 \tilde{\mathbf{a}}_1 + x_2 \tilde{\mathbf{a}}_2 + x_3 \tilde{\mathbf{a}}_3. \tag{14.15}$$

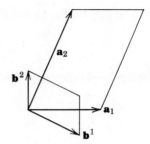

Fig. 2-5. Two-dimensional reciprocal vector systems.

On the other hand, from (14.14),

$$x_1 = \mathbf{y}\tilde{\mathbf{b}}^1, \qquad x_2 = \mathbf{y}\tilde{\mathbf{b}}^2, \qquad x_3 = \mathbf{y}\tilde{\mathbf{b}}^3. \tag{14.16}$$

Hence

$$\tilde{\mathbf{a}}_i \tilde{\mathbf{b}}^k = \delta_{ik}. \tag{14.17}$$

15. CURVILINEAR COORDINATES

Three-Dimensional General Coordinates

In three-dimensional space general coordinates can be introduced by the vector equation

$$\mathbf{r} = \mathbf{i} \cdot x(q_1, q_2, q_3) + \mathbf{j} \cdot y(q_1, q_2, q_3) + \mathbf{k} \cdot z(q_1, q_2, q_3), \tag{15.1}$$

x, y, z being specified functions of the general coordinates q_1, q_2, q_3. If two of the latter coordinates are kept constant, we have an equation for the coordinate curves as shown in Fig. 2-6. On the other hand, if a single coordinate is kept constant, the vector $\mathbf{r}$ describes curved surfaces that may be called coordinate surfaces.

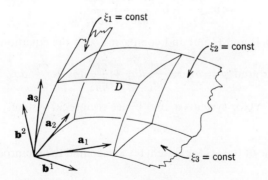

Fig. 2-6. Variable basic vectors a_1, a_2, a_3, and their reciprocals b^1, b^2, b^3 by curvilinear coordinates.

The tangent vectors of the coordinate curves,

$$\mathbf{a}_1 = \frac{\partial \mathbf{r}}{\partial q_1} = \mathbf{i}\frac{\partial x}{\partial q_1} + \mathbf{j}\frac{\partial y}{\partial q_1} + \mathbf{k}\frac{\partial z}{\partial q_1}, \qquad \mathbf{a}_2 = \frac{\partial \mathbf{r}}{\partial q_2}, \qquad \mathbf{a}_3 = \frac{\partial \mathbf{r}}{\partial q_3}, \qquad (15.2)$$

defines a parallelepiped, whose volume is

$$D = \mathbf{a}_1(\mathbf{a}_2 \times \mathbf{a}_3) = \begin{vmatrix} \frac{\partial x}{\partial q_1} & \frac{\partial y}{\partial q_1} & \frac{\partial z}{\partial q_1} \\ \frac{\partial x}{\partial q_2} & \frac{\partial y}{\partial q_2} & \frac{\partial z}{\partial q_2} \\ \frac{\partial x}{\partial q_3} & \frac{\partial y}{\partial q_3} & \frac{\partial z}{\partial q_3} \end{vmatrix} = \frac{\partial(x, y, z)}{\partial(q_1, q_2, q_3)}, \qquad (15.3)$$

which, when multiplied by dq_1, dq_2, dq_3, represents an infinitesimal volume element in space. The volume D is called the functional determinant of the transformation (15.1). The infinitesimal displacement of $\mathbf{r}$ as a function of dq_1, dq_2, dq_3, is

$$d\mathbf{r} = \mathbf{a}_1 \, dq_1 + \mathbf{a}_2 \, dq_2 + \mathbf{a}_3 \, dq_3. \qquad (15.4)$$

On the other hand, the normal vector of the coordinate surfaces are

$$\mathbf{b}^1 = \text{grad } q_1 = \mathbf{i}\frac{\partial q_1}{\partial x} + \mathbf{j}\frac{\partial q_1}{\partial y} + \mathbf{k}\frac{\partial q_1}{\partial z}, \qquad \mathbf{b}^2 = \text{grad } q_2,$$

$$\mathbf{b}^3 = \text{grad } q_3. \qquad (15.5)$$

These vectors are suitable for expressing the gradient of any scalar ψ,

$$\text{grad } \psi = \frac{\partial \psi}{\partial q_1} \text{grad } q_1 + \frac{\partial \psi}{\partial q_2} \text{grad } q_2 + \frac{\partial \psi}{\partial q_3} \text{grad } q_3$$

$$= \mathbf{b}^1 \frac{\partial \psi}{\partial q_1} + \mathbf{b}^2 \frac{\partial \psi}{\partial q_2} + \mathbf{b}^3 \frac{\partial \psi}{\partial q_3}. \qquad (15.6)$$

Multiplying (15.4) and (15.6) and comparing with the equation

$$d\mathbf{r} \text{ grad } \psi = d\psi = \frac{\partial \psi}{\partial q_1} dq_1 + \frac{\partial \psi}{\partial q_2} dq_2 + \frac{\partial \psi}{\partial q_3} dq_3, \qquad (15.7)$$

we see that the vector systems $\mathbf{a}_i$ and $\mathbf{b}^i$ are reciprocal systems,

$$\mathbf{a}_i \mathbf{b}^k = \delta_{ik}. \qquad (15.8)$$

Hence, by means of the vector system $\mathbf{b}^i$, we can find the reciprocal functional determinant

$$\frac{\partial(q_1, q_2, q_3)}{\partial(x, y, z)} = \mathbf{b}^1(\mathbf{b}^2 \times \mathbf{b}^3) = |\mathbf{b}_1, \mathbf{b}_2, \mathbf{b}_3| = \frac{1}{D}. \qquad (15.9)$$

Divergence and the Laplacian in Curvilinear Coordinates

We have already learned how to express the gradient in curvilinear coordinates. It can be expressed by the aid of the generalized operator

$$\nabla = \mathbf{b}^1 \frac{\partial}{\partial q_1} + \mathbf{b}^2 \frac{\partial}{\partial q_2} + \mathbf{b}^3 \frac{\partial}{\partial q_3}. \tag{15.10}$$

For the divergence, however, we cannot simply use this operator. We shall have to consider the original definition of the divergence as the outward vector flux from a volume element as referred to the unit volume.

Consider, for instance, the flux in the direction of $\mathbf{b}^1$. It is proportional to

$$\mathbf{F}(\mathbf{a}_2 \times \mathbf{a}_3)\, dq_2\, dq_3 = \mathbf{F}\mathbf{b}^1\, D dq_2\, dq_3 \tag{15.11}$$

and not only to the product $dq_2\, dq_3$. After calculating the flux we have to divide by the volume D. This leads to the following expression for the divergence of a vector

$$\operatorname{div} \mathbf{F} = \frac{1}{D} \sum_m \frac{\partial}{\partial q_m} D \mathbf{b}^m\, \mathbf{F} = \frac{1}{D} \nabla^1\, D\mathbf{F},$$

$$\nabla^1 = \frac{\partial}{\partial q_1} \mathbf{b}^1 + \frac{\partial}{\partial q_2} \mathbf{b}^2 + \frac{\partial}{\partial q_3} \mathbf{b}^3. \tag{15.12}$$

In particular, the Laplacian of a scalar is

$$\nabla^2 \psi = \operatorname{div}\operatorname{grad} \psi = \frac{1}{D} \sum_m \frac{\partial}{\partial q_m}\left(D \sum_n g^{mn} \frac{\partial \psi}{\partial q_n}\right), \quad g^{mn} = \mathbf{b}^m \mathbf{b}^n. \tag{15.13}$$

Calculation of D^2

Consider the square line element

$$(d\mathbf{r})^2 = (\mathbf{a}_1\, dq_1 + \mathbf{a}_2\, dq_2 + \mathbf{a}_3\, dq_3)^2 = \sum_{ij} g_{ij}\, dq_i\, dq_j,$$

$$g_{ij} = \mathbf{a}_i \mathbf{a}_j = \frac{\partial x}{\partial q_i}\frac{\partial x}{\partial q_j} + \frac{\partial y}{\partial q_i}\frac{\partial y}{\partial q_1} + \frac{\partial z}{\partial q_i}\frac{\partial z}{\partial q_j}. \tag{15.14}$$

Since

$$D = \begin{vmatrix} \dfrac{\partial x}{\partial q_1} & \dfrac{\partial y}{\partial q_1} & \dfrac{\partial z}{\partial q_1} \\ \dfrac{\partial x}{\partial q_2} & \dfrac{\partial y}{\partial q_2} & \dfrac{\partial z}{\partial q_2} \\ \dfrac{\partial x}{\partial q_3} & \dfrac{\partial y}{\partial q_3} & \dfrac{\partial z}{\partial q_3} \end{vmatrix}, \tag{15.15}$$

56 / The Mathematical Foundation of Physics

it is seen by taking the sum products of the rows by determinant multiplication that
$$D^2 = |g_{ij}|. \tag{15.16}$$
Similarly,
$$(\nabla \psi)^2 = \left(\mathbf{b}^1 \frac{\partial \psi}{\partial q_1} + \mathbf{b}^2 \frac{\partial \psi}{\partial q_2} + \mathbf{b}^3 \frac{\partial \psi}{\partial q_3}\right)^2 = \sum_{ij} g^{ik} \frac{\partial \psi}{\partial q_1} \frac{\partial \psi}{\partial q_j} \tag{15.17}$$
and
$$D^{-2} = |g^{ij}|,$$
$$g_{ij} = \mathbf{b}^i \mathbf{b}^j = \frac{\partial q_i}{\partial x}\frac{\partial q_j}{\partial x} + \frac{\partial q_i}{\partial y}\frac{\partial q_j}{\partial y} + \frac{\partial q_i}{\partial z}\frac{\partial q_j}{\partial z}. \tag{15.18}$$

The Laplacian in Curvilinear Orthogonal Coordinates

Equation (15.13) takes a much simpler form if the vectors $\mathbf{a}_i$ and $\mathbf{b}^i$ are mutually orthogonal. In this case the nondiagonal elements of $(d\mathbf{r})^2 (\nabla \psi)^2$ and D^2 vanish. We put, for simplicity,
$$g_{ii} = a_i^2, \qquad g^{ii} = b_i^2 = a_i^{-2}, \qquad D = a_1 a_2 a_3 \tag{15.19}$$
$$\nabla^2 \psi = \frac{1}{D}\frac{\partial}{\partial q_1}\frac{D}{a_1^2}\frac{\partial \psi}{\partial q_1} + \frac{\partial}{\partial q_2}\frac{D}{a_2^2}\frac{\partial \psi}{\partial q_2} + \frac{\partial}{\partial q_3}\frac{D}{a_3^2}\frac{\partial \psi}{\partial q_3}. \tag{15.20}$$

Spherical Coordinates

Cartesian coordinates are transformed into spherical coordinates r, ϑ, φ, by the equations
$$\begin{aligned} x &= r \sin \vartheta \cos \varphi \\ y &= r \sin \vartheta \sin \varphi \\ z &= r \cos \vartheta. \end{aligned} \tag{15.21}$$

The coordinate planes $r = \text{const}$, $\vartheta = \text{const}$, $\varphi = \text{const}$ are spheres, cones of opening angle ϑ around the z-axis, and planes through the z-axis. From (15.2) we obtain the three fundamental vectors

$$\mathbf{a}_1 = \frac{\partial \mathbf{r}}{\partial r} = \mathbf{i} \sin \vartheta \cos \varphi + \mathbf{j} \sin \vartheta \sin \varphi + \mathbf{k} \cos \vartheta = \mathbf{i}_r,$$

$$\mathbf{a}_2 = \frac{\partial \mathbf{r}}{\partial \vartheta} = \mathbf{i} r \cos \vartheta \cos \varphi + \mathbf{j} r \cos \vartheta \sin \varphi - \mathbf{k} r \sin \vartheta = r \mathbf{i}_\vartheta, \tag{15.22}$$

$$\mathbf{a}_3 = \frac{\partial \mathbf{r}}{\partial \varphi} = -\mathbf{i} r \sin \vartheta \sin \varphi + \mathbf{j} r \sin \vartheta \cos \varphi = r \sin \vartheta \, \mathbf{i}_\varphi,$$

where $\mathbf{i}_r$, $\mathbf{i}_\vartheta$, $\mathbf{i}_\varphi$ are orthonormal vectors in radial, meridional, and equatorial directions.

Hence
$$d\mathbf{r}^2 = dr^2 + r^2 d\vartheta^2 + r^2 \sin^2\vartheta \, d\varphi^2$$

$$(\nabla \psi)^2 = \left(\frac{\partial \psi}{\partial r}\right)^2 + \frac{1}{r^2}\left(\frac{\partial \psi}{\partial \vartheta}\right)^2 + \frac{1}{r^2 \sin^2\vartheta}\left(\frac{\partial \psi}{\partial \varphi}\right)^2,$$

$$a_1 = 1 \qquad a_2 = r \qquad a_3 = r \sin\vartheta$$

$$D = a_1 a_2 a_3 = r^2 \sin\vartheta, \tag{15.23}$$

and

$$\nabla^2 \psi = \frac{1}{r^2}\frac{\partial}{\partial r} r^2 \frac{\partial \psi}{\partial r} + \frac{1}{r^2 \sin\vartheta}\frac{\partial}{\partial \vartheta}\left(\sin\vartheta \frac{\partial \psi}{\partial \vartheta}\right) + \frac{1}{r^2 \sin^2\vartheta}\frac{\partial^2 \psi}{\partial \varphi^2} \tag{15.24}$$

Prolate and Oblate Rotational Elliptic Coordinates

Consider the confocal rotational ellipsoids and hyperboloids

$$\frac{x^2 + y^2}{\xi^2 - 1} + \frac{z^2}{\xi^2} = a^2,$$

$$-\frac{x^2 + y^2}{1 - \eta^2} + \frac{z^2}{\eta^2} = a^2, \tag{15.25}$$

with foci in $x = y = 0$, $z = \pm a$, where ξ and η are called prolate elliptic coordinates, the third coordinate being as usual the rotational angle φ around the z-axis. The Cartesian coordinates as expressed in these elliptic coordinates read

$$x = a\sqrt{\xi^2 - 1}\sqrt{1 - \eta^2} \cos\varphi,$$

$$y = a\sqrt{\xi^2 - 1}\sqrt{1 - \eta^2} \sin\varphi, \tag{15.26}$$

$$z = a\xi\eta,$$

$$1 \leq \xi \leq \infty, \qquad -1 \leq \eta \leq 1, \qquad 0 \leq \varphi \leq 2\pi.$$

From these equations we obtain by differentiation

$$d\mathbf{r}^2 = dx^2 + dy^2 + dz^2$$

$$= a^2\left[\frac{\xi^2 - \eta^2}{\xi^2 - 1} d\xi^2 + \frac{\xi^2 - \eta^2}{1 - \eta^2} d\eta^2 + (\xi^2 - 1)(1 - \eta^2)\, d\varphi^2\right]$$

$$(\nabla \psi)^2 = \frac{1}{a^2}\left[\frac{\xi^2 - 1}{\xi^2 - \eta^2}\left(\frac{\partial \psi}{\partial \xi}\right)^2 + \frac{1 - \eta^2}{\xi^2 - \eta^2}\left(\frac{\partial \psi}{\partial \eta}\right)^2 + \frac{(\partial \psi/\partial \varphi)^2}{(\xi^2 - 1)(1 - \eta^2)}\right]$$

$$D = a^3(\xi^2 - \eta^2). \tag{15.27}$$

Applying (15.16) we obtain

$$\nabla^2 \psi = \frac{1}{a^2(\xi^2 - \eta^2)} \cdot \left[\frac{\partial}{\partial \xi} (\xi^2 - 1) \frac{\partial \psi}{\partial \xi} + \frac{\partial}{\partial \eta} (1 - \eta^2) \frac{\partial \psi}{\partial \eta} \right.$$

$$\left. + \left(\frac{1}{\xi^2 - 1} + \frac{1}{1 - \eta^2} \right) \frac{\partial^2 \psi}{\partial \varphi^2} \right]. \quad (15.28)$$

The oblate elliptic coordinates differ from the above with respect to the rotational axis of the ellipsoids and hyperboloids, which is now the minor axis of the ellipsoids. The transformations are

$$x = a\sqrt{\xi^2 + 1}\sqrt{1 - \eta^2} \cos \varphi$$
$$y = a\sqrt{\xi^2 + 1}\sqrt{1 - \eta^2} \sin \varphi \quad (15.29)$$
$$z = a\xi\eta.$$

In this case we obtain the expressions

$$d\mathbf{r}^2 = dx^2 + dy^2 + dz^2$$
$$= a^2 \left[\frac{\xi^2 + \eta^2}{\xi^2 + 1} d\xi^2 + \frac{\xi^2 + \eta^2}{1 - \eta^2} d\eta^2 + (\xi^2 + 1)(1 - \eta^2) d\varphi^2 \right]$$

$$(\nabla \psi)^2 = \frac{1}{a^2} \left[\frac{\xi^2 + 1}{\xi^2 + \eta^2} \left(\frac{\partial \psi}{\partial \xi} \right)^2 + \frac{1 - \eta^2}{\xi^2 + \eta^2} \left(\frac{\partial \psi}{\partial \eta} \right)^2 + \frac{(\partial \psi/\partial \varphi)^2}{(\xi^2 + 1)(1 - \eta^2)} \right]$$

$$D = a^3(\xi^2 + \eta^2) \quad (15.30)$$

and by application of (15.16)

$$\nabla^2 \psi = \frac{1}{a^2(\xi^2 + \eta^2)} \left[\frac{\partial}{\partial \xi} (\xi^2 + 1) \frac{\partial \psi}{\partial \xi} + \frac{\partial}{\partial \eta} (1 - \eta^2) \frac{\partial \psi}{\partial \eta} \right.$$

$$\left. + \left(-\frac{1}{\xi^2 + 1} + \frac{1}{1 - \eta^2} \right) \frac{\partial^2 \psi}{\partial \varphi^2} \right]. \quad (15.31)$$

The oblate elliptic coordinates are useful when dealing with circular plates or circular apertures in acoustics, or in the scattering theory of light and particles. In atomic theory the prolate elliptic coordinates are of greater importances, as for instance in the theory of diatomic molecules.

Parabolic Coordinates

Another set of coordinates with axial symmetry which is used frequently also is that of parabolic coordinates.

Consider the equations
$$r = p + q, \quad z = p - q, \quad (15.32)$$
$$0 \leq p \leq \infty, \quad 0 \leq q \leq \infty,$$

which represent rotational paraboloids with respect to the z-axis. Adding the rotational angle φ, we obtain the transformation from Cartesian to parabolic coordinates p, q, φ, by the equations

$$x = 2\sqrt{pq} \cos \varphi,$$
$$y = 2\sqrt{pq} \sin \varphi, \quad (15.33)$$
$$z = p - q.$$

The square-differential line element is here

$$dr^2 = dx^2 + dy^2 + dz^2 = \frac{p+q}{p} dp^2 + \frac{p+q}{q} dq^2 + 4pq \, d\varphi^2. \quad (15.34)$$

Hence

$$(\nabla \psi)^2 = \frac{p}{p+q}\left(\frac{\partial \psi}{\partial p}\right)^2 + \frac{q}{p+q}\left(\frac{\partial \psi}{\partial q}\right)^2 + \frac{1}{4pq}\left(\frac{\partial \psi}{\partial \varphi}\right)^2. \quad (15.35a)$$

$$D = 2(p + q). \quad (15.35b)$$

Again applying (15.16), we obtain

$$\nabla^2 \psi = \frac{1}{p+q}\left[\frac{\partial}{\partial p}\left(p \frac{\partial \psi}{\partial p}\right) + \frac{\partial}{\partial q}\left(q \frac{\partial \psi}{\partial q}\right) + \left(\frac{1}{4p} + \frac{1}{4q}\right)\frac{\partial^2 \psi}{\partial \varphi^2}\right]. \quad (15.36)$$

It should be noted that the Laplacian has a fairly similar appearance in a set of coordinates ξ, η, φ, which are sometimes used under the name of parabolic coordinates.

If we make the substitution

$$2p = \xi^2 \quad 2q = \eta^2 \quad 0 \leq \xi, \quad \eta \leq \infty, \quad (15.37)$$

we find

$$\nabla^2 \psi = \frac{1}{\xi^2 + \eta^2}\left[\frac{1}{\xi}\frac{\partial}{\partial \xi}\xi\frac{\partial \psi}{\partial \xi} + \frac{1}{\eta}\frac{\partial}{\partial \eta}\eta\frac{\partial \psi}{\partial \eta} + \left(\frac{1}{\xi^2} + \frac{1}{\eta^2}\right)\frac{\partial^2 \psi}{\partial \varphi^2}\right]. \quad (15.38)$$

Parabolic coordinates are useful in connection with a combined central and linear field as, for instance, in the quantum mechanics of the Stark effect.

CHAPTER 3

Linear Transformation. Extremum and Variational Problems

16. THE GENERALIZED VECTOR SPACE. ROTATION OF COORDINATE SYSTEMS

Vectors in the n-Dimensional Space

Quite formally we may write **x** for an arbitrary number of scalars x_1, $x_2, \cdots, x_n$, and call it a vector. For $n = 2$ and $n = 3$ the vectors may be visualized in a plane or in space. If $n > 3$ we have no such illustration, but we may invent something called the n-dimensional space with very much the same properties as our three-dimensional space. For instance, we may think of n mutually orthogonal directions which may be chosen for the direction of axis of a generalized Cartesian coordinate system. To fix the idea, the coordinate systems of several particles (even though in ordinary sense the same) may be thought of as forming a six-, nine-, and twelve-dimensional coordinate system. In physical problems introducing normal coordinates for transforming kinetic (and by vibrational problems sometimes also potential) energy into its normal form, we may consider them coordinates in a manifold space. Obviously there is an infinite number of Cartesian coordinates in the n-dimensional space; otherwise it would be essentially different from two- and three-dimensional space. Hence the same vector **x** may be specified by quite a different set of scalars $x'_1, x'_2, \cdots, x'_n$, referring to another coordinate system.

The Scalar Products of n-Dimensional Vectors

We define the scalar products of two vectors **x** and **y** by the equation

$$\mathbf{xy} = \sum_{i=1}^{n} x_1 y_1 \qquad (16.1)$$

and the magnitude $|\mathbf{x}|$ of a vector **x** by

$$|\mathbf{x}|^2 = \mathbf{xx} = \mathbf{x}^2 = \sum_i x_i^2. \qquad (16.2)$$

62 / The Mathematical Foundation of Physics

Adding the equation

$$\mathbf{x'y'} = \sum_i x'_i y'_i$$

as expressed by the components of the vectors in any coordinate system, these definitions imply that the coordinate systems should be orthogonal or Cartesian.

There is a very interesting and often useful inequality for scalar products in n-dimensional space called *Schwarz's inequality*. Consider $(\mathbf{x} - \lambda \mathbf{y})^2$ which is positive for any value of λ,

$$\mathbf{x}^2 - 2\lambda \mathbf{xy} + \lambda^2 \mathbf{y}^2 \geqq 0. \tag{16.3}$$

Minimizing the expression with respect to λ, we obtain

$$\mathbf{x}^2 - \frac{(\mathbf{xy})^2}{\mathbf{y}^2} \geqq 0 \quad \text{or} \quad (\mathbf{xy})^2 \geqq \mathbf{x}^2 \mathbf{y}^2, \tag{16.4}$$

which is Schwarz's inequality.

The inequality is subject to important generalizations. For instance it can easily be proved that, for two functions $f(x)$ and $g(x)$ in a given region $a \leqq x \leqq b$,

$$(f, g)^2 \leqq (f, f)(g, g), \quad (y, z) = \int_a^b y(x) z(x) \, dx. \tag{16.5}$$

Dividing the coordinate space into n equal parts and considering $f_i = f(x_i)$ the coordinates of a n-dimensional vector, we have by Schwarz's inequality

$$\left(\sum_i f_i g_i \right)^2 \leqq \sum_i f_i^2 \cdot \sum_i g_i^2, \tag{16.6}$$

which as $n \to \infty$ becomes equivalent to (16.5). This suggests that the number of dimensions (n) of a manifold space may also be taken to be infinitely large without changing the main properties of the space. This space is called the space of functions or the *Hilbert space*.

We consider an example. In the region $0 \leqq x \leqq \pi$ we write

$$f(x) = \sum_{m=1}^{\infty} f_m \sin mx, \quad g(x) = \sum_{m=1}^{\infty} g_m \sin mx. \tag{16.7}$$

By integration

$$(f, g) = \sum_{m=1}^{\infty} f_m g_m, \quad (f, f) = \sum_{m=1}^{\infty} f_m^2, \quad \text{etc.} \tag{16.8}$$

Hence, in virtue of (16.5)

$$\left(\sum_m f_m g_m \right)^2 \leqq \sum_m f_m^2 \cdot \sum_m g_m^2 \tag{16.9}$$

Linear Transformation. Extremum and Variational Problems

which is in fact Schwarz's inequality for vectors with components f_m and g_m in a space with infinitely high dimensions. The functions $\sin mx$ are said to be orthogonal functions, $\int \sin mx \sin m'x \, ds = 0$, $m \neq m'$. They may be considered some sort of coordinate axis forming a coordinate system, in which the function $f(x)$ interpreted as a vector has the components f_m.

Unit Vectors and Orthonormal Systems

A unit vector **e** is defined by $\mathbf{e}^2 = 1$. Consider a system of n vectors obeying the equations

$$\mathbf{e}_i \mathbf{e}_k = \delta_{ik}; \tag{16.10}$$

they are said to form an orthonormal system. Clearly the vector **x** may be written

$$\mathbf{x} = \sum_{i=1}^{n} x_i \mathbf{e}_i, \tag{16.11}$$

and correspondingly

$$\mathbf{x} = \sum_{i=1}^{n} x'_i \mathbf{e}'_i \tag{16.12}$$

in another orthonormal system.

Multiplying (16.10) by e_k, we get

$$\mathbf{x} \mathbf{e}_k = x_k, \tag{16.13}$$

which means that vector components are given by the same kind of scalar products as in three-dimensional space.

Obviously the transition of the vector **x** from (16.11) to (16.12) means a rotation in the generalized sense of the coordinate system. Multiplying (16.12) by $\mathbf{e}_i$, we obtain

$$x_i = \sum_{k=1}^{n} (\mathbf{e}_i \mathbf{e}'_k) x'_k = \sum_{k=1}^{n} S_{ik} x'_k = \mathbf{S}_i \mathbf{x}'. \tag{16.14}$$

Here $\mathbf{S}_i$ and $\mathbf{x}'$ are vectors (as expressed in the same coordinate system as **x**) with components $S_{ik} = (\mathbf{e}_i \mathbf{e}'_k)$ and x'_k. The vector $\mathbf{S}_i$ corresponds to the vector $\mathbf{e}_i$ as expressed in the dashed system. The n vectors $\mathbf{S}_i$, hence, form an orthonormal system

$$\mathbf{S}_i \mathbf{S}_k = \mathbf{e}_i \mathbf{e}_k = \delta_{ik}. \tag{16.15}$$

They may be taken to form a tensor or a matrix

$$S = \begin{pmatrix} S_{11} & S_{12} & \cdots & S_{1n} \\ S_{21} & S_{22} & \cdots & S_{2n} \\ \cdot & & & \\ \cdot & & & \\ \cdot & & & \\ S_{n1} & S_{n2} & \cdots & S_{nn} \end{pmatrix} \tag{16.16}$$

64 / The Mathematical Foundation of Physics

which define a rotation in the n-dimensional space and is called a *unitary matrix*.

Since from (16.11) multiplying by e'_i we obtain

$$x'_i = \sum_{k=1}^{n}(\mathbf{e}_k\mathbf{e}'_i)x_k = \tilde{\mathbf{S}}_i\mathbf{x}, \tag{16.17}$$

the vectors $\tilde{\mathbf{S}}_i$ define the reciprocal transformation having the same orthonormal properties as $\mathbf{S}_i$. Their components are S_{ki}, hence they define a matrix $\tilde{S}$, which is obtained by transposition of S. Using the matrix in either succession, the result is an identity transformation. This can be written as

$$S\tilde{S} = \tilde{S}S = 1 \quad \text{or} \quad \tilde{S} = S^{-1} \tag{16.18}$$

which is the definition of a unitary matrix. It can be shown that the determinants of the matrices are 1 or -1, corresponding to transition from a right-hand to either a right-hand or a left-hand coordinate system in three-dimensional space.

17. DETERMINANTS AND LINEAR EQUATIONS

The volume in n-Dimensional Space

Consider a system of n basic vectors $\mathbf{a}_i$ with components a_{ik} with respect to a coordinate system defined by a set of orthonormal vectors $\mathbf{e}_1, \mathbf{e}_2, \cdots, \mathbf{e}_n$. As in ordinary space, we shall think of the vectors $\mathbf{a}_i$ as defining a definite volume in the n-dimensional space which we denote by $(\mathbf{a}_1, \mathbf{a}_2, \cdots, \mathbf{a}_n)$, the order of the vectors being of importance. We shall write

$$D(\mathbf{a}_1, \mathbf{a}_2, \cdots, \mathbf{a}_n) = |a_{ik}| = \begin{vmatrix} a_{11} & a_{12} & \cdots & a_{1n} \\ a_{21} & a_{22} & \cdots & a_{2n} \\ \vdots & & & \\ a_{n1} & a_{n2} & \cdots & a_{nn} \end{vmatrix}, \tag{17.1}$$

where the latter expression is known as the determinant of the tensor or matrix a defined by the n basic vectors. We shall also use the transposed matrix $\tilde{a}$ defined by the transposed system of vectors $\tilde{\mathbf{a}}_i$, with components $\tilde{a}_{ik} = a_{ki}$, and we prove that

$$D(\mathbf{a}_1, \mathbf{a}_2, \cdots, \mathbf{a}_n) = |a_{ik}| = |a_{ki}| = D(\tilde{\mathbf{a}}_1, \tilde{\mathbf{a}}_2, \cdots, \tilde{\mathbf{a}}_n). \tag{17.2}$$

For definition of the volume we have three postulates:

1. *Definition of Unity of Volume*

$$D(\mathbf{e}_1, \mathbf{e}_2, \cdots, \mathbf{e}_n) = 1 \tag{17.3}$$

by the given order of unit vectors.

2. *Additivity Law*

$$D(\mathbf{a}_1, \cdots, \mathbf{a}_i + \mathbf{b}_i, \cdots, \mathbf{a}_n) = D(\mathbf{a}_1, \cdots, \mathbf{a}_i, \cdots, \mathbf{a}_n)$$
$$+ D(\mathbf{a}_1, \cdots, \mathbf{b}_i, \cdots, \mathbf{a}_n). \tag{17.4}$$

3. *Zero Volume for Coplanar Vectors*

$$D(\mathbf{a}_1, \cdots, \mathbf{a}_i, \cdots, \mathbf{a}_n) = 0 \quad \text{if} \quad \sum_i \mu_i \mathbf{a}_i = 0. \tag{17.5}$$

From the latter it follows that $|a_{ik}| = 0$ if $\mathbf{a}_i = 0$ or $\mathbf{a}_i = \mathbf{a}_k$.

Combining Postulates 2 and 3, we find that

$$D(\cdots, \mathbf{a}_i - \mathbf{a}_i, \cdots) = D(\cdots, \mathbf{a}_i, \cdots) + D(\cdots, -\mathbf{a}_i, \cdots) = 0, \tag{17.6}$$

which means that the sign of the determinant is changed by changing the sign of a single vector $\mathbf{a}_i$. By the same postulates

$$D(\mathbf{a}_1 + \lambda \mathbf{a}_i, \cdots, \mathbf{a}_i, \cdots) = D(\mathbf{a}_1, \cdots, \mathbf{a}_i, \cdots) \quad i \neq 1. \tag{17.7}$$

Hence we may add to the vector $\mathbf{a}_1$ or any other vector $\mathbf{a}_2, \mathbf{a}_3, \cdots$ another of the vectors multiplied by an arbitrary factor λ without changing its value.

Repeating this process, we may prove that interchange of any of two vectors has the effect of merely changing the sign of the volume,

$$D(\cdots, \mathbf{a}_i, \cdots, \mathbf{a}_k, \cdots) = D(\cdots, \mathbf{a}_i, \cdots, \mathbf{a}_i + \mathbf{a}_k, \cdots)$$
$$= D(\cdots, -\mathbf{a}_k, \cdots, \mathbf{a}_i + \mathbf{a}_k, \cdots)$$
$$= D(\cdots, -\mathbf{a}_k, \cdots, \mathbf{a}_i, \cdots)$$
$$= -D(\cdots, \mathbf{a}_k, \cdots, \mathbf{a}_i, \cdots). \tag{17.8}$$

If λ is an integer, $\lambda \mathbf{a}_1 = \mathbf{a}_1 + \mathbf{a}_1 + \cdots$, the repeated application of the additivity law yields

$$D(\cdots, \lambda \mathbf{a}_i, \cdots) = \lambda D(\cdots \mathbf{a}_i \cdots). \tag{17.9}$$

Now let p and q be integers,

$$qD\left(\cdots \frac{p}{q} \mathbf{a}_i \cdots\right) = D(\cdots, p\mathbf{a}_i, \cdots) = pD(\cdots, \mathbf{a}_i, \cdots). \tag{17.10}$$

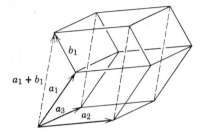

Fig. 3-1. Illustration of the additivity law.

On dividing by q this proves that (17.9) holds for any rational number $\lambda = p/q$.

In these simple rules we recognize the majority of well-known results from the theory of determinants. We now apply the additivity law together with the above rules to identify our definition with the conventional definition of a determinant $|a_{ik}|$ as referred directly to the elements a_{ik}. (See Fig. 3-1.)

We shall first have to note that

$$D(\mathbf{e}_{k_1}, \mathbf{e}_{k_2}, \cdots, \mathbf{e}_{k_n}) = (-1)^P, \tag{17.11}$$

P being the number of elementary permutations (interchange of two vectors) which leads from the succession $1, \cdots, n$, to $k_1, \cdots, k_n$, of the vector $\mathbf{a}_i$. Since $(-1)^P = \pm 1$, the number P need not be uniquely given apart from being odd or even. The minimum number of elementary permutations is obtained by interchanging first $\mathbf{e}_1$ and $\mathbf{e}_{k_1}$, next $\mathbf{e}_2$ and $\mathbf{e}_{k_2}$ and so on. Therefore, the minimum number can be larger than n. For each $\mathbf{e}_{k_i}$ at its normal place the minimum number of elementary permutations is reduced by one, and for each pair $k_i = j$, $k_j = i$, the number is reduced by two. It is easily seen that, for a given permutation, P can never take both positive and negative values.

Now expressing $\mathbf{a}_i$ as $\sum_k a_{ik}\mathbf{e}_k$, we obtain the formula

$$D(\mathbf{a}_1, \cdots, \mathbf{a}_n) = \sum_{k_1 \cdots k_n} D(\mathbf{e}_{k_1} \cdots \mathbf{e}_{k_n}) a_{1k_1} a_{2k_2} \cdots a_{nk_n}$$

$$= \sum_{i_1 \cdots i_n} D(\mathbf{e}_{k_1} \cdots \mathbf{e}_{k_n}) a_{i_1 1} a_{i_2 2} \cdots a_{i_n n}$$

$$= D(\tilde{\mathbf{a}}_1, \cdots, \tilde{\mathbf{a}}_n), \tag{17.12}$$

if

$$P_{k_1 \cdots k_n} = P_{i_1 \cdots i_n}, \tag{17.13}$$

$i_1, \cdots, i_n$, denoting the row number of the elements after rearranging $k_1, \cdots, k_n$ into its natural order of succession. The result is obvious, since the rearrangements of the $k_1, \cdots, k_m$ is obtained by the same permutation of elements a_{ik} as the arrangement of $1, \cdots, n$ into $i_1, \cdots, i_n$.

Linear Transformation. Extremum and Variational Problems | 67

Development of Determinants in Terms of Minors. Reciprocal Vectors

According to the additivity law (17.4) a determinant is a linear function of the components of any of its basic vectors. Hence it can be written as a scalar product of a basic vector, say, $\mathbf{a}_i$ and another vector $\mathbf{A}_i$,

$$D(\mathbf{a}_1, \cdots, \mathbf{a}_n) = \mathbf{a}_i \mathbf{A}_i = \sum_k a_{ik} A_{ik}. \tag{17.14}$$

The components A_{ik} of the vector $\mathbf{A}_i$ are determinants of the $(n-1)$th order and are called *minor determinants* or simply *minors*. Strictly spoken, these are minors of the first order. There are minors of higher orders which we need not discuss here, reserving the word minors for those of the first order.

It is easily shown that the vector $\mathbf{A}_i$ is orthogonal to any of the basic vectors $\mathbf{a}_k$, except the vector $\mathbf{a}_i$. This is because

$$\mathbf{a}_k \mathbf{A}_i = D(\cdots, a_k, \cdots, a_k \cdots) = 0, \quad i \neq k, \tag{17.15}$$

according to (17.5). Hence the vector systems $\mathbf{a}_i$ and $\mathbf{A}_i/D$ are reciprocal systems. Writing for brevity

$$\mathbf{a}^i = \frac{\mathbf{A}_i}{D}, \quad a^{ik} = \frac{A_{ik}}{D}, \tag{17.16}$$

we have

$$\mathbf{a}_i \mathbf{a}^k = \delta_{ik}. \tag{17.17}$$

According to (17.16) the reciprocal system of vectors $\mathbf{a}^i$ exists only if $D = 0$.

Solution of Inhomogeneous Linear Equations

Consider the systems of inhomogeneous linear equations

$$\sum_{k=1}^{n} a_{ik} x_k = y_i, \quad i = 1, 2, \cdots, n, \tag{17.18}$$

or, for brevity,

$$\mathbf{a}_i \mathbf{x} = y_i, \tag{17.19a}$$

or

$$\mathbf{y} = \sum_{k=1}^{n} x_k \tilde{\mathbf{a}}_k. \tag{17.19b}$$

Provided $D = 0$, so that $\mathbf{a}^i$ exists, the system of 17.19a or 17.19b can be solved and the solution can be written

$$\mathbf{x} = \sum_{i=1}^{n} y_i \mathbf{a}^i, \tag{17.20a}$$

or

$$x_k = \mathbf{y} \tilde{\mathbf{a}}^k = \sum_{i=1}^{n} y_i a^{ik}. \tag{17.20b}$$

This is easily proved by inserting, say, (17.20a) in (17.19a). As a rule we may therefore say that the solutions (17.20) are obtained from (17.20) simply by interchanging the vectors **x** and **y** and replacing $\mathbf{a}_i$ by $\mathbf{a}^i$ or $\tilde{\mathbf{a}}_i$ by $\tilde{\mathbf{a}}^i$.

The result may also be written in the more familiar form

$$x_k = \frac{D(\tilde{\mathbf{a}}_1, \cdots, \mathbf{y}, \cdots, \tilde{\mathbf{a}}_n)}{D(\mathbf{a}_1, \cdots, \mathbf{a}_k, \cdots, \mathbf{a}_n)}, \tag{17.21}$$

as the quotient of two determinants in which the nominator is obtained from the determinant D by using for elements of the kth column the components of the vector **y**.

Homogeneous Linear Equations

If $\mathbf{y} = 0$, Eq. 17.19 are said to be homogeneous,

$$\sum_{k=1}^{n} a_{ik} x_k = \mathbf{a}_i \mathbf{x} = 0, \qquad \sum_{k=1}^{n} x_k \tilde{\mathbf{a}}_k = 0. \tag{17.22}$$

Then in general $\mathbf{x} = 0$ is the only solution. If any other solution exists, that is, if any component x_k differs from zero, there will be a linear relation between the transposed vectors $\tilde{\mathbf{a}}_k$. These vectors then are coplanar and their determinant is zero,

$$D(\tilde{\mathbf{a}}_1, \cdots, \tilde{\mathbf{a}}_n) = D(\mathbf{a}_1, \cdots, \mathbf{a}_n) = 0, \tag{17.23}$$

The vectors $\mathbf{a}_i$ are also coplanar. This is a necessary condition for the solubility of a system of homogeneous linear equations.

The solution of (17.22) is easily found. Since the vector **x** need only be orthogonal to any of the vectors $\mathbf{a}_i$, we may write

$$\mathbf{x} = c\mathbf{A}_i \tag{17.24}$$

since, according to (17.17), $\mathbf{a}_k \mathbf{x} = c \mathbf{a}_k \mathbf{A}_i = 0$ when $i \neq k$, and

$$\mathbf{a}_i \mathbf{x} = cD = 0. \tag{17.25}$$

Only, in the case that all minors A_{ik} of the first order vanish, the solution (17.24) is of no use and we have to enter into a more elaborate treatment of the problem, which we shall omit here. In this case we may say, with a view to the three-dimensional geometry, that the vectors $\mathbf{a}_i$ are at least "*collinear.*"

Inhomogeneous Linear Equations with Vanishing Determinant

If $D = 0$ and $\mathbf{y} \neq 0$, the system of equations (17.19) has in general no solution. There is, however, one exception: according to (17.19b), if **y** is coplanar with the coplanar vectors $\tilde{\mathbf{a}}_i$. This means that there is a linear relation between **y** and $n - 1$ of the vector $\tilde{\mathbf{a}}_i$ expressing the solutions.

Linear Transformation. Extremum and Variational Problems / 69

Since the vectors $\tilde{\mathbf{A}}_i$ formed by minors $\tilde{A}_{ik} = A_{ki}$ of the determinant are orthogonal to the vector $\tilde{a}_i$, the properties of the solution $\mathbf{y}$ is most easily expressed by stating that it is orthogonal to any $\tilde{\mathbf{A}}_i$,

$$\mathbf{y}\tilde{\mathbf{A}}_i = 0. \tag{17.26}$$

In the case of a symmetric determinant with $\tilde{\mathbf{a}}_i = \mathbf{a}_i$ and, $\tilde{\mathbf{A}}_i = \mathbf{A}_i$, (17.26) is equivalent to $\mathbf{y}\mathbf{A}_i = 0$, meaning that $\mathbf{y}$ is orthogonal to the solution $\mathbf{x} = c\mathbf{A}_i$ of the homogeneous system of equations.

Numerical Evaluation of Determinants

The determination of the numerical values of determinants of higher order is not an easy task. According to the common definition of a determinant of the nth order, we shall have to compute $n!$ products with n factors, which together with additions of the products necessitates $(n + 1)!$ operations.

Another and shorter procedure might be to orthogonalize the fundamental vectors successively by writing

$$D(\mathbf{a}_1, \cdots, a_n) = D(\mathbf{a}_1, \lambda_{21}\mathbf{a}_1 + \mathbf{a}_2, \cdots, \lambda_{n1}\mathbf{a}_1 + \lambda_{n2}\mathbf{a}_2 + \cdots + \mathbf{a}_n) \tag{17.27}$$

and determining the $\tfrac{1}{2}n(n + 1)$ coefficients from orthogonality considerations. The determinant then is the product of the absolute values of these vectors with the sign $+$ or $-$.

Another very similar method frequently used in practice, with approximately the same number of operations, consists of a successive reduction of the order of the determinant. This can be done by replacing the vectors a_i in the original determinant by

$$\mathbf{b}_i = \mathbf{a}_i - \frac{a_{i1}}{a_{11}} \mathbf{a}_1, \tag{17.28}$$

which are orthogonal to the first unit vector $\mathbf{e}$, that is, $\mathbf{e}_1 \mathbf{b}_i = 0$ because of

$$\mathbf{a}_1 \mathbf{e}_1 = a_{11}, \quad \mathbf{a}_i \mathbf{e}_1 = a_{i1}. \tag{17.29}$$

18. MATRIX ALGEBRA

Linear Transformation and Matrices

In the system of equations

$$\sum a_{ik}x_k = \mathbf{a}_i \mathbf{x} = y_i, \tag{18.1}$$

we may consider the vector $\mathbf{x}$ and its components x_i given quantities. Then we can determine the vector $\mathbf{y}$ or its components y_i. We therefore say that the above system of equations defines a linear transformation. This transformation is defined by the n basic vector $\mathbf{a}_i$ of the system that may be united in a

tensor
$$a = \mathbf{e}_1 \cdot \mathbf{a}_1 + \cdots + \mathbf{a}_n \cdot \mathbf{a}_n = \sum_i \mathbf{e}_i \cdot \mathbf{a}_i, \qquad (18.2)$$

$\mathbf{e}_i$ being a set of orthonormal or unit vectors. By this tensor we may write (18.1) in the form of a single equation

$$a\mathbf{x} = \mathbf{y}. \qquad (18.3)$$

The quadratic scheme of tensor components

$$(a_{ik}) = \begin{vmatrix} a_{11} & a_{12} & \cdots & a_{1n} \\ a_{21} & a_{22} & \cdots & a_{2n} \\ \cdot & & & \\ \cdot & & & \\ \cdot & & & \\ a_{n1} & a_{n2} & \cdots & a_{nn} \end{vmatrix} \qquad (18.4)$$

is usually called a matrix. Sometimes we shall even use the same notation for it as for the tensor, $a = (a_{ik})$.

A linear transformation is uniquely determined by its matrix. By composite linear transformations we shall have to perform certain operations with the matrices themselves; this we shall call *matrix algebra*. Since a matrix defines a certain operation, physical quantities defined by *operators* may sometimes be expressible by matrices.

Composite Transformations. Multiplication of Matrices

If two linear transformations are applied in succession, the result is expressible by a single transformation. If

$$\mathbf{y} = a\mathbf{u}, \quad \mathbf{u} = b\mathbf{x}, \qquad (18.5)$$

we write

$$\mathbf{y} = a(b\mathbf{x}) = (ab)\mathbf{x}, \qquad (18.6)$$

denoting the matrix of the resulting linear transformation by ab, which is called the product matrix of a and b. The elements of this matrix are easily found by the equations

$$y_i = \sum_j a_{ij} u_j, \quad u_j = \sum_k b_{jk} x_k. \qquad (18.7)$$

Inserting u_j in the first equation, we obtain

$$y_i = \sum_k (ab)_{ik} x_k \qquad (18.8)$$

with

$$(ab)_{ik} = \sum_j a_{ij} b_{jk}. \qquad (18.9)$$

Linear Transformation. Extremum and Variational Problems / 71

This is the important formula for matrix multiplication, and the rule is that we have to take the product sum of the ith row of a and the kth column of b in order to obtain the ikth element of the product matrix. By means of the basic vectors of the matrices we may also write

$$(ab)_{ik} = \mathbf{a}_i \tilde{\mathbf{b}}_k. \qquad (18.10)$$

In general the commutative law is not valid for matrix multiplications, which is easily seen by writing

$$(ba)_{ik} = \tilde{\mathbf{a}}_k \mathbf{b}_i. \qquad (18.11)$$

Even if the factor matrices are symmetric, that is, $\tilde{a}_k = a_k$ and $\tilde{b}_k = b_k$, (18.10) and (18.11) are different, since usually $\mathbf{a}_i \mathbf{b}_k \neq \mathbf{a}_k \mathbf{b}_i$.

According to (18.10)

$$(ab)_{ik} = \tilde{\mathbf{b}}_k \mathbf{a}_i = (\tilde{b}\tilde{a})_{ki} \qquad (18.12)$$

or, in matrix notation,

$$ab = (\widetilde{\tilde{b}\tilde{a}}), \qquad (18.13)$$

whereas for symmetric matrices a and b

$$ab = ba. \qquad (18.14)$$

The associative law $a(bc) = (ab)c$ is, however, easily shown to be valid.

Product Determinants

The multiplication of determinants must be expected to follow the same rules as that of matrices. If that is true, however, there are several ways of performing a determinant multiplication, owing to the fact that, in determinants, rows and columns may be interchanged without altering the value of the determinant. Therefore in a product determinant we may take for elements sum products of rows and rows, columns and columns, or, of rows and columns of either of the factor determinants. Expressed by equation, we find that

$$(|a| \cdot |b|)_{ik} = \mathbf{a}_i \tilde{\mathbf{b}}_k = \tilde{\mathbf{a}}_i \mathbf{b}_k = \mathbf{a}_i \mathbf{b}_k = \tilde{\mathbf{a}}_i \tilde{\mathbf{b}}_k. \qquad (18.15)$$

To see that the first expression $\mathbf{a}_i \tilde{\mathbf{b}}_k$ is true consider the transformation

$$\mathbf{y} = \sum_k x_k \mathbf{a}_k. \qquad (18.16)$$

By this transformation the unit vectors $\mathbf{e}_1, \cdots, \mathbf{e}_n$ in the x-space are transformed into the vectors $\tilde{\mathbf{a}}_1, \cdots, \tilde{\mathbf{a}}_n$ in the y-space, causing a unit volume element to change into a volume $D(\tilde{\mathbf{a}}_1, \cdots, \tilde{\mathbf{a}}_n) = |a|$.

To see that this relative change of volume is independent of its shape, hence applicable to a parallelepiped of the vectors $\mathbf{b}$ with volume $|b|$, we write

$$\mathbf{b}_i = \sum_k b_{ik} \mathbf{e}_k. \qquad (18.17)$$

These vectors transform by (18.12) into the vectors

$$c_i = \sum_k b_{ik}\tilde{a}_k. \tag{18.18}$$

Now if we use the additivity theorem repeatedly, $|c|$ appears as a sum of determinants that are either $\pm a$ or 0, however multiplied by products of n coefficients b_{ik}, in which all the indices k as well as i are different. Mathematically we obtain

$$|c| = \sum_{k_1,\cdots,k_n} (\pm 1) b_{1k_1} b_{2k_2} \cdots b_{nk_n} \cdot |a| = |b| \cdot |a|, \tag{18.19}$$

the sign of the product term, as is easily controlled, being in accordance with the definition of the determinant $|b|$.

Inverse Transformations and Reciprocal Matrices

A matrix with only diagonal terms different from zero is termed *diagonal*. If the diagonal terms are all 1, we have the *unit matrix* which we may denote by δ, its elements $\delta_{ii} = 1$, $\delta_{ik} = 0$ for $i \neq k$, corresponding to the frequently used Kronecker δ_{ik} symbol. In cases where confusion with the sign of variation δ is possible it might be advisable to use the notation e, but we shall rather prefer the notation E, since that has already been used for the three-dimensional unit tensor and for the unit quadratic form. The matrix δ provides an *identity transformation* and hence corresponds to the factor 1 in ordinary algebra, $\delta a = a$, and so on.

Two successive transformations resulting in an identity transformation are said to be mutually *inverse transformations*, their matrices being *reciprocal matrices*. By a given matrix a, we denote its reciprocal by a^{-1}. The reciprocal matrix exists only if $|a| \neq 0$. From

$$\mathbf{y} = a\mathbf{x}, \quad \mathbf{x} = a^{-1}\mathbf{y}, \tag{18.20}$$

we obtain, by eliminating $\mathbf{x}$ or $\mathbf{y}$,

$$aa^{-1} = a^{-1}a = \delta, \tag{18.21}$$

or

$$\sum_j a_{ij} a_{jk}^{-1} = \delta_{ik}. \tag{18.22}$$

Comparing with the definition of reciprocal vector systems in the n-dimensional space

$$\mathbf{a}_i \mathbf{a}^k = \sum_j a_{ij} a^{kj} = \delta_{ik}, \quad \mathbf{a}^k = \frac{\mathbf{A}_k}{D}, \tag{18.23}$$

we obtain

$$a_{ik}^{-1} = a^{ki} = \frac{A_{ki}}{D}. \tag{18.24}$$

Linear Transformation. Extremum and Variational Problems | 73

The special kind of transformations called *orthogonal*, or sometimes *unitary*, are characterized by the following properties of their basic vectors $\mathbf{S}_i$,

$$\mathbf{S}_i \mathbf{S}_k = \delta_{ik}, \tag{18.25}$$

and this means that the reciprocal and the transposed matrices of S are identical

$$S^{-1} = \tilde{S}, \tag{18.26}$$

as already stated in (16.18).

19. QUADRATIC FORMS AND THEIR STATIONARY VALUES

Quadratic Forms with Symmetric Coefficients

A quadratic function of a set of variables $x_1, \cdots, x_n$ can always be written in the form with symmetric coefficients,

$$A(x, x) = \sum_{i,k=1}^{n} a_{ik} x_i x_k, \qquad a_{ik} = a_{ki}, \tag{19.1}$$

This is called a *quadratic form*. It may be said to be defined by a symmetric matrix a or (a_{ik}); it can also be interpreted as the scalar product of a symmetric tensor a and the vector x, both as pre- and postfactor,

$$A(x, x) = x(ax) = (xa)x = xax. \tag{19.2}$$

The *unit quadratic form* is written

$$E(x, x) = \sum_i x_i^2 = x^2. \tag{19.3}$$

It should be noted at once that there is a very important and frequently used generalization of quadratic forms leading to what is called Hermitian forms. In these forms the variables may have complex values; the Hermitian forms themselves, however, are real as may be seen from the definition

$$A(x^*, x) = x^* a x = \sum_{i,k} x_i^* a_{ik} x_k, \qquad a_{ik} = a_{ki}^*, \tag{19.4}$$

the asterisk denoting conjugate complex quantities.

Because of the real nature of the forms, Hermitian theory is fairly similar to that of ordinary quadratic forms. For this reason, and also because we have not yet discussed the theory of complex numbers, we shall deal only with ordinary quadratic forms in this section.

The Eigenvalues of a Quadratic Form

A basic problem in the theory is to determine extremum values, whether maximum, minimum, or simply stationary of a quadratic form when the variables are subject to the condition that they are components of a unit vector or that the value of the unit quadratic form is 1. Mathematically we write

$$A(x, x) = \mathbf{x}a\mathbf{x} = \text{extr.} \tag{19.5a}$$

by

$$E(x, x) = \mathbf{x}^2 = 1. \tag{19.5b}$$

This problem could also be solved by looking for extremum values of the quotient

$$\lambda = \frac{A(x, x)}{E(x, x)}, \tag{19.6}$$

leaving aside the restriction on **x**, since both A and E are quadratically proportional to the magnitude of **x**. By ordinary derivation we obtain the conditions for stationary values

$$\frac{\partial \lambda}{\partial x_i} = \frac{1}{E}\left(\frac{\partial A}{\partial x_i} - \frac{A}{E}\frac{\partial E}{\partial x_i}\right) = 0, \quad i = 1, \cdots, n, \tag{19.7}$$

or

$$\frac{\partial A}{\partial x_i} - \lambda \frac{\partial E}{\partial x_i} = 0, \tag{19.8}$$

when reintroducing the so-far unknown extremum—or eigenvalue—parameter λ.

Inserting A and E in (19.8), we obtain the system of equations

$$\sum_k a_{ik} x_k - \lambda x_i = 0, \quad \text{or} \quad a\mathbf{x} - \lambda \mathbf{x} = 0. \tag{19.9}$$

The system can be solved only if its determinant vanishes,

$$|a_{ik} - \lambda \delta_{ik}| = \begin{vmatrix} a_{11} - \lambda & a_{12} & \cdots & a_{1n} \\ a_{21} & a_{22} - \lambda & \cdots & a_{2n} \\ \vdots & & & \\ a_{n1} & a_{n2} & \cdots & a_{nn} - \lambda \end{vmatrix} = 0. \tag{19.10}$$

This provides an algebraic equation of the nth order for determination of the unknown n stationary values of λ. These values which we denote by $\lambda_1, \lambda_2, \cdots, \lambda_n$, are called the *eigenvalues* of the quadratic form $A(x, x)$; the corresponding vectors are called *eigenvectors*.

We are now going to show that all eigenvalues are real. Consider the equation
$$ax = \lambda x, \tag{19.11}$$
or
$$\sum_k a_{ik} x_k = \lambda x_i, \tag{19.12}$$
for any eigenvalue and eigenvector. If λ and $\mathbf{x}$ are complex, then, conjugating (19.12) or (19.11) and remembering that the a_{ik} are real and symmetric, we obtain
$$a\mathbf{x}^* = \lambda^* \mathbf{x}^*. \tag{19.13}$$
If we multiply (19.11) and (19.13) by $\mathbf{x}^*$ and $\mathbf{x}$ and subtract,
$$(\lambda - \lambda^*)\mathbf{x}^*\mathbf{x} = 0. \tag{19.14}$$
Since $\mathbf{x}^*\mathbf{x}$ is a real (and positive) nonvanishing quantity, we find that $\lambda = \lambda^*$, which means that any eigenvalue λ is real.

Even in the case of Hermitian forms, this result holds. By the same procedure we obtain
$$(\lambda - \lambda^*)\mathbf{x}^*\mathbf{x} = \mathbf{x}^* a \mathbf{x} - \mathbf{x} \tilde{a} \mathbf{x}^* \tag{19.15}$$
because of $a^* = \tilde{a}$, and the right-hand side is zero.

Orthogonality of Eigenvectors

If $\mathbf{x}_i$ and $\mathbf{x}_k$ are two different eigenvectors with eigenvalues λ_i and λ_k, we may use the same procedure as in (19.11), (19.13), and (19.14) and obtain
$$(\lambda_i - \lambda_k)\mathbf{x}_i \mathbf{x}_k = 0, \tag{19.16}$$
which proves that
$$\mathbf{x}_i \mathbf{x}_k = 0 \quad \text{if} \quad \lambda_i \neq \lambda_k. \tag{19.17}$$

If $\lambda_i = \lambda_k$, $\mathbf{x}_i$ and $\mathbf{x}_k$ have the same eigenvalue and also, owing to the linearity of the equations, any combination of $\mathbf{x}_i$ and $\mathbf{x}_k$ as, for instance, $\mathbf{x}_i + \mathbf{x}_k$ and $\mathbf{x}_i - \mathbf{x}_k$. These or other mutually orthogonal solutions may be taken for new standard solutions after being renormalized. The same applies to cases where three or more different and linearly independent solutions belong to the same eigenvalue. Therefore, quite generally, we may assume a full set of eigenvectors to exist, which may be characterized as an *orthonormal* set of vectors obeying the conditions
$$\mathbf{x}_i \mathbf{x}_k = \delta_{ik}. \tag{19.18}$$

Transformation on the Principal Axis

Denote the above n eigenvectors by $\mathbf{S}_i$,
$$a\mathbf{S}_k = \lambda_k \mathbf{S}_k, \quad \mathbf{S}_i a \mathbf{S}_k = \lambda_k \delta_{ik}. \tag{19.19}$$

Applying the orthogonal transformation

$$\mathbf{x} = \sum_i y_i \mathbf{S}_i = \tilde{S}\mathbf{y}, \qquad (19.20)$$

we obtain

$$A(x, x) = \mathbf{x}a\mathbf{x} = \sum \lambda_i y_i^2, \qquad E(x, x) = \mathbf{x}\mathbf{x} = \sum_i y_i^2, \qquad (19.21)$$

which proves that the transformation S transforms the quadratic form A on its so-called *principal axis*. The coefficients in this purely quadratic form are the eigenvalues of the form A.

By means of the transformation matrix

$$S = (S_{ik}) = \begin{pmatrix} S_{11} & S_{12} & \cdots & S_{1n} \\ S_{21} & S_{22} & \cdots & S_{2n} \\ \vdots & & & \\ S_{n1} & S_{n2} & \cdots & S_{nn} \end{pmatrix} \qquad (19.22)$$

we may also write for (19.20)

$$\mathbf{x} = \mathbf{y}S = \tilde{S}\mathbf{y}. \qquad (19.23)$$

The quadratic form A then becomes

$$A(x, x) = (\mathbf{y}S)a(\tilde{S}\mathbf{y}) = \mathbf{y}(Sa\tilde{S})\mathbf{y} = \sum_i \lambda_i y_i^2, \qquad (19.24)$$

and this means that $Sa\tilde{S}$ has becomes a diagonal matrix

$$Sa\tilde{S} = \begin{pmatrix} \lambda_1 & 0 & \cdots & 0 \\ 0 & \lambda_2 & \cdots & 0 \\ \vdots & & & \\ 0 & & \cdots & \lambda_n \end{pmatrix} \qquad (19.25)$$

Hence matrices also may be transformed on their main axis and be said to possess definite eigenvalues.

Simultaneous Transformations

It is possible to transform simultaneously two quadratic forms A and B on diagonal form, provided one of them is essentially positive or *positive definite*, however not in general by an orthogonal transformation.

To see this, imagine the positive definite form, say, B transformed on its principal axis by an orthogonal transformation, changing

$$B(x, x) = \sum_{i,k=1}^{n} \sum_{i,k=1}^{n} b_{ik} x_i x_k \qquad (19.26)$$

into

$$B(y, y) = \sum_i b_i y_i^2, \qquad (19.27)$$

b_i being the positive eigenvalues of the form. Next, by the nonorthogonal transformation $z_i = \sqrt{b_i}\, y_i$ we transform it into the unit form

$$B(z, z) = \sum_i z_i^2. \qquad (19.28)$$

By these successive transformations

$$A(x, x) = \sum_{i,k} a_{ik} x_i x_k \qquad (19.29)$$

is transformed into

$$A(z, z) = \sum_{i,k} a_{ik} z_i z_k, \qquad (19.30)$$

which by a second orthogonal transformation, leaving $B(z, z)$ formally unchanged, transforms $A(z, z)$ on its new main axis.

Obviously, this composite procedure solves the problem of finding extremum values of $A(x, x)$ by the restriction of the variables that $B(x, x) = 1$. All the above transformations, therefore, can be made in a single step similar to that outlined above, leading to the system of equations

$$\frac{\partial A}{\partial x_i} - \lambda \frac{\partial B}{\partial x_i} = 0, \qquad i = 1, \cdots, n. \qquad (19.31)$$

The extremum values or eigenvalues are found from the determinant equation

$$|a_{ik} - \lambda b_{ik}| = \begin{vmatrix} a_{11} - \lambda b_{11} & a_{12} - \lambda b_{12} & \cdots & a_{1n} - \lambda b_{1n} \\ a_{21} - \lambda b_{21} & a_{22} - \lambda b_{22} & \cdots & a_{2n} - \lambda b_{2n} \\ \vdots & & & \\ a_{n1} - \lambda b_{n1} & a_{n2} - \lambda b_{n2} & \cdots & a_{nn} - \lambda b_{nn} \end{vmatrix} = 0. \qquad (19.32)$$

This equation also has n roots all of which are positive.

Invariants of a Quadratic Form

The roots λ_1 to λ_n of (19.10) are invariant with respect to orthogonal transformations of A or its matrix a. In the determinant equation written in the form

$$(\lambda_1 - \lambda) \cdots (\lambda_n - \lambda) = c_0 - c_1\lambda + \cdots + (-1)^{n-1}c_{n-1}\lambda^{n-1} + (-1)^n\lambda^n = 0 \quad (19.33)$$

the n coefficients c_0 to c_{n-1} are also invariant. The first of these invariants is the determinant of the matrix a,

$$c_0 = |a| = D. \quad (19.34)$$

The next in importance is c_{n-1} or the *spur*, which is the sum of the diagonal elements,

$$c_{n-1} = \text{sp}(a) = \sum_i a_{ii}. \quad (19.35)$$

Solution of Inhomogeneous Eigenvalue Equations

Consider the system of inhomogeneous equations

$$a\mathbf{x} - \lambda\mathbf{x} = \mathbf{y}, \quad (19.36)$$

the eigenvalues and eigenvectors λ_i and $\mathbf{x}_i$ of the corresponding homogeneous equations being considered known. Equations of this type frequently appear in the treatment of forced oscillations of a vibrating system, whether mechanical or electrical, λ_i usually denoting a square frequency of the system.

To solve (19.36) we expand the solution in terms of the eigenvectors

$$\mathbf{x} = \sum_i c_i \mathbf{x}_i. \quad (19.37)$$

Inserting in (19.36), we obtain

$$\sum_i c_i(\lambda_i - \lambda)\mathbf{x}_i = \mathbf{y}. \quad (19.38)$$

Multiplying by one of the eigenvectors and observing that $\mathbf{x}_i\mathbf{x}_k = \delta_{ik}$, we find the result

$$c_i = \frac{\mathbf{x}_i \mathbf{y}}{\lambda_i - \lambda}, \quad \mathbf{x} = \sum_i \frac{\mathbf{x}_i \mathbf{y}}{\lambda_i - \lambda} \mathbf{x}_i. \quad (19.39)$$

The result is valid for any $\lambda \neq \lambda_i$, $i = 1, \cdots, n$. In the case of λ converging toward one of the eigenvalues, the solution $\mathbf{x}$ becomes infinitely large and, hence, cannot exist. There is only one exception: If at the same time the product $\mathbf{x}_i\mathbf{y}$ vanishes, that is, if the vector $\mathbf{y}$, representing, say, a force causing the forced vibration, is orthogonal to the particular eigenvector $\mathbf{x}_i$, then a solution exists.

This may be interpreted physically so that an outer force acting on a vibration system is not allowed to have the same frequency or be in *resonance* with one of the eigenfrequencies without breaking up the system, unless it is orthogonal to the corresponding eigenvibration, in the sense that it does not at all disturb this particular eigenvibration.

20. VECTORS IN A SPACE OF INFINITE DIMENSIONS. ORTHOGONAL SYSTEMS OF POLYNOMIALS AND FUNCTIONS

Vibration of a Continuum as a Limiting Case of Vibrating Mass Points

In the preceding considerations of n-dimensional vectors, n may be any number however large. Therefore nothing prevents us from generalizing the results already obtained also for the case of $n \to \infty$. It is necessary, however, to keep an eye on the mathematical consequences and the physical meaning of such a process.

Consider, for instance, the physical process of a continuous vibrating string. We may approach the problem by thinking of it as consisting of n equal parts with a mass $m = \rho d$, ρ being the mass density and d the length of each part of the string. Again, we may in the first approximation treat the motion of the different parts of the string as a motion of mass points which are tied together by the elastic forces of the string. These forces may be represented, say, by weightless springs connecting the mass points and of a certain tension p. With respect to longitudinal motion these forces at either side of the mass points may be taken to produce equilibrium for any part of the string. With respect to side motion, however, the result is different.

Referring to Fig. 3-2, we find the transversal forces to be $(u_{k-1} - u_k)p/d$ and $(u_{k+1} - u_k)p/d$, u_k being the side displacement of the kth mass point. Hence the equation of motion for this part of the string reads

$$\rho \frac{d^2 u_k}{dt^2} = \frac{p(u_{k+1} - 2u_k + u_{k-1})}{d^2}. \tag{20.1}$$

Before attempting to determine the motion of this n-particle system, we observe that in the limiting case of $n \to \infty$ or $d \to 0$ the above equation turns

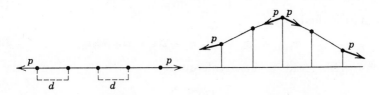

Fig. 3-2. Vibrating string as a limiting case of vibrating mass points.

into the partial differential equation for the continuous string

$$\rho \frac{\partial^2 u}{\partial t^2} = p \frac{\partial^2 u}{\partial x^2}, \tag{20.2}$$

when denoting by $u(x, t)$ the side displacement of the string as a function of time and the length coordinate of the string.

Of particular importance now are the synchronous or eigenvibrations of the system. These are obtained by putting

$$u_k = v_k \sin(\omega t + \eta) \quad \text{respectively,} \quad u = v \sin(\omega t + \eta), \tag{20.3}$$

ω and N being the angular frequency and the phase of a particular eigenvibration. Substituting in (20.1) and (20.2), we obtain

$$\rho \omega^2 \, d^2 v_k + p(v_{k+1} - 2v_k + v_{k-1}) = 0 \tag{20.4}$$

and

$$\rho \omega^2 v + p \frac{d^2 v}{dx^2} = 0. \tag{20.5}$$

As a matter of fact, the latter differential equation is more easily handled than the difference equation (20.4) of the n-body problem. For instance, in the case of a string of length a with fixed ends, the solutions of (20.5) may be written

$$v_m(x) = C_m \sin \frac{m \pi x}{a}, \tag{20.6}$$

where m is any positive integer $m = 1, 2, \cdots, \infty$. The corresponding eigenfrequency is

$$\omega_m = m \left(\frac{p}{\rho}\right)^{1/2} \frac{\pi}{a}. \tag{20.7}$$

We shall now see that the n-particle system (20.1) and (20.4) is vibrating in much the same way. Denoting by x_k the length coordinate of the kth particle, we have by definition $x_{k+1} = x_k + d$ and $x_{k-1} = x_k - d$.

Assuming the side displacement of the particles also, this time given by a sine function

$$v_k = C \sin \beta x_k, \tag{20.8}$$

we can easily find that

$$v_{k+1} - 2v_k + v_{k-1} = -2(1 - \cos \beta d) v_k, \tag{20.9}$$

which by (20.5) leads to the eigenfrequency

$$\omega = \left(\frac{p}{\rho}\right)^{1/2} \cdot \frac{2}{d} \sin \beta \frac{d}{2}. \tag{20.10}$$

Linear Transformation. Extremum and Variational Problems | 81

As to the boundary conditions, which are here also those of a string with fixed ends, we are obviously free to choose the coordinate of the first point. We might choose x_1 at any point between 0 and d, but to be more definite we shall choose

$$x_1 = \tfrac{1}{2}d, \quad x_n = (n - \tfrac{1}{2})d, \tag{20.11}$$

thus obtaining a more symmetric picture of the motion. Putting

$$d = \frac{a}{n} \tag{20.12}$$

and observing that the boundary conditions necessitate as earlier that

$$\beta = \frac{m\pi}{a}, \tag{20.13}$$

we find the frequency of the mth vibration to be

$$\omega_m = \left(\frac{p}{\rho}\right)^{1/2} \frac{2n}{a} \sin\frac{m\pi}{2n}. \tag{20.14}$$

For very large n, these frequencies are not very different from those of the continuous string. For a fixed number of particles n, however large, the maximum number of eigenvibrations is, however, only equal to n. Putting $m = n + 1, n + 2, \cdots$, we get only a repetition of the frequencies ω_{m-1}, $\omega_{m-2}, \cdots$. Continuing up to $m = 2n + 1, 2n + 2, \cdots$, it is true that we get formally different frequencies, namely, those of the opposite sign. This, however, only changes the sign of the time factor and adds multiples of 2π to the argument of the sine function (20.10); consequently, the eigenvibrations are the same as before.

The four possible eigenvibrations of four mass points representing a substitute for the continuous string are illustrated in Fig. 3-3.

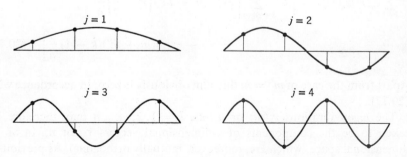

Fig. 3-3. Eigenvibrations of four mass points as compared with the four first eigenvibrations of a continuous homogeneous string.

It is well known that the eigenvibrations of the continuous homogeneous string are mutually orthogonal, in the sense that

$$\int_0^a v_m(x) v_{m'}(x)\, dx = 0, \qquad m' \neq m. \tag{20.15}$$

By the suitable choice of C_m, the functions v_m may be made what are called orthonormal functions with the additional property that

$$\int_0^a v_m^2(x)\, dx = 1. \tag{20.16}$$

It is interesting to note that the same is true for the above eigenvibrations of the n mass points when replacing integrals by sums, that is,

$$\sum_{k=1}^n v_k(m) v_k(m') = 0, \qquad m' \neq m. \tag{20.17}$$

Writing according to (20.8), (20.11), (20.12), and (20.13)

$$v_k(m) = C \sin m\left(k + \frac{1}{2}\right)\frac{\pi}{n}, \tag{20.18}$$

we find

$$v_k(m) v_k(m') = \frac{1}{2} C^2 \left[\cos(m - m')\left(k - \frac{1}{2}\right)\frac{\pi}{n} - \cos(m + m')\left(k - \frac{1}{2}\right)\frac{\pi}{n} \right] \tag{20.19}$$

and hence by summation

$$\sum_{k=1}^n v_k(m) v_k(m') = \frac{1}{4} C^2 \sum_{k=-n+1}^n \left[\exp i(m - m')\left(k - \frac{1}{2}\right)\frac{\pi}{n} - \exp i(m + m')(k - 1)\frac{\pi}{n} \right]. \tag{20.20}$$

Apart from the case of $m' = m$ this sum obviously is zero, in accordance with (20.17).

The eigenvibrations $v_k(m)$ or $u_k(m)$, $k = 1, 2, \cdots, n$ may therefore be taken to be the components of n-dimensional vectors $\mathbf{v}_m$ or $\mathbf{u}_m$ in an n-dimensional space, which are, moreover, mutually orthogonal. As previously demonstrated they are easily generalized into vectors in a space of infinite dimensions, which is often called the Hilbert space or the *function space*.

Orthogonal Polynomials

In the power series of a function

$$y(x) = c_0 + c_1 x + c_2 x^2 + \cdots \qquad (20.21)$$

in a given region, for the sake of definiteness, say, in the region $-1 \leq x \leq 1$ for the real variable x, we may think of the powers x^m as vectors in a Hilbert space and of $y(x)$ as another vector in the same space which is defined by the above coefficients c_n. Equation (20.21) expresses the fact that an arbitrary function or vector $y(x)$ can be represented by a linear combination of *coordinate vectors* or *coordinate functions*, in the above case simply the positive integral powers of the independent variable x.

These powers are not, however, the most suitable coordinate functions in general because they are lacking the important property of mutual orthogonality. This inconvenience can easily be removed, however, by replacing the base powers by linear combinations called *orthogonal polynomials*. In the above case of the fundamental region -1 to 1 for the variable x, it is easily seen that the following set of polynomials up to $n = 5$,

$$P_0(x) = 1, \qquad\qquad P_1(x) = x,$$
$$P_2(x) = \tfrac{3}{2}x^2 - \tfrac{1}{2}, \qquad P_3(x) = \tfrac{5}{2}x^3 - \tfrac{3}{2}x, \qquad (20.22)$$
$$P_4(x) = \tfrac{35}{8}x^4 - \tfrac{15}{4}x^2 + \tfrac{3}{8}, \qquad P_5(x) = \tfrac{63}{8}x^5 - \tfrac{35}{4}x^3 + \tfrac{15}{8}x,$$

possess the property of mutual orthogonality

$$\int_{-1}^{+1} P_n(x) P_m(x)\, dx = 0, \qquad \text{for} \quad n \neq m. \qquad (20.23)$$

(See Fig. 3-4.)

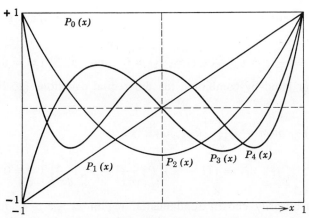

Fig. 3-4. The first four Legendre polynomials.

84 / The Mathematical Foundation of Physics

These functions are known as *Legendre polynomials* and play an important role in mathematical physics, particularly in potential theory. They are not normalized in the same way as the sine function of the vibrating strings, their normalization integral being, as we shall see later,

$$\int_{-1}^{+1} [P_n(x)]\, dx = \frac{2}{2n+1}. \tag{20.24}$$

Instead, they are normalized by the simple condition that $P_n(1) = 1$, where, because of the symmetric properties of the functions, $P_n(-1) = (-1)^n$.

A great many properties of the Legendre functions can be derived easily from the simple definition

$$P_n(x) = \frac{1}{2^n n!} \frac{d^n}{dx^n} (x^2 - 1)^n. \tag{20.25}$$

Putting, for instance,

$$u_n(x) = \frac{1}{2^n n!} (x^2 - 1)^n, \tag{20.26}$$

we very easily find the differential equation

$$(1 - x^2) u_n' + 2nx u_n = 0, \tag{20.27}$$

and, by $n + 1$ successive derivations, the differential equation for the polynomials

$$\left[(1 - x^2) \frac{d^2}{dx^2} - 2x \frac{d}{dx} + n(n+1) \right] P_n(x) = 0. \tag{20.28}$$

The property of mutual orthogonality is derived easily from both (20.25) and (20.28). By partial integration for $n > m$,

$$\int_{-1}^{+1} P_n(x) P_m(x)\, dx = \int_{-1}^{+1} u_n^{(n)} u_m^{(m)}\, dx = \int_{-1}^{+1} u_n u_m^{(m+n)}\, dx = 0, \tag{20.29}$$

since

$$u_n^{(n-1)}(\pm 1) = u_n^{(n-2)}(\pm 1) = \cdots = 0 \quad \text{and} \quad u_m^{(m+n)}(x) = 0.$$

On the other hand by combining the differential equations (20.28) for P_n and P_m,

$$[n(n+1) - m(m+1)] \int_{-1}^{+1} P_n P_m\, dx$$

$$= \left[(1 - x^2) \left(P_n \frac{dP_m}{dx} - P_m \frac{dP_n}{dx} \right) \right]_{-1}^{+1} = 0. \tag{20.30}$$

In any arbitrarily chosen fundamental region $x_0 \leq x \leq x_1$ for a real variable x, we are able to construct such orthogonal polynomials. We may further generalize the polynomials by means of a positive density function $\rho(x)$ according to the condition

$$\int_{x_0}^{x_1} \rho(x) y_n(x) y_m(x)\, dx = 0, \qquad n \neq m, \tag{20.31}$$

the meaning of which is that $z_n(x) = \sqrt{\rho(x)} y_n(x)$ are the true orthogonal functions. Very often, also, such generalized orthogonal polynomials obey a differential equation. However, for a finite region the density function then must be of the form

$$\rho(x) = (x - x_0)^\alpha (x_1 - x)^\beta, \tag{20.32}$$

and the differential equation is

$$\frac{d}{dx}[(x - x_0)^{\alpha+1}(x_1 - x)^{\beta+1}] + \lambda_n (x - x_0)^\alpha (x_1 - x)^\beta y_n = 0. \tag{20.33}$$

with special *eigenvalues* λ_n for any of the *eigenfunctions* y_n. This leads to a type of polynomials of the so-called hypergeometrical type, which will be dealt with later on.

Equation (20.33) is a special type of the Sturm-Liouville eigenvalue problem

$$\left[\frac{d}{dx}\left(p(x)\frac{d}{dx}\right) + \lambda \rho(x)\right] y = 0, \tag{20.34}$$

with positive density and tension functions $\rho(x)$ and $p(x)$. With the density function $\rho = (1 - x^2)^m$, for instance, the resulting polynomials may be written

$$y_n(x) = P_n^{(m)}(x) = \frac{d^m}{dx^m} P_n(x) \tag{20.35}$$

and are known as *generalized Legendre polynomials*. The corresponding true orthogonal functions in the region $-1 \leq x \leq 1$,

$$P_n^m(x) = (1 - x^2)^{m/2} P_n^{(m)}(x), \tag{20.36}$$

are called *generalized Legendre functions* or *spherical harmonics*.

For an infinite region, say, $0 \leq x \leq \infty$, we may also define sets of orthogonal polynomials, however always in the sense of orthogonality with a density function $\rho(x)$ which makes the product integrals convergent. If the density function is taken to be $\rho(x) = e^{-x}$, we easily find some of the first corresponding orthogonal polynomials to be

$$\begin{aligned} L_0(x) &= 1, & L_1(x) &= 1 - x, \\ L_2(x) &= 2 - 4x + x^2, & L_3(x) &= 6 - 18x + 9x^2 - x^3. \end{aligned} \tag{20.37}$$

These polynomials, whose general expression is

$$L_n(x) = e^x \frac{d^n}{dx^n}(x^n e^{-x}) = \sum_{k=0}^{n} \binom{n}{k} \frac{n!}{k!}(-x)^k \qquad (20.38)$$

are known as *Laguerre polynomials*. They may be generalized in very much the same way as the Legendre polynomials, for instance, by introducing the density function $\rho(x) = x^m e^{-x}$. The corresponding orthogonal polynomials are denoted by $L_n{}^m(x)$ and may be taken to be the mth derivatives of the ordinary Laguerre polynomials.

The differential equation for the generalized Laguerre polynomials reads

$$\left[\frac{d}{dx}(x^{m+1} e^{-x})\frac{d}{dx} + (n-m)\right] L_n{}^m(x) = 0. \qquad (20.39)$$

It might be immediately obtained from (20.33) by the quite plausible process of replacing $(x_1 - x)^\beta$, or rather $(1 - x/x_1)^\beta$, by e^{-x} as $x_1 \to \infty$ (with $\beta = x_1$) and by observing that the exponent of the highest power of x in $L_n{}^m(x)$ is $n - m$.

Returning for a moment to the Legendre polynomials, we want to state that they are not so foreign to the eigenfunctions of the vibrating homogeneous string as might be felt at first sight. Putting for instance

$$x = \cos \vartheta, \qquad 0 \leq \vartheta \leq \pi, \qquad (20.40)$$

the differential equation (20.28) turns into

$$\left[\frac{d^2}{d\vartheta^2} + \frac{\cos \vartheta}{\sin \vartheta}\frac{d}{d\vartheta} + n(n+1)\right] P_n(\cos \vartheta) = 0, \qquad (20.41)$$

whose first successive solutions corresponding to (20.22) are

$P_0(\cos \vartheta) = 1,$

$P_1(\cos \vartheta) = \cos \vartheta,$

$P_2(\cos \vartheta) = \tfrac{1}{4}(3 \cos 2\vartheta + 1),$

$P_3(\cos \vartheta) = \tfrac{1}{8}(5 \cos 3\vartheta + 3 \cos \vartheta),$ (20.42)

$P_4(\cos \vartheta) = \tfrac{1}{64}(35 \cos 4\vartheta + 20 \cos 2\vartheta + 9),$

$P_5(\cos \vartheta) = \tfrac{1}{128}(63 \cos 5\vartheta + 35 \cos 3\vartheta + 30 \cos \vartheta),$

or, in general,

$$P_n(\cos \vartheta) = \frac{1}{2^{2n}} \sum_{k=0}^{n} \binom{2k}{k}\binom{2n-k}{n-k} e^{(2k-n)i\vartheta}. \qquad (20.43)$$

See Fig. 3-5.

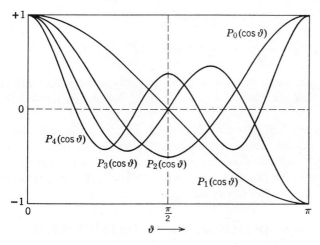

Fig. 3-5. The first four cosine-like functions $P_n(\cos\vartheta)$.

On the other hand, the functions
$$y_n(\vartheta) = \cos n\vartheta, \qquad n = 0, 1, \cdots, \tag{20.44}$$
which are solutions of the equations
$$\left(\frac{d^2}{d\vartheta^2} + n^2\right) y_n(\vartheta) = 0, \tag{20.45}$$
with boundary conditions $y_n'(0) = y_n'(\pi) = 0$, and which form an orthogonal system of functions in the region $0 \leq \vartheta \leq \pi$, may be expressed as polynomials of the independent variable $x = \cos\vartheta$, $-1 \leq x \leq 1$. These polynomials
$$y_n(\cos\vartheta) = T_n(x) = \cos(n \text{ arc cos } x) \tag{20.46}$$
are called *Tschebyscheff polynomials*. The first few of them read

$$T_0(x) = 1, \qquad\qquad T_1(x) = x$$
$$T_2(x) = 2x^2 - 1, \qquad T_3(x) = 4x^3 - 3x \tag{20.47}$$
$$T_4(x) = 8x^4 - 8x^2 + 1, \quad \text{etc.}$$

Here the normalization is the same as that used for Legendre polynomials, namely, $T_n(\pm 1) = (\pm 1)^n$. It should be noted, however, that in the conventional definition of the polynomials the coefficient of the highest power x^n is taken to be 1.

The above polynomials (20.47) are mutually orthogonal with the density function $\rho(x) = (1-x^2)^{-1/2}$. It is obvious, therefore, that they can be

88 / The Mathematical Foundation of Physics

represented by a suitably normalized expression

$$T_n(x) = \frac{(n-1)!}{(2n-1)!} 2^{n-1}(x^2-1)^{1/2} \frac{d^n}{dx^n}(x^2-1)^{n-1/2}, \qquad (20.48)$$

since these can easily be proved to be mutually orthogonal with the above $\rho(x)$. Dropping the factor $2n-1$, we have the conventional definition of $T_n(x)$.

The Tschebyscheff polynomials may be generalized by differentiations in the very same manner as Legendre polynomials. In this way we obtain systems of polynomials that are mutually orthogonal in the region $-1 \leq x \leq 1$ with the density function $\rho(x) = (1-x^2)^{m-1/2}$. These polynomials are, however, usually called *Gegenbauer polynomials*.

21. EXPANSION OF FUNCTIONS BY ORTHONORMAL SYSTEMS

The Components of a Function on the Axis of an Orthonormal System by Least Square Deviation

Consider a region $x_0 \leq x \leq x_1$ of the variable x and two, apart from certain restrictions, arbitrary functions $y(x)$ and $z(x)$. Dropping for convenience the given limits x_0 and x_1 of the region, we write for the definite product integral

$$(y, z) = \int yz\, dx. \qquad (21.1)$$

Considering y and z vectors in a Hilbert or function space, we may take (y, z) to be their scalar product.

Assuming a set of orthonormal functions $y_1, y_2, \cdots, y_n, \cdots$, that is, with the properties

$$(y_n, y_n) = 1, \qquad (y_n, y_m) = 0 \quad \text{for } n \neq m, \qquad (21.2)$$

we may ask how to approximate in the very best way any given function y by means of the finite expression

$$y \approx c_1 y_1 + c_2 y_2 + \cdots + c_n y_n. \qquad (21.3)$$

A definite answer can be given by requiring the difference square integral to have a minimum value, that is,

$$\int \left(y - \sum_{k=1}^{n} c_k y_k\right)^2 dx = \min. \qquad (21.4)$$

Evaluating the integral, we obtain

$$(y, y) - 2\sum_{k=1}^{n} c_k(y, y_k) + \sum_{k=1}^{n} c_k^2 = \min, \qquad (21.5)$$

Linear Transformation. Extremum and Variational Problems | 89

which by differentiation of the coefficients yields the condition

$$c_k = (y, y_k) \tag{21.6}$$

irrespective of the member of terms n. The corresponding minimum value of the difference square integral is

$$\min = (y, y) - \sum_{k=1}^{n} (y, y_k)^2. \tag{21.7}$$

Completeness of a Functional System

If for any, within certain restrictions, arbitrary function $y(x)$ this minimum tends to zero as $n \to \infty$, that is,

$$(y, y) = \sum_{k=1}^{\infty} (y, y_k)^2, \tag{21.8}$$

the system $y_n(x)$ is said to form a complete functional system. The above equation is therefore called the *completeness relation*. It fails, of course, if only a single one of the functions is lacking. This relation may also be expressed in some other way, say, by stating that for any function $y(x)$ the equation

$$y(x) = \sum_{k=1}^{\infty} (y, y_n) y_n(x) \tag{21.9}$$

is true. Let us here introduce for a moment the improper function $\delta(x)$ now often used in mathematical physics and which will be dealt with later on. Its main property is that the integral

$$\int f(x) \delta(x - x') \, dx \tag{21.10}$$

is equal to $f(x')$ if taken across the point x' however small the range, whereas taken only outside of x' it is zero.

Now if we formally change the integration and summation in (21.8), which is of course strictly permitted only in the case of a finite series, we obtain

$$y(x) = \int y(x') \sum_{n=1}^{\infty} y_n(x') y_n(x) \, dx'. \tag{21.11}$$

Hence the completeness relation may be expressed by the improper equation

$$\sum_{n=1}^{\infty} y_n(x') y_n(x) = \delta(x - x'), \tag{21.12}$$

meaning of course only that a finite series of the left-hand form with increasing upper limit gradually takes the form of what is conventionally called a δ-function.

Equation 21.12 should formally be compared with the orthonormal relation

$$\int y_n(x) y_{n'}(x) \, dx = \delta_{nn'}. \tag{21.13}$$

In the case of orthonormality of the functions y_n with a density function $\rho(x)$, we need only a slight modification of the above equations. Writing in this case

$$(y, z) = \int \rho(x) yz \, dx \tag{21.14}$$

and correspondingly

$$(y, y_k) = \int \rho(x) y y_k \, dx \tag{21.15}$$

for scalar products, we easily see that the minimum requirement

$$\int \rho(x) \left(y - \sum_{k=0}^{n} c_k y_k \right)^2 dx = \min \tag{21.16}$$

leads to the same equations (21.6) to (21.9).

We have already seen that systems of orthonormal polynomials or functions are frequently solutions of a differential equation with an eigenvalue parameter, which is usually called a Sturm-Liouville eigenvalue equation. We shall write this equation in the form

$$(py')' - qy + \lambda \rho y = 0 \tag{21.17}$$

and prove that by suitably chosen boundary conditions or by natural boundary conditions caused by so-called singularities of the equation itself at the boundaries, the solutions are mutually orthogonal.

Consider two different eigenvalues λ_n and λ_m with corresponding eigenfunctions y_n and y_m. We multiply the first equation by y_m, the second by y_n, subtract and integrate between the boundaries x_0 and x_1. The result is

$$(\lambda_n - \lambda_m) \int \rho y_n y_m \, dx = [p(y_n y'_m - y_m y'_n)]_{x_0}^{x_1}. \tag{21.18}$$

The right-hand side of this equation is zero if at the boundaries x_0 and x_1 for positive $p(x)$ any of the functions y_n obey the conditions $y = 0$, $y' = 0$ or $y' + ky = 0$, which are called homogeneous boundary conditions. In the case of $p(x_0) = 0$ or $p(x_1) = 0$, which means that the end points are singular points of the equation, it suffices that y_n is finite at these points. This will be the case only for the special values of the parameter λ, which are here called eigenvalues. For any other values of λ solutions of (21.3) will be infinite at least one of the end points x_0 and x_1.

Linear Transformation. Extremum and Variational Problems | 91

The latter example of natural boundary conditions applies to the above orthogonal systems of polynomials.

It should be mentioned, finally, that for an infinite region to one or to both sides of the coordinate axis of the independent variable the Sturm-Liouville eigenvalue equation may have such a form that the eigenvalues form a continuum. The simplest example is that known from the theory of *Fourier integrals*.

Consider, for instance, the region $0 \leq x \leq \infty$ and the boundary conditions, say, $y(0) = 0$ and $y(\infty)$ being finite. Then, for instance, the equation

$$y'' + \lambda y = 0 \qquad (21.19)$$

has the admitted solution

$$y = \sin \sqrt{\lambda}\, x \qquad (21.20)$$

for any positive value of λ.

In such cases the orthogonal properties of the eigenfunctions are of a more intricate nature. Usually, however, they can be satisfactorily expressed by some improper equation, say, of the form

$$\int y(\lambda, x) y(\lambda', x)\, dx = \delta(\lambda - \lambda'), \qquad (21.21)$$

where both λ and x are continuous variables. In this case the completeness relation is usually written

$$\int y(\lambda, x) y(\lambda, x')\, d\lambda = \delta(x - x'), \qquad (21.22)$$

and this demonstrates even more clearly than before the formal similarity of orthonormality and completeness relations of eigenfunctions.

22. ELEMENTS OF THE CALCULUS OF VARIATION

Statement of Problems. Functions of Functions

The calculus of variation may be thought of as a generalization of the elementary theory of maxima and minima of a function $\varphi(x, y, \cdots)$ of one or more independent variables. The necessary condition for such a function to possess a maximum or a minimum is that any of its first partial derivatives must vanish, that is,

$$\frac{\partial \varphi}{\partial x} = 0, \qquad \frac{\partial \varphi}{\partial y} = 0, \cdots. \qquad (22.1)$$

The fulfillment of these conditions does not, however, suffice to produce either a maximum or a minimum. Even in the case of a single variable, we have to add the conditions $\varphi''(x) > 0$ for a minimum and $\varphi''(x) < 0$ for maximum values of the function. If $\varphi''(x) = 0$, we have only a stationary value of the function and the point x for which $\varphi'(x) = 0$ is called an inflection point. Moreover, whereas the equation $\varphi'(x) = 0$ may have several solutions, $x = x_1, x_2, \cdots$, only one of these points can correspond to an absolute maximum or minimum. The others are points of relative maxima or minima, or simply inflection points. (See Fig. 3-6.)

In the calculus of variation we are also dealing with the determination of maxima or minima, or simply stationary values of certain quantities. The fundamental difference between the calculus of variation and the simpler problem of maxima and minima of functions of one or more variables may

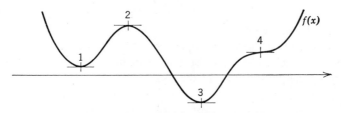

Fig. 3-6. 1 and 2, relative minimum and maximum; 3, absolute minimum; and 4, inflection point.

be expressed as follows. In the calculus of variation we are dealing with quantities that depend on functions of such variables in a given fundamental region of the variables. To express it briefly: We are dealing with *functions of functions* and the aim of the calculus is to determine the functions on which our quantities depend so that the required extremum properties of the said quantities can be realized.

As a very simple example of such a function of functions we mention here the length of a curve $y(x)$ between two given values x_0 and x_1 of the variable x. It is easily seen that this length, which depends on y-values in the whole region, is

$$L(y) = \int_{x_0}^{x_1} \sqrt{1 + y'^2}\, dx. \tag{22.2}$$

We see that the length is given by an integral; it is natural to expect the majority of functions of functions to be expressible by integrals of some kind.

Fermat's Principle

By a similar integral we shall formulate one of the oldest and most famous variation problems which is known as *Fermat's principle*, which states that a ray of light between two points in space travels along what is called the shortest light-path, or in shortest time. The meaning of this is as follows. Let

Linear Transformation. Extremum and Variational Problems | 93

the refraction index be a function $n(x, y, z)$ of the space coordinates, or, for the sake of simplicity, only of two of them $n = n(x, y)$, the direction of propagation always being perpendicular to the z-axis. The velocity of light is then c/n and it travels a unit distance in a time n/c. Hence, the time for traveling any infinitesimal distance ds is proportional to nds. Since the wavelength is $\lambda = \lambda_0/n$, λ_0 being that in empty space, the same is true for the so-called light-path ds/λ, that is, the distance as measured in actual wavelength. Hence, requiring that either light time or light path be a minimum, we are led to consider the variation problem

$$T(y) = \int_{x_0}^{x_1} n(x, y)\sqrt{1 + y'^2}\, dx = \min, \qquad (22.3)$$

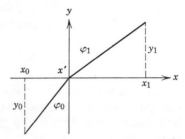

Fig. 3-7. Illustration of the law of refraction.

$y(x)$ being the unknown function to be determined, however with given boundary values $y_0 = y(x_0)$, $y_1 = y(x_1)$.

To solve the problem $n(x, y)$ must be specified. One of the simplest examples is that of only two different homogeneous media on either side of a plane boundary, the refraction indices being n_0 and n_1.

For the sake of simplicity, let the boundary be $y = 0$. Introducing the unknown abscissa x' of the passing ray at the boundary, we have the minimum condition

$$n_1[(x_1 - x_2)^2 + y_1^2]^{1/2} + n_0[(x' - x_0)^2 + y_0^2]^{1/2} = \min, \qquad (22.4)$$

since the path in either medium is rectilinear. By differentiation of the equation with respect to x', and by using the angles φ_1 and φ_0 of Fig. 3-7, we immediately obtain the *Snellius* refraction law

$$n_1 \sin \varphi_1 = n_0 \sin \varphi_0. \qquad (22.5)$$

The Isoperimetric Problem

The isoperimetric problem consists of determining the form of a curve of given length in such a way that the curve encloses a largest possible area.

We shall divide the curve into two equal parts and place the two ends of a half curve on the x-axis, the one at the origin and the other at an unknown point. It is obvious, of course, that the curve is a half-circle. If the problem has a given solution in the upper xy-half-plane, the same symmetrically situated solution must be valid for the lower half-plane. But such a line of symmetry must exist for any direction. Hence the full curve can only be a circle.

Nevertheless, let us formulate the problem mathematically, denoting the length of the curve by πa. Let $y(x)$ be the curve ordinate with $y(0) = 0$, ξ being the x-coordinate of the right-hand end of the curve. Then we have the following variational problem,

$$F(y) = \int_0^\xi L(y)\, dx = \max, \qquad L(y) = \int_0^\xi \sqrt{1 + y'^2}\, dx = \pi a \quad (22.6)$$

with a variable upper limit of the integrals.

Now we introduce the quantity

$$\varphi = \frac{s}{a}, \qquad ds^2 = dx^2 + dy^2, \quad (22.7)$$

where s is the arc length of the curve from the origin and hence φ the corresponding angle as reckoned from the center of the circle which we expect to be the final curve. Then $0 \leq \varphi \leq \pi$ and, writing $y = az$, we are left with the single variational problem

$$F(z) = a^2 \int_0^\pi z\sqrt{1 - z'^2}\, d\varphi = \max \quad (22.8)$$

with the boundary condition $z(0) = z(\pi) = 0$.

Putting $z = \sin \varphi$, we find that the area turns out to be $F = \tfrac{1}{2}\pi a^2$, as it must be. To prove that this is the maximum curve we must know the condition for stationarity of the integral. If the integrand is denoted by $I(z, z')$, the condition is, as we shall discuss more fully later, the Euler differential equation

$$\frac{d}{d\varphi}\frac{\partial I}{\partial z'} - \frac{\partial I}{\partial z} = 0, \quad (22.9)$$

or, as applied to (22.8)

$$\frac{d}{d\varphi}\frac{zz'}{(1 - z'^2)^{1/2}} + \sqrt{1 - z'^2} = 0. \quad (22.10)$$

Performing differentiation and adjusting the equation, we get

$$zz'' + (1 - z'^2) = 0, \quad (22.11)$$

which is equally seen to have the solution $z = \sin \varphi$ with correct boundary values, even though the equation is still of a disagreeable nonlinear type.

The Area of a Surface of Rotation and the Brachystochrone

If a curve $y(x)$ with given end points x_0, y_0 and x_1, y_1 rotates around the x-axis, it will describe a rotational surface of area

$$F(y) = 2\pi \int_{x_0}^{x_1} y\sqrt{1 + y'^2}\, dx. \qquad (22.12)$$

If we require the curve to be of the form that yields a minimum of this area, we obviously have a particular case of Fermat's principle corresponding to $n(x, y) \sim y$. Since, apart from the sign under the square root, the integrand is of the same form as in the preceding section, it is tempting to guess that $y(x)$ is here some hyperbolic function. In fact, using (22.9) with y for z and x for φ, we obtain the Euler equation

$$yy'' - (1 + y^2) = 0. \qquad (22.13)$$

This equation has the following solution, suitable for our purpose,

$$y = a \cosh \frac{x - \xi}{a}, \qquad (22.14)$$

in which the two arbitrary constants a and ξ must be adjusted to the given boundary conditions.

The *brachystochrone* (from Greek *brachy*, short; *chronos*, time) is of historical interest. Jacob Bernoulli used this problem at the end of the 17th century, thus starting the development of the calculus of variation. The brachystochrone is a curve along which a falling body, when forced to follow the curve, passes in shortest possible time from one given point to another. Let $y = y(x)$ be the curve equation and x_0, $y_0 = y(x_0)$ and x_1, $y_1 = y(x_1)$ two fixed points that the body shall have to pass. We shall assume the body to be at rest in x_0, y_0 and not below the point $x_1 y_1$.

It is convenient in this case to let the y-axis point downward and to put $x_0 = y_0 = 0$. Then according to the energy principle the velocity of the body will at any point be $\sqrt{2gy}$, it being understood that there is no friction along the curve. Hence the time required to reach the point x_1, y_1 is

$$T = \int_{x_0}^{x_1} \left(\frac{1 + y'^2}{2gy}\right)^{1/2} dx. \qquad (22.15)$$

Again, we have a particular case of Fermat's principle, this time with $n \sim y^{-1/2}$. The Euler equation is now

$$yy'' + \tfrac{1}{2}(1 + y'^2) = 0 \qquad (22.16)$$

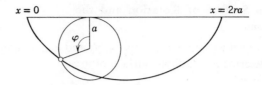

Fig. 3-8. The brachystochrone, a cycloid; $x = a(\varphi - \sin \varphi)$, $y = a(1 - \cos \varphi)$.

as compared to (22.13). Differentiating and integrating, we obtain

$$yy'' + 2y'y'' = 0, \qquad y'^2 y'' = a, \tag{22.17}$$

a being an integration constant. Hence we have the first-order differential equation

$$1 + y'^2 = \frac{2a}{y}, \tag{22.18}$$

or

$$dx = \frac{y\, dy}{[a^2 - (a - y)^2]^{1/2}}. \tag{22.19}$$

Putting

$$y = a(1 - \cos \varphi). \tag{22.20}$$

we obtain

$$dx = a(1 - \cos \varphi)\, d\varphi,$$
$$x = a(\varphi - \sin \varphi). \tag{22.21}$$

Here we have a curve, starting at the origin, which by suitable choice of a can be forced to go through any point x_1, y_1 with positive y_1. An illustration is given in Fig. 3-8; the curve is seen to be a cycloid, φ being the rotation angle of the circle with the fixed point describing the curve.

The Chain Line

The problem of a perfectly flexible chain with fixed ends is mathematically very similar to that of the minimum area of a rotational surface. The chain, of course, will take a shape corresponding to its lowest possible potential energy. This means that as in (22.12) we have to look for the lowest possible value of the integral

$$I = \int_{x_0}^{x_1} y\sqrt{1 + y'^2}\, dx, \tag{22.22}$$

$y(x)$ being the equation of the curve and x_0, y_0 and x_1, y_1 the given end points of the chain. Since the length element is $ds = \sqrt{1 + y'^2}\, dx$, I is obviously proportional to the potential energy of the chain. So far, the problem is mathematically the same as in (22.12). For the chain line, however, we have to add the additional condition

$$\int_{x_0}^{x_1} \sqrt{1 + y'^2}\, dx = s \tag{22.23}$$

because the length s of the chain is given.

The Euler equation (22.22) has already been given in (22.13). To the left-hand expression of this equation we now have to add, according to general rules that will be discussed later, the Euler derivative of (22.23) multiplied by a Lagrange multiplier η [and by $(1 + y'^2)^{3/2}$ as in (22.13)]. The result is the equation

$$(y - \eta)y'' - (1 + y'^2) = 0, \tag{22.24}$$

from which we obtain the curve equation

$$y - \eta = a \cosh \frac{x - \xi}{a}. \tag{22.25}$$

The three arbitrary constants ξ, η, a suffice to satisfy the conditions of the given end points and the given length of the chain.

Maupertuis's Principle or the Principle of Least Action

Maupertuis's principle is nearly as well known as Fermat's principle itself, even though it is formally only a special case of the latter. Maupertuis brought to light the important analogy between geometrical optics and classical mechanics.

The principle of least action may be written

$$\int_a^b mv\, ds = \min, \tag{22.26}$$

where a and b are the end points of the path of a material body with mass m and velocity v as a function of the space coordinates, it being understood that the motion is subject to the energy principle

$$\tfrac{1}{2}mv^2 + v = E, \tag{22.27}$$

even though at that time the general nature of that principle was not fully realized. If $V = V(x, y, z)$ and E a constant, the function $v = v(x, y, z)$ is immediately given by (22.27). To put it in the same form as used above in (22.3) for Fermat's principle we need only specialize for two coordinates x, y and choose one of them, say, x the independent variable. Then (22.26) reads

$$\int_{x_0}^{x_1} \sqrt{2mV(x, y) - E}\, \sqrt{1 + y'^2}\, dx = \min \tag{22.28}$$

with given $y_0 = y(x_0)$, $y_1 = y(x_1)$.

The meaning of the principle is that there exists a natural path of the particle $y(x)$ for which the integral (22.28), or the integral (22.26) with $ds = (dx^2 + dy^2 + dz^2)^{1/2}$, has a minimum value. Any other conceivable path $\bar{y}(x)$ with the same end points would yield a larger value of the integral.

The Euler equation corresponding to the variational principle (22.28) is very foreign to the equations used in classical mechanics. It can be written in the form

$$\frac{y}{1 + y'^2} = -\frac{(\partial V/\partial y) - y'(\partial V/\partial x)}{2(E - V)}. \tag{22.29}$$

However, by introducing the time as an independent variable, writing

$$\frac{d}{dx} = \frac{1}{\dot{x}}\frac{d}{dt}, \tag{22.30}$$

and using

$$-\frac{\partial V}{\partial y} = m\ddot{y}, \quad -\frac{\partial V}{\partial x} = m\ddot{x}, \quad 2(E - V) = m(\dot{x}^2 + \dot{y}^2), \tag{22.31}$$

both the left- and right-hand side of (22.28) can be stated in the form

$$\frac{\dot{x}\ddot{y} - \dot{y}\ddot{x}}{\dot{x}(\dot{x}^2 + \dot{y}^2)}.$$

To bring the principle into contact with more modern principles we start from (22.28), writing $ds = v\,dt$, and obtain

$$\int_{t_0}^{t_1} 2T\,dt = \min, \tag{22.32}$$

where $T = \tfrac{1}{2}mv^2$ denotes the kinetic energy, it being understood that (22.27) is still valid. Then, subtracting the constant left-hand side of (22.27) from the integrand, we obtain the Hamilton principle

$$\int_{t_0}^{t_1} (T - V)\,dt = \min, \tag{22.33}$$

provided t_0 and t_1 have given definite values. This is in accordance with the principle of least action if we assume $x(t)$ and $y(t)$ to have given values at both ends of the time integral, that is, $x(t)$ and $y(t)$ must have given values at $t = t_0$ and $t = t_1$.

23. THE EULER DIFFERENTIAL EQUATIONS OF A VARIATIONAL INTEGRAL WITH BOUNDARY CONDITIONS

The Variation of a Function

Let $y(x)$ be a function that is a solution of a given variational problem as defined by a variational integral taken between given limits of the independent variable x. Next consider a trial function $\bar{y}(x)$ very little different from

Linear Transformation. Extremum and Variational Problems

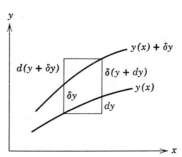

Fig. 3-9. Geometrical interpretation of the equation $\delta(dy) = d(\delta y)$.

$y(x)$. Then we may write

$$\delta y = \bar{y}(x) - y(x) = \varepsilon \eta(x), \tag{23.1}$$

ε being a parameter usually thought of as infinitesimal, whereas $\eta(x)$ may be a function of a finite order of magnitude which may be quite arbitrarily chosen. In this way we have introduced infinitesimal variations of a function y and still kept open the possibility of differentiating the variations. Variation like δy must be strictly distinguished from differentials like $dy = y(x + dx) - y(x)$. We may, however, also form variations of derivatives or differentials of a function. Obviously, according to (23.1)

$$d\bar{y} - dy = \varepsilon \eta' \, dx = (\bar{y}' - y') \, dx,$$

or

$$d(\delta y) = \delta(dy), \tag{23.2}$$

as illustrated geometrically in Fig. 3.9.

In the following we shall use the notation δy and only keep in mind its true interpretation (23.1). For instance, the variation of a function $F(x, y, y', \cdots)$ of a dependent variable y, together with its first and higher derivatives and of the independent variable x as well, now reads

$$\delta F = \frac{\partial F}{\partial y} \delta y + \frac{\partial F}{\partial y'} \delta y' + \cdots, \tag{23.3}$$

considering that $\delta y, \delta y', \cdots$, are all infinitesimal so that higher powers in the variations may be neglected.

We are now in the position of being able to evaluate the variation of an integral

$$I(y) = \int_{x_0}^{x_1} F(y, y', x) \, dx \tag{23.4}$$

in terms of the variation δy, F being, for simplicity and definiteness, a function of the derivative only up to y'. By variation and partial integration, using (23.2),

$$\delta I = \int_{x_0}^{x_1} \left(\frac{\partial F}{\partial y} \delta y + \frac{\partial F}{\partial y'} \delta y' \right) dx = \int_{x_0}^{x_1} \left(\frac{\partial F}{\partial y} - \frac{d}{dx} \frac{\partial F}{\partial y'} \right) \delta y \, dx + \left(\frac{\partial F}{\partial y'} \delta y \right)_{x_0}^{x_1}. \tag{23.5}$$

The Euler Differential Equation

If it is required that the integral I be stationary or that $\delta I = 0$ for any δy, a necessary condition is that the integrand of the right-hand integral (23.5) is zero for any δy, which necessitates that

$$\frac{\delta F}{\delta y} = \frac{\partial F}{\partial y} - \frac{d}{dx}\frac{\partial F}{\partial y'} = 0. \tag{23.6}$$

This is the fundamental equation in the calculus of variations, replacing the simple conditions like $f(x) = 0$ in the theory of maxima and minima or stationary values of ordinary functions by a differential equation for the unknown function $y(x)$, which solves the problem. This is a consequence of the free choice of the variation δy. If the Euler *variational derivative* $\delta F/\delta y$ differs from zero, the integral could by suitable choice of $\delta y = \varepsilon \eta(x)$ be made to attain any value without any chance in general of compensation by the last term.

This being accepted, the Euler equation is still not a sufficient condition. We also have to add the condition

$$\left(\frac{\partial F}{\partial y'} \delta y\right)_{x_0}^{x_1} = 0, \tag{23.7}$$

which enforces some boundary conditions on the function $y(x)$.

Boundary Conditions

The very simplest boundary condition obeying (23.7) is that of prescribed end values y_0 and y_1 of the function $y(x)$. This means that at both ends $\delta y = 0$ and, hence, (23.7) is fulfilled. The other way of obtaining (23.7) is to require that $\partial F/\partial y' = 0$ at both ends. This we shall call a natural boundary condition. For instance, if F is quadratic in y', $y' = 0$ is a natural boundary condition. If the Euler equation has singularities at the ends of the fundamental region, as mentioned under the discussion of eigenvalue problems, and if we deal with second-order equations, we only have to require the solution to be finite, supposing thereby one of the particular solutions near the singular boundary point. For instance, the Legendre polynomials may be defined by a variation principle or a variational integral in which the integrand contains the term $(1 - x^2)y'^2$. The natural boundary condition then becomes $(1 - x^2)y' = 0$, which does not necessitate that $y' = 0$ but only that y' and y are finite. This is the case for the Legendre polynomials. The other solution, however, called *Legendre functions* of the second kind, contains logarithmic terms $y \sim \log (1 + x)/(1 - x)$, $y' \sim (1 - x^2)^{-1}$, making the expression $(1 - x^2)y'$ finite.

Linear Transformation. Extremum and Variational Problems / 101

Several Dependent Variables. Simultaneous Euler Equations

In the case of several unknown functions $y_1(x)$, $y_2(x)$, $\cdots$, $y_n(x)$ of a single independent variable x, it is possible to formulate integral variational principles leading to a determination of the whole set of functions. The equations determining the functions are, of course, simultaneous differential equations. If these are to be second-order equations, we shall have to limit ourselves to functions $F(y_i, y_i', x)$, $i = 1, 2, \cdots, n$, of the unknown function and their first derivatives only, together with, possibly, the independent variable x itself.

Writing

$$I(y_i) = \int_{x_0}^{x_1} F(y_i, y_i', x)\, dx = \text{extr}. \qquad (23.8)$$

with given values of the function y' at $x = x_0$, x_1, or other suitable conditions, $I(y_i)$ must be stationary for variations of any single dependent variable. Hence, the necessary condition for I to be stationary is the fulfillment of the whole set of their equations

$$\frac{\delta F}{\delta y_i} = \frac{\partial F}{\partial y_i} - \frac{d}{dx}\frac{\partial F}{\partial y_i'} = 0, \qquad i = 1, 2, \cdots, n. \qquad (23.9)$$

Several Independent Variables. Partial Differential Equations

Next consider variational integrals in some region G of several independent variables x, y, z, $\cdots$, the boundary of the region being a closed hypergeometric surface g, that is, by two variables x, y a curve, by three variables x, y, z a surface in the ordinary sense, and so on.

Consider now a function $u(x, y, \cdots)$ whose variation we may define by an equation $\delta u = \bar{u} - u = \varepsilon \eta(x, y, \cdots)$, ε being as before an infinitesimal parameter and η in general an arbitrary function of the independent variables. Again it can be shown that the operations of variation and derivation are commutative, irrespective of the fact that we have now several different sorts of derivation.

Denote by u_x, u_y, $\cdots$, the first-order partial derivatives, we want to study the variational problem

$$I(u) = \int_G F(u, u_x, u_y, \cdots, x, y, \cdots)\, dx\, dy = \text{extr}. \qquad (23.10)$$

A necessary condition is that I be stationary, or,

$$\delta I = \int_G \left(\frac{\partial F}{\partial u}\delta u + \frac{\partial F}{\partial u_x}\delta u_x + \frac{\partial F}{\partial y}\delta u_y + \cdots\right) dv = 0, \qquad (23.11)$$

102 / The Mathematical Foundation of Physics

writing for brevity dv for the volume element $dx\,dy\cdots$. Since $\delta u_x = (\partial/\partial x)\,\delta u$, and so forth, we may integrate by parts and obtain

$$\delta I = \int_G \left(\frac{\partial F}{\partial u} - \frac{\partial}{\partial x}\frac{\partial F}{\partial u_x} - \frac{\partial}{\partial y}\frac{\partial F}{\partial u_y} \cdots \right) \delta u\, dv$$

$$+ \int_g \left(\frac{\partial F}{\partial u_x}\frac{\partial x}{\partial n} + \frac{\partial F}{\partial u_y}\frac{\partial y}{\partial n} + \cdots \right) \delta u\, df = 0, \qquad (23.12)$$

df being an element of the surface g and $\partial/\partial n$ denoting differentiation along the normal unit vector $\mathbf{n}$ in ordinary space

$$\mathbf{n} = \mathbf{i}\frac{\partial x}{\partial n} + \mathbf{j}\frac{\partial y}{\partial n} + \mathbf{k}\frac{\partial z}{\partial n}. \qquad (23.13)$$

Requiring (23.12) to be true for any variation δu of the unknown function u, we obtain the *Euler partial differential equation* of the second order

$$\frac{\delta F}{\delta u} = \frac{\partial F}{\partial u} - \frac{\partial}{\partial x}\frac{\partial F}{\partial u_x} - \frac{\partial}{\partial y}\frac{\partial F}{\partial u_y} - \cdots = 0. \qquad (23.14)$$

In addition to this equation we have the boundary condition, for instance $\delta u = 0$ on g, which means that $u(x, y, \cdots)$ is a given function on the boundary. The other possibility is that u obey the natural boundary condition

$$\frac{\partial F}{\partial u_x}\frac{\partial x}{\partial n} + \frac{\partial F}{\partial u_y}\frac{\partial y}{\partial n} + \cdots = 0 \qquad (23.15)$$

which means that the vector with components $\partial F/\partial u_x$, $\partial F/\partial u_y$, $\cdots$, is orthogonal to the normal vector $\mathbf{n}$. For instance, if that part of F which depends on the partial derivatives has the simple form const $(u_x^2 + u_y^2 + \cdots)$, the condition (23.15) is equivalent to

$$u_x \frac{\partial x}{\partial n} + u_y \frac{\partial y}{\partial n} + \cdots = \mathbf{n}\,\text{grad}\,u = \frac{\partial u}{\partial n} = 0. \qquad (23.16)$$

This means that the normal derivative of the function u vanishes all over the surface.

Homogeneous and Linear Euler Equations and Linear Homogeneous Boundary Conditions

If the function F is a homogeneous function in u and its derivatives, the Euler equation will be homogeneous. If F is a quadratic function, the Euler equation will be linear.

In (23.16) we obtained the natural boundary condition $\partial u/\partial n = 0$ from the form $(\text{grad}\,u)^2$ of the integrand. If we add an appropriate surface integral

Linear Transformation. Extremum and Variational Problems | 103

in the variational integral, say,

$$I(u) = \int_G (\text{grad } u)^2 + k \int_a u^2 \, df = \text{extr.} \tag{23.17}$$

k being a positive constant, the natural boundary condition will be the linear homogeneous equation

$$\frac{\partial u}{\partial n} + ku = 0. \tag{23.18}$$

If $k = 0$, we have $\partial u/\partial n = 0$. As $k \to \infty$ the boundary condition approaches $u = 0$. On the other hand, the extremum value of $I(u)$ increases with k. Therefore, $\partial u/\partial n = 0$ is called the weakest and $u = 0$ the *strongest* of the linear homogeneous boundary conditions.

This is a general feature of variational problems, which can be relied upon in the theory of eigenvalues or vibrational problems.

Finally, if in (23.17) we replace u^2 by $(u - u_0)^2$, $u_0(x, y, \cdots)$ being a given function we obtain in the limiting case $k \to \infty$, the inhomogeneous boundary condition $u = u_0$, that is, the function $u(x, y, \cdots)$ is a known function on the boundary g.

These considerations may serve the purpose of demonstrating the internal connection between boundary conditions of different kinds, which may otherwise appear entirely foreign to each other.

24. VARIATIONAL PROBLEMS WITH ADDITIONAL CONDITIONS. EIGENVALUE PROBLEMS

Variational Problems Leading to Linear Differential Equations

As already mentioned, the Euler equations of a variational problem will be linear and homogeneous only if the integrand in the variational integral is a quadratic function of the dependent variable and its derivatives. If in this integrand product terms like

$$2\varphi u(x, y, \cdots) u_x = \varphi \frac{\partial}{\partial x} u^2$$

appear, they can be replaced by $-u^2 \, \partial \varphi / dx$ by partial integration. The general form of the integrand may therefore be taken to be

$$p_{11} u_x^2 + p_{22} u_y^2 + \cdots + 2 p_{12} u_x u_y + \cdots + q u^2,$$

the coefficients p_{ik} and q being given functions of the independent variables.

If p_{ik} were constant, we might, by simple rotation of the coordinate system, bring the integrand to the form

$$p_1 u_x^2 + p_2 u_y^2 + \cdots + q u^2,$$

and finally, by altering the scale for individual coordinates, reach the simple standard form

$$u_x^2 + u_y^2 + \cdots + Qu^2 = (\text{grad } u)^2 + Qu^2. \quad (24.1)$$

The purely quadratic form is also obtainable in the case of variable coefficients p_{ik}; however, we should then have to introduce curvilinear coordinates. Owing to the appreciable difficulties that would arise from such more general problems we shall, therefore, in the case of several independent variables restrict ourselves to the case (24.1). In the case of a single independent variable x, however, we shall use the general form of the integrand

$$py'^2 + qy^2, \quad (24.2)$$

where $p = p(x)$ and $q = q(x)$ are not constants, although known functions of the independent variable.

The Euler equations of (24.1) and (24.2) are

$$(\text{grad } u)^2 - Qu = 0 \quad (24.3)$$

and

$$(py')' - qy = 0. \quad (24.4)$$

Normalization as an Additional Condition

Consider the variational principle

$$I(u) = \int [(\text{grad } u)^2 + Qu^2] \, dv = \text{extr.}, \quad (24.5)$$

where u is subject to the additional condition

$$N(u) = \int u^2 \, dv = 1. \quad (24.6)$$

For further simplification we assume $Q(x, y, \cdots) > 0$ to ensure that both I and N are for any function u positive, that is, *positive definite* integrals.

Equation 24.6 restricts the choice of trial functions to functions with a given norm one. Owing to the quadratic form of the integrands in both (24.5) and (24.6), we may disregard this restriction by writing

$$\lambda = \frac{I}{N} \quad (24.7)$$

for the stationary values of I as defined by (24.5) and (24.6). Hence, assuming $\delta \lambda = 0$, the condition for stationary values may be written

$$\partial(I - \lambda N) = \delta I - \lambda \, \delta N = 0, \quad (24.8)$$

λ being a numerical parameter that is still to be determined.

This obviously is the same as applying the Lagrange method of indefinite multipliers. Writing F and G for the integrand in (24.5) and (24.6), the Euler equation of the problem is

$$\frac{\delta F}{\delta u} - \lambda \frac{\delta G}{\delta u} = 0 \tag{24.9}$$

or

$$\nabla^2 u - Qu + \lambda u = 0. \tag{24.10}$$

Similarly, from the more general one-dimensional problem,

$$I(y) = \int_{x_0}^{x_1} (py'^2 + qy^2)\, dx = \text{extr.} \tag{24.11}$$

by

$$N = \int_{x_0}^{x_1} \rho y^2\, dx = 1 \tag{24.12}$$

we obtain the ordinary differential equation

$$(py')' - qy + \lambda \rho y = 0 \tag{24.13}$$

with undetermined eigenvalue parameter λ.

Equation 24.13 has a particular form usually called *self-adjoint*. Any second-order differential expression $y'' + f(x)y'$ can be brought to self-adjoint form if it is multiplied by a suitable factor p. This function must obey the equations

$$pf = p', \qquad d \log p = f(x)\, dx, \qquad p = e^{\int f\, dx}. \tag{24.14}$$

Any second-order differential equation, therefore, can be derived from a variational problem, which is easily found when the equation has been made self-adjoint.

In addition to (24.13) we shall have to add the condition

$$(py'\, \delta y)_{x_0}^{x_1} = 0 \tag{24.15}$$

so that the variational integral (24.11) can be stationary. Apart from the possibility of $\delta y = 0$ at x_0 and x_1, this leads to the natural boundary condition $y' = 0$ if $p \neq 0$ and $py' = 0$ if $p = 0$ at the same points. In any case, if (24.13) is satisfied by y and λ, we obtain by multiplying by λ and integrating

$$\lambda \int_{x_0}^{x_1} \rho y^2\, dx = \int_{x_0}^{x_1} (py'^2 + qy^2)\, dx, \tag{24.16}$$

in consistency with (24.7), or

$$\lambda = \int_{x_0}^{x_1} (py'^2 + qy^2)\, dx \tag{24.16a}$$

for the normalized function y, in consistency with (24.5).

25. EIGENVALUE PROBLEMS

Eigenvalues and Eigenfunctions

Equations 24.10 and 24.13 are called *eigenvalue equations*. Their solutions obey given boundary conditions only for particular values of the parameter λ, which are called eigenvalues. An eigenvalue problem, therefore, is of a composite nature inasmuch as eigenvalues and corresponding admissible solutions of the differential equation must be determined simultaneously. Assuming such solutions or eigenfunctions to exist, they are bound to have definite properties which we shall now discuss.

Eigenvalues are commonly ordered according to their magnitude. Let us write

$$\lambda = \lambda_1, \lambda_2, \cdots, \lambda_n, \cdots, \tag{25.1}$$

and

$$y = y_1, y_2, \cdots, y_n, \cdots, \tag{25.2}$$

for the eigenvalues and corresponding eigenfunctions. We then have

$$(py_n')' - qy_n + \lambda_n y_n = 0, \quad (py_m')' - qy_m + \lambda_m y_m = 0, \tag{25.3}$$

or, in combining the two equations,

$$(\lambda_n - \lambda_m)\rho y_n y_m = [p(y_n y_m' - y_m y_n')]'. \tag{25.4}$$

As already mentioned in connection with (21.17) and (21.18), this leads by integration to the equation

$$(\lambda_n - \lambda_m)\int_{x_1}^{x_0} \rho y_n y_m \, dx = 0, \tag{25.5}$$

provided the eigenfunction obeys the boundary $py' = 0$ or $y = 0$.

If therefore $\lambda_n \neq \lambda_m$, the corresponding eigenfunctions are mutually orthogonal (with the density function ρ). On the other hand, we are free to multiply any of the eigenfunctions by arbitrary factors, for instance, in order to make them normalized functions. Therefore the whole system of eigenfunctions can be taken to form an orthonormalized system

$$\int_{x_0}^{x_1} \rho y_n y_m \, dx = \delta_{nm}. \tag{25.6}$$

Coincidence of Eigenvalues or Degeneration

In the case of the eigenvalue problem (24.10), as defined by a partial differential equation, we are able to obtain similar results by making use of the *Green theorem* of vector analysis

$$\int u_n \nabla^2 u_m - u_m \nabla^2 u_n \, dv = \int \left(u_n \frac{\partial u_m}{\partial n} - u_m \frac{\partial u_n}{\partial n} \right) df. \tag{25.7}$$

Linear Transformation. Extremum and Variational Problems / 107

By combination of two different eigenvalue equations we obtain the equation

$$(\lambda_n - \lambda_m) \int u_n u_m \, dv = 0 \tag{25.8}$$

provided that on the boundary of the domain of integration the eigenfunction obeys any of the boundary conditions, $u = 0$, $\partial u/\partial n = 0$, or the general homogeneous boundary condition

$$\frac{\partial u}{\partial n} + ku = 0. \tag{25.9}$$

Therefore, here also the eigenfunctions are mutually orthogonal provided that their eigenvalues are different. It may happen, however, that different eigenfunctions u_n and u_m belong to equal eigenvalues $\lambda_n = \lambda_m$.

In this case, owing to the linearity of the differential equation, any linear combination of u_n and u_m is an eigenfunction belonging to the same eigenvalue. If u_n and u_m have the same norm, we may take for example the special combinations $u_n + u_m$ and $u_n - u_m$ as new representative eigenfunctions in order to have an orthogonal set of functions. The same reasoning applies to the cases of three or more eigenfunctions belonging to the same eigenvalue. The process of mutual orthogonalization is always feasible and in connection with a succeeding normalization of the functions this leads to the existence of a full set of orthonormalized eigenfunction of any eigenvalue problem

$$\int u_n u_m \, dv = \delta_{nm}. \tag{25.10}$$

Direct Methods of Solution. The Ritz Method

A great many eigenvalue problems, of course, can be solved by known elementary methods. In general, however, we shall have to expect that this will not be the case and we shall have to rely upon more general numerical methods. It might appear tempting to apply a method similar to that of the integral calculus, namely, to replace an exact solution or curve $y(x)$ by a polygon as indicated in Fig. 3-10. The ordinates of the curve, that is, a discrete set of approximate values of the eigenfunction could then be obtained

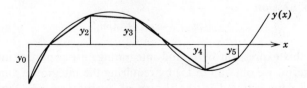

Fig. 3-10. Approximation of a smooth curve by a polygon.

by the ordinary theory for maxima and minima of functions of a finite set of variables.

Because of its complexity, this otherwise very direct and intelligible procedure has been used only infrequently. Far more simple and effective is the method invented by Ritz in 1908 for the general treatment of eigenvalue problems.

The method consists of replacing the exact solution of equations like (24.11), (24.12), or (24.13), whether in the form of variational integrals or of differential equations, by finite expressions

$$y = c_1 u_1 + c_2 u_2 + \cdots + c_n u_n. \tag{25.11}$$

In these expressions the functions $y_n(x)$ are suitably chosen functions, preferably those belonging to orthonormal systems with a density function $\rho(x)$. The coefficients c_i are then determined in such a way that the variational integral (24.11) becomes stationary with respect to variations of the coefficients, which in addition are subject to the condition (24.12). Assuming the orthonormality relation to be valid for the function u_i,

$$\int_{x_0}^{x_1} \rho u_i u_k \, dx = \delta_{ik} \tag{25.12}$$

we obtain

$$I = \sum_{i,k=1}^{n} a_{ik} c_i c_k = \text{extr.}, \qquad N = \sum_{i=1}^{n} c_i^2 = 1, \tag{25.13}$$

where

$$a_{ik} = \int_{x_0}^{x_1} (p u_i' u_k' + q u_i u_k) \, dx. \tag{25.14}$$

In this way we return to a problem that has already been dealt with in the preceding sections, namely, the problem of maxima or minima or stationary values of quadratic forms of a finite set of variables. The equivalent set of linear equations reads

$$\sum_{k=1}^{n} a_{ik} c_k - \lambda c_i = 0, \qquad i = 1, 2, \cdots, n. \tag{25.15}$$

This system could also have been obtained from the differential equation (24.13). Inserting the expression (25.11) in the left-hand side of (24.13) we obtain the expression

$$\sum_{k=1}^{n} c_k [(p u_k')' - q u_k + \lambda \rho u_k]. \tag{25.16}$$

Multiplying this expression by u_i and integrating, using partial integration for the first term, we obtain (25.15) by requiring the integral to vanish.

As already demonstrated in the theory of quadratic forms the possible values of λ, that is, the first n approximate eigenvalues, are obtained from the determinant equation

$$|a_{ik} - \lambda \delta_{ik}| = 0, \quad i, k = 1, 2, \cdots, n. \quad (25.17)$$

If the number n of independent functions u_i in (25.11) gradually increases, the approximate values improve. If the system (25.11) as $n \to \infty$ forms a complete system of functions, the approximate eigenvalues will converge toward the exact eigenvalues.

It is not necessary by the Ritz method that the auxiliary functions $u_i(x)$ be orthogonal, even though it is convenient. When using nonorthogonal functions, we are led to determinant equations of the type (19.32) rather than (19.10) or the above equation (25.17).

Distribution of Approximate Eigenvalues by the Ritz Method

It is highly important to know how approximate eigenvalues in various approximations are related to each other and to the exact eigenvalues when applying the Ritz method. Denote by λ_i the exact eigenvalues and by $\lambda_i^{(n)}$ those that are obtained from a nth-order determinant equation. Then, as will be proved, we have the general law

$$\lambda_i^{(n)} \leqq \lambda_i^{(n-1)} \leqq \lambda_{i+1}^{(n)}. \quad (25.18)$$

If this law is true, we may draw the very important conclusion that not only the lowest approximate eigenvalue, but *all approximate eigenvalues are larger than the corresponding exact eigenvalues*, and that any of them converges towards its final value *monotonously and from the upper side*. This is a result that is known beforehand only for the lowest eigenvalue.

To prove the theorem (25.18) we may imagine the quadratic form I in (25.13) transformed on its principal axis by an orthogonal transformation of the coefficients c_i to a set of new coefficients γ_i, $i = 1, 2, \cdots, n$. We then get the very simple extremum problem

$$I = \sum_{i=1}^{n} \lambda_i^{(n)} \gamma_i^2 = \text{extr.} \quad \text{by} \quad \sum_{i=1}^{n} \gamma_i^2 = 1, \quad (25.19)$$

with the solution

$$I = \lambda_i^{(n)} \quad \text{for} \quad \gamma_i = 1, \quad \gamma_k = 0 \quad \text{for} \quad k \neq i. \quad (25.20)$$

Of course, this means only that, in solving the nth-order secular equation (25.17) and the linear system of (25.15), we have for any approximate eigenvalue λ_i the corresponding approximate eigenfunction

$$v_i = \sum_k c_k u_k, \quad (25.21)$$

the coefficients c_k being different for different λ_i.

110 / The Mathematical Foundation of Physics

The secular equation when we use the system of function v_i instead of u_i, of course, is

$$D^{(n)}(\lambda) = [\lambda_1^{(n)} - \lambda]\{[\lambda_2^{(n)} - \lambda] \cdots [\lambda_n^{(n)} - \lambda]\} = 0, \qquad (25.22)$$

and we shall now try to find out what its form will be when descending to the $(n-1)$th approximation by dropping the last function u_n in the expansion (25.11). This means that we have to put $c_n = 0$.

Now in the inverse orthogonal transformation, that is, from γ_i to c_i, we shall have a linear equation

$$c_n = \alpha_1\gamma_1 + \alpha_2\gamma_2 + \cdots + \alpha_n\gamma_n = 0 \qquad (25.23)$$

which causes a relation between the coefficients γ_i. This means that we no longer write for an approximate solution

$$y = \sum_i \gamma_i v_i \qquad (25.24)$$

with a free choice of γ_i without risking the appearance of the "forbidden" u_n. It can be avoided only by accepting the restriction (25.7). With this relation, however, it can be fairly easily shown that the secular equation turns into

$$D^{(n-1)}(\lambda) = [\lambda_1^{(n)} - \lambda][\lambda_2^{(n)} - \lambda] \cdots [\lambda_n^{(n)} - \lambda] \cdot$$

$$\cdot \frac{\alpha_1^2}{\lambda_1^{(n)} - \lambda} + \frac{\alpha_2^2}{\lambda_2^{(n)} - \lambda} + \cdots + \frac{\alpha_n^2}{\lambda_n^{(n)} - \lambda} = 0, \qquad (25.25)$$

$D^{(n-1)}(\lambda)$ being a polynomial of the $(n-1)$th degree.

Assume first all coefficients α_i except α_n to be zero. Then $\gamma_n = 0$, which means that the suppression of u_n in (25.11) involves the suppression of v_n in (25.24), or, that $v_n = u_n$. The other coefficients γ_1 to γ_{n-1} now being left free, we are bound to obtain from (25.25) the approximate eigenvalues $\lambda = \lambda_1^{(n)}$ to $\lambda_{n-1}^{(n)}$. Similarly, if we take any of the other α_i to be the only coefficient different from zero, the result must be that all λ-values are unchanged except λ_i which is lacking, in perfect conformity with (25.12). Since our considerations are purely mathematical, it does not matter that only the first of the above cases corresponds to neglecting u_n.

So far we do not know, however, whether the particular form of the nominators in (25.12) is correct. We only know that any of them must vanish with the corresponding α_i. Next we consider the case of all coefficients, except two, say, α_i and α_k which are zero. Then only the particular eigenvalues $\lambda_i^{(n)}$ and $\lambda_k^{(n)}$ are changed and united into one. According to (25.22) this value will be

$$\lambda = \frac{\lambda_i^{(n)}\gamma_i^2 + \lambda_k^{(n)}\gamma_k^2}{\gamma_i^2 + \gamma_k^2} = \text{extr.} \qquad \text{with} \qquad \alpha_i\gamma_i + \alpha_k\gamma_k = 0. \qquad (25.26)$$

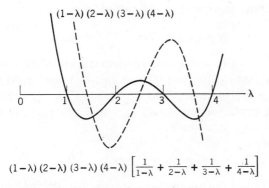

Fig. 3-11. Monotonous depression of eigenvalues by the Ritz method.

Eliminating γ_i, then γ_k also disappears, which leaves us with

$$\lambda = \frac{\lambda_i^{(n)} \alpha_k^2 + \lambda_k^{(n)} \alpha_i^2}{a_k^2 + \alpha_i^2}, \tag{25.27a}$$

or

$$[\lambda_i^{(n)} - \lambda](\lambda_k^{(n)} - \lambda)\left[\frac{\alpha_i^2}{\lambda_i^{(n)} - \lambda} + \frac{\alpha_k^2}{\lambda_k^{(n)} - \lambda}\right] = 0.$$

This ensures that our equation (25.25) is true. In the general case in which all coefficients α_i are different from zero, we now find that the determinant $D^{(n-1)}(\lambda)$ changes its sign as we pass from a λ-value $\lambda_i^{(n)}$ to $\lambda_{i+1}^{(n)}$. Hence there will be a root of (25.25) in between and, since the number of roots is $n-1$, the relation (25.18) has been proved. The result is geometrically illustrated in Fig. 3-11 for the fourth and third approximations of an eigenvalue problem.

The Minimum Properties of Eigenvalues

We shall now prove another very important theorem. Consider a function y subject to the condition

$$\int_{x_0}^{x_1} \rho y_i y \, dx = 0, \quad i = 1, 2, \cdots, m. \tag{25.28}$$

Then

$$I(y) = \int_{x_0}^{x_1} (py'^2 + qy^2) \, dx \geq \lambda_{m+1}, \quad \text{if} \quad \int_{x_0}^{x_1} \rho y^2 \, dx = 1. \tag{25.29}$$

To prove this consider an expression

$$y = c_1 u_1 + c_2 u_2 + \cdots + c_m u_m, \tag{25.30}$$

which by the above treatment can be turned into

$$y = \gamma_1 v_1 + \gamma_2 v_2 + \cdots + \gamma_m v_m, \qquad (25.31)$$

v_1 to v_m being mutually orthogonal approximate eigenfunctions.

Assume now that v_1 to v_m are the true eigenfunctions y_1 to y_m, hence $\lambda_1^{(n)}$ to $\lambda_m^{(n)}$ are the true eigenvalues λ_1 to λ_m. Then $\lambda_{m+1}^{(n)}$ to $\lambda_n^{(n)}$ are all larger than or equal to the corresponding values λ_{m+1} to λ_n; they are all larger than or equal to λ_{m+1}.

Now assume

$$\gamma_1 = \gamma_2 = \cdots = \gamma_m = 0, \qquad (25.32)$$

which does not alter the approximate eigenvalues $\lambda_i^{(n)}$, except that the first m of them have disappeared. It follows that the expression

$$y = \sum_{i=m+1}^{n} \gamma_i v_i, \qquad (25.33)$$

which obeys (25.28) yields an approximate minimum value $\lambda_{m+1}^{(n)} \geq \lambda_{m+1}$. But there is no restriction to the number n or to the functions u_i or v_i, except that from $i = m + 1$ all v_i are orthogonal to $y_1, y_2, \cdots, y_n$. Hence, the theorem (25.29) has been proved.

26. NUMBER AND DISTRIBUTION OF EIGENVALUES

Comparison with Simpler Eigenvalue Equations

When studying the number of eigenvalues of a certain order of magnitude, it is advisable to rewrite the equation

$$(py')' - qy + \lambda \rho y = 0 \qquad (26.1)$$

in a more suitable form. This can be done by different means, namely, by changing either the dependent or the independent variable. If we put

$$y = fz, \qquad dx = g\, d\xi \qquad \text{or} \qquad \frac{d}{d\xi} = \frac{d}{dx}, \qquad (26.2)$$

z, ξ being new dependent and independent variables, respectively, the transformed equation may be brought to the self-adjoint form

$$\frac{d}{d\xi}\left(\frac{p}{g} f^2 \frac{dz}{d\xi}\right) - \left(qgf^2 - f\frac{d}{d\xi}\frac{p}{g}\frac{df}{d\xi}\right)x + \lambda\rho gf^2 z = 0. \qquad (26.3)$$

With $g = 1$, $\xi = x$, that is, by changing only the dependent variable, we obtain by putting $f = \rho^{-1/2}$ and $f = p^{-1/2}$ the two variants

$$\frac{d}{dx}\left(\frac{p\,dz}{\rho\,dx}\right) - \left(\frac{q}{\rho} - \rho^{-1/2}\frac{d}{dx}p\frac{d}{dx}\rho^{-1/2}\right)z + \lambda z = 0, \tag{26.4}$$

$$\frac{d^2z}{dx^2} - \left(\frac{q}{p} - p^{-1/2}\frac{d}{dx}p\frac{d}{dx}p^{-1/2}\right)z - \lambda\frac{\rho}{p}z = 0, \tag{26.5}$$

that is, with constant density or tension functions, respectively. If, on the other hand, we put $f = 1$ or $y = z$ and $g = p$, changing only the independent variable, we obtain the equation

$$\frac{d^2y}{d\xi^2} - qpy + \lambda\rho py = 0 \tag{26.6}$$

which is similar to (26.5).

Any of these transformations may be useful for particular purposes. For our present purpose, however, we want a still simpler equation with constant tension as well as density function. This can be attained by putting simultaneously

$$\frac{p}{g}f^2 = 1, \qquad \rho g f^2 = 1, \qquad \text{or} \qquad f = (\rho p)^{-1/4}, \qquad g = \left(\frac{p}{\rho}\right)^{1/2}. \tag{26.7}$$

In this way we obtain the equation

$$\frac{d^2z}{d\xi^2} - Qz + \lambda z = 0, \tag{26.8}$$

with

$$Q = \frac{q}{\rho} - (\rho p)^{-1/4}\frac{d}{d\xi}\left[(\rho p)^{1/2}\frac{d}{d\xi}(\rho p)^{-1/4}\right], \tag{26.9}$$

and

$$z = (\rho p)^{1/4}y, \qquad \xi = \int_{x_0}^{x}\left(\frac{\rho}{p}\right)^{1/2}dx \tag{26.10}$$

the limits of the region for the variable ξ now being

$$0 \quad \text{and} \quad a = \int_{x_0}^{x_1}\left(\frac{\rho}{p}\right)^{1/2}dx. \tag{26.11}$$

Equation 26.8 is now easily comparable with

$$\frac{d^2z}{d\xi^2} + \lambda z = 0 \tag{26.12}$$

whose eigenvalues and eigenfunctions for the same region are easily found to be

$$\lambda_n = n^2 \frac{\pi^2}{a^2}, \qquad z_n = \sin n \frac{\pi}{a} \xi, \qquad (26.13)$$

if the boundary conditions are $z = 0$.

The Total Number of Eigenvalues

From the exact solution of (26.12) we see that there is an infinite number of eigenvalues. This can also be seen from geometrical considerations of the curve $z(\xi)$ as illustrated in Fig. 3.12.

For positive λ the solution of (26.12) is always a curve with curvature of opposite sign of the ordinate z and, has its concave side in the direction of the ξ-axis. Sooner or later it is therefore forced to cut the axis, and this process is repeated, producing an infinite number of zeros for the function z. The higher λ, the more dense become the zeros. Hence we obtain solutions with zero points at the boundaries $\xi = 0$ and $\xi = a$ and a number of zeros $0, 1, \cdots,$ $n - 1, \cdots$ in between. The number of solutions obeying the given boundary conditions, therefore, is infinite.

But the same arguments apply to (26.8) if Q is finite in the region $0 \leq \xi \leq a$. The only difference is that the curves, particularly for small λ, are not so regular as those drawn for (26.12) in Fig. 3-12. Hence any eigenvalue equation of the form (26.1) with finite p, q, and ρ in a finite region $x_0 \leq x \leq x_1$ have an infinite number of eigenvalues.

Asymptotic Behavior of Eigenvalues

Not only the number but also the magnitude of the eigenvalues of (26.8) and (26.12) may be compared. The two sets of eigenvalues necessarily are of the same order of magnitude. As $n \to \infty$ and $\lambda_n \to \infty$ the difference can even be neglected, since the function Q plays an unimportant role in comparison to λ when λ becomes very large, provided Q has only finite values. We may, therefore, use the first formula (26.13) as an asymptotic expression for the

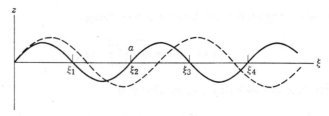

Fig. 3-12. Integral curves for different equations (26.8) and (26.12) or for one of them by different λ.

Linear Transformation. Extremum and Variational Problems | 115

eigenvalues of (26.8), writing

$$\lim_{n\to\infty} \frac{n^2}{\lambda_n} = \frac{a^2}{\pi^2} = \frac{1}{\pi^2}\left[\int_{x_0}^{x_1}\left(\frac{\rho}{p}\right)^{\frac{1}{2}} dx\right]^2. \tag{26.14}$$

A more accurate formula is obtainable by inserting z_n of (26.13), considered as an approximate eigenfunction, into the variational integral corresponding to (26.8),

$$\int_0^a \left[\left(\frac{dz}{d\xi}\right)^2 + Qz^2\right] d\xi \quad \text{by} \quad \int_0^a z^2 \, d\xi = 1. \tag{26.15}$$

We then obtain for large n the more accurate expression

$$\lambda_n \approx n^2 \frac{\pi^2}{a^2} + \bar{Q}, \tag{26.16}$$

$\bar{Q}$ denoting the mean value of Q in the region $0 \leq \xi \leq a$.

27. COMPLETENESS PROPERTIES FOR SOLUTIONS OF EIGENVALUE EQUATIONS

Proof of the Completeness of a System

The eigenfunctions of an ordinary *Sturm-Liouville equation* (24.1) in a finite interval $x_0 < x < x_1$ form a complete system of orthogonal or, if wanted, orthonormal functions in the sense of (20.16).

Consider an arbitrary function $f(x)$ in the above fundamental region with components

$$c_k = (f, y_k) = \int_{x_0}^{x_1} \rho f y_k \, dx \tag{27.1}$$

on the axis defined by the eigenfunctions y_k if $f(x)$, like y_k, is considered a vector.

We now expand $f(x)$ in terms of the first n functions y_k and write for the mean square error

$$a_n^2 = \int_{x_0}^{x_1} \rho \left(f - \sum_{k=1}^n c_k y_k\right)^2 dx = (f,f) - \sum_{k=1}^n c_k^2. \tag{27.2}$$

Next we consider the normalized function

$$y = \frac{1}{a_n}\left(f - \sum_{k=1}^n c_k y_k\right), \tag{27.3}$$

which, obviously, is orthogonal to the first n eigenfunctions y_k. Inserted into the variational integral I, it therefore must give a value of the integral not smaller than $n + 1$,

$$\int_{x_0}^{x_1} (py'^2 + qy^2)\, dx = \frac{1}{a_n^2}\left[\int_{x_0}^{x_1} (pf'^2 + gf^2) - \sum_{k=1}^{n} \lambda_k c_k^2\right] \geq \lambda_{n+1}. \quad (27.4)$$

Provided that the bracket expression is finite, as it must be since it is positive and smaller than the first finite positive integral, the quantity a_n^2 must tend to zero as $n \to \infty$ in order that the whole expression above (or at least not below) the ever-increasing λ_n be on the order of magnitude n^2. Hence it follows from (27.2) that

$$(f,f) = \sum_{n}^{\infty} c_n^2, \qquad f = \sum_{n=1}^{\infty} c_n y_n, \qquad c_n = (f, y_n). \quad (27.5)$$

Completeness of Systems in an Infinite Region

Consider the eigenvalue equation

$$\left(\frac{d}{dx} x \frac{d}{dx} - \frac{x}{4} + \lambda\right) y = 0 \quad (27.6)$$

in the infinite region $0 \leq x \leq \infty$. Substituting

$$y = e^{-x/2} L(x), \quad (27.7)$$

we obtain

$$\left[x \frac{d^2}{dx^2} + (1 - x) \frac{d}{dx} + \lambda - \frac{1}{2}\right] L = 0 \quad (27.8)$$

which with

$$\lambda = \lambda_n = n + \tfrac{1}{2}, \qquad n = 0, 1, \cdots, \infty, \quad (27.9)$$

is the differential equation for the nth Laguerre polynomial $L_n(x)$ in (20.38). Equation 27.6 can be obtained from the variational problem

$$\lambda = I = \int_0^\infty \left(xy'^2 + \frac{x}{4} y^2\right) dx = \text{extr}, \qquad N = \int_0^\infty y^2\, dx = 1, \quad (27.10)$$

by the additional condition that competing functions y must be finite and quadratically integrable in the whole infinite region. The eigenfunctions solving this problem are according to (27.7)

$$y_n = e^{-x/2} L_n(x). \quad (27.11)$$

These eigenfunctions are easily seen by ordinary methods to be mutually orthogonal. In fact, all of the above considerations made on the basis of the calculus of variations are valid. For instance, a function y orthogonal to the first y_1 up to y_n and subject to the condition $N = 1$ can never yield a value of the variational integral smaller than λ_{n+1}.

Therefore, using the methods of (27.2), (27.3), and (27.4) and keeping in mind that $\lambda_n = n + \tfrac{1}{2} \to \infty$, it follows that the eigenfunctions (27.11) constitute a complete system of orthogonal functions.

As in the case of a finite region it may be proved that eigenfunctions of more complicated eigenvalue equations constitute complete systems of functions. We shall consider, for instance, the equation

$$\left[\frac{d}{dx} x \frac{d}{dx} - \frac{x}{4} - q(x) + \lambda\right] y = 0 \qquad (27.12)$$

or the equivalent variational problem

$$\lambda = I = \int_0^\infty \left[xy'^2 + \frac{x}{4} y^2 + q(x) y^2 \right] dx = \text{ext}, \qquad N = \int_0^\infty y^2\, dx = 1, \quad (27.13)$$

$q(x)$ having, for instance, only positive values. If $q(x)$ is proportional to x or if it is a constant, the problem can be solved simply as (27.6) or (27.10). We shall assume that $q(x)$ is of a more general character, however, that the function is finite. If it has a maximum value q_m in the region of integration, then it is obvious that the nth eigenvalue is smaller than $n + \tfrac{1}{2} + q_n$. On the other hand the eigenvalue of (27.12) or (27.13) must be larger than $n + \tfrac{1}{2}$. We therefore know the asymptotic behavior of the eigenvalues λ_n of (27.12) and (27.13). The point is that $\lambda_n \to \infty$ as $n \to \infty$, and this ensures as in the cases of (27.6) and (27.10) the completeness of the system of eigenfunctions of (27.12) and (27.13). Since (27.12) comprises a fairly wide class of eigenvalue equations, we see that the general feature of the eigenfunction of a Sturm-Liouville problem is to constitute complete systems of functions, whether in finite or infinite regions.

Noncomplete Systems of Eigenfunctions

We shall now demonstrate by an example that the ordinary eigenfunctions of an eigenvalue equation may form an incomplete system.

Considering the equations for Laguerre functions, we find

$$\left(\frac{d}{dx} x \frac{d}{dx} - \frac{x}{4} + n + \frac{1}{2}\right) y_n(x) = 0,$$

$$y_n(x) = e^{-x/2} L_n(x). \qquad (27.14)$$

Changing to the variable ξ and writing

$$x = \frac{\xi}{n + \tfrac{1}{2}}, \qquad (27.15)$$

we obtain the equations

$$\left[\frac{d}{d\xi}\xi\frac{d}{d\xi} - \frac{\xi}{(2n+1)^2} + 1\right]y_n\left(\frac{\xi}{n+\frac{1}{2}}\right) = 0, \qquad n = 0, 1, \cdots \quad (27.16)$$

which obviously correspond to the eigenvalue equation

$$\left(\frac{d}{d\xi}\xi\frac{d}{d\xi} + \lambda\xi + 1\right)z(\xi) = 0 \qquad (27.17)$$

with

$$\lambda_n = -\frac{1}{(2n+1)^2}, \qquad z_n = y_n\left(\frac{\xi}{n} + \frac{1}{2}\right). \qquad (27.18)$$

Here we have an eigenvalue problem in the infinite region $0 \leq x \leq \infty$, still with an infinite number of eigenvalues $n = 0, 1, \cdots, \infty$. The eigenvalues are not, however, condensing at infinity as in the former problems, but at a finite value, the point $\lambda = 0$. Here our general procedure for proving the completeness of the system breaks down, the true reason for this being that the system is, in fact, *incomplete*. Beside the discrete set of negative eigenvalues there are also positive eigenvalues λ, forming what should be called a continuous spectrum.

Assume λ to be positive. Then for very large values of ξ (27.17) is very similar to the equation

$$\left(\frac{d^2}{d\xi^2} + \lambda\right)z(\xi) = 0, \qquad (27.19)$$

which is said to be the asymptotic form of (27.17) as $\xi \to \infty$. The solutions of (27.19) are $\cos\sqrt{\lambda}\,\xi$ and $\sin\sqrt{\lambda}\,\varepsilon$, which are said to be asymptotic solutions of (27.17). These asymptotic solutions for positive λ have a permissible form, being finite as $\xi \to \infty$, in contrast to asymptotic solutions $\exp +(-\lambda)^{\frac{1}{2}}\xi$ when λ is negative. Only the other form $\exp -(-\lambda)^{\frac{1}{2}}\xi$ of the asymptotic solution is permissible and this will be the form only by the discrete eigenvalues (27.18).

Independent Proof of the Completeness Relation

It is possible in certain cases of particularly simple eigenvalue equations to establish a direct proof of the completeness relation in the form (21.12) and avoid the variational theorems. This means that usually we are able to find some simpler equation suitable for comparison with another given eigenvalue equation. For instance, given the equation

$$z'' - q(x)z + \lambda z = 0 \qquad (27.20)$$

with boundary conditions, say, $y = 0$ at the ends of a fundamental region, say, $0 \leq x \leq \pi$. We compare the equation with the simpler equation

$$y'' + \lambda y = 0 \qquad (27.21)$$

with the solutions

$$\lambda = \lambda_n = n^2, \qquad y = y_n = \left(\frac{2}{\pi}\right)^{1/2} \sin nx, \qquad n = 1, 2, \cdots, \infty, \quad (27.22)$$

assuming the system y_n to be complete. As to the function $q(x)$ we assume it to be finite in the above region, hence, that

$$\left| \int_0^\pi q(x) y_n z_m \, dx \right| \leq M, \qquad (27.23)$$

that is, an upper limit of the value of the integral exists which is independent of n and m.

We are now justified in expanding an eigenfunction z_m in terms of the eigenfunctions y_n,

$$z_m = \sum_n y_n c_{nm}, \qquad c_{nm} = \int_0^\pi y_n z_m \, dx. \qquad (27.24)$$

As a result of the asymptotic behavior of the eigenvalues $\lambda_m \approx m^2$ it is possible to show by means of the differential equations for y_n and z_m that, for sufficiently high m,

$$|c_{nm}| < \frac{M}{|m^2 - n^2|}. \qquad (27.25)$$

From the orthonormality relations

$$\int_0^\pi z_m z_{m'} \, dx = \delta_{mm'}, \qquad \int_0^\pi y_n y_{n'} \, dx = \delta_{nn'} \qquad (27.26)$$

and (27.24) it follows then that

$$\sum_{n=1}^\infty c_{nm} c_{nm'} = 0, \quad m \neq m', \qquad \sum_{n=1}^\infty c_{nm}^2 = 1. \qquad (27.27)$$

On the other hand, we may write

$$y_n = \sum_{m=1}^\infty c_{nm} z_m \qquad (27.28)$$

provided that

$$\int_0^\pi y_n^2 \, dx = \sum_{m=1}^\infty c_{nm}^2 = 1. \qquad (27.29)$$

This will be the completeness relation for the system z_m if valid for any n, since an arbitrary function $f(x)$ can be expanded in terms of y_n.

120 / The Mathematical Foundation of Physics

From (27.25) and (27.27) we now obtain

$$1 - \sum_{n=1}^{N} c_{nm}^2 = \sum_{n=N+1}^{\infty} c_{nm}^2 < \sum_{n=N+1}^{\infty} \frac{M^2}{(n-m)^2(n+m)^2} < \frac{C}{N^2}. \quad (27.30)$$

On the other hand, the positive integral

$$\int_0^\pi \left(y_n - \sum_{m=1}^{\infty} c_{nm} z_m \right)^2 dx = 1 - \sum_{m=1}^{N} c_{nm}^2 \geq 0 \quad (27.31)$$

must tend to zero as $N \to \infty$ in order that the system z_m be complete.

In place of the right-hand expression (27.31) we calculate the double sum

$$\sum_{n=1}^{N} \left(1 - \sum_{m=1}^{N} c_{nm}^2 \right) = \sum_{m=1}^{N} \left(1 - \sum_{n=1}^{N} c_{nm} \right) \leq \sum_{m=1}^{N} \frac{C}{N^2} = \frac{C}{N}. \quad (27.32)$$

This proves that any of the left-hand brackets and, hence, the right-hand side of (27.31) tend to zero as $N \to \infty$, which is the proof of the completeness relations for the system z_m.

CHAPTER 4

Theory of Functions

28. COMPLEX NUMBERS

Complex Numbers as Vectors

Complex numbers were invented by Gauss and Argand at the end of the 18th century. At the same time or even earlier the same idea was put forward by the Norwegian Caspar Wessel in a paper which had the misfortune of remaining unknown for a long time. He used the geometrical means of illustration usually called an Argand diagram or the Gauss plane. In this plane a complex number is given by a point (x, y) or its two coordinates x and y. We shall call it the complex plane and we shall write for the complex number

$$z = x + iy. \tag{28.1}$$

It is said to have a *real* and an *imaginary* component x and y, but it often will be useful to employ the notations Re and Im and write

$$x = \text{Re}\, z, \quad y = \text{Im}\, z. \tag{28.2}$$

The number $z = x$ is said to be a *real* number situated on the *real axis* and $z = iy$ is said to be purely *imaginary* and lying on the *imaginary axis* of the complex plane. The numbers

$$z = x + iy \quad \text{and} \quad z^* = x - iy \tag{28.3}$$

are said to be *complex conjugate* numbers.

The complex numbers, of course, could be treated as two-dimensional vectors in the notation **z** instead of z. For very particular reasons the use of this procedure is very limited. Whereas vectors can be transformed by rotations, there is no sense in transforming in the complex plane, say, the real into the imaginary axis. We shall therefore use the vector notation only for the purpose of identifying important theorems of the vector analysis with those of the imaginary theory of numbers.

Operations with Complex Numbers

The fundamental operations with complex number are the same as for ordinary or real numbers: addition and subtraction, multiplication and division. Either of them is easily generalized, subtraction and division, of course, being defined as inverse operations of addition and multiplication. The addition of complex numbers is understood to consist of the separate addition of real and imaginary components of the numbers. Hence, in the vector picture, the addition and subtraction of two numbers can be expressed by vector equations

$$z_1 + z_2 = \mathbf{z}_1 + \mathbf{z}_2, \qquad z_1 - z_2 = \mathbf{z}_1 - \mathbf{z}_2. \tag{28.4}$$

The procedure of multiplication is not as simple. The concept of the imaginary unit i originates with the desire to perform the operation $\sqrt{-1}$. If that is accomplished by an extension of the concept of numbers there are no more barriers for any root operation, whether square or cube or any other. We shall see that the passage from real to complex numbers is ideally accomplished by simply introducing the multiplication rule

$$i \cdot i = i^2 = -1, \tag{28.5}$$

together with the interpretation of any imaginary number yi as product of the imaginary unit and the real number y. Adopting any of the rules of commutation, association, and distribution on which the analysis of real number is based, we may write for the result of multiplication

$$z_1 z_2 = (x_1 + iy_1)(x_2 + iy_2) = x_1 x_2 - y_1 y_2 + i(x_1 y_2 + y_1 x_2). \tag{28.6}$$

In the vector notation this is, curiously enough,

$$z_1 z_2 = \mathbf{z}_1^* \mathbf{z}_2 + i(\mathbf{z}_1^* \times \mathbf{z}_2) = \mathbf{z}_2^* \mathbf{z}_1 + i(\mathbf{z}_2^* \times \mathbf{z}_1) \tag{28.7}$$

if by the above expressions we mean, respectively, the scalar product and the normal component of the vector product. Hence there is even some arbitrariness in vector illustration, which provides a less elegant notation than the direct use of $z_1 z_2$.

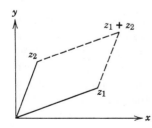

Fig. 4-1. Addition of complex numbers.

Complex Numbers in Polar Coordinates

For the purpose of multiplication and division there is, however, another and extremely useful illustration by means of polar coordinates r and z. Writing

$$x = r \cos \varphi \qquad y = r \sin \varphi \qquad (28.8)$$

$$r = \mod z = |z| = \sqrt{x^2 + y^2} \qquad (28.9)$$

is called the *modulus* or *absolute value* and

$$\varphi = \arg z = \arctan \frac{y}{x} \qquad (28.10)$$

the *argument* of the complex number z.

If expressed in polar coordinates, the result of multiplication is

$$z_1 z_2 = r_1 r_2 [\cos(\varphi_1 + \varphi_2) + i \sin(\varphi_1 + \varphi_2)], \qquad (28.11)$$

the modulus being multiplied and the arguments added. Therefore division by a number also means division by its modulus and the subtraction of its argument

$$\frac{z_1}{z_2} = \left(\frac{r_1}{r_2}\right)[\cos(\varphi_1 - \varphi_2) + i \sin(\varphi_1 - \varphi_2)]. \qquad (28.12)$$

Even with no knowledge of the detailed properties of the exponential function, this would suggest that we introduce such a function and write

$$z_1 = r_1 e^{i\varphi_1}, \qquad z_2 = r_2 e^{i\varphi_2},$$
$$z_1 z_2 = r_1 r_2 e^{i(\varphi_1 + \varphi_2)}, \qquad (28.13)$$
$$\frac{z_1}{z_2} = \left(\frac{r_1}{r_2}\right) e^{i(\varphi_1 - \varphi_2)}.$$

(See Figs. 4-2a and 4-2b.)

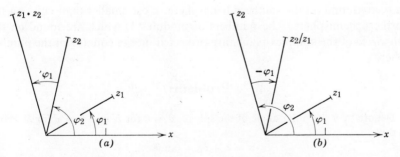

Fig. 4-2. (*a*) Multiplication of complex numbers. (*b*) Division by a complex number.

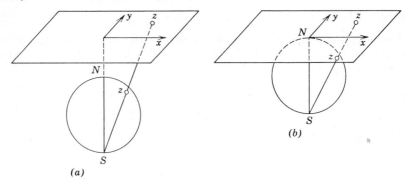

Fig. 4-3. (a) The number sphere. (b) A special number sphere. $\xi^2 + \eta^2 + (\xi + \frac{1}{2})^2 = \frac{1}{4}$.

The Numbers Zero and Infinity. The Unit Circle and the Number Sphere

Another illustration that may be of interest is the number sphere. This sphere may be discussed whether infinity, as represented, say, by a circle of an infinitely increasing radius, is a line or a point. In geometry it is a line. In the theory of complex numbers, if it is to be a single number, it is a point. (See Fig. 4-3a.)

This is most neatly illustrated by projecting the point of the complex plane on a sphere, say, beneath the plane and for the sake of convenience of diameter 1 and touching the complex plane at $z = 0$. This particular number sphere is found to the right in Fig. 4-3b.

For brevity we may talk of the north and south pole N and S of the sphere. The points of the complex plane are now projected on the sphere by lines drawn from the points to the south pole S. It is seen immediately that the process is unique. The zero is already found at N, and any point at sufficiently large distance from zero, hence with sufficiently large modulus, approaches the single point S on the sphere, which is now said to represent a single number termed *infinity* or $z = \infty$. Numbers of modulus larger than unity are found on the southern hemisphere, those smaller than unity on the northern hemisphere. The numbers of modulus 1, which are found on the *unit circle* of the complex plane, are projected on the equator of the number sphere.

Problems

1. Denote by ϑ the polar angle of the sphere, $\vartheta = 0$ for N and $\vartheta = \pi$ for S. Prove that

$$z = x + iy = \tan \vartheta \, e^{i\varphi},$$

hence that any complex of given modulus is on a given latitude circle on the sphere.

2. Denote by ξ, η, ζ the coordinates of a point on the sphere. Find the transformations from x, y to ξ, η, ζ. Prove that the equation for any circle in the plane transforms into the equation of a plane in the ξ, η, ζ space, hence, that the circle is projected on that circle on the sphere which is cut by the plane.

29. ANALYTIC FUNCTIONS

Definition of an Analytic Function

A complex function of the two components x and y of a complex variable $z = x + iy$ is said to be a function of z and is written

$$u(z) = F(x, y) + iG(x, y), \tag{29.1}$$

where F and G are real functions. We are mainly interested in a special class of function called *analytic functions* of the complex variable z, which are defined as follows. Let F and G be continuous functions of x and y. Then $u(z)$ is an analytic function if, and only if, $u(z)$ has a unique and continuous derivative

$$u'(z) = \lim_{\Delta z \to 0} \frac{u(z + \Delta z) - u(z)}{\Delta z} = \lim_{\Delta z \to 0} \frac{\Delta u}{\Delta z}. \tag{29.2}$$

In the complex plane Δz may tend to zero in different ways, along the real or the imaginary axis or along a line making an arbitrary angle with the real axis. Expanding Δu in terms of Δx and Δy up to the first power, we have

$$\Delta u = \left(\frac{\partial F}{\partial x} + i\frac{\partial G}{\partial y}\right)\Delta x + \left(\frac{1}{i}\frac{\partial F}{\partial y} + \frac{\partial G}{\partial y}\right)i\,\Delta y. \tag{29.3}$$

From this expression we obtain two different expressions for the derivative

$$\Delta u'(z) = \frac{\partial F}{\partial x} + i\frac{\partial G}{\partial x} = \frac{1}{i}\frac{\partial F}{\partial y} + \frac{\partial G}{\partial y}, \tag{29.4}$$

whose real and imaginary parts must be equal

$$\frac{\partial F}{\partial x} = \frac{\partial G}{\partial y}, \qquad \frac{\partial G}{\partial x} = -\frac{\partial F}{\partial y}. \tag{29.5}$$

These are known as the *Cauchy-Riemann equations*.

Eliminating one of the real functions, we find that either of them, and also the complex function, obey the Laplace equation

$$\nabla^2 F = 0, \quad \nabla^2 G = 0, \quad \nabla^2 u = 0, \quad \nabla^2 = \frac{\partial^2}{\partial x^2} + \frac{\partial^2}{\partial y^2}. \tag{29.6}$$

The simplest example of a continuous function that is not analytic is $z^* = x - iy$. In the direction of the real axis its derivative is 1; along the imaginary axis it is -1.

Construction of Analytic Functions

If $u(x)$ is any continuous complex function of the single variable x, it can be shown that, provided that $u(z)$ can be calculated from the definition of the function, it is an analytic function. If $u'(x)$ is the derivative of the function $u(x)$ in the ordinary sense, $u'(z)$ is simply that function with the argument x replaced by z.

The argument holds for any positive or negative power z^n of the variable z with an integral exponent n. This follows simply from the definition of multiplication and division. It holds also for any polynomial of z, irrespective of the degree of the highest power. Hence it must hold for any infinite power series in z, as long as the series is convergent.

Since by repeated use of multiplications any power of a variable z can be found, there is no difficulty in defining the square root, the cube root, and so on. Hence there is no limit for definitions of powers with rational exponents, and powers with irrational exponents can be defined by limiting processes using rational exponents. As we shall see, care must be taken, however, in defining what are called the singular points of an analytic function in which the function ceases to be analytic. For the powers of z with nonintegral exponents the singular points are $z = 0$ and $z = \infty$.

In polar coordinates any power of a complex variable

$$z = r(\cos \varphi + i \sin \varphi) \qquad (29.7)$$

can be expressed in a very simple way. We have

$$z^n = r^n(\cos n\varphi + i \sin n\varphi) \qquad (29.8)$$

which means that the nth power of the variable is obtained by taking the nth power of the modulus and by multiplying the argument by n. This applies, of course, only to real n. For nonintegral real n the argument of z^n increases by 2π if the variable point z encircles the origin. Hence, for the same point z in the complex plane, z^n will be a many-valued function except in the case of the integral n. Many-valued functions can be made single-valued by cuts between singular points of the function. The meaning of the cuts is that they shall not be traversed by the variable z.

We can obtain analytic functions by solving the Laplace equations $\nabla^2 F = 0$ and $\nabla^2 G = 0$ in different sets of coordinates in which the equation is separable. If we choose Cartesian coordinates and write

$$F(x, y) = X(x)Y(y), \qquad (29.9)$$

we must have

$$\frac{\nabla^2 F}{F} = \frac{X''}{X} + \frac{Y''}{Y} = 0. \qquad (29.10)$$

Writing
$$\frac{X''}{X} = -\frac{Y''}{Y} = k^2, \qquad (29.11)$$

where k^2 is a separation constant independent of either x and y, we have four possible independent solutions

$$e^{kx}\cos ky, \qquad e^{kx}\sin ky,$$
$$e^{-kx}\cos ky, \qquad e^{-kx}\sin ky,$$

which can be used for either $F(x, y)$ or $G(x, y)$. From the first-order Cauchy-Riemann equation it is found that the combination must be

$$\begin{aligned} F(x, y) &= e^{kx}\cos ky, & G(x, y) &= e^{kx}\sin ky, \\ F(x, y) &= e^{-kx}\cos ky, & G(x, y) &= e^{-kx}\sin ky, \end{aligned} \qquad (29.12)$$

Using the formalism,
$$\cos ky \pm i \sin ky = e^{\pm iky}, \qquad (29.13)$$

we obtain two different analytic functions which may be written

$$u(z) = e^{kz} \quad \text{and} \quad u(z) = e^{-kz}. \qquad (29.14)$$

Replacing in (29.11) k^2 by $-k^2$, we obtain by the same procedure two analytic functions

$$u(z) = e^{ikx-ky} = e^{ikz}, \qquad u(z) = e^{-ikx+ky} = e^{-ikz}. \qquad (29.15)$$

In polar coordinates the Laplace equation reads

$$\left(\frac{\partial^2}{\partial r^2} + \frac{1}{r}\frac{\partial}{\partial r} + \frac{1}{r^2}\frac{\partial^2}{\partial \varphi^2}\right)u = 0 \qquad (29.16)$$

with solutions

$$u(z) = r^n e^{in\varphi} = z^n, \qquad u(z) = r^{-n}e^{-in\varphi} = z^{-n}, \qquad (29.17)$$
$$u(z) = r^n e^{-in\varphi}, \qquad u(z) = r^{-n}e^{in\varphi}. \qquad (29.17a)$$

The expressions (29.17) are analytic functions, whereas (29.17a) are not. This is most easily proved by using the Cauchy-Riemann equations as expressed in polar coordinates

$$\frac{\partial F}{\partial r} = \frac{1}{r}\frac{\partial G}{\partial \varphi}, \qquad \frac{\partial G}{\partial r} = -\frac{1}{r}\frac{\partial F}{\partial \varphi}. \qquad (29.18)$$

The proof is easy and may be left for the reader. Considering the two first

128 / The Mathematical Foundation of Physics

functions of (29.17) and (29.17a) with

$$F = r^n \cos n\varphi, \quad G = r^n \sin n\varphi,$$
and
$$F = r^n \cos n\varphi, \quad G = -r^n \sin n\varphi,$$
(29.19)

it is easily seen from (29.18) that only in the first case is $u(z) = F + iG$ an analytic function.

The solutions of the Laplace equation in polar coordinates provide analytic functions that are powers of the complex variable z.

There is no difficulty in obtaining analytic functions in the form of powers of z with complex exponents n. For instance,

$$z^{in} = (re^{i\varphi})^{in} = (e^{\log r + i\varphi})^{in} = e^{-n\varphi + in \log r} \quad (29.20)$$

is an analytic function of z. From

$$F = e^{-n\varphi} \cos(n \log r), \quad G = e^{-n\varphi} \sin(n \log r) \quad (29.21)$$

this is easily controlled by (29.18).

Complex Integrals and Cauchy's Theorem

When defining complex integrals, we shall have to ensure that they present natural generalizations of the integrals of a real variable. Such integrals may then be considered complex integrals taken along some region of the real axis in the complex plane.

First of all, we must define what is called the path of integration. A curve in the complex plane can be given by the equations $x = x(t), y = y(t), t$ being some parameter. The differential dz on the curve then is

$$dz = dx + i\, dy = [x'(t) + iy'(t)]\, dt. \quad (29.22)$$

Usually only the end points a and b of the curve, corresponding to some values $t = t_0$ and $t = t_1$ of the independent variable t, are given by the integration, the path of integration being thought of as a curve that could be defined as stated above. For analytic functions, however, small deviations of the path usually do not matter, and for this reason the path is usually very inaccurately specified.

In general we therefore write for a complex integral

$$\int_a^b u(z)\, dz = \int_a^b (F + iG)(dx + i\, dy) = \int_a^b \{F\, dx - G\, dy + i(G\, dx + F\, dy)\}. \quad (29.23)$$

This integral which contains a real and an imaginary term is not equivalent to any integral in vector analysis, although either of them may be so. Introducing the infinitesimal vectors $d\mathbf{s}$ and $d\mathbf{n}$ corresponding to the equally large

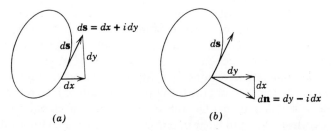

Fig. 4-4. (a) Definition of the curve element $d\mathbf{s}$. (b) Definition of the normal element $d\mathbf{n}$.

line and normal elements $dz = dx + i\,dy$ and $dy - i\,dx$, we may write

$$\int_a^b u(z)\,dz = \int_a^b [\mathbf{u}^* \cdot d\mathbf{z} + i(\mathbf{u}^* \times d\mathbf{z})] = \int_a^b (\mathbf{u}^* \cdot d\mathbf{s} + i\mathbf{u}^* \cdot d\mathbf{n}). \quad (29.24)$$

For a closed curve it follows from the Stokes and the two-dimensional Gauss theorem that

$$\oint u(z)\,dz = \oint (\operatorname{curl} \mathbf{u}^* + i \operatorname{div} \mathbf{u}^*)\,df$$

$$= \int \left[-\frac{\partial G}{\partial x} - \frac{\partial F}{\partial y} + i\left(\frac{\partial F}{\partial x} - \frac{\partial G}{\partial y}\right) \right] df, \quad (29.25)$$

df being a surface element of the complex plane and the last integral being taken over the whole region inside the closed curve.

If now at any point of that region the function $u(z)$ is analytic, it follows from the Cauchy-Riemann equation that the integrand vanishes identically and the integral is zero. Hence, for functions that are analytic in the whole region,

$$\oint u(z)\,dz = 0, \quad (29.26)$$

and the Cauchy-Riemann equation may be written in the vector notation as

$$\operatorname{div} \mathbf{u}^* = 0, \quad \operatorname{curl} \mathbf{u}^* = 0. \quad (29.27)$$

Equation (29.26) is often referred to as the Cauchy theorem. (See Figs. 4-4a and 4-4b.)

Deformation of Path of Integration

From the Cauchy theorem it follows that the complex integral depends only on the end points a and b of the path of integration and not on the path itself, provided that the integrand is considered to be in an analytic region. The path of integration can also be continuously displaced. The result that a

130 / The Mathematical Foundation of Physics

complex integral has the same value whether taken along fully drawn or dotted curves of a figure is obtained by applying the Cauchy theorem to the region between the two curves on the supposition that in this region the integrand is an analytic function.

Usually the nonanalytic regions of a function are points that are called *singular points*. By the process of deformation of a path of integration it must not surpass a singular point. On the other hand, if a closed path of integration—corresponding to what in particular is called a contour integral—encircles a singular point, the path may be deformed so as to follow either some fairly distant curve as the curve C of Fig. 4-5 or some curve very near to the point as the small circle c of the same figure. The value of the integral will be the same in either case.

The Cauchy Integral Theorem

Consider an analytic function $u(z)$ and the contour integral

$$\oint \frac{u(z)}{z-a} dz$$

in which the integrand is made artificially nonanalytic in the point $z = a$ on dividing $u(z)$ by $z - a$. The path of integration of the contour integral must be taken in some specified direction which, if not particularly mentioned is the positive or counterclockwise direction. It can easily be shown that the value of the integral is given by the equation

$$u(a) = \frac{1}{2\pi i} \oint \frac{u(z)}{z-a} dz. \tag{29.28}$$

To obtain this result we deform the path of integration as shown in Fig. 4-5 from the given arbitrary curve C to a very small circle with its center at $z = a$. For an infinitely small circle the function $u(z)$ may be treated as a constant whose value is $u(a)$.

On the other hand, writing $z - a = \rho e^{i\varphi}$, with a constant ρ that may be taken arbitrarily small, we have $dz = \rho e^{i\varphi} i\, d\varphi$. Hence the value of the integral

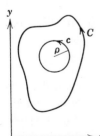

Fig. 4-5. The Cauchy integral theorem.

is given by

$$u(a) \cdot \frac{1}{2\pi} \int_0^{2\pi} d\varphi = u(a). \tag{29.29}$$

The theorem (29.28) is also due to Cauchy. To distinguish it from the former theorem we shall call it the Cauchy *integral* theorem. It states the fairly unusual result that the value of an analytic function at any point of a closed region is uniquely determined by its values on the boundary of the region.

It is very interesting to learn that the same is true for any of the derivatives of the function. Write (29.28) as

$$u(z) = \frac{1}{2\pi i} \oint \frac{u(\zeta)}{\zeta - z} d\zeta \tag{29.30}$$

with the auxiliary variable ζ, so that $u(z)$ appears as a function of the variable z within the inside region of the closed curve. Then, for the difference of the function at two neighboring points z and $z + \Delta z$, we have without any approximations

$$\frac{\Delta u}{\Delta z} = \frac{1}{2\pi i} \oint \frac{u(\zeta) \, d\zeta}{(\zeta - z)(\zeta - z - \Delta z)}. \tag{29.31}$$

When passing to the limit $\Delta z \to 0$, this yields

$$\frac{du}{dz} = \frac{1}{2\pi i} \oint \frac{u(\zeta) \, d\zeta}{(\zeta - z)^2}, \tag{29.32}$$

a result which could as well have been obtained by direct derivation of the integrand in (29.30) with respect to the variable z considered as a parameter.

Obviously we may therefore continue that process of derivation and obtain higher derivative of $u(z)$ in the form

$$u^{(n)}(z) = \frac{n!}{2\pi i} \oint \frac{u(\zeta) \, d\xi}{(\zeta - z)^{n+1}}. \tag{29.33}$$

This means that an analytic function not only has a unique first-order derivative, but the whole sequence of higher derivatives do exist and are unique.

There is another and very important theorem in the theory of analytic functions which is called the *Liouville theorem* and is easily obtained from (29.30).

Assume the function $u(z)$ to have no singular points. Then the Liouville theorem states that this function is a constant. The point is that in this case the path of integration of the contour integral (29.30) may be deformed into a very distant curve. If this curve is, say, a circle of radius R, its circumference is $2\pi R$, that is, of the order of magnitude R.

132 / The Mathematical Foundation of Physics

On the other hand, if the function $u(z)$ has no singular point and, hence, is continuous and bounded everywhere, we see directly from (29.32) that its first derivative must be zero. Since $u(\zeta)$ is bounded and the denominator becomes of the order of magnitude R^2, the integral itself is of the order of magnitude R^{-1}, that is, its limiting value is zero. But the integral is not changed by the deviation of its path of integration; hence it is identically zero. This implies that $u(z)$ itself is a constant.

The result is different if $u(z)$ is allowed to have a singularity at infinity. Such functions having no singularity except at $z = \infty$ are called integral functions. The simplest example of an integral function is an integral power of the variable z,

$$u(z) = z^n. \tag{29.34}$$

Another example is the infinite series

$$u(z) = e^z = 1 + z + \frac{z^2}{1 \cdot 2} + \cdots + \frac{z^n}{n!} + \cdots. \tag{29.35}$$

It is easily understood that any infinite power series with integral exponents, which converges for any value of the variable z, is an integral function.

The reason why the Liouville theorem is restricted to functions with $z = \infty$ as an ordinary point is that otherwise the supposition of $u(z)$ being finite everywhere does not hold and, hence, the integral (29.32) need not be zero. This is most easily seen from the example (29.34).

Connection with Vector Analysis

Despite the ideal simplicity of the above results, we shall not miss the opportunity of looking for connecting lines that lead to the fundamental results already known from vector analysis. This is because we may profit by being able to understand the subject on the basis of methods that are familiar to us from other branches of theoretical physics. We shall find at the end that nothing is gained for the technique of calculation. On the contrary, it is vector analysis that may profit from the methods of complex number theory, and there is even some advantage in knowing well the relation between the theories.

We shall begin by studying a Laplacian vector field **v** in a closed region, the x, y-plane, that is, a field obeying the equations. On the boundary curve we shall use the notations $d\mathbf{s}$ for an ordinary vector curve element and $d\mathbf{n}$ for the corresponding normal element as already shown in Figs. 4-4a and 4-4b. The field shall have to obey the equations

$$\text{div } \mathbf{v} = 0, \quad \text{curl } \mathbf{v} = 0. \tag{29.36}$$

Theory of Functions | 133

As already stated, this vector cannot be represented by the analytic function $u(z)$ but rather with $u^*(z)$. Hence we shall write (29.30)

$$v_x - iv_y = \frac{1}{2\pi i} \oint \frac{\mathbf{v}\,ds + i\mathbf{v}\,dn}{\xi - x - i(\eta - y)}, \qquad (29.37)$$

ξ and η being the coordinates of the curve element $d\mathbf{s}$ or $d\mathbf{n}$. Since the operation of division with a vector is not defined in vector analysis, we shall have to make the denominator real by multiplying with $(\zeta - z)^*$. Introducing for the denominator the abbreviation r, we may write

$$v_x - iv_y = \frac{1}{2\pi} \oint \frac{r^*(\mathbf{v}\,dn - i\mathbf{v}\,ds)}{|r|^2}$$

$$= \frac{1}{2\pi} \oint \frac{(v_x - iv_y)(\mathbf{r}\,dn - i\mathbf{r}\,ds)}{r^2}. \qquad (29.38)$$

In this equation we have a mixture of notations, but it serves the purpose of leading the way to the corresponding vector equation. The products $\mathbf{v}\,d\mathbf{n}, \cdots,$ $\mathbf{r}\,d\mathbf{s}$, as well as r^2, are scalar products in the meaning of vector theory. The integral itself is still a complex number. The identity of the two right-hand expressions can easily be controlled and is simply a consequence of the commutation law for complex multiplication as applied to the complex numbers $v_x - iv_y$ and r^*.

By performing the multiplication in the nominator of the last expression (29.38), separate expressions for v_x and v_y are obtained and these expressions if united into a vector $\mathbf{v}$ in the meaning of vector theory are found to be

$$\mathbf{v} = \frac{1}{2\pi} \oint \frac{\mathbf{v}(\mathbf{r}\,d\mathbf{n}) + (\mathbf{k} \times \mathbf{v})(\mathbf{r}\,d\mathbf{s})}{r^2}, \qquad (29.39)$$

$\mathbf{k}$ being the unit vector normal to the x, y-plane.

We now start from the opposite side, namely, that of the vector theory, and in a slightly more general way. Let the field be a more general one with

$$\text{div } \mathbf{v} = \rho, \quad \text{curl } \mathbf{v} = \mathbf{j}. \qquad (29.40)$$

It appears that it will be advantageous to make use of the Laplacian vector $\mathbf{r}/r^2$ with the properties

$$\text{div } \frac{\mathbf{r}}{r^2} = 0 \quad \text{curl } \frac{\mathbf{r}}{r^2} = 0. \qquad (29.41)$$

This applies to any vector $\mathbf{r}$ from a fixed to a variable point. We shall consider here ζ, η the variable point and x, y the fixed point, the operators div and curl referring to ξ and η. For brevity we shall write r_x for $\xi - x$, and so forth.

By very simple calculations we find that

$$\operatorname{div}\frac{r_x\mathbf{v}}{r^2} - \operatorname{curl}\frac{r_y\mathbf{v}}{r^2} = \frac{r_x\rho}{r^2} - \frac{r_y\mathbf{j}}{r^2},$$
$$\operatorname{div}\frac{r_y\mathbf{v}}{r^2} + \operatorname{curl}\frac{r_x\mathbf{v}}{r^2} = \frac{r_y\rho}{r^2} + \frac{r_x\mathbf{j}}{r^2}. \tag{29.42}$$

This simple result is obtained only by the special combination of div and curl on the left-hand side of the equations.

Applying the Gauss and Stokes theorems to the left-hand side of either of the equations, and collecting the result in a single vector equation, we obtain for a bounded region with surface element df

$$\mathbf{v} + \frac{1}{2\pi}\int\left(\frac{\mathbf{r}}{r^2}\rho + \frac{\mathbf{k}\times\mathbf{r}}{r^2}j\right)df$$
$$= \frac{1}{2\pi}\oint\left(\frac{\mathbf{r}}{r^2}(\mathbf{v}\,d\mathbf{n}) + \frac{\mathbf{k}\times\mathbf{r}}{r^2}(\mathbf{v}\,d\mathbf{s})\right)$$
$$= \frac{1}{2\pi}\oint\left(\frac{\mathbf{v}}{r^2}(\mathbf{r}\,d\mathbf{n}) + \frac{\mathbf{k}\times\mathbf{v}}{r^2}(\mathbf{r}\,d\mathbf{s})\right). \tag{29.43}$$

Here the left-hand term $\mathbf{v}$ originates from a contour integral around the point $\mathbf{r} = 0$ which must be avoided in the surface integrals in the Gauss and Stokes formulas. For a very small circle around this point, $\mathbf{r}\,d\mathbf{n}/r^2 = d\varphi$, φ being the polar angle, and hence

$$\lim_{r\to 0}\frac{1}{2\pi}\oint\frac{\mathbf{v}}{r^2}(\mathbf{r}\,d\mathbf{n}) = \mathbf{v}. \tag{29.44}$$

The second term of this contour integral is zero.

In the case of a Laplacian vector with $\rho = 0, j = 0$, (29.39) is derived from (29.43), thus establishing the connection between the vector and complex number theory. Despite the formal elegance of vector theory it is evident that it cannot compete at all with the theory of complex numbers and analytic functions. The reason for this is that division is feasible with complex numbers and it is unknown in vector theory.

30. POWER SERIES FOR ANALYTIC FUNCTIONS

Expansion at a Regular Point

An analytic function $u(z)$ with all its derivatives is uniquely determined at any point in a given region by the values of the function on the boundary curve; conversely, these values are uniquely determined by the values of the function and its derivatives at a single point of the region.

Let a be such a point, and, since the function $u(z)$ is analytical, it is what is called a *regular point* of the function. By repeated use of the auxiliary equation

$$\frac{1}{\zeta - z} = \frac{1}{\zeta - a - (z - a)} = \frac{1}{\zeta - a} + \frac{z - a}{(\zeta - z)(\zeta - a)},$$

the factor $1/(\zeta - z)$ in the Cauchy integral theorem can be replaced by

$$\frac{1}{\zeta - z} = \frac{1}{\zeta - a} + \frac{z - a}{(\zeta - a)^2} + \cdots + \frac{(z - a)^n}{(\zeta - a)^{n+1}} + \frac{(z - a)^{n+1}}{(\zeta - a)^{n+1}(\zeta - z)}. \tag{30.1}$$

Using (29.26), (29.28), (29.32), and (29.33) with z replaced by a, we obtain

$$u(z) = u(a) + u'(a)(z - a) + \cdots + \frac{u^{(n)}(a)}{n!}(z - a)^n + R_{n+1}, \tag{30.2}$$

with

$$R_{n+1} = \frac{1}{2\pi i} \oint \frac{u(\zeta)(z - a)^{n+1}}{(\zeta - z)(\zeta - a)^{n+1}} d\zeta. \tag{30.3}$$

In the expression (30.2) the function $u(z)$ is approximated by a polynomial up to the nth power in $z - a$. The accuracy of the approximate expression increases or decreases according to whether R_{n+1} decreases or increases. If R_{n+1} definitely decreases, tending to zero as $n \to \infty$, expression (30.2) can be replaced by an infinite series which is an exact expression for the analytic function $u(z)$.

If $u(z)$ is an analytic function, hence bounded on the contour, $|u(z)| < M$, M being a given positive number, the condition for $R_{n+1} \to 0$ and $n \to \infty$ is that $|z - a| < |\xi - a|$. If therefore z is inside a circle with its center at a and tangential to the contour curve at its nearest point, the condition is fulfilled. On the other hand, the boundary curve can be displaced outward in any direction until it is stopped by some singularity of the function $u(z)$, that is, some singular point of the function in which it ceases to be an analytic function. Therefore, within a circle with its center at a and not touching the nearest singular point of $u(z)$, the infinite series

$$u(z) = u(a) + u'(a)(z - a) + u''(a)\frac{(z - a)^2}{1 \cdot 2} + \cdots \tag{30.4}$$

represents correctly the analytic function $u(z)$. It has a definite sum and is said to be convergent within its *circle of convergence*, which is the circle through the nearest singular point of $u(z)$. Its radius is called the radius of convergence for the power series (30.4). (See Fig. 4-6.)

136 / The Mathematical Foundation of Physics

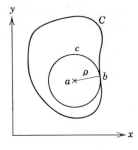

Fig. 4-6. Illustration of the circle of convergence of an analytic function with its radius of convergence ρ as the distance to the nearest singular point b.

Some Examples of Power Series

Consider first the nth power $u(z) = z^n$ of the variable z with a positive integral n. From

$$\Delta u = (z + \Delta z)^n - z^n = \Delta z[(z + \Delta z)^{n-1} + (z + \Delta z)^{n-2}z + \cdots + z^{n-1}]$$

(30.5)

it follows, making $z \to 0$,

$$\frac{d(z^n)}{dz} = nz^{n-1}.$$

(30.6)

The higher derivatives are easily obtained by the same formula and are formally the same as those of the power of a real variable. According to (30.4) the power series of the function in a point a is

$$z^n = a^n + \binom{n}{1}a^{n-1}(z-a) + \binom{n}{2}a^{n-2}(z-a)^2 + \cdots + (z-a)^n, \quad (30.7)$$

where $\binom{n}{k} = n!/[k!\,(n-k)!]$ are the Newton binomial coefficients. Equation (30.7), of course, is the binomial formula extended to complex variables. With a finite number of terms the series converges for any value of z. This means that the function $u(z) = z^n$ has no singularity except at $z = \infty$.

The function $u(z) = z^{-n}$ with integral n can be defined by the equation $u(z)z^n = 1$ from which, be repeated differentiation of the equation, any derivative of the function can be found. Defining $u(z) = z^{1/2}$ by the equation $u^2 = z$, and so on, it is possible to find the derivatives of any power of z as $u(z) = z^n$ with arbitrary n, and the result can always be expressed by formally the same equation as (30.6).

The same result can be obtained as easily in polar coordinates, writing

$$u(z) = z^n = r^n(\cos n\varphi + i \sin n\varphi) \quad (30.8)$$

or, for brevity,

$$u(z) = z^n = r^n e^{in\varphi}, \quad z = re^{i\varphi}. \quad (30.8a)$$

To obtain the derivative we may differentiate only with respect to r, putting $dz = e^{i\varphi} dr$. The result is

$$\frac{du}{dz} = nr^{n-1}e^{i(n-1)\varphi} = nz^{n-1}. \tag{30.9}$$

Differentiation with respect to φ yields the same value. Hence we see that for powers of z with arbitrary exponents n—even if the exponents are complex numbers—any higher derivative can be obtained in the ordinary way of differentiation as known from real variables.

From the above result the power series for an arbitrary power $u = z^n$ at any point a of the complex plane can be deduced,

$$u(z) = z^n = [a + (z - a)]^n = \sum_{m=0}^{\infty} \binom{n}{m} a^{n-m}(z - p)^m, \tag{30.10}$$

where

$$\binom{n}{m} = \frac{n(n-1)\cdots(n-m+1)}{1 \cdot 2 \cdots m} = \frac{n!}{m!(n-m)!} \tag{30.10a}$$

are generalized Newton binomial coefficients. The meaning of $n!$ and $(n-m)!$ for nonintegral n, of course, needs an explanation. They must be defined by the gamma function; however, their ratio is immediately seen from the expression (30.10a).

As to the radius of convergence of the series it must be $\rho = |a|$ since, apart from $z = \infty$, $z = 0$ is the only and, hence, the nearest singular point of z^n. The series therefore converges if $|z - a| < |a|$. In the case of the positive nonintegral n, one should not select z^n as an analytic function merely on the basis of its zero value at $z = 0$. As we shall see, it is not a single-valued function at this point, and its nonanalytic character is easily revealed by its higher derivatives becoming infinitely large at $z = 0$.

By a slight modification of (30.10) we arrive at the power series of the expression $(1 + z/a)^n$. Replacing $z - a$ by z and dividing by a^n, we have

$$u(z) = \left(1 + \frac{z}{a}\right)^n = \sum_{n=0}^{\infty} \binom{n}{m}\left(\frac{z}{a}\right)^m. \tag{30.11}$$

The radius of convergence of this series still is $\rho = |a|$, the singular point being at $z = -a$. If a is replaced by n, and we make $n \to \infty$, $u(z) \to e^z$ according to the definition of the exponential function of a real variable. The limiting values of the expansion coefficients are

$$\lim_{n \to \infty} \frac{n(n-1)\cdots(n-m+1)}{m! \, n^m} = \frac{1}{m!}. \tag{30.12}$$

Hence

$$\lim_{n \to \infty} \left(\frac{1+z}{n}\right)^n = \sum_{m=0}^{\infty} \frac{z^m}{m!} = u(z). \tag{30.13}$$

Defining $u(z) = e^z$ by the differential equation $u' = u$ together with $u(0) = 1$, we obtain $u^{(n)}(0) = 1$ for any of the higher derivatives and according to our general expression (30.4) the expansion of e^z in power series of z must be given by (30.13). The radius of convergence of the left-hand expression (30.13) is $\rho = n$ for real positive n. As $n \to \infty$, $\rho \to \infty$ and the only singular point of e^z is $z = \infty$. However, by the limiting process the character of this singularity has been changed. For reasons that will be explained in more detail later, the point $z = \infty$ is called an irregular singular point.

The function $u(z) = e^{iz}$ quite naturally must be defined by the differential equation $u' = iu$. Its power series is obtained from (30.13) by replacing z by iz. For real z it can be separated into a real and an imaginary power series. Writing in general

$$e^{iz} = \cos z + i \sin z, \qquad (30.14)$$

we may define the functions $\cos z$ and $\sin z$ by the power series

$$\cos z = \sum_{m=0}^{\infty} \frac{(-z^2)^m}{(2m)!}, \qquad \sin z = \sum_{m=0}^{\infty} \frac{z(-z^2)^m}{(2m+1)!}. \qquad (30.15)$$

Both series, of course, are convergent throughout the complex plane. The functions e^z, e^{-z}, e^{iz}, e^{-iz}, and $\cos z$ are all *integral functions*.

$$\begin{aligned} u'' + u = 0, & \quad u(0) = 1, \quad u'(0) = 0 \text{ for } u(z) = \cos z, \\ u'' + u = 0, & \quad u(0) = 0, \quad u'(0) = 1 \text{ for } u(z) = \sin z \end{aligned} \qquad (30.16)$$

which are valid for the functions cosine and sine of a real variable also lead easily to the same expansions (30.15). This, of course, gives a quite definite meaning to our earlier abbreviation of $\cos \varphi + i \sin \varphi$ into $e^{i\varphi}$.

We shall now turn to logarithmic functions which are widely different from the above integral functions. We have as a rule two singular points, $z = 0$ and $z = \infty$, for the elementary function $\log z$; they resemble the single powers z^n of a variable. However, the character of the singular points is different, the logarithmic functions becoming infinite at any singular point, although to a fainter degree than any power z^n and z^{-n} with positive real n in the $z = \infty$ and $z = 0$.

The function $\log z$ may be defined in a number of ways. The essential point is that we want it to be the inverse of the exponential function

$$z = e^u, \qquad u = \log z. \qquad (30.17)$$

Since we are here entering the field of somewhat more intricate problems, we shall mention some general theorems for analytic functions which should be kept in mind. For instance, if $u(z)$ and $v(z)$ are analytic functions $u(z)v(z)$ is an analytic function. The same is true for $u(z)/v(z)$, provided that $v(z) \neq 0$.

The definition (30.17) introduces another class of analytic function as generally defined by the implicit equations

$$u(z) = \zeta, \quad v(\zeta) = z,$$

or   (30.18)

$$v(u(z)) = z.$$

If $v(\zeta)$ is an analytic function of ζ, we have

$$\Delta u = \Delta \zeta, \quad \Delta z = v'(\zeta) \Delta \zeta.$$

Hence

$$u'(z) = \lim_{\Delta z \to 0} \frac{\Delta v}{\Delta z} = \frac{1}{v'(\zeta)} \qquad (30.19)$$

is a unique expression for $u'(z)$, and $u(z)$ is an analytic function.

Applying this formula to the definition (30.17), we obtain

$$u(z) = \log z, \quad u'(z) = \frac{1}{z}, \qquad (30.20)$$

from which any higher derivative of the logarithm can be obtained.

Since $z = 0$ is a singular point of $\log z$, this does not lead to any power series in z. We shall therefore as an example displace the singular point of the function to, say, $z = -1$, and study the function $u(z) = \log(1+z)$.

We may choose to evaluate the higher derivatives of $\log(1+z)$ from repeated differentiation of $u' = 1/(1+z)$, or, we may use the power series for the derivative,

$$u'(z) = 1 - z + z^2 - z^3 + \cdots \qquad (30.21)$$

and determine by integration the power series

$$u(z) = \log(1+z) = z - \frac{z^2}{2} + \frac{z^3}{3} - \frac{z^4}{4} + \cdots. \qquad (30.22)$$

From the functional equation $\log z_1 z_2 = \log z_1 + z_2$, which is a consequence of the definition (30.17), we may construct power series of more complex analytic functions as, for instance,

$$u(z) = \frac{1}{2} \log \frac{1+z}{1-z} = z + \frac{z^3}{3} + \frac{z^5}{5} + \cdots. \qquad (30.23)$$

Whereas the series (30.22) apart from $z = \infty$ has only one singular point in $z = -1$, (30.23) has the singular points $z = \pm 1$. The radius of convergence in either case is $\rho = 1$.

It may be of interest to consider a similar function with singularities in $z = \pm i$,

$$u(z) = \frac{1}{2i} \log \frac{1+iz}{1-iz} = z - \frac{z^3}{3} + \frac{z^5}{5} - \cdots, \qquad (30.24)$$

obtained by replacing z by iz in (30.23) and dividing by i. Its derivative is, whether applied to the left- or the right-hand side,

$$u'(z) = \frac{1}{1+z^2} = 1 - z^2 + z^4 - \cdots ; \qquad (30.25)$$

the function therefore must be identified with the function arc tan z. Hence we write

$$\text{arc tan } z = \frac{1}{2i} \log \frac{1+iz}{1-iz}. \qquad (30.26)$$

31. CONVERGENCE OF INFINITE SERIES

Absolute and Conditional Convergence

It is possible, as we have seen, to tell whether or not a power series in a complex variable is convergent if the series is an expansion of a known function with given singularities. In general, however, we shall have to judge directly from the known terms of a series whether it converges or not. It will therefore be useful to study first the general rules of convergence for any series.

An infinite series of complex terms

$$z_1 + z_2 + \cdots + z_n + \cdots = \sum_{n=1}^{\infty}(x_n + iy_n) \qquad (31.1)$$

is known to be convergent if for any positive integer p

$$|z_{n+1} + \cdots + z_{n+p}| < \varepsilon, \qquad \text{when } n > N(\varepsilon). \qquad (31.2)$$

In this inequality ε is a positive number that may be chosen arbitrarily small. Then $N(\varepsilon)$ will be a number depending only on ε and increasing with decreasing ε. The statement is simply that the series has a definite limit and that the error in breaking off the series at n can be made arbitrarily small for sufficiently large n. The statement, of course, also necessitates that either of the real or imaginary series that can be obtained from (31.1) be convergent.

According to the geometrical addition of complex numbers

$$|z_{n+1} + \cdots + z_{n+p}| \leq |z_{n+1}| + \cdots + |z_{n+p}|. \qquad (31.3)$$

Hence, if

$$|z_{n+1}| + \cdots + |z_{n+p}| \leq \varepsilon, \qquad \text{when } n > N(\varepsilon), \qquad (31.4)$$

the series $\sum_n z_n$ is convergent. The inequality (31.4) states that the modulus $\sum_n |z_n|$ is convergent. If that is true, the series $\sum_n z_n$ is said to be *absolutely convergent*. If, on the other hand, the modulus series does not converge, the series $\sum_n z_n$ may still converge. It is then said to be only conditionally convergent.

Interchange of Terms in Conditionally Convergent Series

Conditionally convergent series are convergent only under definite restrictions. Whereas in an absolutely convergent series any of its terms may be interchanged without restriction to their numbers, in a conditionally convergent series a systematic interchange of terms is not allowed. To see the reason for this it suffices to consider a real series, say $\sum_n x_n$ in (31.1). If the series is absolutely convergent, we may sum up separately the positive and negative part of the series, since either of the sums must exist. Hence the interchange of terms can no longer alter these partial sums. In the case of a conditionally convergent series, either of the two sums must be divergent, the convergence of the original series being effected by intermixing positive and negative terms. Hence, if we interchange positive and negative terms in a systematic manner by an infinite process, the sum of the series must change.

Consider, for instance, the series

$$S = 1 - \frac{1}{2} + \frac{1}{3} - \frac{1}{4} + \cdots = \frac{1}{1 \cdot 2} + \frac{1}{3 \cdot 4} + \cdots. \qquad (31.5)$$

From this series we may construct another series with, say, two positive terms between successive negative terms

$$S' = 1 + \tfrac{1}{3} - \tfrac{1}{2} + \tfrac{1}{5} + \tfrac{1}{7} - \tfrac{1}{4} + \cdots. \qquad (31.6)$$

This new series can easily be summed up by dividing negative terms into two equal parts and interchanging one of them with the foregoing term, which obviously does not change the sum,

$$\begin{aligned}S' &= 1 - \tfrac{1}{4} + \tfrac{1}{3} - \tfrac{1}{4} + \tfrac{1}{5} - \tfrac{1}{8} + \tfrac{1}{7} - \tfrac{1}{8} + \cdots \\ &= S + \tfrac{1}{2} - \tfrac{1}{4} + \tfrac{1}{6} - \tfrac{1}{8} + \cdots = S + \tfrac{1}{2}S = \tfrac{3}{2}S. \end{aligned} \qquad (31.7)$$

If, on the other hand, we allow for two negative terms between successive positive terms, the result is

$$\begin{aligned} S'' &= 1 - \tfrac{1}{2} - \tfrac{1}{4} + \tfrac{1}{3} - \tfrac{1}{6} - \tfrac{1}{8} + \cdots \\ &= \tfrac{1}{2} - \tfrac{1}{4} + \tfrac{1}{6} - \tfrac{1}{8} + \cdots = \tfrac{1}{2}S. \end{aligned} \qquad (31.8)$$

Comparison Series

If from some definite term the following terms of a series have smaller moduli than a corresponding series with only positive terms which is known to be convergent, then the series is absolutely convergent. A series used for this purpose is termed a comparison series. The simplest one is the geometric series

$$a(1 + k + k^2 + \cdots + k^n + \cdots) = \frac{a}{1 - k} \quad \text{for} \quad k < 1. \qquad (31.9)$$

This leads to the criterion of absolute convergence

$$\lim_{n \to \infty} \sqrt[n]{|z_n|} = k, \quad k < 1, \tag{31.10}$$

for any series. It is not necessary, however, that the values $\sqrt[n]{|z_n|}$ condense around a single value k as $n \to \infty$. There may be several k-values. Hence, the criterion should rather be written

$$\lim \sqrt[n]{|z_n|} < k, \quad k < 1,$$

as $n \to \infty$.

In the case of a single k-value we may also write

$$\lim_{n \to \infty} \left| \frac{z_{n+1}}{z_n} \right| = k, \quad k < 1, \tag{31.11}$$

as the condition for absolute convergence.

In all the above cases of $k < 1$, the convergence must be characterized as fairly good. The most important and highly interesting case therefore is $k = 1$. To find whether in this case a series is still convergent we shall choose for comparison the very useful series

$$\int_0^a \frac{dx}{(1-x)^{1-\varepsilon}} = [1 - (1-a)^\varepsilon] = a + \frac{1-\varepsilon}{1 \cdot 2} a^2 + \frac{(1-\varepsilon)(2-\varepsilon)}{1 \cdot 2 \cdot 3} a^3 + \cdots, \tag{31.12}$$

putting $a = 1$. The series

$$1 + \frac{1-\varepsilon}{1 \cdot 2} + \frac{(1-\varepsilon)(2-\varepsilon)}{1 \cdot 2 \cdot 3} + \cdots \tag{31.13}$$

obviously is convergent and has the sum $1/\varepsilon$ if $\varepsilon > 0$. If $\varepsilon \leq 0$, the series is divergent.

The ratio of two consecutive terms is

$$\frac{n-1-\varepsilon}{n} = 1 - \frac{1+\varepsilon}{n}. \tag{31.14}$$

Hence, we find that if for any series

$$\lim_{n \to \infty} \left| \frac{z_{n+1}}{z_n} \right| = 1 - \frac{1+\varepsilon}{n} + O\left(\frac{1}{n^2}\right), \tag{31.15}$$

or, more stringently,

$$\lim_{n \to \infty} n \left(1 - \left| \frac{z_{n+1}}{z_n} \right| \right) = 1 + \varepsilon, \quad \varepsilon > 0, \tag{31.16}$$

then the series is absolutely convergent. In the limiting case of $\varepsilon = 0$ the series may still be conditionally convergent.

We shall use this theorem for deducing the convergence criterion of the hypergeometric series

$$F(a, b, c, x) = 1 + \frac{ab}{1 \cdot c} x + \frac{a(a+1)b(b+1)}{1 \cdot 2 \cdot c(c+1)} x^2 + \cdots \qquad (31.17)$$

in the dubious case of $x = 1$.

For two consecutive terms the ratio of their absolute values is

$$\left| \frac{(n+a)(n+b)}{(n+1)(n+c)} \right| = \left| 1 - \frac{1+c-a-b}{n} + O\left(\frac{1}{n^2}\right) \right|$$

$$= 1 - \frac{1 + \operatorname{Re}(c - a - b)}{n} + O\left(\frac{1}{n^2}\right). \qquad (31.18)$$

This shows that the condition for absolute convergence of the hypergeometric series with the value $x = 1$ of the variable is

$$\operatorname{Re}(c - a - b) > 0. \qquad (31.19)$$

From (31.10) and (31.11) it follows that the radius of convergence ρ of a power series $\sum_n c_n(z-a)^n$ is given by

$$\frac{1}{\rho} = \lim_{n \to \infty} \sqrt[n]{|c_n|} = \lim_{n \to \infty} \left| \frac{c_{n+1}}{c_n} \right|. \qquad (31.20)$$

On the circle of convergence itself the series may be convergent or divergent. From (31.12) and (31.13) it follows that the series is absolutely convergent all over the circle of convergence if

$$\lim_{n \to \infty} \left| \frac{c_{n+1}}{c_n} \right| = \frac{1}{\rho}\left(1 - \frac{1+\varepsilon}{n}\right), \qquad \varepsilon > 0. \qquad (31.21)$$

In the region $-1 < \varepsilon < 0$ the series is conditionally convergent at any point of the circle apart from its singular points. For $\varepsilon = 0$ the series is divergent. The transition from absolute to conditional convergence takes place at $\varepsilon = 0$ and may be illustrated by logarithmic functions as

$$\log \frac{1}{1-z} = z + \frac{z^2}{2} + \frac{z^3}{3} + \cdots. \qquad (31.22)$$

For $z = -1$ and for any other z with $|z| = 1$ the series is conditionally convergent. However, for $z = 1$ we get the divergent series $1 + \frac{1}{2} + \frac{1}{3} + \cdots$.

In the case of $\varepsilon = -1$ the divergence may be illustrated by means of the geometric series

$$(1-z)^{-1} = 1 + z + z^2 + \cdots \qquad (31.23)$$

which is divergent for any z with modulus $|z| = 1$.

Uniform Convergence

There still remains one important concept of convergence to be mentioned, which is termed *uniform convergence*. It relates to series whose terms are functions of a real or complex variable. To fix the idea consider a power series in the complex variable z or $z - a$. Take first a region, say, within the circle of radius $\rho/2$, ρ being the circle of convergence. We choose a positive value ε, say, equal to $\frac{1}{16}$ of the absolute value of the first term. Then the modulus of the sum of the fifth and following terms will be smaller than ε, provided that z is in the region of the above circle. If ε is, say, $\frac{1}{256}$ times the first term, we may write $N(\frac{1}{256}) = 9$, meaning the modulus of the sum of the ninth and following term is smaller than ε. The point is that the number $N(\varepsilon)$, telling where we have to begin in order to have a remaining sum smaller than ε, must depend only upon ε and not on the variable z. This we express by saying that the series is uniformly convergent in the region $|z|$ and $|z - a| < \frac{1}{2}\rho$.

For points outside that circle we have no more $N(\frac{1}{16}) = 5$, $N(\frac{1}{256}) = 9$, and so on. However, choosing again a circle of radius, say, 0.9ρ, it is possible to state that the remaining sum is smaller than any arbitrarily chosen ε, provided that we calculate this sum for terms above some number $N(\varepsilon)$ depending only on ε. In this case we may perhaps have $N(\frac{1}{16}) = 25$ and $N(\frac{1}{256}) = 50$, but the statement is of the same nature. Hence we define a uniformly convergent series $s_1(z) + s_2(z) + \cdots + s_n(z) + \cdots$ of a complex variable to be uniformly convergent in a given region if

$$\left| \sum_{p=1}^{\infty} s_{n+p}(z) \right| < \varepsilon, \quad \text{when} \quad n < N(\varepsilon), \tag{31.24}$$

if the number $N(\varepsilon)$ depends only on ε and not on the variable z.

An infinite series of terms that are continuous functions of a real or complex variable represents a continuous function in the region in which the series is uniformly convergent. If the convergence is nonuniform, the function will be discontinuous. (See Fig. 4-7.)

A power series, hence, is uniformly convergent in a region well inside the circle of convergence. It may or it may not be uniformly convergent on the circle of convergence itself. The theory of Fourier series yields numerous examples of series that are not uniformly convergent.

To see the point concerning nonuniform series representing discontinuous functions we shall consider only one example, the series

$$S(x) = x^2 + \frac{x^2}{1 + x^2} + \cdots + \frac{x^2}{(1 - x^2)^n} + \cdots. \tag{31.25}$$

For any real value of x this series is absolutely convergent, even for $x = 0$, since then $S(0) = 0$. For $x \neq 0$ the series can be summed up like a geometric

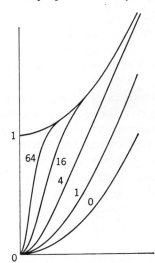

Fig. 4-7. Nonuniform convergence 1, 4, 16, $\cdots$ for various partial sums $y_n = 1 + x^2 - 1/(1 + x^2)^{n-1}$. Limiting curve $y = 1 + x^2$.

series, the result being

$$S(x) = \frac{x^2}{1 - 1/(1 + x^2)} = 1 + x^2. \qquad (31.26)$$

This apparently is a continuous function. However, for $x = 0$ in which case the summation does not apply, we find directly $S(0) = 1$. Hence the function represented by the series (31.25) is discontinuous at the point $x = 0$.

This result can be foreseen from the convergence properties of the series. The remaining sum after the nth term is

$$r_n(x) = \sum_{p=1}^{\infty} S_{n+p}(x) = \frac{x^2}{(1 + x^2)^{n+1}} + \cdots$$

$$= \frac{S(x)}{(1 + x^2)^{n+1}} = \frac{1}{(1 + x^2)^n}. \qquad (31.27)$$

By a given x the remaining sum can be made arbitrarily small by choosing n large. Hence the series is absolutely convergent.

On the other hand, however large we may have chosen n, the remaining sum cannot be kept below a given positive value by variation of x, except by excluding a smaller region around the point $x = 0$. This proves that the convergence of the series is nonuniform in the whole region of real x-values.

32. ANALYTIC CONTINUATION OF POWER SERIES

Different Power Series of an Analytic Function

In Sections 29 and 30 we explored the lines of thought that are peculiar to what may be called the *geometric* theory of functions as specially represented by Riemann. Another method, usually thought of as the more modern and stringent, was forwarded by Weierstrass. According to this method analytic functions are preferably defined directly by given explicit power series.

For convenience denote generally by $P(z - a)$ a power series in $z - a$. An analytic function $u(z) = P(z - a)$ is then defined only within the convergence circle of $P(z - a)$. It is possible, however, to form by a new power series, the so-called *analytic continuation* defined by the first series, and this analytic continuation usually extends to a new region of the independent complex variable. For this reason the definition of analytic functions by a single power series proves to be as general as any other definition based on the "inner properties" of the function, that is, on the knowledge of its singular points, zeros, and so forth.

By means of the definition

$$u(z) = P(z - a), \qquad |z - a| < \rho, \tag{32.1}$$

it is possible to calculate the numerical value of the analytic function $u(z)$, together with any of its first and higher derivatives at a point b situated inside the circle of convergence. This can be used for expressing the same function by another power series

$$u(z) = P(b - a) + P'(b - a)(z - a)$$
$$+ P''(b - a)\frac{(z - a)^2}{1 \cdot 2} + \cdots = P_1(z - b). \tag{32.2}$$

The new power series $P_1(z - b)$ may have a larger or a smaller radius of convergence; it may extend outside the convergence circle of $P(z - a)$ as indicated in Fig. 4-8. In the common region of the circles the two series are bound to identical values of $u(z)$ and all its derivatives. In the region of the second circle outside of the first circle of convergence $P_1(z)$ provides the analytic continuation of $u(z)$ as defined by $P(z - a)$.

To make the point clearer consider the functions z^m and e^z whose power series in the points $z = a$ and $z = 0$, respectively, we have given in (30.10)

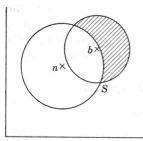

Fig. 4-8. Circles of convergence for power series $P(z - a) = P_1(z - b)$. In the dark region $P_1(z - b)$ provides the analytic continuation of $P(z - a)$.

and (30.13). The power series in $(z - b)$ are readily found and read

$$z^m = \sum_{n=0}^{\infty} \binom{m}{n} b^{m-n}(z - b)^n, \qquad (32.3)$$

$$e^z = e^b \sum_{n=0}^{\infty} \frac{(z - b)^n}{n!} = e^b \cdot e^{z-b}. \qquad (32.4)$$

Owing to the infinite radius of convergence of the series (30.13) for e^z, the series (32.4) is obtainable for any point b. The transition from (30.10) to (32.3) is possible only if $|b - a| \leq |a|$. Hence, for arbitrary n, $|a|$ is the circle of the series (30.10).

It is interesting to note that it is not permitted to place b on the circle of convergence itself. It is true that the series (30.16)

$$z^m = \sum_{n=1}^{\infty} \binom{m}{n} a^{m-n}(z - a)^n \qquad (32.5)$$

is absolutely convergent all over the circle of convergence for real $n > 0$, and for $0 > n > -1$ it still will be conditionally convergent over the whole circle apart from the point $z = a$. Nevertheless the higher derivatives are not. Hence we might possibly be able to calculate z^m, together with a few of its derivatives but not all of them. This is necessary in order to obtain the expansion in power series at a point b on the circle of convergence, even though we know that such an expansion must exist. At any inside point, however near to the convergence circle, it is possible to determine the expansion in a power series and even to pass over to points *on* the convergence circle by limiting considerations.

Similar considerations are valid for a great many functions as $\log(1 + z)$, arc tan z, and so forth, as defined by the power series. We shall therefore look upon such a series from a different point of view, namely, by changing the independent variable.

The above considerations are concerned with linear transformations $\zeta = z - b$ of the variable. Let us study the more general transformation

$$\zeta = \frac{z - 1}{z + 1}, \qquad z = \frac{1 + \zeta}{1 - \zeta}, \qquad (32.6)$$

and apply it to the function

$$\log z = \log \frac{1 + \zeta}{1 - \zeta} = 2\left(\zeta + \frac{1}{3}\zeta^3 + \frac{1}{5}\zeta^5 + \cdots\right). \qquad (32.7)$$

This series converges if $|\zeta| < 1$, which means that Re $z = x > 0$ or that z is situated in the right half of the complex plane. We may also write

$$\log z = \log(-1) + \log \frac{1 + (1/\zeta)}{1 - (1/\zeta)} = \log(-1) + 2\left(\frac{1}{\zeta} + \frac{1}{3}\frac{1}{\zeta^3} + \frac{1}{5}\frac{1}{\zeta^5} + \cdots\right) \tag{32.8}$$

and this series converges if $|\zeta| > 1$, and this condition corresponds to Re $z = x < 0$ or that z is in the half-plane to the left of the imaginary axis. Hence, we have in (32.7) and (32.8) expansions defining the function $\log z$ over the entire complex plane, provided that we can assign a definite meaning to the quantity $\log(-1)$.

This can be attained by requiring the two power series to be identical at some definite point, say, $\zeta = i$, $z = i$. Inserting $\zeta = i$ in (32.7) and (32.8),

$$\log(-1) = 2i\left(1 - \frac{1}{3} + \frac{1}{5} - \cdots\right) - \frac{2}{i}\left(1 - \frac{1}{3} + \frac{1}{5} - \cdots\right) = 4i\frac{\pi}{4} = i\pi. \tag{32.9}$$

If, however, we require the series to agree at $\zeta = -i$, $z = -i$, the result will be different, namely,

$$\log(-1) = -2i\frac{\pi}{4} + \frac{2}{i}\frac{\pi}{4} = -i\pi. \tag{32.10}$$

This proves that $\log z$ is not a single-valued function. As will be shown later, the function $\log z$ can be made unique only by making a cut in the complex plane from $z = 0$ to $z = \infty$, for instance, along the real positive axis, the meaning of the word "cut" being that we are not allowed to pass. On either side of this cut there will be a constant difference of $2\pi i$ in the single-valued function. If such a cut is not made, the function $\log z$ is a many-valued function acquiring an additional term $2\pi i$ each time the variable z encircles the point $z = 0$.

Many-Valued Functions

A suitable example for studying the main properties of many-valued functions is the elementary function $(z - a)^n$ with a singular point in $z = a$ or simply $u(z) = z^n$. Introducing the modulus

$$u(z) = \rho^n(\cos n\varphi + i \sin n\varphi)^n = \rho^n e^{in\varphi},$$

$$\rho = |z|, \qquad \varphi = \arctan\frac{y}{z}. \tag{32.11}$$

If the point z of the complex variable encircles the singular point $z = 0$ counterclockwise, that is, in the positive sense, φ increases by 2π and the function is multiplied by the factor

$$e^{2\pi i n} = \cos 2\pi n + i \sin 2\pi n. \tag{32.12}$$

If n is a real irrational number, the function may acquire an infinity of different values for the same value of the variable z. If on the other hand the exponent is real and rational, $n = p/q$, p and q being mutually prime integers; $u(z) = z^n$ will be multiplied by $e^{2\pi i p}$ and will return to its original value if z encircles the origin q times. The various values of $u(z)$ are said to belong to q different *branches* of the function. The most frequent case of many-valued functions is that of $q = 2$ with only two branches.

The logarithmic function $\log z$, as already discussed above, is defined by the equation

$$z = e^{\log z}. \tag{32.13}$$

From this equation we immediately see that the multiple of $2\pi i$ can be added to the logarithm without changing the value of z. Hence, the many-valued character of $\log z$ is caused by the periodic nature of e^z. Writing $z = re^{i\varphi}$

$$\log z = \log r + i\varphi + 2\pi n, \quad n = \cdots, -2, -1, 0, 1, 2, \cdots. \tag{32.14}$$

Hence the function $\log z$ has an infinite number of branches. To make it single-valued, as well as any other function like z^n, we shall have to fix a definite value for the function at a given point, say, at a point on the real x-axis and, next, other values at the same point by cuts in the z-plane, as mentioned above. For instance, both for $\log z$ and z^n we may limit the region of the argument φ to 2π, for instance, from $-\pi$ to π. This can be achieved by a cut between the singular points $z = 0$ and $z = \infty$, say, along the negative real axis as illustrated in Fig. 4-9. In a similar manner the function

$$u(z) = \left(\frac{z-a}{z-b}\right)^n \tag{32.15}$$

can be made single-valued by a cut between the singular points a and b as shown in Fig. 4-10. If the variable z encircles both points a and b, the function (32.15) regains its original value, even though $(z-a)^n$ and $(z-b)^n$ do not.

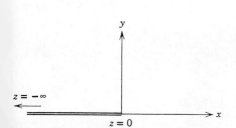

Fig. 4-9. The many-valued functions z^n with nonintegral n and $\log z$ can be made single-valued by a cut from $z = 0$ to $z = \infty$.

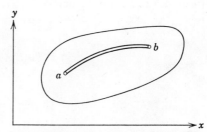

Fig. 4-10. Uniqueness of the function $u(z) = \left(\dfrac{z-a}{z-b}\right)^n$ as obtained by a cut between the two singular points.

This is valid for any real or complex value of n, the numerator and denominator both being multiplied by $e^{2\pi i n}$, the first owing to the singular point a and the latter owing to the singular point b.

33. COMPLEX INTEGRATION AND CONTOUR INTEGRALS

Evaluation of Definite Integrals by Complex Integration

It is frequently possible to transform definite integrals of functions of real variables into complex integral or contour integrals, that is, integrals taken along a closed curve in the complex plane. The usefulness of this procedure rests on the fact that the path of integration may be deformed without changing the value of the integral. Hence it is possible to look for simpler ways of actually calculating the integral, the main aim frequently being to contract the path of integration around the simplest singular points of the functions.

Consider the analytic and the *meromorphic* functions $u(z)$ and $f(z)$ in the point a,

$$u(z) = c_0 + c_1(z - a) + \cdots + c_n(z - a)^n + \cdots \qquad (33.1)$$

and

$$f(z) = \frac{u(z)}{(z-a)^{m+1}} = C_{-m-1}(z-a)^{-m-1} + \cdots + C_{-1}(z-a)^{-1} + C_0 + \cdots$$

$$C_{-m-1} = c_0, \qquad C_{-1} = c_m, \qquad C_n = c_{n+m+1}, \qquad (33.2)$$

with positive integral m. Since $u(z)$ is an analytic function in a, $f(z)$ is a function with a pole of $(m+1)$th order containing the negative powers $(z-a)^{-m-1}$ up to $(z-a)^{-1}$ and called a meromorphic in contrast to *holomorphic* or analytic. The coefficient of the power $(z-a)^{-1}$, $C_{-1} = c_m$, is called the residue of the function in the point a. According to the Cauchy formula (29.33) for higher derivatives of a function $u(z)$, putting $u^{(m)}(a) = m \cdot c_m$

$$C_{-1} = c_m = \frac{1}{2\pi i} \oint \frac{u(z)\,dz}{(z-a)^{m+1}}, \qquad (33.3)$$

or

$$2\pi i C_{-1} = \oint f(z)\,dz. \qquad (33.4)$$

The Theorem of Residues

Equation 33-4 is the theorem of residues for meromorphic functions. The residue of $f(z)$ in the point a is called C_{-1}, and the contour integral around this point is $2\pi i$ times the residue C_{-1}.

Theory of Functions | 151

If a function $f(z)$ has a pole of the first or higher orders in $z = \infty$, we may for sufficiently large values of $|z|$ expand the function in an infinite series of negative powers of z

$$f(z) = c_n z^n + \cdots + c_1 z + c_0 + c_{-1} z^{-1} + \cdots. \tag{33.5}$$

By the transformation $z = \zeta^{-1}$ changing the singular point $z = \infty$ into $\zeta = 0$ the residuum of $f(z) = f(1/\zeta)$ apparently is c_1. It is, however, the coefficient c_{-1} that determines the value of the integral around $z = \infty$ or $\zeta = 0$, the reason for this being the form of the differential $dz = -\zeta^{-2} d\zeta$. In fact,

$$f(z) \, dz = (c_n \zeta^{-n-2} + \cdots + c_1 \zeta^{-3} + c_0 \zeta^{-2} + c_{-1} \zeta^{-1} + \cdots) \, d(-\zeta). \tag{33.6}$$

If the path of integration is in the positive sense of a circle of large radius (z) in the z-plane, the path in the ζ-plane is a small circle around $\zeta = 0$ in the negative sense. Hence the value of the contour integral is $2\pi i c_{-1}$,

$$\oint f(z) \, dz = -\oint f\left(\frac{1}{\zeta}\right) \frac{d\zeta}{\zeta^2} = 2\pi i c_{-1}, \tag{33.7}$$

and c_{-1} may therefore be called the residue of (z) in $z = \infty$ even though c_1 and not c_{-1} is the residue of $f(1/\zeta)$ in $\zeta = 0$.

Contour Integrals

We shall not enter too much into details in describing the method of complex integration for definite integrals. Some few examples may serve to illustrate the main features of the method.

Consider first the very simple definite integral $I = \int_0^\pi d\varphi = \pi$. It can be transformed into a contour integral by the following procedure. Write $\varphi = \arccos z$ and

$$I = \int_{-1}^{+1} \frac{dz}{(1 - z^2)^{1/2}} = \frac{1}{2} \int_{-1}^{+1} \frac{dz}{(1 - z^2)^{1/2}} - \frac{1}{2} \int_{+1}^{-1} \frac{dz}{(1 - z^2)^{1/2}}. \tag{33.8}$$

Because the square root $\sqrt{1 - z}$ is multiplied by $e^{i\pi} = -1$ when encircling the point $z = 1$ and $\sqrt{1 + z}$ is multiplied by the same factor when z encircles $z = -1$, a cut between the singular or branching points 1 and -1 makes the integrand single-valued. Fixing its value by $\sqrt{z^2 - 1}$ real on the real axis to the right of $z = 1$, we may take the two separate integrals (33.8) to be parts of a contour integral

$$I = \frac{1}{2i} \oint \frac{dz}{(z^2 - 1)^{1/2}} \tag{33.9}$$

on the lower and upper side of the real axis, respectively, the path of integration encircling the points ± 1 and closely following the cut between the

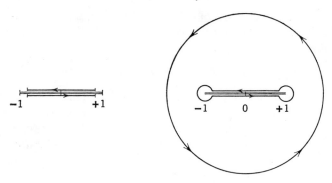

Fig. 4-11. Contour integral around a branching cut, the contour being contracted around the pole $z = \infty$.

points. This is illustrated in Fig. 4-11, the cut and the original two paths of (33.8) being given at the left, the first path of complex integration together with with a deformed path at the right.

The integrand of (33.9) has no other singular point than ± 1 except at infinity $z = \infty$. Therefore the path of integration can be deformed in any way, except that it must remain a closed curve, for instance, into a circle large enough so that it permits an expansion of the inverse square root (33.9) into a series of negative powers in z,

$$(z^2 - 1)^{-1} = z^{-1} + \tfrac{1}{2}z^{-3} + \cdots \tag{33.10}$$

with residue 1 at infinity. Hence the value of the integral is $I = 2\pi i/2i = \pi$. This may appear a very modest result of lengthy considerations. Nevertheless, the method is a vigorous one as seen when applied to more intricate integrals.

For instance, if in a definite integral in the region $0 \leq \varphi \leq \pi$, we take the integrand to be $\sin^{2m} \varphi \cos^{2n} \varphi$, m and n being positive integers, the integral, although elementary, requires some few integrations by part until its actual value can be safely predicted. Applying the above method, however, we at once find

$$I_{nm} = \int_0^\pi \sin^{2m} \varphi \cos^{2n} \varphi \, d\varphi = \frac{(-1)^m}{2i} \oint (z^2 - 1)^{m-\frac{1}{2}} z^{2n} \, dz. \tag{33.11}$$

If $|z| > 1$, we may expand the integrand as follows in powers of $1/z$:

$$(z^2 - 1)^{m-1} z^{2n}$$
$$= z^{2m+2n-1} \left(\frac{1-1}{z^2}\right)^{m-\frac{1}{2}} = z^{2m+2n-1} - \left(m - \frac{1}{2}\right) z^{2m+2n-3} + \cdots$$
$$+ (-1)^m \frac{(m-\frac{1}{2})(m-\frac{3}{2}) \cdots \frac{1}{2} \cdot \frac{1}{2} \cdot \frac{3}{2} \cdots (n-\frac{1}{2})}{1 \cdot 2 \cdots (n+m)} z^{-1} + \cdots. \tag{33.12}$$

The last term contains the residue at $z = \infty$ and determines the value of the integral, which is

$$I_{nm} = \pi \frac{\frac{1}{2} \cdot \frac{3}{2} \cdots (m - \frac{1}{2}) \cdot \frac{1}{2} \cdot \frac{3}{2} \cdots (n - \frac{1}{2})}{1 \cdot 2 \cdots (n + m)}. \tag{33.13}$$

By means of this result, more complicated integrals as

$$I(a) = \int_0^\pi \frac{d\varphi}{1 - a \cos \varphi} = \pi \left(1 + \frac{1}{2} a^2 + \frac{\frac{1}{2} \cdot \frac{3}{2}}{1 \cdot 2} a^4 + \cdots \right) = \frac{\pi}{(1 - a^2)^{1/2}} \tag{33.14}$$

and

$$J(a) = \int_0^\pi \frac{\sin^2 \varphi \, d\varphi}{1 - a \cos \varphi} = \pi \left(\frac{\frac{1}{2}}{1} + \frac{\frac{1}{2} \cdot \frac{1}{2}}{1 \cdot 2} a^2 + \frac{\frac{1}{2} \cdot \frac{1}{2} \cdot \frac{3}{2}}{1 \cdot 2 \cdot 3} a^4 + \cdots \right)$$

$$= \frac{\pi}{a^2} [1 - (1 - a^2)^{1/2}] \tag{33.15}$$

for $a < 1$ can be evaluated fairly easily. However, in this case also, direct complex integration is the shortest way to the goal. For instance, the first integral may be written

$$I(a) = \frac{1}{2i} \oint \frac{dz}{(z^2 - 1)^{1/2}(1 - az)} = -\frac{1}{2ia} \oint \frac{dz}{(z^2 - 1)^{1/2}(z - 1/a)}. \tag{33.16}$$

Here the integrand has a pole at $z = 1/a$ and one at infinity. Therefore the path of integration around the cut from $z = -1$ to $z = 1$ can be deformed so as to contract finally around the pole $z = 1/a$. The result is easily seen to be the same as in (33.14).

The integral (33.15) $J(a)$ takes the form

$$J(a) = \frac{1}{2ia} \oint \frac{\sqrt{z^2 - 1}}{z - 1/a} dz \tag{33.17}$$

and the integrand has a pole at infinity, the residuum of which is easily seen to be $1/a$ by expansions of the denominator, $(z - 1/a)^{-1} = z^{-1} + z^{-2}/a + \cdots$. Adding the contribution from the pole $z = 1/a$, we obtain the result

$$J(a) = \frac{\pi}{a} \left[\frac{1}{a} - \left(\frac{1}{a^2} - 1\right)^{1/2}\right], \tag{33.18}$$

which is the same as (33.15). (See Fig. 4-12.)

Another example of a contour integral, which played an important role in the old quantum theory of atomic orbits, is

$$I_r = \oint \left(-A + \frac{2B}{r} - \frac{C}{r^2}\right)^{1/2} dr, \quad B^2 > AC. \tag{33.19}$$

154 / The Mathematical Foundation of Physics

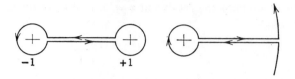

Fig. 4-12. Complex integrations by poles at $z = 1/a$ and $z = \infty$ of the integrand.

By the additional condition the radicand is positive, and hence the integrand rational only in a region $r_{\min} \leq r \leq r_{\max}$. The integral is taken from $r_{\min}$ to $r_{\max}$ with positive sign, and back to $r_{\min}$ with negative sign of the square root, which corresponds to the contour integral

$$I_r = \oint \left(A - \frac{2B}{z} + \frac{C}{z^2} \right)^{\frac{1}{2}} dz, \tag{33.20}$$

the square root being positive on the real z-axis to the right of $z = r_{\max}$. The integrand has two poles, at $z = 0$ and $z = \infty$. The residuum at the first is $-\sqrt{C}$ since the square root is negative to the left of $z = r_{\min}$, the negative sign being counterbalanced by the negative direction of integration. Hence, the contribution from this point is $2\pi i \cdot i\sqrt{C}$. At infinity the residuum is found by expansion in powers of $1/z$ to be $-B\sqrt{A}$. Multiplying by $i \cdot 2\pi i = -2\pi$, we obtain the result

$$I_r = 2\pi \left(\frac{B}{\sqrt{A} - \sqrt{C}} \right), \tag{33.21}$$

which is the famous radial phase integral of the Bohr-Sommerfeld theory of a one-electron atom.

Since it would require too much space here to work out a representative selection of examples of important contour integrals, we have only presented the fundamental ideas. More will be learned from the subsequent use of contour integrals used in definitions of analytic functions.

34. THE FACTORIAL OR GAMMA FUNCTION

Definition of the Gamma Function

Even though the gamma function of Euler and Gauss is a more intricate function than any of the functions to be discussed in the next chapter, we shall present this function here because it will be needed when studying the particular functions as defined by differential equations. The study of the gamma function will also afford some useful applications of contour integrals.

The main properties of the gamma function are that it has poles of the first order for negative integral values of the independent variable z and that it obeys the functional equation

$$\prod(z) = z\prod(z-1), \tag{34.1}$$

$\prod(z)$ being the notation introduced by Gauss as different from Eulers notation $\Gamma(z) = \prod(z-1)$. As a consequence of (34.1), choosing $\prod(0) = 1$, the Gauss gamma function equals for positive integers n the factorials $n!$,

$$\prod(n) = n! \tag{34.2}$$

and appears as a natural generalization of the factorials for nonintegral values of the variable. It is therefore also called the *factorial* function.

It is to be expected that various definitions of analytic functions obeying the functional equation (34.1) and having an infinite set of correct values as in (34.2) for all positive integers will serve as a correct definition of the gamma function. A definition of that kind is provided by the *Euler integral* of the so-called *second kind*,

$$\prod(n) = \int_0^\infty e^{-z} z^n \, dz, \tag{34.3}$$

denoting for comparison with (34.2) the independent variable by n.

The above integral is convergent; hence the function is defined for any complex value of the variable n, provided that $\operatorname{Re} n > -1$, that is, if $n+1$ is at the right of the imaginary axis. For $\operatorname{Re} n < -1$ with $n+1$ at the left of the imaginary axis, the definition may be completed by means of the functional equation (34.1). Integrating by parts, we easily find from (34.3) that

$$\prod(n) = n\int_0^\infty e^{-z} z^{n-1} \, dz = n\prod(n-1). \tag{34.4}$$

Since from (34.3) we find $\prod(0) = 1$, it follows that the definition (34.3) conforms with both (34.2) and (34.1) or (34.4).

If n is not a positive integer, the definition (34.3) can be replaced by a contour integral as illustrated in Fig. 4-13. The path of integration is so chosen to start from $+\infty$ on the upper side of the real axis, encircling the origin in the positive sense and returning to infinity beneath the real axis.

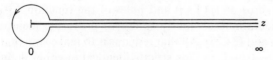

Fig. 4-13. Path of integration by definition of the gamma function.

We write the integral

$$\Pi(n) = \frac{1}{e^{\pi i n} - e^{-\pi i n}} \oint_{+\infty+i\varepsilon}^{+\infty-i\varepsilon} e^{-z}(-z)^n \, dz. \tag{34.5}$$

The meaning of the use of the negative argument in $(-z)^n$ is that for real n the integrand is taken to be real on the negative real z-axis. This ensures that the integral itself is real for real n.

In following the small circle of Fig. 4-12 from the negative z-axis clockwise to the upper side of the positive axis, the integrand is multiplied by $e^{-\pi i n}$ and by $e^{\pi i n}$ if following the circle counterclockwise to the lower side. This ensures that the integrals (34.3) and (34.5) are identical in the region $\operatorname{Re} n > -1$ in which the integral (34.3) has been proved to exist. The integral (34.5) in contrast to (34.3) exists for any complex value of n apart from the poles of the function, $n = -1, -2, \cdots$. This is due to the exclusion of the point $z = 0$ from the path of integration and the converging factor e^{-z} at $z = +\infty$. Moreover, the functional equation (34.4) is also valid. Hence, (34.5) provides a definition of the gamma function for any complex value of its argument.

Alternative Definition of the Gamma Function

Another definition of the gamma function by complex integration is provided by the contour integral

$$\frac{1}{\Pi(n)} = \frac{1}{2\pi i} \oint_{+\infty+i\varepsilon}^{+\infty-i\varepsilon} e^{-z}(-z)^{-n-1} \, d(-z). \tag{34.6}$$

Integration by parts proves that this definition conforms also with the functional equation (34.4) in the form

$$\frac{1}{\Pi(n)} = \frac{1}{2\pi i n} \oint e^{-z}(-z)^{-n} = \frac{1}{n\,\Pi(n-1)}. \tag{34.7}$$

On the other hand for $n = 0$, or any positive integer n, the ends of the contour curve at $z = \infty$ may be united, because of the common value of the integrand on either side of the real axis, and the path of integration contracted into a circle around $z = 0$. This yields $\Pi(0) = 1$ and also $\Pi(n) = n!$ either directly or in virtue of (34.6).

If n is a negative integer, the path of integration may be contracted in the same manner yielding the value zero for the integral (34.6). Hence, $n = -1$, $-2, \cdots$, are zeros of $1/\Pi(n)$ and poles of the function $\Pi(n)$ itself.

These considerations obviously suffice to prove the identity of the definitions (34.26) and (34.27). All that remains is to make sure that the nature of the poles $n = -1, -2, \cdots$, is strictly identical in either definition. If that, however, is taken for granted, the difference of the two possibly different functions (34.5) and (34.6) is an analytic function throughout the complex plane including the point $z = \infty$. According to *Liouville's theorem*, as inferred from (29.30) and (29.32), such a function is a constant and therefore, since the difference is zero in one point $n = 0$ or $n = 1, 2, \cdots$, it is zero everywhere.

It should be noted that the definition (34.6) is usually written in the more precise form

$$\frac{1}{\prod(n)} = \frac{1}{2\pi i} \oint_{-\infty-i\varepsilon}^{-\infty+i\varepsilon} e^z z^{-n-1} \, dz, \tag{34.8}$$

which is readily obtained from (34.6) replacing $(-z)$ by z. The form (34.6) has been preferred here for the sake of easier comparison with (34.5).

The definitions (34.5) and (34.6) can be used for deducing important formulas in the theory of the gamma function. For instance, replacing n by $-n$ in (34.7) and comparing with (34.5), we obtain

$$\frac{1}{\prod(-n)} = \frac{1}{2\pi i n} \oint e^{-z}(-z)^n \, dz = \frac{2i \sin n\pi}{2\pi i n} \prod(n)$$

or

$$\prod(n) \prod(-n) = \frac{n\pi}{\sin n\pi}, \tag{34.9}$$

which is the *reflection formula* for the gamma function.

The Gamma Function Expressed by Infinite Products

Consider the function

$$\prod_m (n) = \int_0^m \left(1 - \frac{z}{m}\right)^m z^n \, dz = m^{m+1} \int_0^1 (1-t)^m t^n \, dt$$

$$= \frac{m^{n+1} m!}{(n+1)(n+2) \cdots (n+1+m)} \tag{34.10}$$

for $\operatorname{Re} n > -1$. Making $m \to \infty$, we obtain

$$\lim_{m \to \infty} \prod_m (n) = \prod(n) = \lim_{m \to \infty} \frac{m^{n+1} m!}{(n+1) \cdots (n+1+m)}. \tag{34.11}$$

This is the fundamental definition of the gamma function as first used by Euler. Replacing m by $m + 1$, in (34.11), and observing that $(m+1)^2/(n+1+m)(n+2+m) \to 1$ as $m \to \infty$ and further that $m+1 = 2 \cdot \frac{3}{2} \cdots (m+1)/m$, we obtain the infinite product

$$\prod(n) = \prod_{m=1}^{\infty} \frac{(1+1/m)^n}{1+n/m}. \tag{34.12}$$

Again, observing that

$$\sum_{m=1}^{N-1}\left(1+\frac{1}{m}\right)^n = N^n = e^{n\log N} \qquad (34.13)$$

and using the definition

$$C = \lim_{N\to\infty}\left(1+\frac{1}{2}+\cdots+\frac{1}{N}-\log N\right) = 0.577\,215\,665\cdots \qquad (34.14)$$

for the *Euler constant* C, we obtain the modified infinite product

$$\prod(n) = e^{-Cn}\prod_{m=1}^{\infty}\frac{e^{n/m}}{1+n/m} \qquad (34.15)$$

which was first given by Weierstrass.

The above definitions of the gamma function are all general and not restricted by the condition $\mathrm{Re}\,n > -1$.

The Euler Integral of the First Kind

Consider the integral

$$\prod(n) = \int_0^\infty e^{-z}z^n\,dz = 2\int_0^\infty e^{-x^2}x^{2n+1}\,dx \qquad (34.16)$$

and next the product

$$\prod(n)\prod(m) = 4\int_0^\infty\int_0^\infty e^{-(x^2+y^2)}x^{2n+1}y^{2m+1}\,dx\,dy \qquad (34.17)$$

for $\mathrm{Re}\,n > -1$ and $\mathrm{Re}\,m > -1$. Introducing polar coordinates r and φ, we find that the double integral becomes

$$\prod(n)\prod(m) = 2\int_0^\infty e^{-r^2}r^{2n+2m+3}\,dr \cdot 2\int_0^{\pi/2}\cos^{2n+1}\varphi\,\sin^{2m+1}\varphi\,d\varphi. \qquad (34.18)$$

The first right-hand side integral is $\prod(n+m+1)$ as seen by comparison with (34.16). For the second integral we write $\xi = \cos\varphi$ and next $\eta = \xi^2$ and obtain

$$2\int_0^1 \xi^{2n+1}(1-\xi^2)^m\,d\xi = \int_0^1 \eta^n(1-\eta)^m\,d\eta = \frac{\prod(n)\prod(m)}{\prod(n+m+1)}. \qquad (34.19)$$

This formula is very frequently used and is known as the *Euler integral of the first kind*.

If $\mathrm{Re}\,n < -1$ or $\mathrm{Re}\,m < -1$, or both, the integral (34.19) may be a contour integral encircling one or both of the singular points $\eta = 0$ and $\eta = 1$.

In the latter case in which the integral (34.19) is divergent at both ends, hence requiring a definition by contour integrals, the form of the integral and the path of integration require special treatment, therefore we shall omit this study here.

The Logarithmic Derivative of the Gamma Function

For the logarithmic derivative of $\prod(n)$ we have the notation

$$\Psi(n) = \frac{d}{dn} \log \prod (n). \tag{34.20}$$

Using for $\prod(n)$ the Weierstrass product, we obtain

$$\Psi(n) = -C + \sum_{m=1}^{\infty} \left(\frac{1}{m} - \frac{1}{m+n} \right). \tag{34.21}$$

The right-hand sum may be written

$$\int_0^\infty \frac{e^{-s} - e^{-(n+1)s}}{1 - e^{-s}} ds = \sum_{m=1}^{\infty} \left(\frac{1}{m} - \frac{1}{m+n} \right) \tag{34.22}$$

and the constant $-C$

$$\int_0^\infty \left(\frac{e^{-s}}{s} - \frac{e^{-s}}{1 - e^{-s}} \right) ds = \lim_{N \to \infty} \int_0^\infty \left(\frac{e^{-s} - e^{-Ns}}{s} - \frac{e^{-s} - e^{-(N+1)s}}{1 - e^{-s}} \right) ds$$

$$= \lim_{N \to 0} \left[\int_\varepsilon^{N\varepsilon} \frac{e^{-t}}{t} dt - \left(1 + \frac{1}{2} + \cdots + \frac{1}{N} \right) \right] = -C. \tag{34.23}$$

Adding (34.22) and (34.23), we have the integral representation of $\Psi(n)$,

$$\Psi(n) = \int_0^\infty \left[\frac{e^{-s}}{s} - \frac{e^{-(n+1)s}}{1 - e^{-s}} \right] ds. \tag{34.24}$$

The Duplication and Multiplication Formulas

Consider the formula

$$\frac{\prod(n) \prod (n - \frac{1}{2})}{\prod(2n) \prod(-\frac{1}{2})} = \frac{1}{2^{2n}}, \tag{34.25}$$

which is easily seen to be true for any positive integer or half-integral number. With some care it is seen to be true also for negative integral and half-integral numbers. Having no singular point, except certainly at infinity, it is an integral function and it is highly probable that it is the exponential function $e^{-2n \log 2}$.

160 / The Mathematical Foundation of Physics

To be certain we calculate the logarithmic derivative of the left-hand expression by (34.21), getting

$$\sum_{m=1}^{\infty}\left(\frac{2}{m+2n}-\frac{1}{m+n}-\frac{1}{m+\frac{1}{2}-\frac{1}{2}}\right)$$

with the particular value $-2(1-\frac{1}{2}+\frac{1}{3}-\frac{1}{4}+\cdots)=-2\log 2$ for $n=0$. The above series converges for any value of n, even for negative integral and half-integral numbers. It is therefore analytic throughout the complex plane inclusive of $n=\infty$. Hence it is a constant and its value is $-2\log 2$ in conformity with (34.25).

Writing $\prod(-\frac{1}{2})=\sqrt{\pi}$, which is easily obtained from the reflection formula (34.9), we have the conventional form of the *duplication* formula

$$\prod(n)\prod(n-\tfrac{1}{2})=2^{-2n}\sqrt{\pi}\prod(2n). \tag{34.26}$$

Consider next the formula

$$\frac{\prod(n)\prod[n-(1/m)]\cdots\prod[n-(m-1)]/m}{\prod(-1/m)\cdots\prod[-(m-1)/m]}=\frac{\prod(mn)}{m^{mn}}, \tag{34.27}$$

m being a positive integer. This formula holds for any positive integer n and furthermore for any number in between that differs from the integers by multiple of $1/m$. The left-hand expression has no poles for real negative number and hence is an integral function that can be nothing else but the right-hand side of (34.27), even though we shall not prove this by computing the logarithmic derivative. Equation 34.27 will provide the conventional *multiplication theorem* if the left-hand denominator can be computed.

To accomplish this partially we apply the reflection formula for symmetric factors

$$\prod\left(-\frac{1}{m}\right)\prod\left(-1+\frac{1}{m}\right)=\frac{\pi}{\sin(\pi/2)},$$

$$\prod\left(-\frac{2}{m}\right)\prod\left(-1+\frac{2}{m}\right)=\frac{\pi}{\sin(2\pi/2)},\cdots,\prod\left(-\frac{1}{2}\right)=\sqrt{\pi}. \tag{34.28}$$

If now

$$\frac{1}{\sin(\pi/m)}\cdot\frac{1}{\sin(2\pi/m)}\cdots=\left(\frac{2^{m-1}}{m}\right)^{\frac{1}{2}}, \tag{34.29}$$

the multiplication formula becomes

$$\prod(n)\prod\left(n-\frac{1}{m}\right)\prod\left(n-\frac{2}{m}\right)\cdots\prod\left(n-1+\frac{1}{m}\right)$$

$$=\left(\frac{(2\pi)^{m-1}}{m}\right)^{\frac{1}{2}}\frac{\prod(mn)}{m^{mn}}. \tag{34.30}$$

The proof of (34.29) is as follows. Squaring the equations, we may write

$$\sin\frac{\pi}{m}\sin 2\frac{\pi}{m}\cdots\sin\frac{m-1}{2}\frac{\pi}{m}=\frac{m}{2^{m-1}} \qquad (34.31)$$

since

$$\sin^2 k\frac{\pi}{m}=\sin k\frac{\pi}{m}\sin(m-k)\frac{\pi}{m}.$$

In (34.29) the sines go up to $\sin[(m-1)/2](\pi/m)$ for odd and to $\sin \pi/2 = 1$ for even m, so that in the latter case $\sin^2 \pi/2 = \sin \pi/2$. In either case, therefore, (34.29) and (34.31) are equivalent.

From the product

$$x^{2m}-1=(x^2-1)\left[x^2-\exp\left(-i\frac{2\pi}{m}\right)\right]\cdots x^2-\exp\left[-i(m-1)\frac{2\pi}{m}\right], \qquad (34.32)$$

multiplying by $x^{-1} \cdot x^{-1} \exp[i(\pi/m)] \cdots x^{-1} \exp[i(m-1)\pi/m] = x^{-m} \cdot i^{m-1}$ and replacing x by $\exp(ix)$, it follows that

$$\frac{\exp(imx)-\exp(-imx)}{i}=\frac{\exp(ix)-\exp(-ix)}{i}$$

$$\cdot\frac{\exp[i(x+\pi/m)]-\exp\{-i[x+(\pi/m)]\}}{i}\cdots$$

$$\frac{\exp[i(x+(m-1)\pi/m)]-\exp[-i(x+(m-1)\pi/m)]}{i}, \qquad (34.33)$$

or

$$\frac{\sin mx}{2^{m-1}\sin x}=\sin\left(x+\frac{\pi}{m}\right)\sin\left(x+2\frac{\pi}{m}\right)\cdots\sin\left(x+(m-1)\frac{\pi}{m}\right) \qquad (34.34)$$

which as $x \to 0$ yields (34.31).

35. THE DELTA FUNCTION AND THE COMPLETENESS OF SYSTEMS OF FUNCTIONS

The Concept of the Delta Function

In the theory of functions, particularly in the orthonormal system of eigenfunctions, we frequently need a symbol generalizing the Kronecker symbol δ_{ik}. The new symbol may be said to be a function of two continuous variables, in contrast to the discrete variables i and k of δ_{ik}, although a highly discontinuous or improper function. Whereas δ_{ik} may be taken to represent the unit matrix (by an infinite number of elements), the new symbol which we

162 / The Mathematical Foundation of Physics

denote by $\delta(x, x')$, or $\delta(x - x')$ corresponding to $\delta_{ik} = \delta_{0,i-k}$, may be said to represent some sort of a *continuous unit matrix*.

The improper function $\delta(x - x')$ was introduced by Dirac by the definitions

$$\int_{x-\varepsilon}^{x+\varepsilon} f(x')\, \delta(x - x')\, dx' = f(x), \qquad (35.1a)$$

$$\delta(x - x') = 0 \quad \text{when} \quad x \neq x', \qquad (35.1b)$$

ε being an arbitrarily small quantity. If $f(x') = 1$,

$$\int_{x-\varepsilon}^{x+\varepsilon} \delta(x - x')\, dx' = 1, \qquad (35.2)$$

meaning that we are considering functions or curves of an area above the axis of the independent variable. Moreover, this area is concentrated within the limits $x - \varepsilon$ and $x + \varepsilon$ of the variable x'. Outside of this region the function is zero, according to the Dirac definition. A more satisfactory definition would be

$$\int_{x''-\varepsilon}^{x''+\varepsilon} \delta(x - x')\, dx = \begin{cases} 1 & \text{when } x'' = x, \\ 0 & \text{when } x'' \neq x, \end{cases} \qquad (35.3)$$

comprising both equations (35.1) in a somewhat different way.

The meaning of the modified definition (35.3) will appear from Figs. 4-14*a* and 4-14*b*, the first representing some kind of a Gauss curve, which might naturally be thought of by the Dirac definition, the second being an

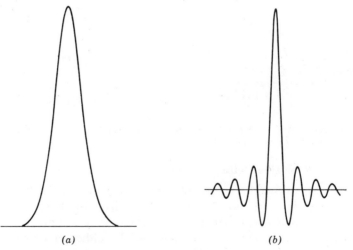

Fig. 4-14. (*a*) The δ-function in the form of a deep Gauss curve. (*b*) The δ-function from Fourier integrals.

oscillating curve with finite amplitudes outside of the central peak with infinite height in the limiting case. The latter type is what we usually encounter with actual delta functions.

The true meaning of the over-simplified definition (35.3), of course, can become clear only by some additional considerations. Consider a function $\delta_a(x)$ that is rapidly changing at $x = 0$, and maybe for any value of x. Then if

$$\lim_{a \to \infty} \int_{x'-\varepsilon}^{x'+\varepsilon} \delta_a(x)\, dx = \begin{cases} 1 & \text{for } x' = 0, \\ 0 & \text{for } x' \neq 0, \end{cases} \quad (35.4)$$

we shall think of $\delta_a(x)$ as a delta function by ever-increasing a and write for convenience

$$\lim_{a \to \infty} \delta_a(x) = \delta(x). \quad (35.5)$$

This is pure convention and aims at competing with the mathematical language of corresponding processes for continuous functions. This convention being accepted, functions like

$$\delta_a(x) = (a/\pi)^{1/2} e^{-ax^2} \quad (35.6)$$

or

$$\delta_a(x) = \frac{\sin ax}{\pi x}, \quad (35.7)$$

which are those illustrated in Figs. 4-14a and 4-14b may serve the purpose of representing delta functions $\delta(x)$.

Orthogonality of Functions with a Continuous Parameter

Consider the complex function

$$y(\omega, x) = (2\pi)^{-1/2} e^{i\omega x} \quad (35.8)$$

of a real variable in the region $-\infty \leq x \leq \infty$ and of a parameter ω in the corresponding region of real values. Next we calculate the integral

$$\int_{-a}^{+a} y^*(\omega', x) y(\omega, x)\, dx = \frac{\sin a(\omega - \omega')}{\pi(\omega - \omega')} = \delta_a(\omega - \omega'). \quad (35.9)$$

If we make a very large, the integral assumes more and more the shape of an improper δ-function. We express this by writing conventionally

$$\int y^*(\omega', x) y(\omega, x)\, dx = \delta(\omega - \omega'), \quad (35.10)$$

it being understood that the integration refers to the whole fundamental region of the variable x. This is the orthonormality relation of property-normalized eigenfunctions with continuous eigenvalues.

It is evident that by corresponding considerations we may write

$$\int y^*(\omega, x')y(\omega, x)\, d\omega = \delta(x - x'). \tag{35.11}$$

In Fourier analysis an arbitrary function, apart from certain requirements of continuity and integrability, may be expressed by a Fourier integral

$$f(x) = (2\pi)^{-1/2} \int g(\omega)e^{i\omega x}\, d\omega, \tag{35.12}$$

where

$$g(\omega) = (2\pi)^{-1/2} \int f(x)e^{-i\omega x}\, dx \tag{35.13}$$

is called the *Fourier transform* of $f(x)$. Inserting (35.13) in the right-hand side of (35.12), we obtain

$$\frac{1}{2\pi}\iint f(x')e^{i\omega(x-x')}\, dx\, d\omega = \int f(x')\,\delta(x-x')\, dx' = f(x), \tag{35.14}$$

where the limits of x and ω have first to be reduced, say, to $\pm a$ and $\pm b$ so that the succession of integration may be changed. The integration in ω then yields a δ-function in the above sense and, going to the limiting case of $a \to \infty$, $b \to \infty$, the result will be that of the right-hand side of (35.14).

Equation 35.11 on which the result (35.14) is based, and which is analogous to (35.9) and (35.10), is the completeness relation for the functional system (35.8).

The δ-Function of Fourier Series

Consider the system of functions

$$y_n(x) = (2\pi)^{-1/2} e^{inx}, \quad n = -\infty, \cdots, -1, 0, 1, \cdots, \infty, \tag{35.15}$$

in a region of given length 2π, say, $\pi - \pi < x < \pi$ and obeying the orthonormality relation

$$\int_0^{2\pi} y_m(x)y_n(x)\, dx = \delta_{nm}. \tag{35.16}$$

We now calculate the finite geometric series

$$\sum_{m=-N}^{N} y_n^*(x')y_n(x) = \frac{1}{2\pi}\sum_n e^{in(x-x')} = \frac{\sin(N+\tfrac{1}{2})(x-x')}{2\pi \sin\tfrac{1}{2}(x-x')} = \delta_N(x-x'). \tag{35.17}$$

For a very small argument $x - x'$ the denominator differs very little from $\pi(x - x')$ and hence $\delta_N(x - x')$ represents a δ-function for large N, just as well as (35.7) for large a, since in the definition of a δ-function we have referred to integrations over a very narrow region 2ε.

Owing to its construction, however, the $\delta_N(x - x')$ is a periodic function of period 2π in x or x' as easily seen also from the right-hand side of (35.17). The function therefore has periodic peaks of area 1, both at $x = x'$ and at distances an integral number of 2π from that point. This does not matter, however, since the fundamental region has the length of 2π. Therefore (35.17) may serve as a δ-function, which we express by writing

$$\sum_{n=-\infty}^{\infty} y_n^*(x')y_n(x) = \delta(x - x'). \tag{35.18}$$

This is the completeness relation for the system of functions (35.15).

CHAPTER 5

Differential Equations and Particular Functions

36. SOLUTION OF DIFFERENTIAL EQUATIONS BY POWER SERIES

Classification of Differential Equations

Differential equations are equations with one or more unknown functions of one or several independent variables, together with derivatives of the functions of the first or higher orders and, finally, known functions of the independent variables.

This is a very wide definition. We shall at once simplify by considering only *ordinary* differential equations in which only one independent variable enters. If there is more than one independent variable, the equations are called *partial* differential equations.

If we have more than one unknown function or dependent variable, we must have a corresponding number of equations, which are then called *simultaneous* differential equations.

We shall in the following sections deal mainly with ordinary differential equations for a single unknown function of a real or complex variable. The *order* of the equation is defined as the order of the highest derivative entering the equation. In physics and in many branches of the theory of functions second-order equations play a predominant role. A differential equation of the second order may usually be written in the explicit form

$$u''(z) = F(u, u', z). \tag{36.1}$$

The equation (36.1) may be homogeneous or inhomogeneous in u, u' and u''. Again we must distinguish between *linear* and *nonlinear* equations in the unknown function u and its derivatives. Nonlinear equations are as a rule extremely more difficult to solve than linear equations. To sum up, partial, simultaneous, high-order and nonlinear equations are all difficult to handle for particular reasons. The simplest, and therefore as a rule the most important in practice, are ordinary linear equations. We shall frequently find them difficult enough. A *linear homogeneous* differential equation of the second order may be written

$$\left[\frac{d^2}{dz^2} + p(z)\frac{d}{dz} + q(z)\right]u(z) = 0, \qquad (36.2)$$

p and q being given functions.

A fundamental property of this equation is that if $u_1(z)$ and $u_2(z)$ are two differential solutions,

$$u(z) = c_1 u_1 + c_2 u_2 \qquad (36.3)$$

is also a solution and, as a matter of fact, since it contains two arbitrary integration constants c_1 and c_2, it is the general solution.

If on the right-hand side of (36.2) we find an arbitrary function $f(z)$, the equation becomes *inhomogeneous*, although still *linear*. If $u_0(z)$ is any particular solution of this equation,

$$u(z) = u_0 + c_1 u_1 + c_2 u_2 \qquad (36.4)$$

is the general solution. Similar results are obtained for linear equations of higher order.

The Wronskian of Second-Order Equations

We have already seen under the heading of eigenvalue problems how a Sturm-Liouville equation can be used in various ways on a much simpler form. The same is true in the above case. Transforming only the dependent variable by

$$u(z) = e^{-\frac{1}{2}\int p\, dz} v(z), \qquad (36.5)$$

we obtain the standard form

$$v'' + Qv = 0 \qquad (36.6)$$

of the homogeneous and correspondingly

$$v'' + Qv = f \qquad (36.7)$$

for the inhomogeneous equation with

$$Q = q - \tfrac{1}{4}p^2 - \tfrac{1}{2}p'. \qquad (36.8)$$

Let v_1 and v_2 be two different solutions of the homogeneous equation,

$$v_1'' + Qv_1 = 0,$$
$$v_2'' + Qv_2 = 0. \qquad (36.9)$$

Multiplying by v_2 and v_1 and subtracting, we obtain

$$(v_1 v_2' - v_1' v_2)' = 0. \tag{36.10}$$

The above determinant is called the *Wronskian* of the solutions. It is a constant which by suitable normalization of the solutions, or rather their product $u_1 u_2$, can be given any arbitrarily chosen value. The value 1, of course, is convenient in a great many cases. Hence we write for the Wronskian of standard solutions

$$v_1 v_2' - v_1' v_2 = 1. \tag{36.11}$$

In the case of the simple equation $v'' = 0$, or $v'' + v = 0$, $v_1 = 1$, $v_2 = z$ or $v_1 = \cos z$, $v_2 = \sin z$ are standard solutions obeying the relation (36.11).

Combining (36.7) with either of (36.9) in the same way, we obtain

$$(v_1 v' - v_1' v)' = v_1 f, \qquad v_1 v' - v_1' v = \int_a^z v_1 f \, dz,$$
$$(v_2 v' - v_2' v)' = v_2 f, \qquad v_2 v' - v_2' v = \int_b^z v_2 f \, dz. \tag{36.12}$$

The Wronskians of the general solution v, and either of v_1 and v_2, are here given by integrals containing two arbitrary constants a and b. Multiplying by v_2 and v_1 and subtracting, using the normalization (36.11), we obtain

$$v = v_2 \int_a^z v_1 f \, dz + v_1 \int_z^b v_2 f \, dz. \tag{36.13}$$

This means that if the solution of the homogeneous equation is known, the general solution of the inhomogeneous equation containing two arbitrary constants a and b can be found immediately by quadratures.

If only one of the solutions v_1 and v_2 is known, the other can be found from the constant Wronskian, writing for (36.11)

$$\left(\frac{v_2}{v_1}\right)' = \frac{1}{v_1^2}, \tag{36.14}$$

which by integration yields

$$v_2 = v_1 \int_a^z \frac{dz}{v_1^2}. \tag{36.15}$$

Putting for instance $v_1 = \cos z$, $a = 0$, we obtain

$$v_2 = \cos z \int_0^z \frac{dz}{\cos^2 z} = \cos z \cdot \tan z = \sin z. \tag{36.16}$$

170 / The Mathematical Foundation of Physics

From this result it follows that we should be able to obtain the general solution from only one of the solutions v_1 and v_2. Starting from the first of (36.12)

$$\left(\frac{v}{v_1}\right)' = \frac{1}{v_1^2} \int_a^z v_1 f \, dz, \qquad (36.17)$$

which yields by another integration the general solution

$$v = v_1 \int_b^z \frac{1}{v_1^2} \int_a^z v_1(\zeta) f(\zeta) \, d\zeta \, dz \qquad (36.18)$$

with two integration constants.

Solution by Power Series. The Homogeneous Equation in a Regular Point

The above results clearly demonstrate that the central problem in the theory is to find the solutions of the homogeneous equations. For this reason it is highly important to have a general method of solution, and this is provided by the use of power series.

Writing as usual $v(z) = P(z - a)$ for a power series in the region of a point a, we consider first the case that a is quite an ordinary or *regular* point, which for convenience we first take to be $a = 0$.

If in (36.6) the function $Q(z)$ can be expanded in an ordinary power series

$$Q(z) = Q_0 + Q_1 z + Q_2 z^2 + \cdots, \qquad (36.19)$$

the solution of the equation can be expanded correspondingly,

$$v(z) = c_0 + c_1 z + c_2 z^2 + \cdots. \qquad (36.20)$$

Inserting (36.20) in (36.6), we get

$$\sum_{n=2}^{\infty} n(n-1) c_n z^{n-2} + \left(\sum_{n=0}^{\infty} Q_n z^n\right)\left(\sum_{n=0}^{\infty} c_n z^n\right) = 0, \qquad (36.21)$$

or, comparing for a single power,

$$(n+1)(n+2) c_{n+2} + \sum_{k=0}^{n} Q_{n-k} c_k = 0. \qquad (36.22)$$

Hence the coefficient c_{n+2} can be calculated from the n first coefficients in (36.20). The choice of the first two of them, c_0 and c_1, provides a set of integration constants. Choosing $c_0 = 1$, $c_1 = 0$, we obtain a standard solution $v_1(z)$ and another $v_2(z)$ by choosing, say, $c_0 = 0$, $c_1 = 1$. These solutions then obey (36.11) and the general solution is

$$v(z) = a_1 v_1(z) + a_2 v_2(z). \qquad (36.23)$$

If, for instance, we put $Q(z) = 1$ in (36.19), then (36.21) and (36.22) lead to the solution

$$v(z) = c_0\left(1 - \frac{z^2}{1\cdot 2} + \frac{z^4}{1\cdot 2\cdot 3\cdot 4} - \cdots\right) + c_1\left(z - \frac{z^3}{1\cdot 2\cdot 3} + \cdots\right)$$
$$= c_0 \cos z + c_1 \sin z. \tag{36.24}$$

The solution (36.23) is valid within the region of convergence of the series which will be the same as that of the $Q(z)$ series (36.19), namely, the circle of convergence going through the nearest singular point of the function $Q(z)$. The simplest way of proving this statement is to expand the solution in a series

$$v(z) = v_0(z) + v_1(z) + v_2(z) + \cdots \tag{36.25}$$

and requiring successive functions to obey the equations

$$v_0'' = 0,$$
$$v_n'' + Qv_{n-1} = 0, \qquad n = 1, 2, \cdots. \tag{36.26}$$

The solutions of these equations are

$$v_0 = c_0 + c_1 z, \qquad v_n = -\int_0^z \left[\int_0^z Q(\zeta) v_{n-1}(\zeta)\, d\zeta\right] dz. \tag{36.27}$$

Applying this to the case of $Q = 1$, we obtain for successive solutions the successive powers of z corresponding to (36.24). This series has an infinite radius of convergence corresponding to $Q(z) = 1$ having no singularities.

To prove the convergence of (36.25) assume

$$|Q(z)| \leq M^2 \qquad \text{and} \qquad |c_0| \leq \alpha, \qquad |c_1| \leq \alpha M, \tag{36.28}$$

the first inequality being valid within a circle of radius ρ somewhat smaller than the radius of convergence. It follows that

$$|v_1(z)| \leq \alpha\left(\frac{M^2 |z|^2}{1\cdot 2} + \frac{M^3 |z|^3}{1\cdot 2\cdot 3}\right). \tag{36.29}$$

Continuing the process of integration, keeping in mind that the absolute value of a product is always smaller than the product of absolute values of the factor, we obtain the general result

$$|v_n(z)| \leq \alpha\left[\frac{(M\rho)^{2n}}{(2n)!} + \frac{(M\rho)^{2n+1}}{(2n+1)!}\right]. \tag{36.30}$$

It follows by summation that

$$|v(z)| \leq \alpha e^{M\rho}, \tag{36.31}$$

hence the series (36.25) converges better than the convergent series for the exponential function $e^{M\rho}$ within the circle of convergence.

Power Series at Singular Points

Singular points of $Q(z)$ are called singular points of (36.6) or the corresponding (36.2). Singular points may be grouped into two classes, *regular* and *irregular* singular points. In (36.2) $Q(z)$ has no singularity other than a pole that is not larger than the second order.

Using a suitable transformation, again at $z = 0$, we write

$$Q(z) = Q_{-2}z^{-2} + Q_{-1}z^{-1} + Q_0 + \cdots = \sum_{n=-2}^{\infty} Q_n z^n. \tag{36.32}$$

It appears that the solution must now be written in the form of a power series in a more general sense

$$v(z) = z^\rho P(z) = \sum_{n=0}^{\infty} c_n z^{\rho+n} \tag{36.33}$$

and the coefficient of the expansion must obey the system of equations

$$(n + \rho)(n - 1 + \rho)c_n + \sum_{k=0}^{n} Q_{n-2-k} c_k = 0. \tag{36.34}$$

Here, in general, only the coefficient c_0 can be chosen freely, provided that

$$\rho(\rho - 1) + Q_{-2} = 0 \quad \text{or} \quad \rho = \binom{\rho_1}{\rho_2} = \frac{1}{2} \pm \left(\frac{1}{4} - Q_{-2}\right)^{1/2}. \tag{36.35}$$

For convenience we may write

$$\begin{aligned} Q_{-2} &= -m(m+1), & \rho_1 &= m+1, \\ \rho_2 &= -m, & \rho_1 - \rho_2 &= 2m+1. \end{aligned} \tag{36.36}$$

If the latter difference of the two initial exponents is not an integer, that is, if m is not integral or half-integral, it is possible for any of the exponents to calculate $c_1, c_2, \cdots$ in the expansion (36.33) by (36.34) from the chosen value of c_0. Choosing for instance $c_0 = 1$ and denoting the solutions by v_1 and v_2, we have the general solution

$$v(z) = a_1 v_1(z) + a_2 v_2(z) \tag{36.37}$$

with two integration constants a_1 and a_2.

Logarithmic Solutions

If the exponent difference in (36.36) is a positive integer, the solution $v_1(z)$ can still be found as above. Equation 36.34 fails, however, to determine the $(2m + 1)$th coefficient for $v_2(z)$, that is, the coefficient of $z^{\rho_2+2m+1} = z^\rho$. This is due to the vanishing coefficient $(2m + 1 + \rho_2)(2m + \rho_2) + Q_{-2} = \rho_1(\rho_1 - 1) + Q_{-2}$ of c_{2m+1} in (36.34).

We begin by writing

$$v_1 = z^{m+1}(1 + c_1 z + c_2 z^2 + \cdots),$$
$$v_1^{-2} = z^{-2m-2} + C_{-2m-1} z^{-2m-1} + \cdots + C_{-1} z^{-1} + \cdots \quad (36.38)$$

and use (36.15) for determining $v_2(z)$. The result is a series in integral or half-integral powers beginning with $-z^{-m/2m+1}$, together with an additive term $C_{-1} \log z \cdot v_1(z)$. The coefficients of the series from z^{-m} to z^m are uniquely determined. The coefficient of z^{m+1}, however, can be freely chosen, and this means of course only that the solution $v_1(z)$ with an arbitrary factor may be added or subtracted.

Let us see now how this can be found directly by putting

$$v_2(z) = A \log z \cdot v_1(z) + w_2(z). \quad (36.39)$$

Inserting the expression in (36.6), we determine w_2 by the inhomogeneous equation

$$w_2'' + Q w_2 + A \left(\frac{2}{z} v_1' - \frac{1}{z^2} v_1 \right) = 0. \quad (36.40)$$

Expanding now w_2 in a power series beginning with $-z^{-m/(2m+1)}$ in order to satisfy (36.11), we determine the coefficients up to the power z^m in the usual way. Inserting the power z^{m+1} in the expansion of w, however, we find that the coefficient may be chosen arbitrarily and that the equation must be satisfied by choosing a definite value of A. The following coefficients in the expansion of $w_2(z)$ are then uniquely determined apart from the said arbitrariness in the coefficient of z^m.

The problem of the character of the solutions in an ordinary singular point is not fully discussed without mentioning that of a pole of the first order,

$$Q(z) = Q_{-1} z^{-1} + Q_0 + \cdots \quad (36.41)$$

where the initial exponents are $\rho_1 = 1$, $\rho_2 = 0$. It is easily seen that a solution $v = v_1(z)$ exists, starting with the first power of z. The other solution $v_2(z)$ cannot be expressed by an ordinary power series. It is easily found that we shall have to add the term $A \log z \cdot v_1(z)$. Hence this case obviously is only a special case of the problem just described with logarithmic second solution $v_2(z)$ due to integral exponent difference $\rho_1 - \rho_2$.

Irregular Singular Points

Considering the very strong restriction on what are called regular singular points, it is to be expected that irregular singular points will in general provide tremendous difficulties for determining the solutions of an equation. This is true. However, there are also various kinds of irregular singular points, and we shall preferably be concerned with those that are less complicated. The common feature of such points is that no solution of the differential equation exists in the form of a power series at that point.

Consider the simple equation

$$v''(z) + v(z) = 0 \qquad (36.42)$$

which has no singular point except at infinity $z = \infty$. This, however, is an irregular or essential singular point. If we try to expand the solution in a series of negative powers z^{-1}, we shall find that this method does not work. Introducing the new independent variable $\zeta = z^{-1}$, (36.42) transforms into

$$\zeta^2 \frac{d}{d\zeta} \zeta^2 \frac{dv}{d\zeta} + v = 0 \quad \text{or} \quad v'' + \frac{2}{\zeta} v' + \frac{v}{\zeta^4} = 0. \qquad (36.43)$$

The solutions of this equation are known and may be written $v = e^{\pm iz} = e^{\pm i/\zeta}$. Obviously they cannot be expanded in positive powers of ζ. This is a consequence of the coefficient of v in (36.43) having as strong an irregularity as a pole of the fourth order.

To study the character of the solution $v = e^{i/\zeta}$ in $\zeta = 0$, put $\zeta = \pm i\varepsilon$, ε being small. This yields the values $e^{\pm 1/\varepsilon}$ of v, meaning that at one side of the point $\zeta = 0$ the function v is extremely large, to the opposite side extremely small. Moreover, following a small circle around the point, the function is violently oscillating.

Asymptotic Solutions

In spite of the fact that the solutions of (36.42) do not exist in the form of a power series in $1/z$, we are able to express them by means of the exponential function. By less simple equations with irregularities of the same nature, as for instance the Bessel equation,

$$\frac{d^2}{dz^2} + \frac{1}{z} \frac{d}{dz} + 1 - \frac{m^2}{z^2} u = 0, \qquad (36.44)$$

the exponential functions $e^{\pm iz}$ are very approximately solutions of the equation for large values of $|z|$. Such approximate solutions in the neighborhood of a singular point are called *asymptotic solutions*. The name is not reserved for irregular singular points only. At the regular singular point $z = 0$ of (36.6) z^{ρ_1} and z^{ρ_2} are asymptotic solutions.

$$z^{\rho_1} + c_1 z^{\rho_1 + 1}$$

is still an asymptotic solution, however, of higher accuracy. In this case the asymptotic solution may turn into the exact solution by continuing the series. The series, whether infinite or finite, converges.

Also, at an irregular singular point the first rough asymptotic solution can be amended by expansion in series. Formally we may even find infinite series.

These series are divergent, however, in a special way called semiconvergence. This means that, even though the series as a whole must be rejected, some few or sometimes a great many terms can be used to give a fairly good asymptotic description of the function. The name semiconvergence of a divergent asymptotic series comes from the fact that, to begin with, the successive terms are decreasing in a manner similar to those of convergent series. Later on, the terms are gradually and, finally, violently increasing. The cutting of the series after the minimum terms yields bad results, cutting off before fairly good results, and cutting off at minimum terms yields the best and sometimes excellent results.

In the above case of the Bessel equation (36.44) the poorest asymptotic solution $e^{\pm iz}$ may be split off. Writing

$$u(z) = e^{\pm iz}v(z), \tag{36.45}$$

we obtain from (36.44) the differential equation

$$\frac{d^2}{dz^2} \pm 2i\frac{d}{dz} + \frac{1}{z}\frac{d}{dz} \pm \frac{i}{z} - \frac{m^2}{z^2}v = 0, \tag{36.46}$$

and a very simple formal solution of this equation exists in the form of a power series in z^{-1} beginning with the power $z^{-\frac{1}{2}}$. The series breaks off for half-integral m. In any other case the series is infinite and as already stated divergent but in a semiconvergent manner.

Beside the concept of asymptotic solutions we also frequently use the phrase the *asymptotic form* of a differential equation. Consider, for instance, the Bessel equation (36.44), assuming $u(z)$ and its derivatives to be of the same order of magnitude. For large values of $|z|$, then, the second and the fourth term are small and, hence, of secondary importance. The asymptotic form therefore is

$$\left(\frac{d^2}{dz^2} + 1\right)u = 0 \tag{36.47}$$

with asymptotic solutions $u(z) = e^{\pm iz}$, as first pointed out.

There is, however, some difference in the order of magnitude of the smaller terms, the first of them being small only of the order z^{-1}. For this reason it is advisable to use the transformation $u = z^{-\frac{1}{2}}v(z)$ and obtain an equation of the form (36.6), namely,

$$\frac{d^2}{dz^2} + 1 - \frac{m^2 - \frac{1}{4}}{z^2}v = 0, \tag{36.48}$$

for which

$$v'' + v = 0 \tag{36.49}$$

176 / The Mathematical Foundation of Physics

is a much better asymptotic form. Hence $u(z) = z^{-1/2}e^{\pm z}$ is a more correct asymptotic solution of Bessel's equation than $u(z) = e^{\pm iz}$. In the case of $m = \frac{1}{2}$ the first form even yields the exact solution. Even for higher half-integral m the solution becomes exact by the addition of $m - \frac{1}{2}$ higher powers of $1/z$.

37. ARRANGEMENT OF DIFFERENTIAL EQUATIONS ACCORDING TO THEIR SINGULAR POINTS

Differential Equations with Not More than Two Singular Points

Consider the very simplest differential equation of the second order and its solution

$$\frac{d^2v}{dz^2} = 0, \quad v(z) = c_0 + c_1 z. \tag{37.1}$$

The equation has no singularities at all for finite values of z. However, as seen from the solution there is a singular point at infinity. Using the $\zeta = z^{-1}$, we obtain from (37.1)

$$\frac{d^2v}{d\zeta^2} + \frac{2}{\zeta}\frac{dv}{d\zeta} = 0, \quad v = \frac{c_1}{\zeta} + c_0. \tag{37.2}$$

Again, (37.2) may be transformed into

$$\frac{d^2w}{d\zeta^2} = 0, \quad w = c_1 + c_0\zeta \tag{37.3}$$

by the transformation $v = \zeta^{-1}w$. This means that the singularity can be removed from $z = \infty$ but then it reappears at $z = 0$. Hence a differential equation of the second order with no singularities does not exist.

In the following we shall return to the original standard form

$$\left(\frac{d^2}{dz^2} + p\frac{d}{dz} + q\right)u = 0 \tag{37.4}$$

for our second-order differential equation. If then $p(z)$ has a pole of the first and $q(z)$ one of the second order in a point a, this point is a regular singular point. If

$$p(z) = \frac{p_{-1}}{x} + p_0 + \cdots, \quad q(z) = \frac{q_{-2}}{z^2} + \frac{q_{-1}}{z} + \cdots, \tag{37.5}$$

the equation determining the initial exponents of the solutions reads

$$\rho(\rho_{-1} + p_{-1}) + q_{-2} = 0. \tag{37.6}$$

If $q_{-2} = 0$ one of the exponents is zero. This can always be achieved by splitting off, say, a factor $(z - a)^{\rho_1}$ from the solution of (37.4) by writing $u = (z - a)^{\rho_1} v$. We shall use this repeatedly in the following in order to simplify our equations.

Passing to equations with only two regular singularities of which one is $z = a$ and the other $z = \infty$, we may write

$$\left(\frac{d^2}{dz^2} + \frac{1 - \beta}{z - a} \frac{d}{dz} \right) u = 0 \qquad (37.7)$$

with initial exponents 0 and β in a. At infinity, using $\zeta = z^{-1}$ as the independent variable the initial exponents are 0 and $-m$. The singular point $z = \infty$, of course, can be transformed to a finite point, say, by the transformation $z = 1/(\zeta - b)$, which yields the equation

$$\frac{d^2 u}{d\zeta^2} + \left[\frac{1 - \beta}{\zeta - b - 1/a} + \frac{1 + \beta}{\zeta - b} \right] \frac{du}{d\zeta} = 0. \qquad (37.8)$$

The singular points $z = a$ and $z = \infty$ are now transformed into $\zeta = b + 1/a$ and $\zeta = b$ and the initial exponents are as before 1, m and 1, $-m$, respectively.

Confluence of Singular Points

Second-order equations with only two ordinary singular points are therefore too elementary to be of much interest unless one of the points is an irregular singularity. An irregular singularity, however, may be thought of as produced by confluence of two regular singular points and hence makes the equation less elementary than a corresponding ordinary point. Selecting for instance in (37.7) $a \to \infty$ putting $\beta = 2ia$, we obtain the equation

$$\left(\frac{d^2}{dz^2} - 2i \frac{d}{dz} \right) u = 0 \qquad (37.9)$$

or

$$\left(\frac{d^2}{dz^2} + 1 \right) v = 0 \qquad (37.10)$$

by the substitution $u = e^{iz} v$. This is in accord with the solution $u = 1$ and $u = (1 - z/a)^{1+2ia}$ becoming $u = 1$ and $u = e^{2iz}$ as $a \to \infty$, corresponding to $v = e^{\pm iz}$.

Differential Equations with Three or More Singular Points

We shall now introduce a differential equation with an arbitrary number $n + 1$ of regular singular points $a_1, a_2, \cdots, a_n$ together with $z = \infty$. By suitable transformations the initial exponents may be taken to be 0 and β_ν at

178 / The Mathematical Foundation of Physics

a_ν. Then we have the equation

$$\left(\frac{d^2}{dz^2} + \sum_{\nu=1}^{n} \frac{1-\beta_\nu}{z-a_\nu}\frac{d}{dz} + \frac{A_{n-2}z^{n-2} + \cdots + A_0}{\prod_{\nu=1}^{n}(z-a_\nu)}\right)u = 0. \quad (37.11)$$

The asymptotic form of this equation as $z \to \infty$ is

$$\left[\frac{d^2}{dz^2} + \left(n - \sum_\nu \beta_\nu\right)\frac{1}{z}\frac{d}{dz} + \frac{A_{n-2}}{z^2}\right]u = 0, \quad (37.12)$$

proving that $z = \infty$ is a regular singular point with initial exponents of $1/z$ given by the equation

$$\rho^2 + \left[\sum_\nu \beta_\nu - (n+1)\right]\rho + A = 0. \quad (37.13)$$

Many of the equations of mathematical physics can be obtained from (37.12) by starting with three singular points and making two of them join into a single irregular singularity by confluence. It has been found, however, that a better review can be given by starting with five special singularities of exponent difference $\frac{1}{2}$. The corresponding equation reads

$$\left[\frac{d^2}{dz^2} + \sum_{\nu=1}^{4}\frac{\frac{1}{2}}{z-a_\nu}\frac{d}{dz} + \frac{Az^2 + Bz + C}{(z-a_1)(z-a_2)(z-a_3)(z-a_4)}\right]u = 0. \quad (37.14)$$

It is easily seen that the confluence of any two points, say, $a_3 = a_4$ produces a regular singular point that may have any exponent difference.

In (37.14) $z = \infty$ has an exponent difference $\frac{1}{2}$ only if $a = \frac{3}{16}$. For any other value of A the difference has another value and $z = \infty$ is an ordinary singular point. By arbitrary A (37.14) may therefore be taken to correspond to an equation with six singular points of exponential difference $\frac{1}{2}$ and to five if the singular point a_4 is left out. This equation is directly the well-known Lamé's equation associated with ellipsoidal harmonics. Mathieu's equation which originates from problems associated with the elliptic cylinder has two singularities of difference $\frac{1}{2}$ and a third irregular singularity caused by the confluence of three points of the above sort.

Both Lamé's and Mathieu's equations are too difficult to be dealt with here. Therefore, the consideration of singular points of exponent difference $\frac{1}{2}$ loses much of its interest. It is true that the Legendre and Hermite equations may be written in a form containing a singular point of that sort. It can be removed, however, by introducing $z^{1/2}$ as the independent variable.

The Bessel equation also can be deduced from an equation with five original singularities of exponent difference $\frac{1}{2}$. It is, however, a specialized confluent hypergeometric equation, as the Legendre equation is a specialized hypergeometric equation. These equations can be obtained only from an original equation with six singular points. Leaving aside the very difficult Lamé's and Mathieu's equations, we shall prefer to base our considerations on an equation with three ordinary singular points as obtained from (37.11), putting $n = 2$, or from (37.14) by arbitrary A, putting $a_3 = a_1$ and $a_4 = a_2$. From (37.11) we then obtain the most suitable standard form

$$\left[\frac{d^2}{dz^2} + \left(\frac{1-\alpha}{z-a} + \frac{1-\beta}{z-b}\right)\frac{d}{dz} + \frac{A}{(z-a)(z-b)}\right]u = 0. \quad (37.15)$$

The Hypergeometric Equation. Legendre's Equation

Equation 37.15 is the general hypergeometric equation. It is usual to place the singular points a and b at $z = 0$ and $z = 1$. Moreover, the initial exponents at $z = \infty$ are denoted by a and b and those at $z = 0$ by 0 and $1 - c$. The equation then becomes the well-known standard hypergeometric equation

$$\left\{z(1-z)\frac{d^2}{dz^2} + [c - (a+b+1)z]\frac{d}{dz} - ab\right\}u = 0, \quad (37.16)$$

of which one of the solutions is the hypergeometric series

$$u(z) = F(a, b, c, z) = 1 + \frac{a \cdot b}{1 \cdot c}z + \frac{a(a+1) \cdot b(b+1)}{1 \cdot 2 \cdot c(c+1)}z^2 + \cdots. \quad (37.17)$$

If in (37.15) we write $a = 0$, $b = 1$, $\alpha = \frac{1}{2}$, $\beta = -m$, $A = -\frac{1}{4}(n-m) \times (n+m+1)$,

$$\left[\frac{d^2}{dz^2} + \left(\frac{1}{2z} + \frac{m+1}{z-1}\right)\frac{d}{dz} - \frac{(n-m)(n+m+1)}{4z(z-1)}\right]u = 0 \quad (37.18)$$

we obtain on introducing $z = \xi^2$ the Legendre equation

$$\left[(1 - \xi^2)\frac{d^2}{d\xi^2} - (2m+2)\xi\frac{d}{d\xi} + n(n+1) - m(m+1)\right]u = 0 \quad (37.19)$$

of which one solution is the mth derivative of the Legendre polynomial $P_n(\xi)$,

$$u = P_n^{(m)}(\xi). \quad (37.20)$$

This, however, is only another hypergeometric function of the variable ξ. Writing $a = 1$, $b = -1$, and $\alpha = \beta = -m$, $A = -n(n+1) + m(m \pm 1)$, and finally $z = \xi$ in (37.15), we obtain (37.19).

The Confluent Hypergeometric Equation

If in (37.15) we put $a = 0$, $\alpha = -m$, making b, β, and $A \to \infty$ keeping β/b and A/b constant, we obtain the confluent hypergeometric equation

$$\left[z \frac{d^2}{dz^2} + (m + 1 - z) \frac{d}{dz} + n \right] u = 0 \qquad (37.21)$$

by suitable choice of β/b and A/b. If m and n are positive integers, this is the differential equation for the Laguerre polynomials $L^m_{n+m}(z)$ which is the nth derivative of $L_{n+m}(z)$.

Making the substitution

$$u(z) = z^{-m/2} e^{-z/2} v(z), \qquad (37.22)$$

we obtain the equation

$$\left[\frac{d^2}{dz^2} + \frac{1}{z} \frac{d}{dz} - \frac{m^2}{4z^2} - \frac{1}{4} + \frac{n + \frac{1}{2}(m + 1)}{z} \right] v(z) = 0 \qquad (37.23)$$

which is the differential equation for the associated Laguerre functions.

Putting $n + \frac{1}{2}(m + 1) = 0$, replacing m by $2m$ and z by $2iz$, we have the Bessel equation

$$\left(\frac{d^2}{az^2} + \frac{1}{z} \frac{d}{dz} - \frac{m^2}{z^2} + 1 \right) J_m(z) = 0. \qquad (37.24)$$

Again in (37.21), putting $m = -\frac{1}{2}$ and writing $\frac{1}{2}(n + \frac{1}{2})$ for n, we obtain the transformation $z = \xi^2$

$$\left[\frac{d^2}{d\xi^2} - 2\xi \frac{d}{d\xi} + (2n + 1) \right] H_n(\xi) = 0 \qquad (37.25)$$

which is the equation for the nth Hermite polynomial.

The above considerations suffice to prove that hypergeometric and confluent hypergeometric functions cover a very wide range of functions needed in mathematical physics, the Legendre, Bessel, and Hermite functions being only special modifications of the former.

38. HYPERGEOMETRIC FUNCTIONS

The Hypergeometric Series

The solution of the hypergeometric equation (37.16) can be expressed in a great variety of ways. In (37.17) we have already given the expansion of one of the solutions in the singular point $z = 0$. The expansion of the second solution may be found by the following consideration.

The initial exponents of power series at $z = 0, 1, \infty$ are 0 and $1 - c$, 0 and $a + b - c$, a and b, respectively. To obtain the second solution at $z = 0$, we remove the factor z^{1-c}, writing $u = z^{1-c}v$, and consider the equation for v. This does not alter the initial exponents at $z = 1$. However, at infinity it raises the exponents by $1 - c$ to $a + 1 - c$ and $b + 1 - c$, and at $z = 0$ they are now 0 and $c - 1$. Hence the second solution is

$$u = z^{1-c}F(a + 1 - c, b + 1 - c, 2 - c, z). \tag{38.1}$$

The expansions (37.17) and the above equation are valid simultaneously only for nonintegral c.

At $z = 1$ and $z = \infty$, using $\zeta = \frac{1}{2}$ or $\zeta = 1/(1 - z)$, similar expressions may be found for either of two independent solutions that are in general different from the solutions at $z = 0$. Moreover, since the various circles of convergence are passing through the nearest second singular point, the regions of convergence are different for the various types of series. In addition alternative expressions may be found by transformations using $z/(1 - z)$ and $(1 - z)$ as new variables. We are not interested here in listing all these solutions and we shall therefore return to the standard hypergeometric series $F(a, b, c, z)$ and study some of its properties.

The ratio of successive coefficients of the series is

$$\frac{c_{n+1}}{c_n} = \frac{(n + a)(n + b)}{(n + 1)(n + c)} = 1 - \frac{1 + c - a - b}{n} + O\left(\frac{1}{n^2}\right). \tag{38.2}$$

Therefore the radius of convergence is 1, as is known beforehand from the position of the second singular point at $z = 1$. According to convergence theorems it follows from (2) that the hypergeometric series is absolutely convergent if Re $(c - a - b) > 0$.

Integral Representation of the Hypergeometric Series

By means of gamma functions, as denoted by factorials, the hypergeometric series may be written

$$F(a, b, c, z) = \frac{(c - 1)!}{(a - 1)!\,(b - 1)!} \sum_{n=0}^{\infty} \frac{(n - 1 + a)!\,(n - 1 + b)!}{n!\,(n + 1 + c)!} z^n. \tag{38.3}$$

Introduce the Euler integral of the first kind,

$$\frac{(n - 1 + b)!\,(c - 1 - b)!}{(n - 1 + c)!} = \int_0^1 s^{n-1+b}(1 - s)^{c-1-b}\,ds, \tag{38.4}$$

assuming for the sake of simplicity that the integral is convergent so that the replacement of (38.4) by a contour integral is not needed. Inserting (38.4) in (38.3), interchanging the integration and summation, and noting that

$$\sum_{n=0}^{\infty} \frac{(n - 1 + a)!}{n!} s^n z^n = (a - 1)!\,(1 - sz)^{-a}, \tag{38.5}$$

we obtain the result

$$F(a, b, c, z) = \frac{(c-1)!}{(c-1-b)!(b-1)!}$$

$$\times \int_0^1 s^{-1+b}(1-s)^{c-1-b}(1-sz)^{-a}\, ds. \quad (38.6)$$

The Value of $F(a, b, c, 1)$

Again, putting $z = 1$, (38.6) turns into a Euler integral of the first kind and we obtain the result

$$F(a, b, c, 1) = \frac{(c-1)!(c-1-a-b)!}{(c-1-a)!(c-1-b)!} \quad (38.7)$$

provided that none of the above steps of modification invalidates the requirements of convergence. Of course, if that is to be true, the values of a, b, and c separately are very much restricted. The procedure may be satisfactorily amended in this respect, however, by using convergent contour integrals.

The formula (38.7), which is an important and useful one, however, is correct and is valid if Re $(c - a - b) > 0$.

Legendre Functions of the First Kind

Consider the Legendre differential equation

$$\left[(1-z^2)\frac{d^2}{dz^2} - 2z\frac{d}{dz} + n(n+1)\right]P_n(z) = 0 \quad (38.8)$$

together with that obtained by m derivations,

$$\left[(1-z^2)\frac{d^2}{dz^2} - (2m+2)z\frac{d}{dz} + n(n+1) - m(m+1)\right]P_n^{(m)}(z) = 0, \quad (38.9)$$

and finally the equation for

$$P_n^m(z) = (1-z^2)^{m/2}P_n^{(m)}(z), \quad (38.10)$$

namely,

$$\left[(1-z^2)\frac{d^2}{dz^2} - 2z\frac{d}{dz} - \frac{m^2}{1-z^2} + n(n+1)\right]P_n^m(z). \quad (38.11)$$

For integral n, the $P_m(z)$ are ordinary, and the $P_n^m(z)$ associated, Legendre functions forming for any m by $n \geq m$ complete orthogonal systems of functions. These functions are solutions of the Legendre equations (38.8) and (38.11) which are finite only at the third singular point $z = 1$. They are normalized by the condition $P_n(\pm 1) = (\pm 1)^n$.

Differential Equations and Particular Functions | 183

Explicit expressions and definitions of polynomials $P_n(z)$ have been given in the section on orthogonal polynomials [(20.22) to (20.28); (20.41) and (20.42)].

In order to prove readily various useful formulas we shall repeat the definition

$$P_n(z) = u_n^{(n)}(z), \tag{38.12a}$$

$$u_n = \frac{1}{2^n n!}(z^2 - 1)^n. \tag{38.12b}$$

By successive differentiations of (38.12b) or (20.27) we obtain the equations (38.8) and (38.9). To obtain other formulas we shall combine derivatives of u_n, u_{n+1} and u_{n-1}. Using the relations

$$u'_{n+1} - xu_n = 0,$$
$$xu'_{n+1} - u_n = (2n+2)u_{n+1}, \tag{38.13}$$

we obtain by $n + 1 + m$ differentiations

$$P_{n+1}^{(m+1)} - zP_n^{(m+1)} = (n+1+m)P_n^{(m)},$$
$$zP_{n+1}^{(m+1)} - P_n^{(m+1)} = (n+1-m)P_{n+1}^{(m)}. \tag{38.14}$$

Multiplying the equation by 1 and z and subtracting, or, by z and 1 subtracting and replacing $n+1$ by n, we have

$$(1 - z^2)P_n^{(m+1)} = (n+1+m)zP_n^{(m+1)} - (n+1-m)P_{n+1}^{(m)},$$
$$(1 - z^2)P_n^{(m+1)} = (n+m)P_{n-1}^{(m)} - (n-m)zP_n^{(m+1)}. \tag{38.15}$$

Subtracting, we have the recurrence formula

$$(2n+1)xP_n^{(m)} = (n+1-m)P_{n+1}^{(m)} + (n+m)P_{n-1}^{(m)}. \tag{38.16}$$

Again, we replace $n+1$ by n in the last formula (38.14) and add to the first; then we have

$$P_{n+1}^{(m+1)} - P_{n-1}^{(m+1)} = (2n+1)P_n^{(m)}, \tag{38.17}$$

which is also a useful formula.

From (38.15) we may also partly eliminate $zP_n^{(m+1)}$,

$$(1-z^2)P_n^{(m+1)} - 2mxP_n^{(m)} = \frac{(n+m)(n+1-m)}{2n+1}[P_{n-1}^{(m)} - P_{n+1}^{(m)}]$$
$$= (n+m)(n+1-m)P_n^{(m-1)}, \tag{38.18}$$

by (38.17), and this is the differential equation for $P_n^{(m-1)}$.

Equations (38.14) to (38.17) and the differential equation (38.9) or (38.18) form a consistent system of formulas that are needed frequently.

The normalization integral

$$\int_{-0}^{+1}[P_n{}^m(z)]^2\,dz = \int_{-1}^{+1}(1-z^2)^m[P_n^{(m)}(z)]^2\,dz = \frac{2}{2n+1}\frac{(n+m)!}{(n-m)!} \quad (38.19)$$

is also easily obtained from the definition (38.12) and (38.10). Writing $P_n^{(m)}(z) = u_n^{(n+m)}(z)$ for one of the $P_n^{(m)}$ in (38.18), observing that the highest power for the remaining factor is given by

$$(1-z^2)^m P_n^{(m)}(z) = \frac{1}{2^n n!}\frac{(2n)!\,(-1)^m z^{n+m}}{(n-m)!}, \quad (38.20)$$

and integrating $n+m$ times by part, we obtain the result

$$\int_{-1}^{+1}[P_n{}^m(z)]^2\,dz = \frac{(2n)!}{2^{2n}(n!)^2}\frac{(n+m)!}{(n-m)!}\int_{-1}^{+1}(1-z^2)^n\,dz$$

$$= \frac{1}{2^{2n}}\frac{(n+m)!}{(n-m)!}\int_{-1}^{+1}(1+z)^{2n} = \frac{2}{2n+1}\frac{(n+m)!}{(n-m)!} \quad (38.21)$$

Integral Representation of $P_n(z)$

From the definition (38.12), which is called *Rodrigues' formula*, we may by the Cauchy theorem pass directly to *Schläfli's integral* for $P_n(z)$,

$$P_n(z) = \frac{1}{2\pi i}\oint^{(+z)}\frac{(s^2-1)^n}{2^n(s-z)^{n+1}}\,ds, \quad (38.22)$$

denoting by $(+z)$ that the integral be taken counterclockwise around z. Integrals of the same form, however, with a modified path of integration may be used for definition of other kinds of Legendre functions.

If the integral (38.21) is taken on a circle of given radius, we may obtain the well-known integral representation due to Laplace,

$$P_n(z) = \frac{1}{\pi}\int_0^{\pi}(z+\sqrt{z^2-1}\cos\varphi)^n\,d\varphi. \quad (38.23)$$

It is obtained by writing

$$s = z + \sqrt{z^2-1}\,e^{i\varphi}, \quad -\pi \leq \varphi \leq \pi, \quad (38.24)$$

which means that the path of integration is a circle of radius $\sqrt{|z^2-1|}$. Because of

$$s^2 - 1 = 2\sqrt{z^2-1}\,e^{i\varphi}(z+\sqrt{z^2-1}\cos\varphi), \quad s-z = \sqrt{z^2-1}\,e^{i\varphi},$$

$$ds = i\sqrt{z^2-1}\,e^{i\varphi}\,d\varphi.$$

The result is

$$P_n(z) = \frac{1}{2\pi} \int_{-\pi}^{\pi} (z + \sqrt{z^2 - 1} \cos \varphi)^n \, d\varphi, \qquad (38.25)$$

which is equivalent to (38.23). The integral (38.25) has the advantage of being particularly well suited for a general definition of the function $P_n(z)$ when n is unrestricted.

Generating Function for Legendre Polynomials

Consider a point on the z-axis of a Cartesian coordinate system at a distance a from the origin. Let r_1 be its distance to a variable point x, y, z. Then

$$r_1 = (r^2 - 2ra \cos \vartheta + a^2)^{1/2} \qquad (38.26)$$

as expressed in polar coordinates r, ϑ, φ, the distance being independent of φ. As is well known the reciprocal distance $1/r_1$ satisfies the Laplace equation $\nabla^2(1/r_1) = 0$.

If $r > a$ we must have an expansion of $1/r_1$ in descending powers of r, that is, in ascending powers of a/r,

$$\frac{1}{(r^2 - 2ra \cos \vartheta + a^2)^{1/2}} = \sum_{n=0}^{\infty} P_n(\cos \vartheta) \frac{a^n}{r^{n+1}}, \qquad (38.27)$$

$P_n(\cos \vartheta)$ being so far unknown coefficients depending only on ϑ. What can be said at once, however, is that their values at $\vartheta = 0$ and $\vartheta = \pi$ are the same as for Legendre polynomials, $P_n(\pm 1) = (\pm 1)^n$. This can be seen immediately by putting $\cos \vartheta = \pm 1$. Considering the Laplace operator in spherical coordinates and omitting the differential operator in φ which is of no use here, that is, writing

$$\nabla^2 = \frac{1}{r^2} \left(\frac{\partial}{\partial r} r^2 \frac{\partial}{\partial r} + \frac{1}{\sin \vartheta} \frac{\partial}{\partial \vartheta} \sin \vartheta \frac{\partial}{\partial \vartheta} \right) \qquad (38.28)$$

it is easily seen from the right-hand side of (38.27) that the coefficients $P_n(\cos \vartheta)$ must obey the Laplace equation

$$\left[\frac{d^2}{d\vartheta^2} + \frac{\cos \vartheta}{\sin \vartheta} \frac{d}{d\vartheta} + n(n+1) \right] P_n(\cos \vartheta) = 0, \qquad (38.29)$$

which is the differential equations of the above-defined Legendre polynomials.

Hence, writing $\cos \vartheta = x$, $a/s = t$, it is seen from (38.27) that the Legendre polynomials may be defined by a generating function $\Psi(x, t)$, by the equation

$$\Psi(x, t) = \frac{1}{(1 - 2xt + t^2)^{1/2}} = \sum_{n=0}^{\infty} P_n(x) t^n, \qquad t < 1. \qquad (38.30)$$

186 / The Mathematical Foundation of Physics

The same result may be obtained from Schläfli's integral by the following procedure. We multiply the polynomials in (38.22) (with $z = x$ and $s = z$) successively by the power t^n of a parameter $t < 1$ and calculate the infinite sum

$$\sum_{n=0}^{\infty} P_n(x) t^n = \frac{1}{2\pi i} \sum_{n=0}^{\infty} \oint \frac{(z^2-1)^n t^n}{2^n (z-x)^{n+1}} dz = \frac{1}{2\pi i} \oint \frac{2 dz}{t(z_1-z)(z-z_2)},$$

$$\begin{Bmatrix} z_1 \\ z_2 \end{Bmatrix} = \frac{1}{t} \pm \left(\frac{1}{t^2} - \frac{2x}{t} + 1 \right)^{1/2}. \qquad (38.31)$$

As $t \to 0$, $z_1 \to \infty$, and $z_2 \to x$. The path of integration which is a closed curve encircling the point $z = x$ has been chosen so as to ensure the convergence of the above series. Afterward it may be contracted to a small circle around $z = x$. This yields by the Cauchy theorem

$$\sum_{n=0}^{\infty} P_n(x) t^n = \frac{1}{t(z_1 - z_2)}, \qquad (38.32)$$

which is seen to be the definition (38.30).

Differentiating (38.30) with respect to x, we see that the derivatives $P_n^{(m)}(x)$ of $P_n(x)$ are defined by a generating function as follows:

$$\frac{1 \cdot 3 \cdots (2m-1)}{(1 - 2xt + t^2)^{m+1/2}} = \sum_{n=m}^{\infty} P_n^{(m)} t^{n-m}. \qquad (38.33)$$

Gegenbauer's Polynomials

The Gegenbauer polynomials usually denoted by $C_n^\nu(x)$ may be defined by a generating function in a similar way as the Legendre polynomials,

$$(1 - 2xt + t^2)^{-\nu} = \sum_{n=0}^{\infty} C_n^\nu(x) t^n, \qquad t < 1. \qquad (38.34)$$

In this way the polynomials are defined for any positive value of ν, comprising for the half-integral ν the Legendre polynomials. For $\nu = 0$ the definition fails.

The properties of the polynomials can be obtained from studying the above generating function. As in the case of the Legendre polynomials this is not, however, the easiest way. We shall be satisfied, therefore, by stating only the differential equation for the generating function

$$\left[(1-x^2) \frac{\partial^2}{\partial x^2} - (2\nu+1)x \frac{\partial}{\partial x} + t^2 \frac{\partial^2}{\partial t^2} + (2\nu+1)t \frac{\partial}{\partial t} \right]$$
$$\times (1 - 2xt + t^2)^{-\nu} = 0, \qquad (38.35)$$

which is fairly easily controlled. The differential equation for the polynomials, hence, are

$$\left[(1-x^2)\frac{d^2}{dx^2} - (2\nu+1)x\frac{d}{dx} + n(n+2\nu)\right]C_n^\nu(x) = 0. \quad (38.36)$$

Introducing $x = \cos \varphi$, we get

$$\left[\frac{d^2}{d\varphi^2} + 2\nu\frac{\cos \varphi}{\sin \varphi}\frac{d}{d\varphi} + n(n+2\nu)\right]C_n^\nu(\cos \varphi) = 0, \quad (38.37)$$

and, finally,

$$\left[\frac{d^2}{d\varphi^2} - \frac{\nu(\nu-1)}{\sin^2 \varphi} + (n+\nu)^2\right]\sin^\nu \varphi C_n^\nu(\cos \varphi) = 0. \quad (38.38)$$

In the case of integral ν inclusive of $\nu = 0$, we shall prefer to deal with the orthogonal function of (38.38) which forms a consistent system for all integrals ν. For convenience we shall redefine them by writing

$$\left[\frac{d^2}{d\varphi^2} - \frac{m(m-1)}{\sin^2 \varphi} + n^2\right]u_n^m(\varphi) = 0,$$

$$n = m, m+1, \cdots, \infty, \quad (38.39)$$

the lower index of the Gegenbauer polynomials corresponding to $n - m$. It is possible then to write

$$u_n^{m+1} = -\frac{1}{(n^2 - m^2)^{1/2}} \sin^m \varphi \frac{d}{d\varphi} \sin^{-m} \varphi u_n^m, \quad (38.40a)$$

$$u_n^m = \frac{1}{(n^2 - m^2)^{1/2}} \sin^{-m} \varphi \frac{d}{d\varphi} \sin^m \varphi u_n^{m+1}. \quad (38.40b)$$

Introducing (38.40a) into (38.40b), we obtain the result (38.39). Inserting (38.40b) in (38.40a), we need only change the sign of m in (38.39) to obtain the equation for u_n^{m+1}, which on the other hand is obtainable from (38.39) by writing $m + 1$ for m. It is also easily seen by partial integration, using first (38.40a) and then (38.40b) that u_n^m and u_n^{m+1} have the same normalization integrals.

In the region $0 \leq x \leq \pi$ we now have the following systems of successive orthonormal systems based on the Gegenbauer (and the Tschebyscheff) polynomials

$$u_0 = \frac{1}{\pi^{1/2}}, \quad u_n^0(\varphi) = \left(\frac{2}{\pi}\right)^{1/2} \cos n\varphi, \quad n = 1, 2, \cdots, \quad (38.41a)$$

$$u_n^1(\varphi) = \left(\frac{2}{\pi}\right)^{1/2} \sin n\pi, \quad n = 1, 2, \cdots, \quad (38.41b)$$

$$u_n^2 = \frac{(2/\pi)^{1/2}}{(n^2-1)^{1/2}}\left(\frac{\sin n\varphi \cos \varphi}{\sin \varphi} - n \cos n\varphi\right), \quad n = 2, 3, \cdots, \quad \text{etc.}$$

$$(38.41c)$$

The Delta Function of Gegenbauer Polynomials

In the case of Fourier series and integrals we obtained expressions for δ-like functions which makes it possible to see directly how the expansion of a function generates the function. If the Gegenbauer polynomials are complete systems of functions, the above orthonormal systems must possess similar delta functions. Observing that $2 \cos n\varphi \cos n\varphi' = \cos n(\varphi - \varphi') + \cos n(\varphi + \varphi')$, and so forth, it is easily found by summation of geometric series in $e^{i(\varphi-\varphi')}$ and $e^{-i(\varphi-\varphi')}$ that

$$\delta_N^0 = \sum_{n=0}^{N} u_n^0(\varphi) u_n^0(\varphi') = \frac{\sin(N+\tfrac{1}{2})(\varphi-\varphi')}{2\pi \sin \tfrac{1}{2}(\varphi-\varphi')} + \frac{\sin(N+\tfrac{1}{2})(\varphi+\varphi')}{2\pi \sin \tfrac{1}{2}(\varphi+\varphi')}, \tag{38.42a}$$

$$\delta_N^1 = \sum_{n=1}^{N} u_n^1(\varphi) u_n^1(\varphi') = \delta_N(\varphi - \varphi') - \delta_N(\varphi + \varphi'), \tag{38.42b}$$

where the meaning of δ_N is seen from (38.42a). The functions (38.42a) and (38.42b), of course, are the delta functions of the cosine and sine Fourier series.

Apparently the latter function in (38.42a) and (38.42b) is of no use. Replacing, for instance, the exact expansion of an arbitrary function $f(\varphi)$ by the approximate expression

$$f_N(\varphi) = \sum_{n=0}^{N} f_n u_n^0(\varphi), \quad f_n = \int_0^{\pi} f(\varphi') u_n^0(\varphi') \, d\varphi', \tag{38.43a}$$

we have

$$f_N(\varphi) = \int_0^{\pi} f(\varphi') \, \delta_N^0(\varphi, \varphi') \, d\varphi' \tag{38.43b}$$

and in the limiting case $N \to \infty$ the integral containing $\delta_N(\varphi + \varphi')$ vanishes. There is an exception, namely, $\varphi = 0$, in which case the integration in φ' goes only half through the delta functions. In this case $\delta_N^0 = \delta_N(-\varphi') + \delta_N(\varphi') = 2\delta_N(\varphi')$ and $\delta_N^1 = \delta_N(-\varphi') - \delta_N(\varphi') = 0$, and so for the system (38.43a) we obtain $f_N(0) \to f(0)$, whereas in the case (38.43b) $f_N(0) \to 0$ as $N \to \infty$. This means that the system (38.43a) of cosine functions is adapted for symmetric functions in φ, and the system (38.43b) of sine functions is adapted only to antisymmetric functions with zero value $f(0) = 0$. The cosine and sine series of a given function $f(\varphi)$ in the region $0 \leq \varphi \leq \pi$ can repeat the function outside of this region only in a symmetric or an antisymmetric way. Therefore, if $f(0) \neq 0$ and $f(\pi) \neq 0$, the sine series will contain a discontinuity at $\varphi = 0$ and $\varphi = \pi$, which we shall discuss later. It should also be noted that the additional terms $\pm \delta_N(\varphi + \varphi')$ in (38.42a) and (38.42b) are necessary in order to make the expressions have the correct symmetry and boundary in φ and φ' according to their definitions.

To learn in connection with Gegenbauer functions real facts about the delta functions, and thus of the way in which orthonormal series expansions can reproduce a given function, we shall have to use functional systems of higher m. We may, of course, use the expression (38.41c) and the following and sum up directly products like $u_n{}^m(\varphi)u_n{}^m(\varphi')$ but this proves fairly soon to be a troublesome procedure. Defining the δ-like function in either case by

$$\delta_N{}^m = \sum_{n=m}^{N} u_n{}^m(\varphi)u_n{}^m(\varphi'), \tag{38.44}$$

we find it easier to use the relation

$$\left(\frac{d}{d\varphi} + m\frac{\cos\varphi}{\sin\varphi}\right)\delta_N^{m+1} = \left(-\frac{d}{d\varphi'} + m\frac{\cos\varphi'}{\sin\varphi'}\right)\delta_N{}^m, \tag{38.45}$$

the correctness of which can be controlled by the definition (38.40). It is not difficult to prove, without using any explicit expressions as in (38.41), that

$$\delta_N{}^2 = \delta_N{}^0 - \frac{1}{\sin\varphi \sin\varphi'}\left[\frac{u'_N(\varphi)u'_N(\varphi')}{N} + \frac{u'_{N+1}(\varphi)u'_{N+1}(\varphi')}{N}\right] \tag{38.46a}$$

$$\delta_N{}^3 = \delta_N{}^1 - \frac{1}{\sin\varphi \sin\varphi'}\left[\frac{u_N(\varphi)u_N{}^2(\varphi')}{N} + \frac{u^2_{N+1}(\varphi)u^2_{N+1}(\varphi')}{N+1}\right], \tag{38.46b}$$

the additional terms becoming less and less important as N becomes large. Again, the obvious duty of the additional nonsignificant terms is to produce the correct symmetry in φ and φ' of $\delta_N{}^2$ and $\delta_N{}^3$ according to their construction in (38.42).

The δ-like functions $\delta_N{}^m$ may even be found for higher m but we shall not continue our investigation, since our aim has been one of principle rather than of a practical nature.

The Completeness of Legendre Polynomial Systems

It is fairly obvious that Gegenbauer polynomials $C_n{}^\nu(x)$ with nonintegral ν must have delta functions of similar nature as the explicit expression by integral ν. It might have been of some interest to have an explicit expression for half-integral ν corresponding to the Legendre polynomials $P_n{}^m(x)$. These appear not to be obtainable by an elementary procedure as used above and we shall omit further discussion on this point. The completeness of the polynomial systems $P_n^{(m)}(x)$ which requires the existence of delta functions

$$\sum_{n=m}^{\infty} \frac{2n+1}{2}\frac{(n-m)!}{(n+m)!} P_n{}^m(x)P_n{}^m(x') = \delta(x-x') \tag{38.47}$$

190 / The Mathematical Foundation of Physics

can, however, be readily proved by the methods of Chapter 3, Section 27, in the latter case by comparing with the nearest system of Gegenbauer's polynomials.

It is obvious, however, from the expressions of cosine functions in terms of Legendre polynomials,

$$1 = P_0(\cos \vartheta), \quad \cos \vartheta = P_1(\cos \vartheta),$$

$$\cos 2\vartheta = \tfrac{4}{3}P_2 - \tfrac{1}{3}P_0, \quad \cos 3\vartheta = \tfrac{8}{5}P_3 - \tfrac{3}{5}P_1, \quad \text{etc.}, \quad (38.48)$$

which may be continued infinitely, that any function $f(\vartheta)$ which can be expanded in an infinite series of cosine functions in the region $0 \leq \vartheta \leq \pi$, can as well be expanded in an infinite series of Legendre polynomials $P_n(\cos \vartheta)$. Hence the set of functions $P_n(\cos \vartheta)$ forms a complete set of functions.

Legendre Functions of the Second Kind

If in Schläfli's integral (38.22) we choose another suitable path of integration, the integral may still be a solution of the Legendre equation (38.8) representing the second solution of the equation. It can be fairly easily shown that

$$\left[\frac{d}{dz}(1-z^2)\frac{d}{dz} + n(n+1)\right]\int_C \frac{(\zeta^2-1)^{n+1}\,d\zeta}{(\zeta-z)^{n+1}} = (n+1)\int_C d\frac{(\zeta^2-1)^{n+1}}{(\zeta-z)^{n+2}}.$$

(38.49)

It suffices therefore for the integral to be a solution that the contour C be chosen so as to make $(\zeta^2 - 1)^{n+1}/(\zeta - z)^{n+2}$ have the same value at both ends or vanish. The first condition is fulfilled in Schläfli's integral for $P_n(z)$, the contour encircling the pole $\zeta = z$ and the above expression returning to its original value. The second condition is fulfilled for the integral used to define the second solution

$$Q_n(z) = \frac{1}{2^{n+1}} \int_{-1}^{+1} \frac{(1-\zeta^2)^n}{(z-\zeta)^{n+1}} \, d\zeta, \quad n = 0, 1, \cdots \quad (38.50)$$

This function is called a *Legendre function of the second kind*.

In order that the function be single-valued we shall have to draw a cut between the singular points $z = \pm 1$, preferably along the real axis. As we shall see, there will then be a difference πi in the functional value on the upper and the lower side of the cut, which is due to the logarithmic character of the singularities. The function is real and positive for real values of $z > 1$.

By n partial integrations the integral (38.50) becomes

$$Q_n(z) = \frac{1}{2} \int_{-1}^{+1} \frac{P_n(\zeta)}{z-\zeta} \, d\zeta, \quad (38.51)$$

Differential Equations and Particular Functions / 191

using the Rodrigues formula for $P_n(\zeta)$. Since $P_n(\zeta) = P_n(z) + [P_n(\zeta) - P_n(z)]$ and the last term is divisible by $(z - \zeta)$, we may also write

$$Q_n(z) = Q_0(z)P_n(z) + R_{n-1}(z), \qquad (38.52)$$

$R_{n-1}(z)$ being a polynomial of the $(n-1)$th degree and

$$Q_0(z) = \frac{1}{2} \log \frac{z+1}{z-1} \qquad (38.53)$$

being the first Legendre function of the second kind.

From the common differential equation of $P_n(z)$ and $Q_n(z)$ we obtain the modified Wronskian determinant

$$(1 - z^2)[Q_n(z)P_n'(z) - P_n(z)Q_n'(z)] = \text{const} = 1. \qquad (38.54)$$

The particular value 1 is found by putting $z = 1$; its value is then the limit of $2(z-1)Q_n'(z)$ or, according to (38.52), of $2(z-1)Q_0'(z)$ as $z \to 1$. But this value is 1 as seen from (38.53).

The function $Q_n(z)$ frequently occurs in expansions of functions of several variables. From the above results and definitions it follows immediately that

$$\frac{1}{z-\zeta} = \sum_{n=0}^{\infty} (2n+1)Q_n(z)P_n(\zeta), \qquad (38.55)$$

provided that z and ζ are real and $z > 1$, $\zeta < 1$. This is seen by considering (38.55) an expansion of $(z - \zeta)^{-1}$ in terms of $P_n(\zeta)$. Then multiplying (38.55) by $P_n(\zeta)$ and integrating, we obtain the result that follows from (38.51) and the square integral (38.19) of $P_n(z)$. The result is true even for complex values of z and ζ, in particular for $|z| > 1$, $|\zeta| < 1$.

Associate Legendre Functions of the Second Kind

There are two definitions of associate Legendre functions of the second kind

$$Q_n{}^m(z) = (1 - z^2)^{\frac{1}{2}m} Q_n^{(m)}(z) \qquad (38.56)$$

due to Hobson. We shall adopt the second expression which is real for real $z > 1$.

It should be noted that in dealing with functions of the same variable we should modify the definition of $P_n{}^m(z)$ correspondingly, using the factor $(z^2 - 1)^{\frac{1}{2}m}$. In a great many applications, however, $P_n{}^m$ and $Q_n{}^m$ are functions of different variables and very frequently in different real regions -1 to 1 and 1 to ∞, respectively, for which we prefer the definitions used here.

From the definition (38.50) it follows immediately that the functions $Q_n(z)$ obey the same recurrence formula $(n+1)Q_{n+1} - (2n+1)zQ_n + nQ_{n-1} = 0$ as the functions $P_n(z)$. This leads to the conclusion that any of the formulas (38.14) to (38.18) are valid when the $P_n^{(m)}(z)$ are replaced by $Q_n^{(m)}(z)$.

39. CONFLUENT HYPERGEOMETRIC FUNCTIONS

Laguerre Polynomials and Orthogonal Functions

We shall start with (37.23)

$$\left[\frac{d^2}{dx^2} + \frac{1}{x}\frac{d}{dx} - \frac{m^2}{4x^2} - \frac{1}{4} + \frac{n + \frac{1}{2}(m + 1)}{x}\right]y = 0 \qquad (39.1)$$

which will be termed the *Laguerre differential equation*. This equation has become of quite extraordinary importance in modern theoretical physics as a basic equation of wave mechanics. The constants n and m may have arbitrary values. In our considerations we shall restrict them to being real and n in general a positive integer. The constant m may be any positive real number, even though in the majority of applications it will also be an integer.

The singular points of the equation are $z = 0$ and $z = \infty$. It is easily seen that the asymptotic solutions in these points are, respectively,

$$y = x^{\pm \frac{1}{2}m} \quad \text{and} \quad y = e^{\pm \frac{1}{2}x}. \qquad (39.2)$$

Writing

$$y(x) = e^{-x/2} x^{m/2} v(x), \qquad (39.3)$$

we are back at the equation (37.21) or

$$xv'' + (m + 1 - x)v' + nv = 0. \qquad (39.4)$$

Solving this equation by a power series

$$v = c_0 + c_1 x + \cdots + c_k x^k + \cdots, \qquad (39.5)$$

we obtain the recurrence formula

$$(k + 1)(k + 1 + m)c_{k+1} + (n - k)c_k = 0. \qquad (39.6)$$

Roughly we may write

$$\lim_{k \to \infty} \frac{c_{k+1}}{c_k} = \frac{1}{k + 1} \qquad (39.7)$$

which is the exact law for the coefficients of the expansion in the power series of e^x. The series (39.5) therefore in general leads to the asymptotic solution $v \sim e^x$ or $u \sim e^{\frac{1}{2}x}$ as $x \to \infty$. In this case it has been of little use to split off the asymptotic solution $e^{-\frac{1}{2}x}$. The solution having the asymptotic character $x^{\frac{1}{2}m}$ at $x = 0$ contains both the asymptotic solutions at $x = \infty$.

There is an exception, however, and this is important for the determination of n in (39.1) if n is taken to be an eigenvalue parameter of the equation considered as an eigenvalue equation with finite solutions both in $x = 0$ and

$x = \infty$. In this case n must be an integer, $n = 0, 1, \cdots, \infty$, and the solution $v = v_n$ of (39.4) is a polynomial of the nth degree.

It is easily seen from the differential equation of the ordinary Laguerre polynomials

$$\left[x\frac{d^2}{dx^2} + (1-x)\frac{d}{dx} + (n+m)\right]L_{n+m}(x) = 0, \qquad (39.8)$$

differentiating m times, that

$$\left[x\frac{d^2}{dx^2} + (m+1-x)\frac{d}{dx} + n\right]\frac{d^m}{dx^m}L_{n+m}(x) = 0 \qquad (39.9)$$

and, hence, that the solution of (39.3) is

$$v_n(x) = L_{n+m}^m(x), \qquad (39.10)$$

the upper index denoting simply the mth derivative of L_{n+m}.

An explicit expression of the nth Laguerre polynomials has been given in (20.38). It can easily be obtained from the recurrence formula (39.6), putting $m = 0$. It may be normalized by putting $(-1)^n$ for the coefficient of the highest power x^m which is the conventional normalization. A better choice would have been $(-1)^n/n!$ corresponding to the value 1 of the constant term. For this reason we shall frequently use the function $L_n(x)/n!$.

To verify the correctness of the definition (23.38) we introduce the function

$$u_n(x) = x^n e^{-x} \qquad (39.11)$$

obeying the equation

$$xu_n' - (n-x)u_n = 0. \qquad (39.12)$$

Differentiating $n + 1$ times the result is

$$xu_n^{(n+2)} + (1+x)u_n^{(n+1)} + (n+1)u_n^{(n)} = 0. \qquad (39.13)$$

On writing $u_n^{(n)} = e^{-x}L_n$ which is equivalent to the definition (22.38) the result is

$$xL_n'' + (1-x)L_n' + nL_n = 0. \qquad (39.14)$$

Hence

$$\frac{L_n(x)}{n!} = e^x \frac{1}{n!}\frac{d^n}{dx^n}(x^n e^{-x}). \qquad (39.15)$$

As in the case of the Legendre polynomials useful recurrence and differential formulas may be obtained by the same procedure. These formulae are as a rule simpler than in the Legendre case and therefore explicit formulas are not so necessary.

Integral Representations and Generating Functions of Laguerre

From (39.15) it follows immediately using Cauchy's integral theorem that $L_n(x)$ is given by a contour integral

$$\frac{L_n(x)}{n!} = \frac{e^x}{2\pi i} \oint \frac{z^n e^{-z}}{(z-x)^{n+1}} \, dz, \tag{39.16}$$

the path of integration encircling the point $z = x$. Multiplying by s^n and summing up the infinite series, we obtain

$$\psi(x, s) = \sum_{n=0}^{\infty} \frac{L_n(x)}{n!} s^n = \frac{e^x}{2\pi i} \oint \frac{e^{-z} \, dz}{2 - x - 2s}, \tag{39.17}$$

provided that $|2s/(z-x)| < 1$. The integrand has now a pole in $z = x/(1-s)$. In this pole $2s/(x-x) = -1$ showing that the pole is well inside the contour. By the theory of residues then

$$\psi(x, s) = \exp[x - (x/1 - s)]. \tag{39.18}$$

Among others this formula may be used for the orthonormality of the orthonormal Laguerre functions

$$y_n = \frac{\exp(-x/2) L_n(x)}{n!}, \quad n = 0, 1, \cdots, \infty. \tag{39.19}$$

From (39.17) and (39.18) we have

$$\sum_{n=0}^{\infty} \sum_{m=0}^{\infty} y_n y_m s^n t^m = e^{-x} \psi(x, s) \psi(x, t) = \exp\{-x[(1 - st)/(1-s)(1-t)]\} \tag{39.20}$$

and by integration

$$\sum_{n,m} s^n t^m \int_0^{\infty} y_n y_m \, dx = \frac{1}{1 - st} = \sum_{n=0}^{\infty} s^n t^n. \tag{39.21}$$

This proves that

$$\int_0^{\infty} y_n y_m \, dx = \delta_{nm}. \tag{39.22}$$

Similarly, differentiating m times (39.17) and (39.18), we obtain the generating function for the derivatives of Laguerre polynomials

$$\sum_{n=0}^{\infty} \frac{(-1)^m L_{n+m}^m(x) s^n}{(n+m)!} = \frac{\exp[-xs/(1-s)]}{(1-s)^{m+1}} = \psi_m(x, s). \tag{39.23}$$

Differential Equations and Particular Functions / 195

Computing

$$\int_0^\infty x^m e^{-x} \psi_m(x,s)\psi_m(x,t)\,dx = \frac{m!}{(1-st)^{m+1}} = \sum_{n=0}^\infty \frac{(n+m)!}{n!} s^n t^n, \quad (39.24)$$

it is also seen that the following functions

$$y_{nm}(x) = x^{m/2} e^{-x/2} \frac{(-1)^m}{(n+m)!} L_{n+m}^m(x), \quad (39.25)$$

called *associate Laguerre functions*, form an orthogonal system

$$\int_0^\infty y_{nm}(x) y_{n'm}(x)\,dx = \frac{(n+m)!}{n!} \delta_{nn'}. \quad (39.26)$$

An illustration of the orthonormal ordinary Laguerre functions with $m=0$ is given in Fig. 5-1.

From (39.25) it is seen that the system $y_{nm+1}(x)$ may be produced from $y_{nm}(x)$ by a differential operator

$$y_{(n-1)(m+1)}(x) = -x^{(m+1)/2} e^{-(x/2)} \frac{d}{dx} x^{-(m/2)} e^{x/2} y_{nm}(x). \quad (39.27)$$

It can be proved that another operator exists that transforms the system $y_{nm+1}(x)$ into the system $y_{nm}(x)$,

$$y_{nm}(x) = \frac{1}{n} x^{-(m/2)} e^{x/2} \frac{d}{dx} x^{(m+1)/2} e^{-(x/2)} y_{(n-1)(m+1)}(x). \quad (39.28)$$

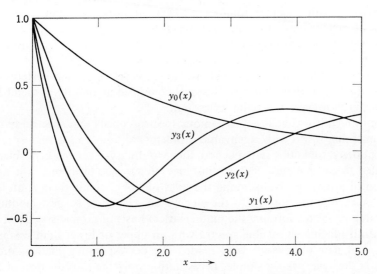

Fig. 5-1. Orthonormal functions $y_n = e^{-x/2} L_n(x)/n!$

This is true if the differential equation (39.1) can be written

$$\left(x^{-(m/2)}e^{x/2}\frac{d}{dx}\,x^{m+1}e^{-x}\frac{d}{dx}\,x^{-(m/2)}e^{x/2} + n\right)y_{nm}(x) = 0 \qquad (39.29)$$

and it is easily controlled that the equation is the same as

$$\left(x\frac{d^2}{dx^2} + x\frac{d}{dx} - \frac{m^2}{4x} - \frac{x}{4} + \frac{m+1}{2} + n\right)y_{nm} = 0. \qquad (39.30)$$

The corresponding equation for $y_{(n-1)(m+1)}(x)$ is obtainable in the same way from (39.27) and (39.28).

By partial integration, using first (39.27) and then (39.28), we obtain

$$\int_0^\infty [y_{(n-1)(m+1)}(x)]^2\, dx = n\int_0^\infty [y_{nm}(x)]^2\, dx \qquad (39.31)$$

or

$$\int_0^\infty y_{nm}^2\, dx = (n+1)\int_0^\infty y_{(n+1)(m-1)}^2\, dx$$

$$= (n+1)\cdots(n+m)\int_0^\infty y_{n+m}^2\, dx = \frac{(n+m)!}{n!} \qquad (39.31a)$$

in accordance with (39.26).

Hydrogen Atomic Functions

So far we learn only about eigenvalue equations leading to eigenvalues λ_n increasing with its index number n toward infinity. We may term this system of eigenvalues an *eigenvalue spectrum*. The fact that the higher eigenvalues are increasing toward infinity may be expressed by saying that the eigenvalues have a condensation point at $\lambda = \infty$.

In atomic spectroscopy it was known at an early stage that systems of spectral lines, or rather of spherical terms, contained an infinity of members condensing, however, at a finite series limit.

Even though it was believed by famous spectroscopists that spectral terms had something to do with eigenvalue problems, although of a very special kind, nobody took the step beyond those ordinary eigenvalue problems presented to us by nature, say, for instance, in acoustic problems. The first solution of atomic problems along the lines of Bohr atomic theory, on the other hand, was too foreign to the eigenvalue scheme to suggest modified eigenvalue problems forming the mathematical basis of atomic spectra. The step of introducing eigenvalue spectra with an upper or lower limit was not taken until new discoveries were made. The corresponding mathematical theory is, however, fairly simple, necessitating only the introduction of an infinite fundamental region of some independent variable.

Consider the differential equation for the Laguerre polynomial $L_{n+l}^{2l+1}(x)$ which can be obtained from (39.9)

$$\left[\frac{d^2}{dx^2} + (2l + 2 - x)\frac{d}{dx} + (n - l - 1)\right]L_{n+l}^{2l+1}(x) = 0. \tag{39.32}$$

Introducing

$$y_n^l(x) = x^l e^{-(x/2)} L_{n+l}^{2l+1}(x), \tag{39.33}$$

we find the equation

$$\left[x\frac{d^2}{dx^2} + 2\frac{d}{dx} - \frac{l(l+1)}{x} - \frac{x}{4} + n\right]y_n^l(x) = 0. \tag{39.34}$$

To transform this equation into an eigenvalue equation of quite a different nature, we need only introduce the new variable $\xi = \tfrac{1}{2}nx$ and write

$$x = \frac{2\xi}{n}, \quad \frac{d}{dx} = \frac{n}{2}\frac{d}{d\xi}, \quad \psi_n^l(\xi) = y_n^l\left(\frac{2\xi}{n}\right) \tag{39.35}$$

to obtain

$$\left[\frac{d^2}{d\xi^2} + \frac{2}{\xi}\frac{d}{d\xi} - \frac{l(l+1)}{\xi^2} - \frac{1}{n^2} + \frac{2x}{\xi}\right]\psi_n^l(\xi) = 0. \tag{39.36}$$

In this equation the quantity

$$\lambda_n = -\frac{1}{n^2}, \quad n = 1, 2, \cdots, \infty \tag{39.37}$$

may be considered the possible values of an eigenvalue parameter λ, and this eigenvalue spectrum has just the character of the stationary states energy spectrum of an atom. The eigenvalues are negative corresponding to the binding energy of an atomic electron as reckoned from its state of freedom far from the atom. The spectrum has increasing terms with a series limit at zero corresponding to zero energy of an unbound electron with no kinetic energy.

We shall not discuss the hydrogen atomic functions in much detail here, since they must be treated extensively in wave mechanics. It should be noted, however, that from (39.36) we easily deduce their orthogonality properties,

$$\int_0^\infty \psi_n^l(\xi)\psi_{n'}^l(\xi)\xi^2 \, d\xi = 0, \quad \text{for} \quad n \neq n' \tag{39.38}$$

which are far from easily seen from explicit expressions of the functions.

If now in the region $0 \leq \xi = \infty$ we try to expand an arbitrary function $f(\xi)$ in terms of the hydrogenic functions, we shall find that this is easily done in principle and that the result is a minimum square deviation from the

198 / The Mathematical Foundation of Physics

function $f(\xi)$. We are not, however, in the position of being able to prove that this deviation is zero and, hence, that the expansion is correct. In fact, the expansion is usually not correct, the reason for this being that the system of hydrogenic functions belonging to the above eigenvalue spectrum is not a complete system. The completeness of the system is obtained only by adding solutions of the equation

$$\left[\frac{d^2}{d\xi^2} + \frac{2}{\xi}\frac{d}{d\xi} - \frac{l(l+1)}{\xi^2} + \lambda + \frac{2}{\xi}\right]\psi = 0 \tag{39.39}$$

for positive values of λ. These solutions are finite both at $\xi = 0$ and $\xi = \infty$ for any positive value of λ, this eigenvalue parameter having then what is called a *continuous eigenvalue spectrum*. Again, this is a feature of known atomic spectra, the series limit of spectral lines being frequently continued by a continuous spectrum.

Hermite's Polynomials and Functions

The above hypergeometric and confluent hypergeometric functions all find their application in elementary problems of wave mechanics. There is one problem, however, in some respects the simplest one, namely, that of the harmonic oscillator, which leads to functions of a somewhat different type called Hermite's polynomials or orthogonal functions.

As a matter of fact these functions are also confluent hypergeometric functions, but of a special type with exponent difference $\frac{1}{2}$ at the origin of the independent variable z. This means that in the Laguerre equation we have to put $m = \pm\frac{1}{2}$. Next we shall find that either of the solutions at $z = 0$ may serve as an eigenfunction for the eigenvalue problem formed by the differential equation for the functions. Finally, introducing a new variable by $z = x^2$, we find that the potential difference in $x = 0$ becomes 1 and the point ceases to be a singular point.

In (39.9) let us put $m = -\frac{1}{2}$ and replace n by $\frac{1}{2}n$, the latter modification meaning simply that even integers n correspond to modified Laguerre polynomials of index $m = -\frac{1}{2}$. For odd n, however, we shall obtain another set of polynomials corresponding to Laguerre polynomials of index $m = \frac{1}{2}$. The equation becomes

$$\left[z\frac{d^2}{dz^2} + \left(\frac{1}{2} - z\right)\frac{d}{dz} + \frac{n}{2}\right]H = 0, \tag{39.40}$$

writing z and H for independent and dependent variables.

Now we change to the variable x by

$$z = x^2, \quad 2\frac{d}{dz} = \frac{1}{x}\frac{d}{dx}, \tag{39.41}$$

and (39.40) becomes

$$\left(\frac{d^2}{dx^2} - 2x\frac{d}{dx} + 2n\right)H_n(x) = 0, \quad n = 0, 1, 2, \cdots, \quad (39.42)$$

putting an index n to the unknown function $H(x)$. This is the differential equation for Hermite's polynomial. In fact, writing $H(x)$ as a power series it is easily seen that this series will stop at the power x^m. Moreover, there will be a two-term recurrence formula for the coefficient so that all powers of the expansion are of the same parity, even or odd, the first kind having a constant and the latter kind the power x for its first term. The polynomials are even or odd corresponding to the parity of n, and of the nth degree.

Writing (39.42) in self-adjoint form, we see that the polynomials form an orthogonal system of functions provided that e^{-x^2} is taken for the density function. This means that the true orthogonal functions are

$$u_n = C_n e^{-x^2/2} H_n, \quad (39.43)$$

the constant C_n being left for the purpose of normalization. Substituting H_n according to (39.43) in (39.42), we obtain

$$\left[\frac{d^2}{dx^2} + (2n + 1) - x^2\right]u_n(x) = 0, \quad n = 0, 1, 2, \cdots. \quad (39.44)$$

An expression for the polynomial $H_n(x)$ in (39.42) is easily found. We write $H_n = e^{x^2} K_n$ and obtain

$$\left(\frac{d^2}{dx^2} + 2x\frac{d}{dx} + 2n + 2\right)K_n = 0 \quad (39.45)$$

and compare with $(d/dx + 2x)e^{-x^2} = 0$ or rather

$$\left(\frac{d^2}{dx^2} + 2x\frac{d}{dx} + 2\right)e^{-x^2} \quad (39.46)$$

when differentiated n times. The identity with (39.45) then proves that

$$H_n(x) = (-1)^n e^{x^2} \frac{d^n}{dx^n} e^{-x^2} \quad (39.47)$$

if it is accepted that the highest power of the polynomials is $< (2x)^n$. The first few of the polynomials are

$$H_0(x) = 1, \qquad H_1(x) = 2x,$$
$$H_2(x) = 4x^2 - 2, \qquad H_3(x) = 8x^3 - 12x, \qquad \text{etc.} \quad (39.48)$$

Using the integral representation

$$H_n(x) = \frac{n!(-1)^n}{2\pi i} \oint \frac{e^{-z^2}}{(z-x)^{n+1}} \tag{39.49}$$

which is equivalent to (39.47) we may easily obtain a generating function

$$\psi(x, s) = \sum_{n=0}^{\infty} \frac{H_n(x) s^n}{n!} = \frac{e^{x^2}}{2\pi i} \oint \frac{e^{-z^2} dz}{(z - x + s)^{n+1}}$$

$$= e^{x^2 - (x-s)^2} = e^{2xs - s^2}. \tag{39.50}$$

This is a useful function for integration purposes. For instance,

$$\sum_{n,m=0}^{\infty} s^n t^m \int_{-\infty}^{+\infty} \frac{e^{-x^2} H_n(x) H_m(x) \, dx}{n! \, m!}$$

$$= \int_{-\infty}^{+\infty} \psi(x, s) \psi(x, t) e^{-x^2} dx$$

$$= \int_{-\infty}^{+\infty} e^{2st - (x-s-t)^2} dx = \sqrt{\pi} \, e^{2st} = \sqrt{\pi} \sum_{n=0}^{\infty} \frac{(2st)^n}{n!}. \tag{39.51}$$

This proves the orthogonality of the system and yields the normalization integrals. For the constant C_n in (39.43) we obviously have to put

$$C_n = \frac{1}{(2^n n! \, \pi^{1/2})^{1/2}} \tag{39.52}$$

in order that (39.43) may be an orthonormal system.

It is obvious from (39.45) and (39.42) that there are two differential operators transforming, respectively, $H_n(x)$ into $H_{n+1}(x)$, and vice versa. These are $-e^{x^2} \frac{d}{dx} e^{-x^2}$ and $1/(2n+2) \, d/dx$. The corresponding transformations for the orthonormal functions are

$$e^{x^2/2} \frac{d}{dx} e^{-(x^2/2)} u_n = \left(\frac{d}{dx} - x\right) u_n = -\sqrt{2(2n+2)} \, u_{n+1},$$

$$e^{-(x/2)} \frac{d}{dx} e^{x^2/2} u_{n+1} = \left(\frac{d}{dx} + x\right) u_{n+1} = \sqrt{2(2n+2)} \, u_n, \tag{39.53}$$

from which by elimination of either u_n or u_{n+1} we obtain their differential equations. By partial integration it also follows from the transformations that two successive functions have the same normalization integral in the fundamental region $-\infty \leq x \leq \infty$.

Differential Equations and Particular Functions / 201

The factor $\sqrt{2(2n+2)}$ in the transformation (39.43) is connected with the famous matrix elements of the harmonic oscillator in quantum mechanics.

Finally it should be stated once more that apart from normalization factors depending on particular definitions

$$H_{2n}(x) = L_{n-\frac{1}{2}}^{-\frac{1}{2}}(x^2), \qquad H_{2n+1} = xL_{n+\frac{1}{2}}^{\frac{1}{2}}(x^2). \tag{39.54}$$

40. BESSEL FUNCTIONS

First Solution of the Bessel Equation

A particular kind of confluent hypergeometric equation is the *Bessel differential equation*

$$\frac{d^2u}{dz^2} + \frac{1}{z}\frac{du}{dz} + \left(1 - \frac{m^2}{z^2}\right)u = 0. \tag{40.1}$$

It is obtained from (39.1) by omitting the last term and then replacing m by $2m$ and x by $2iz$. Because of its predominant role in a great many branches of mathematical physics we shall devote some space to studying with some care the functions defined by these equations known under the common name of *Bessel functions*.

Since we have already dealt with the asymptotic behavior of the solutions of (40.1) in the singular points $z = 0$ and $z = \infty$, and since the method of solving a differential equation by power series is already a familiar one, we shall start by defining the Bessel function $J_m(z)$ by the power series

$$J_m(z) = \sum_{k=0}^{\infty} \frac{(-1)^k(\frac{1}{2}z)^{m+2k}}{k!\,(k+m)!}, \tag{40.2}$$

denoting for simplicity the Gauss gamma function $\prod(k+m)$ by the factorial function $(k+m)!$ even in the case of nonintegral m.

The following differential formulas are now easily obtainable

$$z^{-m}\frac{d}{dz}z^m J_m(z) = J_{m-1}(z),$$
$$z^m\frac{d}{dz}z^{-m}J_m(z) = -J_{m+1}(z). \tag{40.3}$$

Subtracting, we obtain the recurrence formula

$$\frac{2m}{z}J_m(z) = J_{m-1}(z) + J_{m+1}(z). \tag{40.4}$$

Adding, we have the differential formula

$$2J'_m(z) = J_{m-1}(z) - J_{m+1}(z). \tag{40.5}$$

Replacing m by $m-1$ in the last formula (40.3) and inserting $J_{m-1}(z)$ from the first equation we have

$$z^{m-1}\frac{d}{dz}z^{-2m+1}\frac{d}{dz}z^{m}J_{m}(z) = -J_{m}(z), \qquad (40.6)$$

and this is easily seen to be the differential equation (40.1) with $u(z) = J_m(z)$. Alternatively we might have replaced m by $m+1$ in the first equation and obtained by elimination of $J_{m+1}(z)$

$$z^{-m-1}\frac{d}{dz}z^{2m+1}\frac{d}{dz}z^{-m}J_{m}(z) = -J_{m}(z), \qquad (40.7)$$

which is again the differential equation.

Equations (40.3) to (40.5) form, together with (40.1), a consistent system of equations that are useful for various purposes. It is convenient to adopt the same equations when defining the more general class of functions called Bessel functions comprising other solutions of the Bessel equation.

Second Solution

In the case of nonintegral m the second solution with the asymptotic behavior $u(z) \sim z^{-m}$ at $z = 0$ is obtained by simply changing the sign of m in formula (40.2). Hence we may write

$$u_1(z) = J_m(z), \qquad u_2(z) = J_{-m}(z) \qquad (40.8)$$

for the two independent solutions of the Bessel equation.

In the special case of integral m, however, the solutions are no longer independent. Changing the sign of m in (40.2) and replacing in the case of integral $m = n$, k by $k + n$, we obtain

$$J_{-n}(z) = \sum_{k=-1}^{\infty} \frac{(-1)^n(\tfrac{1}{2}z)^{m+2k}}{(k+n)!\,k!}. \qquad (40.9)$$

Since $k!$ is infinite for negative integers, the first m terms vanish and the result is

$$J_{-n}(z) = (-1)^n J_n(z) \qquad (40.10)$$

for integral $m = n$.

We now calculate the value of the generalized Wronskian of the two solutions (40.8), which as a consequence of the form of (40.1) has the form

$$W(u_1, u_2) = z(u_1 u_2' - u_1' u_2), \qquad (40.11)$$

the factor z being necessary in order that the Wronskian be a constant. Putting $z = 0$ in the right-hand expression we need only consider first terms in the expansions of $J_m(z)$ and $J_{-m}(z)$. The result is

$$W[J_m(z), J_{-m}(z)] = -\frac{2m}{m!(-m)!} = -\frac{2}{\pi}\sin m\pi, \qquad (40.12)$$

using the reflection formula for the factorial or gamma function.

It is desirable to replace the second solution $J_{-m}(z)$ by another solution $u_2(z)$ which makes the Wronskian of u_1 and u_2 independent of m. This may be obtained by dividing $J_{-m}(z)$ by the right-hand side of (40.12), which makes $W = 1$, or rather on dividing by $-\sin m\pi$ which makes $W = 2/\pi$.

Bessel Functions of the Second Kind. The Neumann Function $N(z)$

We shall adopt the latter procedure, which is the more commonly used, and which leads to a definition of the Neumann function $N_m(z)$. The function $-J_{-m}(z)/\sin m\pi$, however, becomes infinite for integral m. We therefore subtract the function $J_m(z)$ as multiplied by a suitable factor which has proved to be $-\cot m\pi$. Hence we write

$$N_m(z) = \frac{J_m(z)\cos m\pi - J_{-m}(z)}{\sin m\pi} \qquad (40.13)$$

for the Neumann function.

Let it be known that the asymptotic value of a Bessel function is

$$J_m \to \left(\frac{2}{\pi x}\right)^{1/2} \cos\left(z - \left(m + \frac{1}{2}\right)\frac{\pi}{2}\right) \qquad (40.14)$$

as $z \to \infty$ and that

$$J_{-m}(z) \to \left(\frac{2}{\pi z}\right)^{1/2} \cos\left(z - \left(m + \frac{1}{2}\right)\frac{\pi}{2} + m\pi\right). \qquad (40.15)$$

From the above equations it follows that

$$N_m(z) \to \left(\frac{2}{\pi z}\right)^{1/2} \sin\left(z - \left(m + \frac{1}{2}\right)\frac{\pi}{2}\right) \quad \text{when} \quad z \to \infty. \qquad (40.16)$$

By this definition it is seen that the Bessel and Neumann functions are related to each other in the same way as the cosine and sine functions, the Wronskian being modified only by the factor $2/\pi$.

Bessel Functions of the Third Kind or Hankel Functions

The third kind of Bessel functions are only combinations of those of the first and second kind. They may be taken to correspond to the exponential functions $e^{\pm iz}$. Hence we define them by the equations

$$H_m^{(1)}(z) = J_m(z) + iN_m(z),$$
$$H_m^{(2)}(z) = J_m(z) - iN_m(z). \tag{40.17}$$

It follows from (40.14) and (40.16) that the asymptotic values of the functions as $z \to \infty$ are

$$H_m^{(1,2)}(z) = \left(\frac{2}{\pi z}\right)^{1/2} \exp\left[\pm i\left(z - \left(m + \frac{1}{2}\right)\frac{\pi}{2}\right)\right]. \tag{40.18}$$

Asymptotic Values and Asymptotic Expansions

The simplest way of obtaining asymptotic expansions for large $|z|$ is to transform the power series (40.2) of $J_m(z)$ into a definite integral or contour integral and deform the contour according to particular requirements. If, for instance, we use the duplication formula

$$2^{2k}k!\,\frac{(k-\tfrac{1}{2})!}{(-\tfrac{1}{2})!} = (2k)!\,, \tag{40.19}$$

the rest of the kth coefficient in (40.2) can be expressed by the Euler integral

$$\frac{(m-\tfrac{1}{2})!\,(k-\tfrac{1}{2})!}{(k+m)!} = \int_0^1 (1-t)^{m-1/2} t^{k-1/2}\, dt = \int_{-1}^{+1} (1-s^2)^{m-1/2} s^{2k}\, ds, \tag{40.20}$$

writing $t = s^2$. The series (40.2) may now be summed up into

$$J_m(z) = \frac{(\tfrac{1}{2}z)^m}{(-\tfrac{1}{2})!\,(m-\tfrac{1}{2})!} \int_{-1}^{+1} \cos sz (1-s^2)^{m-1/2}\, ds$$
$$= \frac{(\tfrac{1}{2}z)^m}{(-\tfrac{1}{2})!\,(m-\tfrac{1}{2})!} \int_{-1}^{+1} e^{isz}(1-s^2)^{m-1/2}\, ds. \tag{40.21}$$

The path of integration in the complex s-plane may now be deformed so as to pass from $s = -1$ to $s = i\infty$ and back to $s = 1$. It is tempting then to assume that either of the integrals represent Hankel functions

$$H_m^{(1)}(z) = -\frac{2(\tfrac{1}{2}z)^m}{(-\tfrac{1}{2})!\,(m-\tfrac{1}{2})!} \int_1^{i\infty} e^{isz}(1-s^2)^{m-1/2}\, ds, \tag{40.22a}$$

$$H_m^{(2)}(z) = \frac{2(\tfrac{1}{2}z)^m}{(-\tfrac{1}{2})!\,(m-\tfrac{1}{2})!} \int_{-1}^{i\infty} e^{isz}(1-s^2)^{m-1/2}\, ds. \tag{40.22b}$$

Changing in the latter integral first the sign of i and next of s, we see that the two functions are conjugate complex as are the Hankel functions.

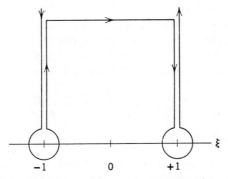

Fig. 5-2. Evaluation of the asymptotic formula for $J_m(z)$.

Of course, this is not enough to prove rigorously that they are Hankel functions. We should have to demonstrate that their imaginary parts are Neumann functions, or that the integrals separately obey the Bessel equation, which in fact they do. We shall be satisfied, however, to study the asymptotic behavior of the above functions stating that it conforms with (40.18).

To that end we introduce in (40.22a) the new variable x by

$$s = 1 + \frac{ix}{z}. \tag{40.23}$$

Noting that $(-\tfrac{1}{2})! = \dfrac{\pi}{2}$ it is found that

$$H_m^{(1)}(z) = \left(\frac{2}{\pi z}\right)^{1/2} \exp\left[i\left(z - \left(m + \frac{1}{2}\right)\frac{\pi}{2}\right)\right]$$
$$\times \frac{1}{(m - \tfrac{1}{2})!} \int_0^\infty \exp(-x)\left(1 + \frac{ix}{2z}\right)^{m - 1/2} dx. \tag{40.24}$$

As $z \to \infty$ this yields the asymptotic value (40.18).

It is possible, however, to improve the asymptotic formula (40.17) by adding more terms corresponding to an expansion of the last factor in (40.24) in a power series in z^{-1}. Since this expansion is valid only up to $x \leq 2|z|$, the resulting infinite series cannot be used and is in fact divergent. We shall write this equation only as a formal solution. It is understood that the series, being only semiconvergent, shall have to be terminated at a convenient place where successive terms are still decreasing or at least not increasing. This being agreed upon, we may write

$$H_m^{(1)}(z) = \left(\frac{2}{\pi z}\right)^{1/2} \sum_{k=0}^{n} \frac{(m - \tfrac{1}{2} + k)!}{k!\,(m - \tfrac{1}{2} - k)!} \frac{\exp\left[i(z - (m + \tfrac{1}{2} - k)\pi/2)\right]}{(2z)^k}, \tag{40.25}$$

the choice of the upper limit n of the series being a matter of convenience.

The expression (40.25) conforms with the formal solution of Bessel equations as obtained by splitting off an asymptotic factor e^{iz} and expanding the other factor of the solution in a power series in z^{-1}.

It should be noted that in the case of the half-integral m, say, $m = n + \frac{1}{2}$, $n = 0, 1, 2, \cdots$, the expression (40.24) is exact, yielding

$$H^{(1)}_{n+\frac{1}{2}}(z) = \left(\frac{2}{\pi z}\right)^{\frac{1}{2}} \sum_{k=0}^{n} \frac{(n+k)!}{k!(n-k)!} \frac{\exp[i(z-(n+1-k)\pi/2)]}{(2z)^k}. \quad (40.26)$$

The corresponding Bessel function is

$$J_{n+\frac{1}{2}}(z) = \left(\frac{2}{\pi z}\right)^{\frac{1}{2}} \sum_{k=0}^{n} \frac{(n+k)!}{k!(n-k)!} \frac{\sin(z-(n-k)\pi/2)}{(2z)^k}. \quad (40.27)$$

In particular we have

$$J_{\frac{1}{2}}(z) = \left(\frac{2}{\pi z}\right)^{\frac{1}{2}} \sin z, \quad J_{\frac{3}{2}}(z) = \left(\frac{2}{\pi z}\right)^{\frac{1}{2}}\left(-\cos z + \frac{\sin z}{z}\right). \quad (40.28)$$

The corresponding Neumann functions are according to (40.12) the Bessel functions of negative half-integral indices,

$$N_{n+\frac{1}{2}}(z) = -(-1)^n J_{-(n+\frac{1}{2})}(z). \quad (40.29)$$

Hence

$$N_{\frac{1}{2}}(z) = -J_{-\frac{1}{2}}(z) = -\left(\frac{2}{\pi z}\right)^{\frac{1}{2}} \cos z,$$

$$N_{\frac{3}{2}}(z) = J_{-\frac{3}{2}}(z) = \left(\frac{2}{\pi z}\right)^{\frac{1}{2}}\left(-\sin z - \frac{\cos z}{z}\right). \quad (40.30)$$

Bessel Functions with Imaginary Arguments

The function $i^{-m} J_m(iz)$ is usually written

$$I_m(z) = \sum_{k=0}^{\infty} \frac{(\frac{1}{2}z)^{m+2k}}{k!(m+k)!}, \quad (40.31)$$

which obeys the differential equation

$$\left[\frac{d^2}{dz^2} + \frac{1}{z}\frac{d}{dz} - \frac{m^2}{z^2} - 1\right] I_m(z) = 0. \quad (40.32)$$

As in the case of ordinary Bessel functions, $I_{-m}(z)$ is a second solution of (40.31). The Wronskian (40.11) of the two functions $I_m(z)$ and $I_{-m}(z)$ is the same as in (40.12) for $J_m(z)$ and $J_{-m}(z)$. Another second solution replacing $I_{-m}(z)$ may be defined by the equation

$$K_m(z) = \frac{\pi}{2} \frac{I_{-m}(z) - I_m(z)}{\sin m\pi}, \quad (40.33)$$

yielding an *m*-independent Wronskian

$$W[I_m(z), K_m(z)] = -1. \tag{40.34}$$

The function $K_m(z)$ is of particular importance in the case of the integral *m*. It is then defined as the limit of expression (40.33).

Integral Representations

In addition to the integral representation already given in (40.21) there is another frequently used alternative. This is found immediately by using for $(m + k)!$ in formula (40.2) the definition of the gamma function by a contour integral

$$\frac{1}{(m+k)!} = \frac{1}{2\pi i} \oint^{(+1)} \frac{e^t \, dt}{t^{m+k+1}}, \tag{40.35}$$

the path of integration in *t* coming from $-\infty$ encircling the origin $t = 0$ in the counterclockwise sense and returning to $-\infty$.

Provided the turn around $t = 0$ is chosen so that $|t| > \frac{1}{2}|z|$ the series (40.2) for $J_m(z)$ may be summed up to form the following contour integral

$$J_m(z) = \frac{1}{2\pi i} \oint^{(+0)} \frac{(\frac{1}{2}z)^m}{t^{m+1}} \exp[t - (z^2/4t)] \, dt = \frac{1}{2\pi i} \oint^{(+0)} \frac{\exp[(z/2)(s-1/s)]}{s^{m+1}} \, ds, \tag{40.36}$$

putting $t = \frac{1}{2}s$.

This is an integral that may be written in a variety of forms and is sometimes taken as the starting point in the theory of Bessel functions. It is particularly well suited in the case of integral *m*, the contour integral being then simply a closed curve around $s = 0$.

If in (40.36) for the integral $m = n$ we write $s = e^{i\varphi}$, we obtain

$$J_n(z) = \frac{1}{2\pi} \int_0^{2\pi} \exp(iz \sin \varphi - in\varphi) \, d\varphi. \tag{40.37}$$

According to the theory of Fourier series, as will be shown later, this means that

$$\exp(iz \sin \varphi) = \sum_{n=-\infty}^{\infty} J_n(z) \exp(in\varphi), \tag{40.38}$$

or, reintroducing $s = e^{i\varphi}$

$$\exp\left[\frac{z}{2}\left(s - \frac{1}{s}\right)\right] = \sum_{n=-\infty}^{\infty} J_n(z) s^n. \tag{40.39}$$

Hence the function $\exp\left[\frac{z}{2}\left(s - \frac{1}{s}\right)\right]$ when expanded in a *Laurent series* may be considered a generating function of the Bessel functions of integral order. These functions are sometimes called *Bessel coefficients*.

Observing that $J_{-n}(z) = (-1)^n J_n(z)$ and equating for real z real and imaginary parts of (40.38) we obtain

$$\cos(z \sin \varphi) = J_0(z) + 2 \sum_{n=1}^{\infty} J_{2n}(z) \cos 2n\varphi,$$

$$\sin(z \sin \varphi) = 2 \sum_{n=1}^{\infty} J_{2n+1}(z) \sin(2n+1)\varphi.$$

(40.40)

This is in accordance with (40.37) as written in the form

$$J_n(z) = \frac{1}{2\pi} \int_0^{2\pi} \cos(z \sin \varphi - n\varphi) \, d\varphi \qquad (40.41)$$

or

$$J_n(z) = \frac{1}{\pi} \int_0^{\pi} \cos(z \sin \varphi) \cos n\varphi \, d\varphi, \qquad n \text{ even,}$$

$$J_n(z) = \frac{1}{\pi} \int_0^{\pi} \sin(z \sin \varphi) \sin n\varphi \, d\varphi, \qquad n \text{ odd.}$$

(40.42)

PART II

Mechanics and Statistics

Introduction

This Part largely follows the same lines of exposition as Parts 1 and 3; the general remarks in these earlier prefaces also apply here.

The author has not avoided detailed mathematical treatments, even though these may at times be too advanced for the average reader. On first perusal the reader may therefore quite safely skip over the most difficult sections; even more so as these tend to treat specialized problems, while the easier sections keep to what one might call the "communal body of knowledge" of all physicists. Although no typographical accentuation of the difference between easy and more difficult subject matter has been made, this circumstance, which leaves the task of differentiation to the reader, may not be wholly detrimental. After all, both the capacity for learning and individual fields of interest may differ widely.

Concerning the first main section of this book, *mechanics*, it may be said that this subject has received a fairly detailed general treatment. More specialized topics, such as the motion of rigid bodies, elasticity, and hydrodynamics, have also been included. We have treated the theory of elasticity in a somewhat independent and novel fashion, by relating the coefficients of elasticity more closely to quadratic forms for the potential energy than to the customary tensorial description of surface forces.

The last chapters, dealing with *thermodynamics*, the *kinetic theory of gases*, and *statistical mechanics*, have been by far the most difficult to write. However, the topics treated have been subjected to a thorough analysis, and most of the basic features of these important disciplines of physics (which in the course of time have undergone such an enormous expansion) have been included within a reasonably sized volume.

CHAPTER 1

Elementary Mechanical Problems

1. HISTORICAL BACKGROUND

The subject of mechanics is one of the basic disciplines of theoretical physics. Among those who have made fundamental contributions to mechanics we find, in antiquity Archimedes of Syracuse (287–212 B.C.), in more recent times Galileo Galilei (1564–1642), Christian Huygens (1629–1695), and indirectly Johannes Kepler (1571–1630) with his discovery of the laws of planetary motion. However, the real founder of mechanics was Isaac Newton (1642–1727). The basic principles he laid down in his book *Principia* (*Philosophiae Naturalis Principia Mathematica*), published in 1687, stood unchallenged right up to 1905, when Albert Einstein put forward a somewhat modified theory of mechanics, in his Special Theory of Relativity.

But mechanics has not only been a basic physical subject; through a couple of centuries it was actually *the basis* of theoretical physics. Through the work of D'Alembert (1717–1783), Leonhard Euler (1707–1783), and Joseph Louis Lagrange (1730–1813), Newtonian mechanics was already approaching the summit of its capacity from the formal point of view. In particular, the "analytic mechanics" of Lagrange was a mighty step forward, based as it was on the introduction of quite general coordinates, with a consequent liberation from more "visual" geometrical considerations. The preface of his book *Mécanique Analytique* (1788) is famous: "Dans cette recherche on ne trouve pas de figures."

We also mention the appearance of certain integral principles in mechanics: the "principle of least action" (Maupertuis 1698–1759) and the related principle of Hamilton (1805–1865). A further perfection of mechanics was attained with Hamilton's canonical equations, and the Hamilton-Jacobi method of solving these equations (Jacobi 1804–1851).

In the following years mechanics achieved outstanding results in astronomy, where we note especially the contributions of Simon Pierre Laplace (1749–1827), with his five-volume work *Mécanique Céleste* (1799–1825). Through its success in astronomy, by the computation of the orbits of known and even undiscovered planets (Neptune), the prestige of mechanics was further enhanced.

Having elucidated these fundamental problems of the macrocosm, mechanics now turned to new problems in the microcosm. In the beginning of the 19th century John Dalton (1766–1844) put forward his atomic hypothesis which was destined to revolutionize physics a century later. At the same time Earl Rumford (1753–1814) and Humphrey Davy (1778–1829) made important contributions to the understanding of the nature of heat. This new insight was in time to lead to not only a complete understanding of the nature of heat, but also the law of conservation of energy, which was postulated by Robert Mayer (1814–1878) and more closely investigated by James Prescott Joule (1818–1889) and Hermann Helmholtz (1821–1894). Through the kinetic theory of gases, founded by Rudolf Clausius (1822–1888) and James Clerk Maxwell (1831–1879), and further developed by Ludwig Boltzmann (1844–1906) and Josiah Willard Gibbs (1839–1903), mechanics also invaded the domains of atomic theory and thermodynamics.

We should also mention that another large physical discipline, optics, was brought under the sway of mechanics after the discovery by Thomas Young (1773–1829) and Augustin Fresnel (1788–1827) of the interference properties of light; this, to be sure, was accomplished only by means of the hypothetical "luminiferous aether" medium.

This epoch of the total domination of mechanics lasted essentially until the year 1864, when Maxwell put forward his revolutionary electromagnetic theory of light. In practice this state of affairs was maintained for yet another couple of decades, according to a kind of natural law of inertia in the development of science. But from the year 1888, with the discovery of electromagnetic waves by Heinrich Hertz (1857–1894), the period of total domination was definitely terminated.

From now on the new notions, conceived earlier by that genius of scientific research, Michael Faraday (1791–1867), and formulated in precise mathematical form by Maxwell in his electromagnetic field equations, made rapid headway; in particular, they furnished the basis of an electron theory of matter, created by a successor of Maxwell and Hertz, Hendrik Antoon Lorentz (1853–1928). This theory was to prove just right for accommodating the flood of new discoveries confronting physicists by the turn of the century. Maxwell's equations and Lorentz's electrodynamics contain the same peculiarities that primarily distinguish Einstein's relativity theory; the term "Lorentz transformations" actually has its origin in electrodynamic theory, and these are equivalent to the transformations of relativity.

The further development of physics in the 20th century has not yet become history. A new theoretical edifice has arisen which has reformulated and modified mechanics in a very special way; this is the so-called *quantum theory*, which is fundamental to modern atomic physics.

Despite the onslaught of new theories during the last 50 or 100 years, Newtonian mechanics still stands as a fundamental pillar of theoretical physics. It has been modified through relativity and quantum mechanics, by the former when the velocities of bodies approach that of light, by the latter when distances and particle masses are of atomic orders of magnitude. Nevertheless, the central position of the theory is fully justified: not only because it is still in practice fully valid for the great majority of natural processes, but even more because of the all-encompassing (and enduring) physical principles it has brought forth, one of the most prominent of which is the principle of conservation of energy.

2. FUNDAMENTAL LAWS OF MECHANICS
THE NEWTONIAN AXIOMS

It is characteristic of Newton's formulation of the fundamental laws of mechanics that they are introduced axiomatically, in the form of three *laws*, with appurtenant definitions and auxiliary theorems.

The Law of Inertia

The so-called *law of inertia*, which had already been stated by Galilei, is formulated as Newton's first law:

Every body continues in its state of rest, or of uniform rectilinear motion, unless it is compelled to change that state by forces impressed upon it.

The insight into natural processes which this law expressed is, from a historical point of view, of extraordinary importance. In contrast to the view held in earlier times, it puts the state of rest and that of uniform rectilinear motion of a body on an equal footing. No external force is necessary in order to maintain such a motion; it continues without change, due to an intrinsic property of all bodies, which we call *inertia*.

Newton made the law more precise by introducing definitions of *quantity of motion* and *mass*. The quantity of motion of a body equals the product of its velocity and the amount of matter it contains. Newton later replaces the "quantity of matter" by the synonymous concept of mass, which then in reality represents a measure of the inertia of the body.

Thus he writes

$$\mathbf{p} = m\mathbf{v} \tag{2.1}$$

where $\mathbf{v}$ is the velocity of the body, m its mass, and $\mathbf{p}$ its quantity of motion, which (like the velocity) is a vector. The law of inertia may then be expressed mathematically by the equation

$$\mathbf{p} = \text{const}, \tag{2.2}$$

assuming that no external forces act on the body.

216 / Mechanics and Statistics

That the law of inertia was far from being self-evident at the time of its conception, may be demonstrated from the fact that Immanuel Kant maintained, as late as 1747, that there are "two kinds of motion," one that ceases after some time, and another that is lasting. By the law of inertia the former are just motions subject to friction, that is, external forces.

Instead of the term quantity of motion for **p** we shall in the following frequently use the word *momentum*, which also signifies the quantity force multiplied by time. The law of inertia is thus equivalent to the law of conservation of momentum for systems of bodies, and identical with this law for a single body.

The Fundamental Equation of Mechanics

The description of the actual motion of a body is found in Newton's second law:

The change of motion is proportional to the motive force impressed, and occurs in the direction of the straight line along which that force is acting.

By "change of motion" Newton undoubtedly meant the time derivative of the quantity of motion. The law may consequently be expressed in mathematical terms as follows:

$$\dot{\mathbf{p}} = \mathbf{F}, \qquad \mathbf{p} = m\mathbf{v}, \tag{2.3}$$

where **F** is the force acting:

$$\frac{d}{dt}(m\mathbf{v}) = \mathbf{F}. \tag{2.4}$$

This law governs the change of momentum of a body due to the action of a force, and may thus be referred to as the *momentum law* or *momentum theorem*. However, a more common designation is the *acceleration law*, this because Newton's second law is generally written as:

$$m\mathbf{a} = \mathbf{F} \quad \text{or} \quad m\dot{\mathbf{v}} = \mathbf{F}, \tag{2.5}$$

with the tacit assumption that the mass of a body is a constant.

One may well wonder at the foresight revealed here, consciously or not, by Newton. For, in the theory of relativity, the acceleration is not always proportional to the acting force, since the mass of a body depends upon its velocity:

$$m = \frac{m_0}{(1 - v^2/c^2)^{1/2}}, \tag{2.6}$$

where c is the velocity of light. In the theory of relativity, therefore, Newton's second law must be employed in the form originally given to it by Newton, as in (2.3) and (2.4).

Newton's Third Law

Newton's third law contains the famous principle of *action and reaction:*

To every action there is always opposed an equal reaction; or, the mutual actions of two bodies upon each other are always equal in magnitude and oppositely directed.

This means that the forces of nature always occur in pairs. The action-reaction principle is just as valid for forces acting at a distance as for those requiring material contact. A pressure is always accompanied by an equal counterpressure. At first glance, only one force appears to be present in the case of a body situated in the earth's gravity field, since it is difficult for us to imagine the reaction of the body on the field itself. However, let us consider the body generating the field, that is, the earth. It is then clear that the body in the field attracts the earth with a force of exactly the same magnitude as the weight of the body.

The action-reaction principle does not affect our treatment of the motion of a single body under the influence of given forces, but only our treatment of complex systems. Also, it may be dispensed with or deduced, if we would accept other related postulates, as for example, that a complex body or system of bodies cannot accelerate its motion by means of internal forces only. By "motion" we mean, in this case, motion of the center of mass.

Force and Work

Before proceeding we should pause for a moment, and consider the concept of *force*, which has been the subject of much discussion. If the concept is to have a clear physical meaning, one must state how forces are to be measured. Quite often they are measured by the acceleration they produce in bodies of known masses, using Newton's second law. If it is defined in such a way, the concept of force loses its independent physical meaning; this was the view advocated by Gustav Kirchhoff (1824–1887), and later by Heinrich Hertz.

However, forces may also be measured directly in other ways, similar in some respects to certain methods used for measuring temperatures. Thus we may compare a spring, as a measuring instrument for forces, to a mercury thermometer. The higher the mercury column rises, the higher the temperature; and similarly, the greater the stretching of the spring, the greater the force acting on it.

We cannot, however, just set the force proportional to the elongation of the spring; this would furnish a purely individual measure of force, just as one particular thermometer would define an individual scale of temperature. What we can do, using a spring, is to verify experimentally whether two given forces (for instance the weights of two different chunks of material) are equal. By adding the forces from a set of several such unit weights, we may then obtain known forces of different magnitudes, and thus calibrate any spring or other force-measuring instrument in a consistent quantitative way. Hence we arrive at a rational scale of forces, much easier than we obtain a rational scale of temperatures in thermodynamics.

The use of such springs or other measuring instrument is not restricted to statical forces alone. We may also imagine that they be attached to moving bodies, thus enabling us to measure the forces acting between these bodies. In this way it is possible, at least in principle, to assign a meaning to the notion of acting force, both for bodies *at rest* and bodies *in motion*.

Quite a different matter is that the unit of force defined by the weight of a body in a given gravity field is not so readily reproducible, since the weight does not depend only on the body, but also on the strength of the field. On the other hand, the mass of a body is an intrinsic property. For this reason the unit of mass is preferred to that of force in the scientific system of units, in line with the fact that forces usually are determined by the acceleration which they impart to bodies of known mass. Taking the unit of mass to be that of a gram (or kilogram) weight, we arrive at a so-called *scientific* or "absolute" unit system, in contrast to the *technical* system. In particular, units of mass, length, and time in the CGS-system are the gram, centimeter, and second, respectively; the unit of force in this system is called the dyne, which is approximately $\frac{1}{981}$ of the weight of a body of unit mass.

Having defined the unit of force, we may also define the concept of *work*, as force multiplied by distance for a body or mass point undergoing a displacement $d\mathbf{r}$; this may be expressed mathematically by

$$dA = \mathbf{F}\, d\mathbf{r}. \tag{2.7}$$

The Parallelogram Theorem

In the *Principia* Newton formulates a fourth law, or rather lemma, which was already known earlier, however; this law states that forces add vectorially, that is, that the net effect of a set of forces $\mathbf{F}_1, \mathbf{F}_2, \cdots$ is that of a single force $\mathbf{F}$, given by

$$\mathbf{F} = \mathbf{F}_1 + \mathbf{F}_2 + \cdots. \tag{2.8}$$

This, of course, is just the parallelogram theorem for two forces acting at the same point: the resultant force is given, in magnitude and direction relatively to the two component forces, by the diagonal of the parallelogram constructed geometrically by these forces.

Coordinate Systems. "Absolute" Space

In the introduction to the *Principia* Newton states that "absolute space, in its own nature and without relation to anything external, remains always similar and immovable."

This assumption is the only really obsolete feature of the Newtonian theory of mechanics, and it is therefore not surprising that Newton himself never actually made use of it.

All motion must be described relatively to some rigid body or geometrically invariant material system, that is, to a *coordinate system*. Our formulation of the Newtonian laws of motion, in particular the second law, in the form

$$m\mathbf{a} = \mathbf{F} \tag{2.9}$$

presupposes that the motion is referred to a certain class of coordinate systems. To a certain extent, the definition of such systems may be deduced from Newton's first law (the law of inertia), since every body that is left to itself, and consequently moves only by its own inertia, may serve as a coordinate system. Such a body will have no acceleration relative to other freely moving bodies, and a body acted on by forces will have the same acceleration relative to all such bodies.

Galilei being the discoverer of the law of inertia, such coordinate systems are generally called *Galilean*. If we want to retain Newton's idea of absolute space, we should then have to picture all Galilean systems moving with uniform velocities relative to this. However, it is a consequence of the laws of motion that there exists no way to detect which of these systems represents (i.e. is at rest relative to) absolute space.

We shall refer to the transformations of coordinates and equations of motion from one Galilean system to another as *Galilean transformations*. They do not change the form of these equations; hence the forces do not depend on the coordinate system of reference either, at least as long as we stick to Galilean systems.

However, the forces will change if we refer the motion to an accelerated coordinate system. Let us assume, for instance, that forces are to be referred to an elevator that is suddenly allowed to fall freely. A spring that formerly was extended by a weight would then just as suddenly recontract; the force of gravity has been "suspended." We experience the same sort of thing on the Earth, which "falls" freely in the solar gravity field; the attraction by the sun on ourselves and other bodies cannot be detected, as long as all motions are referred to the Earth itself.

All this will become more clear later on, when we discuss relative motion. At the present stage we shall only remark that Newton's original formulation of the fundamental law of mechanics,

$$\frac{d}{dt}(m\mathbf{v}) = \mathbf{F}, \tag{2.10}$$

necessitates other transformations than the Galilean, if mass is assumed to vary with velocity as demanded by the theory of relativity. These are the already-mentioned *Lorentz transformations*.

Let us consider two coordinate systems having a velocity v relative to each other in the direction of the x-axis, and call the sets of coordinates in the two systems x, y, z and x', y', z', and the corresponding time variables t and t'. The Galilean transformations may then be written as follows:

$$x' = x - vt, \quad y' = y, \quad z' = z, \quad t' = t, \qquad (2.11)$$

while the Lorentz transformations are

$$x' = \frac{x - vt}{[1 - (v^2/c^2)]^{1/2}}, \quad y' = y, \quad z' = z, \quad t' = \frac{t - xv/c^2}{[1 - (v^2/c^2)]^{1/2}}. \qquad (2.12)$$

These transformations are treated in detail at the end of Vol. II, Part 3 on electromagnetism.

3. FALLING AND OTHER ONE-DIMENSIONAL MOTIONS

Until further notice we shall confine our attention to so-called point mechanics; in this theory a body is represented by a single material point, the motion of which is defined in terms of its three coordinates. Thus we reduce the number of degrees of freedom of the body from six to three, by ignoring any rotations. This is equivalent to regarding the center of mass of the body as a material point, with a mass equaling the total mass of the body, acted on by a force equal to the sum of all forces acting on the component parts of the body.

In point mechanics a system of N bodies is thus assumed to have $3N$ degrees of freedom; by considering only one body, this number is then reduced to three, corresponding to the three coordinates of the material point (or the center of mass).

To start with, however, we shall simplify things even further, by considering one-dimensional problems only: for example, motion in one coordinate direction. A simple example of such problems is falling: here there are no forces acting horizontally, so that this motion is truly one-dimensional. A body may also be constrained to move in such a way that its motion may be described by means of only one variable. An example is that of a pendulum oscillating in a plane about a fixed axis; its position is determined by the angle of deflection only.

General Conditions for Integration of Equations of Motion for One-Dimensional Problems

We consider the equation of motion for a material point (particle) with one degree of freedom:

$$m\dot{v} = F \qquad (3.1)$$

and investigate the general possibilities for integrating this equation. The force may be assumed to vary in divers ways: for example, depend on time, on the position of the particle, or on its velocity. We shall now show that the equation may be integrated analytically (in the sense that solution may be expressed analytically in terms of elementary functions, convergent series, and/or integrals of known functions) in all these three cases, provided that F depends on only one of the quantities mentioned.

1. Time Dependence Only: $F = F(t)$

The equation may be solved by ordinary integration. Integrating once, we get $v = G(t)$; setting $v = \dot{x}$ and integrating again, we get $x = H(t)$. The former expression then contains one constant of integration, while the latter contains two; these constants may be taken to represent the initial position and velocity of the particle.

2. Velocity Dependence Only: $F = F(v)$

In this case we may put $1/\dot{v} = dt/dv$, so that

$$t = \int \frac{m\,dv}{F(v)} = f(v). \tag{3.2}$$

Hence
$$v = G(t), \tag{3.3}$$

and we may proceed as in 1.

3. Position Dependence Only: $F = F(x)$

Multiplying (3.1) by $v = \dot{x}$ and integrating, we get

$$\frac{m}{2}v^2 = E - V(x),$$
$$V = -\int F(x)\,dx, \tag{3.4}$$

where E is a constant of integration. We shall later learn to recognize (3.4), in more complicated forms, as the *energy equation*.

Rearranging (3.4), we find

$$v = \left\{\frac{2}{m}[E - V(x)]\right\}^{1/2},$$
$$t = \int \frac{dx}{\{(2/m)[E - V(x)]\}^{1/2}} = f(x), \tag{3.5}$$

and hence $x = g(t)$.

If the force depends on all the three variables jointly: $F = F(x, v, t)$, the equation of motion will in general not be solvable by analytic integration. A closer investigation shows that this is possible only when the force depends linearly on the position coordinate x and velocity $v = \dot{x}$, for instance,

$$F = -ax - b\dot{x} + f(t). \tag{3.6}$$

This is because linear differential equations are so much easier to solve than nonlinear ones; since the coordinate and velocity are dependent variables, the force must depend linearly on them to ensure linearity in the equations to be solved.

The equation

$$m\ddot{x} + b\dot{x} + ax = f(t), \tag{3.7}$$

corresponding to a force of the type (3.6), may (as is well known) be solved by first finding a particular solution, and then the general solutions of the corresponding homogeneous equation:

$$m\ddot{x} + b\dot{x} + ax = 0. \tag{3.8}$$

The general solutions of the inhomogeneous equation (3.7) are then found by adding the solutions of (3.8) to the particular solution of (3.7).

When a and b are constants, the solutions of the homogeneous equation (3.8) become very simple: either periodic sine- and cosine functions, or exponential functions with real arguments. If $b \neq 0$, products of such functions appear in the solutions. Also, $b = 0$ corresponds, as a rule, to simple nondamped oscillations, while $b \neq 0$ ($b > 0$) represents different types of damped oscillations. When $f(t) \neq 0$, the oscillations are *forced;* while (3.8) describes *free* oscillations.

Some Elementary Examples

1. Free Fall

The driving force on a falling body is its weight, which equals mg, where m is the mass and g the acceleration of gravity. We may then cancel the mass m in the Newtonian equation of motion, obtaining

$$\ddot{z} = g \quad \text{or} \quad \dot{v} = g, \quad v = \dot{z}; \tag{3.9}$$

here the vertical coordinate, taken to increase in the downward direction, is designated by z. This is the simplest form possible for the equation of motion; the force mg is a constant, and hence explicitly independent of time, position, and velocity.

By elementary integration in two steps we obtain:

$$v = v_0 + gt,$$
$$z = z_0 + v_0 t + \tfrac{1}{2}gt^2. \tag{3.10}$$

The constants v_0 and z_0 are constants of integration, which may be adjusted to describe the velocity and position coordinate at the time $t = 0$, if these are known.

2. Fall with Air Resistance

We may here distinguish between two possibilities:

$$F = mg - mkv, \tag{3.11a}$$
$$F = mg - mav^2. \tag{3.11b}$$

In both cases we have a frictional resistance. Physically, (3.11a) may be taken to describe slow motions through the air, by very small and light bodies; while (3.11b) describes the force for large velocities, as attained by heavy massive bodies. In general the frictional resistance of the air will be more correctly described by a combination of a linear and a quadratic term (in the velocity). The integration of the equation of motion for such a compound force does not present any problems in principle; nevertheless we shall simplify matters by treating the two special cases (3.11a) and (3.11b) separately.

From (3.11a) we obtain

$$\dot{v} = g - kv. \tag{3.12}$$

By inspection we see that $v_1 = g/k$ is a particular solution of the inhomogeneous equation (3.12); and since the homogeneous equation $\dot{v} = -kv$ has the solution $v_2 = ce^{-kt}$, a solution with a given initial velocity $v = v_0$ may be written as

$$v = \left(v_0 - \frac{g}{k}\right)e^{-kt} + \frac{g}{k}, \tag{3.13}$$

$v_1 = g/k$ being the limiting velocity attained after a long time of falling.

From (3.11b) it follows that

$$\dot{v} = g - av^2. \tag{3.14}$$

It is possible here to solve this equation by the general method shown by (3.2), but we may also proceed as follows.

A particular solution of (3.14) is the limiting velocity $v_1 = \sqrt{g/a}$. Substituting $v = v_1 + v_2$, we obtain a simpler, though nonlinear, equation in v_2. However, a linear equation may be obtained by replacing v_2 by const/u, for instance, by assuming that

$$v = \left(\frac{g}{a}\right)^{1/2}\left(1 - \frac{2}{u}\right). \tag{3.15}$$

We then obtain an equation in the variable u:

$$\dot{u} = 2\sqrt{ga}\,(u - 1). \tag{3.16}$$

Thus

$$u - 1 = Ce^{2\sqrt{ga}\,t}, \tag{3.17}$$

or

$$v = \left(\frac{g}{a}\right)^{1/2} \frac{Ce^{2\sqrt{ga}\,t} - 1}{Ce^{2\sqrt{ga}\,t} + 1}. \tag{3.18}$$

The solution may be adjusted to accommodate a given initial velocity V_0, by putting

$$C = \left(\left(\frac{g}{a}\right)^{1/2} + v_0\right) \Big/ \left(\left(\frac{g}{a}\right)^{1/2} - v_0\right). \tag{3.19}$$

3. Electric Circuit with Fixed Resistance

In theoretical physics, problems from widely different fields may often be treated mathematically in very much (or even exactly) the same way. This is because the problems generally appear in the form of differential equations, which often are of very similar types, especially when they are linear. An electric circuit with fixed resistance, for instance, corresponds in this respect to the first case previously treated for falling with air resistance (compare (3.11a). The electromotive force E in the circuit will correspond to the weight mg of the body; the current I, to the velocity v; and the amount of charge

$$q = \int_0^t I\,dt \tag{3.20}$$

which has passed a cross section of the circuit, to the distance traveled (i.e. the height of fall) by the body. Moreover, the resistance R (in ohms) of the circuit will correspond to the air resistance mk, and finally the self-induction L to the inertia or mass m of the body. Hence we arrive at the following equation for the circuit:

$$L\frac{dI}{dt} = E - RI, \tag{3.21}$$

with the solution

$$I = \left(I_0 - \frac{E}{R}\right)e^{-(R/L)t} + \frac{E}{R}. \tag{3.22}$$

The saturation current E/R then corresponds to the limiting velocity g/k attained by the body falling in air.

4. FREE AND FORCED ELASTIC OSCILLATIONS

Proper Oscillations and Proper Frequency

The forces occurring in the preceding examples were independent of position and time. As time-dependent forces generally are the most difficult to handle in mechanics we shall first study forces depending only on position. A special type of such forces, leading to harmonic monochromatic oscillations (i.e. with the frequency independent of the amplitude of oscillation) are the elastic or quasielastic force.

Such oscillations may be treated very simply. This is also true when they are forced, that is, driven by external time-dependent forces. The reason for this simplicity is that quasielastic forces lead to linear equations of motion.

We shall still consider only one-dimensional motions, designating the deflection (from the equilibrium position of rest) of an oscillating system by x, whether this be a coordinate for a single particle or, say, the angle of deflection of a pendulum. A quasielastic force is by definition a force that is proportional to the deflection and drives the particle (or, in general, the oscillator) back toward the equilibrium position. For a one-dimensional oscillator we may thus write

$$F = -ax. \tag{4.1}$$

Letting m be the mass of the particle (or the inertia of the oscillator), we obtain

$$m\ddot{x} + ax = 0. \tag{4.2}$$

This linear differential equation with constant coefficients may (as is well known) be solved by substituting

$$x = Ce^{\rho t}, \tag{4.3}$$

which leads to a second-degree equation for ρ with the roots:

$$\rho = \pm i\omega_0, \quad \omega_0 = \left(\frac{a}{m}\right)^{1/2}. \tag{4.4}$$

The two linearly independent solutions are thus exponential functions with a positive and a negative argument, expressible also by sines and cosines:

$$x = c_1 \sin \omega_0 t + c_2 \cos \omega_0 t = A \sin(\omega_0 t + \eta),$$
$$A = \sqrt{c_1^2 + c_2^2}, \quad \eta = \arctan\left(\frac{c_2}{c_1}\right). \tag{4.5}$$

In the second of the above expressions for x, A is referred to as the *amplitude* and η as the *phase;* the motion (4.5) of x is generally designated as *free proper oscillation*. Also, $\omega_0 = 2\pi\nu_0$ is called the *angular frequency* of these oscillations, while $\nu_0 = 1/T$, where $T = 1/\nu_0 = 2\pi/\omega_0$ is the time taken by the oscillator to complete one cycle, is called the *cyclic frequency*. We shall usually just use the word frequency, since it will be clear from the context which type is meant. The terms v_0 and ω_0, representing the frequency exhibited by free oscillation, are often referred to as *proper frequencies* (cyclic and angular, respectively). "Free" or "proper" oscillations derive from the fact that these oscillations keep going, once they have been started, with undiminished amplitude and energy.

We may adjust the two constants of integration in the solution (4.5) to accommodate a given initial position x and initial velocity $\dot{x} = v = v_0$, by writing

$$x = x_0 \cos \omega_0 t + \left(\frac{v_0}{\omega_0}\right) \sin \omega_0 t. \tag{4.6}$$

Forced Oscillations

When the oscillator is acted upon by a time-dependent external force, it will be made to execute forced oscillations, at the same time as its own proper oscillations are set going.

A time-dependent force may, for a limited or unlimited interval of time, be resolved into a Fourier series or Fourier integral, respectively. Thus, when the equations are *linear*, we may always consider each of the monochromatic components of the external force separately. If such a component is characterized by the frequency ω and the amplitude F_0:

$$F = F_0 \sin(\omega t + \eta), \tag{4.7}$$

the equation of motion for an oscillator of proper frequency ω_0 and mass m becomes

$$\ddot{x} + \omega_0^2 x = \left(\frac{F_0}{m}\right) \sin(\omega t + \eta). \tag{4.8}$$

A solution of this inhomogeneous equation is

$$x_1 = \left[\frac{F_0}{m}(\omega_0^2 - \omega^2)\right] \sin(\omega t + \eta). \tag{4.9}$$

The general solution is then obtained by adding free sine and cosine oscillations of arbitrary amplitudes to (4.9). These amplitudes may be chosen in such a way that the initial position and velocity (i.e. x and $\dot{x}$ at $t = 0$) have given values x_0 and V_0; we then find

$$x = x_0 \cos \omega_0 t + \left(\frac{v_0}{\omega_0}\right) \sin \omega_0 t + \left[\frac{F_0}{m}(\omega_0^2 - \omega^2)\right]$$
$$\times \left[\sin(\omega t + \eta) - \sin \eta \cos \omega_0 t - \left(\frac{\omega}{\omega_0}\right) \cos \eta \sin \omega_0 t\right]. \tag{4.10}$$

Resonance

It is evident that (4.10) exhibits two different kinds of oscillation: first, one with the same frequency and phase as the external force, and with amplitude as given for x_1 in (4.9); second, the two independent free oscillations, which keep going with undiminished amplitudes, once they are started. Furthermore it is clear from (4.10) that for a given initial state $x = x_0$, $\dot{x} = v_0$, the amplitudes of $\cos \omega_0 t$ and $\sin \omega_0 t$ are not the same when an external force is acting as when the oscillations are free.

The exact formula (4.10) is very useful in the treatment of an important physical phenomenon, which we call *resonance*. It follows from (4.9) and (4.10) that the amplitude of the forced oscillation becomes infinite when the frequency of the external force approaches the proper frequency ω_0, that is, the system "explodes" after a short time.

To study the oscillations in detail for a limited time, we must therefore know the initial state x_0, v_0 of the oscillator at a given moment of time, say $t = 0$; we shall assume $x_0 = 0$, $v_0 = 0$, so that the two first terms of (4.10) vanish.

Now, as $\omega \to \omega_0$ or $\omega_0 \to \omega$, we obtain a zero-divided-by-zero expression; the correct value of this may then be found by differentiating both the numerator and the denominator with respect to ω or ω_0, and then putting ω equal to ω_0. This leads to

$$x = \left(\frac{F_0}{2m\omega_0}\right)[\cos \eta \sin \omega t - \omega t \cos (\omega t + \eta)]. \quad (4.11)$$

The last term of this equation will increase without limit when $t \to \infty$; the amplitude then increases for each "swing" proportionally to the time, and on the average this increase will be given by

$$x_{\max} = \left(\frac{F_0}{2m\omega}\right)t. \quad (4.12)$$

Electric Oscillatory Circuit with Self-Induction and Capacitance

For electric circuits the capacitance C of a condenser connected into the circuit will correspond to the quasielastic counterforce of an oscillator. The reason for this is clear: When the condenser is charged, an electric counterpotential is set up in the circuit, determined by the potential on the condenser, or for a plate condenser, the potential difference between the plates. This potential is $V = q/C$, where q is the charge on the condenser. The larger the capacitance, the smaller the counterpotential or, if you like, the quasielastic counterforce. The equation of motion for the circuit becomes

$$L\ddot{q} + \left(\frac{1}{C}\right)q = 0. \tag{4.13}$$

The proper frequency and period of oscillation for the circuit are thus

$$\omega_0 = \frac{1}{\sqrt{LC}}, \qquad T = 2\pi\sqrt{LC}. \tag{4.14}$$

The further treatment of forced oscillations and resonance for an electric circuit follows exactly the same lines as those previously described for a mechanical oscillator. (See Fig. 1-1.)

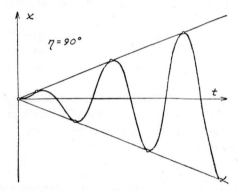

Fig. 1-1. Forced nondamped oscillations with resonance.

5. DAMPED OSCILLATIONS

Variation of Frequency and Amplitude

If an oscillator is subjected to friction, the equation of motion may still be integrated analytically, provided that the frictional force is linear in the velocity $\dot{x} = v$, as in the case previously studied for falling with air resistance. In this case, however, we shall see that the oscillations are *damped*, that is, of *decreasing amplitude*.

For simplicity we put the damping force equal to $-2mk\dot{x}$, enabling us (since the quasielastic term has been written as $ax = m\omega_0^2 x$) to cancel the masses. This leads to the following equation for free damped oscillations

$$\ddot{x} + 2k\dot{x} + \omega_0^2 x = 0. \tag{5.1}$$

This equation may be solved by the same method as before, that is, by substituting (4.3), but the possible values of ρ are now:

$$\rho = -k \pm i\sqrt{\omega_0^2 - k^2}. \tag{5.2}$$

Elementary Mechanical Problems | 229

The two linearly independent solutions will thus have a common factor e^{-kt}, describing the damping of the oscillations; moreover, they will contain sine and cosine terms with a modified proper frequency

$$\bar{\omega}_0 = \sqrt{\omega_0^2 - k^2}. \tag{5.3}$$

The general solution of (5.1), with two constants of integration, may be written

$$x = e^{-kt}(c_1 \sin \bar{\omega}_0 t + c_2 \cos \bar{\omega}_0 t), \tag{5.4}$$

or, if it is adjusted to a given initial state:

$$x = e^{-kt}\left[x_0 \cos \bar{\omega}_0 t + \left(\frac{1}{\bar{\omega}_0}\right)(v_0 + kx_0) \sin \bar{\omega}_0 t\right]. \tag{5.5}$$

(See Fig. 1-2.)

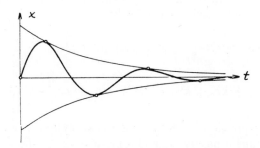

Fig. 1-2. Damped free oscillations.

These damped free oscillations will die out in the course of time, as the mechanical energy of oscillation is "used up" in combating the frictional resistance. How quickly they die out depends on the damping constant. It is customary to describe the damping by the so-called logarithmic decrement δ, which shows how much the logarithm of the amplitude decreases during one period of oscillation T; this is given by

$$\delta = kT = \frac{2\pi k}{\omega_0}. \tag{5.6}$$

Periodic and Aperiodic Oscillations

In the previous considerations we have tacitly assumed $k < \omega_0$. However, if $k = \omega_0$ or $k > \omega_0$, the proper frequency $\bar{\omega}_0$ of (5.3) will become zero, or purely imaginary. This means that the oscillations are no longer periodic; we shall refer to such oscillations as *aperiodic*.

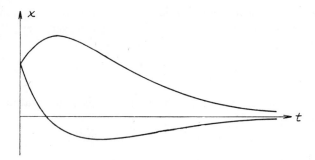

Fig. 1-3. Aperiodic oscillations.

We now investigate the nature of such oscillations for $k > \omega_0$, and in the limit $k = \omega_0$. In the first case, the proper frequency becomes purely imaginary: $\bar{\omega}_0 = ik'$, $k' < k$. We may then replace the trigonometric functions by hyperbolic sines and cosines, which leads to:

$$x = e^{-kt}\left[x_0 \cosh k't + \left(\frac{1}{k'}\right)(v_0 + kx_0) \sinh k't\right]$$

$$= \left(\frac{1}{2}k'\right)[(v_0 + kx_0 + k'x_0)e^{-(k-k')t} - (v_0 + kx_0 - k'x_0)e^{-(k+k')t}]. \quad (5.7)$$

From the last expression it follows that x is always positive if x_0 and v_0 are positive. Assuming a positive x_0 and a negative $v_0 < -(k + k')x_0$, the first (negative) term of (5.7) will dominate for large values of t. Hence we obtain the two possible types of oscillation shown in Fig. 1-3.

The aperiodic oscillations in the limiting case $k = \omega_0$ may be determined by setting $\bar{\omega}_0 = k' = 0$ in (5.5) and (5.7), respectively, thus obtaining

$$x = e^{-kt}[x_0 + (v_0 + kx_0)t]. \quad (5.8)$$

The qualitative nature of the oscillations are as shown in Fig. 1-3. If $v_0 > -kx_0$, only deflections to one side occur, while for $v_0 < -kx_0$ we also get a deflection to the other side.

Forced Damped Oscillations

For forced damped oscillations the mathematical treatment becomes somewhat more complicated, and it is therefore customary to make the calculations easier to follow by introducing complex variables.

To start with, we express the external force in the form:

$$F = F_0 \cos \omega t = \operatorname{Re} F_0 e^{i\omega t}. \quad (5.9)$$

Henceforth we shall take the variables x, $\dot{x}$, and so forth, to have two different meanings, regarding them on the one side as complex quantities, and on the other side as the real parts of these quantities; these real parts then represent the coordinates and physical quantities that are to be determined. (Since this is conventional, we might just as well have chosen to use the imaginary parts for this purpose.)

By this convention, we may write the equation for the oscillation as follows:

$$\ddot{x} + 2k\dot{x} + \omega_0^2 x = \left(\frac{F_0}{m}\right)e^{i\omega t}. \tag{5.10}$$

A particular solution of (5.10) is

$$x = \frac{(F_0/m)e^{i\omega t}}{\omega_0^2 - \omega^2 + 2i\omega k},$$

$$\dot{x} = \frac{(F_0/m)e^{i\omega t}}{2k + (\omega_0^2 - \omega^2)/i\omega}. \tag{5.11}$$

These equations describe the forced oscillations which are permanent. The free oscillations are transient, that is, they are damped and die out quickly. Hence they are of importance only for very special problems; the complete solution and its adjustment to a given initial state does not bring up anything essentially new, compared with the previous treatment of forced oscillations without damping. We shall therefore leave out these transients here.

The special feature distinguishing forced damped oscillations from undamped ones is that the damped oscillations do not exhibit exact resonance, in the sense that the amplitude increases to infinity. Physically this is, of course, easy to understand: the frictional resistance will ultimately "consume" all the energy supplied by the driving force. We shall, however, still use the term resonance to describe the situation when the amplitudes of deflection or velocity achieve their largest possible values. We may also define the resonance frequency as that for which the energy of the oscillator attains its greatest value.

The maximum values of x and $\dot{x}$ are clearly those for which the denominators in (5.11) are as small as possible. Hence we obtain

$$(\omega_0^2 - \omega^2)^2 + 4\omega^2 k^2 = \min, \quad \omega^2 = \omega_0^2 - 2k^2,$$

$$\frac{(\omega_0^2 - \omega^2)^2}{\omega^2} + 4k^2 = \min, \quad \omega^2 = \omega_0^2. \tag{5.12}$$

The maximum energy, however, occurs (as we shall show later) at the average frequency

$$\bar{\omega}_0^2 = \omega_0^2 - k^2. \tag{5.13}$$

232 / Mechanics and Statistics

That this frequency lies between the two resonance frequencies for the deflection and velocity was to be expected; since it is equal to the proper frequency (5.3) for free oscillations, it is natural to define it as *the* resonance frequency of the oscillator.

Amplitude Variation and Phase Shift for Forced Damped Oscillations

We shall now write the solutions (5.11) in a slightly different form, expressing the denominators in terms of their moduli and arguments:

$$x = \left(\frac{F_0}{i\omega Z}\right)e^{i(\omega t + \eta)},$$

$$\dot{x} = \left(\frac{F_0}{Z}\right)e^{i(\omega t + \eta)},$$
(5.14)

where

$$Z = m\left[4k^2 + \frac{(\omega_0^2 - \omega^2)^2}{\omega^2}\right]^{\frac{1}{2}}, \quad \eta = \arctan\frac{\omega_0^2 - \omega^2}{2k\omega} \quad (5.15)$$

and η is the phase shift for the velocity $\dot{x}$. For the deflection x the shift is $\eta - \pi/2$ (since $1/i = e^{-i(\pi/2)}$). In the case of an exact resonance for the velocity ($\omega = \omega_0$) $\dot{x}$ will then oscillate exactly "in step" with the external force while x will be delayed a quarter-period behind. For higher frequencies the delay will increase, and when $\omega \to \infty$ the deflections of the oscillator will be in counterphase with (i.e. one half-period behind) the external force.

Forced Damped Electric Oscillations

For electric circuits the damping is due to the resistance R (in ohms), which gives rise to a counterpotential RI in the circuit. Since $I = \dot{q}$, we obtain the equation of motion:

$$L\ddot{q} + R\dot{q} + \left(\frac{1}{C}\right)q = E_0 e^{i\omega t}, \quad (5.16)$$

E_0 being the amplitude of the potential on the condenser. Using (5.10) and (5.11), we then find

$$\dot{q} = I = \frac{E_0 e^{i\omega t}}{R - i[(1/C\omega) - L\omega]}, \quad (5.17)$$

or, by (5.14) and (5.15):

$$I = (E_0/Z)e^{i(\omega t + \eta)},$$

$$Z = \left[R^2 + \left(\frac{1}{C\omega} - L\omega\right)\right]^{\frac{1}{2}}, \quad \eta = \arctan\frac{[(1/C\omega) - L\omega]}{R}. \quad (5.18)$$

The quantity Z is called the *impedance*, and $1/C\omega - L\omega$ the *reactance*. For high frequencies the reactance attains a large negative value; the phase delay of the current then approaches 90°, and the average work of the external driving force becomes zero. This may also be achieved by including a very large self-induction L in the circuit.

Work of External Force and Energy of Oscillator for Damped Oscillations

We shall now calculate the amount of work

$$A = \int_{x_0}^{x_1} F\,dx = \int_{t_0}^{t_1} F\dot{x}\,dt \tag{5.19}$$

performed by the external force on the oscillator during the time interval from t_0 to t_1. Using (5.10), we find

$$A = \int_{t_0}^{t_1} 2mk\dot{x}^2\,dt + \frac{m}{2}\{\dot{x}^2 + \omega_0^2 x^2\}\big|_{t_0}^{t_1}. \tag{5.20}$$

The first term increases with t_1, since the integrand is positive; this term represents the work performed against the friction. The last term, which we shall designate by $E(t)$, obviously represents the accumulated mechanical energy of the oscillator. The variation of this energy will be very small; hence the work performed by the external force will in the long run be consumed by frictional resistance alone.

Inserting complex variables in

$$E(t) = \frac{m}{2}(\dot{x}^2 + \omega_0^2 x^2) \tag{5.21}$$

and substituting x^2 and $\dot{x}^2$ by xx^* and $\dot{x}\dot{x}^*$, respectively, we obtain the sums of the squares of the real and imaginary parts. Since these represent oscillations with a quarter-period phase shift, it is reasonable to expect small variations in the energy to be "smoothed out." Thus, substituting the expressions (5.11) in (5.21), and dividing by two in order to average the two oscillations, we obtain the average total energy

$$\bar{E} = \left(\frac{F_0^2}{4m}\right) \cdot \frac{\omega_0^2 + \omega^2}{(\omega_0^2 - \omega^2)^2 + 4k^2\omega^2}. \tag{5.22}$$

For $\omega^2 = \omega_0^2$, as well as for $\omega^2 = \omega_0^2 - 2k^2$, this energy is $F_0^2/8mk^2$; while for the average frequency $\bar{\omega}_0^2 = \omega_0^2 - k^2$, it is

$$\bar{E} = \frac{F_0^2}{8mk^2} \cdot \frac{\omega_0^2 - \tfrac{1}{2}k^2}{\omega_0^2 - \tfrac{3}{4}k^2}, \tag{5.23}$$

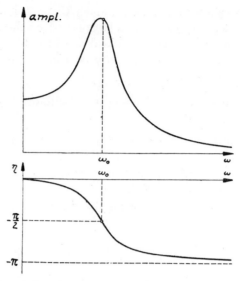

Fig. 1-4. Amplitude and phase for forced damped oscillations as functions of the frequency.

and thus greater. However, differentiation of (5.22) shows that the maximum value occurs at

$$\omega^2 = -\omega_0^2 + 2\omega_0\sqrt{\omega_0^2 - k^2} \qquad (5.24)$$

and then is

$$\bar{E} = \frac{F_0^2}{8mk^2} \cdot \frac{1}{2}\left(1 + \frac{\omega_0}{\sqrt{\omega_0^2 - k^2}}\right). \qquad (5.25)$$

The difference between (5.23) and (5.25) is very small. (See Fig. 1-4.)

6. OTHER OSCILLATION PROBLEMS

The Mathematical Pendulum

We shall now consider a few other simple oscillatory problems, that cannot be described by linear differential equations and thus do not represent strictly harmonic oscillations. These problems must therefore be treated quite differently from those already discussed: by means of the general procedure outlined in (3.4) and (3.5). The simple mathematical and physical pendula are examples of such anharmonic oscillators.

A *mathematical pendulum* may be thought of as consisting of a material sphere of mass m, suspended from a point by a thread of a given length l; the frictional resistance of the air and the weight of the thread are then assumed to be negligible, and the sphere is considered as a material point.

The sphere may also be suspended by a weightless rod, turning on a fixed horizontal axis. The pendulum may then oscillate in a certain plane only, so that the motion becomes one-dimensional, while the thread pendulum may oscillate in any horizontal direction. If the weight of the rod and the extension of the sphere cannot be neglected, we obtain a *physical pendulum*.

Actually the pendulum sphere executes a two-dimensional motion, which is made formally one-dimensional by a certain constraint; hence we should really deduce the one-dimensional equation of motion. However, it is easy to set up this equation right away. We observe that the component of the force in the direction of the motion is $-mg \sin \varphi$, where φ is the angle of deflection, and represent the deflection by $s = l\varphi$; the equation of motion

$$m\ddot{s} = -mg \sin \varphi$$

may then be written as

$$\ddot{\varphi} + \left(\frac{g}{l}\right) \sin \varphi = 0. \qquad (6.1)$$

(See Fig. 1-5.)
For very small deflection angles φ, we may substitute φ for $\sin \varphi$, thus obtaining the familiar equation for harmonic oscillations; the solution of this is

$$\varphi = a \sin(\omega_0 t + \eta), \qquad \omega_0 = \left(\frac{g}{l}\right)^{1/2}, \qquad (6.2)$$

with the proper frequency and period

$$\nu_0 = \frac{1}{2\pi}\left(\frac{g}{l}\right)^{1/2}, \qquad T = 2\pi\left(\frac{l}{g}\right)^{1/2}. \qquad (6.3)$$

Similar equations and solutions may be found for the physical pendulum; however, it is then necessary to determine the center of the mass and the moment of inertia.

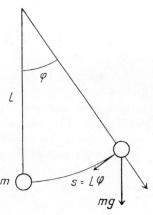

Fig. 1-5. The mathematical pendulum.

For large deflections we must proceed in a different way. Multiplying (6.1) by $\dot\varphi$ and integrating, we obtain

$$\frac{1}{2}\dot\varphi^2 + \left(\frac{g}{l}\right)^{1/2}(\cos\varphi_m - \cos\varphi) = 0. \tag{6.4}$$

Here φ_m is a constant of integration, which clearly represents the maximum deflection of the pendulum, corresponding to the turning point $\dot\varphi = 0$. Inversion of (6.4), followed by another integration, then gives us t as a function of the deflection φ. To simplify matters further we introduce the identities

$$\cos\varphi = 1 - 2\sin^2\frac{\varphi}{2}, \qquad \cos\varphi_0 = 1 - 2\sin^2\frac{\varphi_0}{2}, \tag{6.5}$$

obtaining

$$t = \left(\frac{l}{g}\right)^{1/2}\int_0^t \frac{d\varphi}{[4\sin^2(\varphi_0/2) - 4\sin^2(\varphi/2)]^{1/2}}. \tag{6.6}$$

Thus, for infinitesimal φ_0 and φ ($\varphi < \varphi_0$),

$$t = \left(\frac{l}{g}\right)^{1/2}\int_0^\varphi \frac{d\varphi}{\sqrt{\varphi_0^2 - \varphi^2}} = \left(\frac{l}{g}\right)^{1/2}\arcsin\frac{\varphi}{\varphi_0},$$

$$\varphi = \varphi_0\sin\omega_0 t, \qquad \omega_0 = \left(\frac{l}{g}\right)^{1/2}, \tag{6.7}$$

as in (6.2).

For finite φ_0, on the other hand, (6.6) gives us the time t as an elliptic integral in φ, so that φ becomes an elliptic function of t. Hence the oscillations are no longer strictly harmonic, and the period depends on the maximum deflection φ_0,

$$T = 4\left(\frac{l}{g}\right)^{1/2}\int_0^{\varphi_0}\frac{d\varphi}{[4\sin^2(\varphi_0/2) - 4\sin^2(\varphi/2)]^{1/2}}. \tag{6.8}$$

On inspection of (6.8), one finds that the period increases with the deflection and approaches infinity for $\sin(\varphi_0/2) = 1$, $\varphi_0 = \pi$. This means that a rigid pendulum will stop in the upside-down position in an unstable equilibrium. If its energy of oscillation is further increased, it will go right around, that is, perform a nonuniform rotational motion.

The period may be calculated as follows: The substitution

$$x = \frac{\sin\varphi/2}{\sin\varphi_0/2} = \sin u \tag{6.9}$$

leads to
$$\frac{d\varphi}{2\sin\varphi_0/2} = \frac{dx}{\cos\varphi/2} = \frac{dx}{[1-x^2\sin^2(\varphi_0/2)]^{1/2}} \ ;$$
hence
$$T = 4\left(\frac{l}{g}\right)^{1/2}\int_0^1 \frac{dx}{\{[1-x^2\sin^2(\varphi_0/2)](1-x^2)\}^{1/2}}$$
$$= \left(\frac{l}{g}\right)^{1/2}\int_0^{2\pi} \frac{du}{[1-\sin^2(\varphi_0/2)\sin^2 u]^{1/2}}. \tag{6.10}$$

Expanding the denominator in the last integral, and introducing the well-known mean values of powers of sines

$$\frac{1}{(1-s^2)^{1/2}} = 1 + \frac{1}{2}s^2 + \frac{1\cdot 3}{2\cdot 4}s^4 + \cdots, \quad \overline{\sin^2 u} = \frac{1}{2}, \quad \overline{\sin^4 u} = \frac{1\cdot 3}{2\cdot 4}, \cdots,$$

we finally obtain

$$T = 2\pi\left(\frac{l}{g}\right)^{1/2}\left[1 + \left(\frac{1}{2}\right)^2\sin^2\frac{\varphi_0}{2} + \left(\frac{1\cdot 3}{2\cdot 4}\right)^2\sin^4\frac{\varphi_0}{2} + \cdots\right]. \tag{6.11}$$

The Cycloid Pendulum

It is now of some interest to investigate how the mathematical pendulum should be modified, in order to make the oscillations strictly harmonic. This problem was solved earlier by Huygens, who also described special methods of suspension, by means of which the solution may be realized in practice.

We shall here treat the problem in a purely analytic way and assume that the pendulum sphere is constrained to follow a certain curve, the determination of which is now our task. We may, for instance, assume that a sphere rolls on a concave cylindrical surface; the problem is then to find the form of the cross section of this surface for the harmonic motion of the sphere. It is to be noted that the rotational energy imparted to the sphere by this rolling motion will not influence the character of the motion: this energy is, just like the translational energy, proportional to the square of the velocity. Thus the inertia of the sphere with respect to rotation (because of its moment of inertia) will manifest itself only by an apparently small increase in its mass.

So, we imagine a new curve instead of the circle described by the rigid pendulum, and designate the arc length by s. In the coordinate system chosen in Fig. 1-6 the tangential component of the force of gravity will be $-mg\,dy/ds$, and the equation of motion for the pendulum becomes

$$\ddot{s} + g\frac{dy}{ds} = 0. \tag{6.12}$$

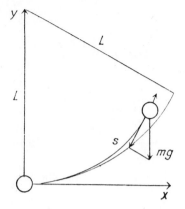

Fig. 1-6. The cycloid pendulum.

In order to obtain the same equation as for the mathematical pendulum in the case of small deflections, we put

$$\frac{dy}{ds} = \frac{s}{l}, \qquad y = \frac{s^2}{2l}. \qquad (6.13)$$

Utilizing $dx^2 + dy^2 = ds^2$, we then find

$$\frac{dx}{ds} = \left[1 - \left(\frac{s}{l}\right)^2\right]^{1/2}$$

or, substituting $s = l \sin \varphi$,

$$dx = l \cos^2 \varphi \, d\varphi,$$

$$x = \frac{l}{2}\left(\varphi + \frac{1}{2}\sin 2\varphi\right), \qquad y = \frac{l}{4}(1 - \cos 2\varphi). \qquad (6.14)$$

Writing $l = 4a$ and $2\varphi = u$, we then obtain the well-known parametric representation of the cycloid

$$x = a(u + \sin u), \qquad y = a(1 - \cos u). \qquad (6.15)$$

As is well known, a cycloid is generated by a point on a circle rolling along a straight line. In order to produce the cycloid needed for our construction of the ideal harmonic pendulum, the circle must roll on the horizontal straight line of Fig. 1-7. The radius of the circle is a, and the "effective length" of the

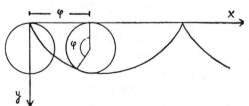

Fig. 1-7. The cycloid.

pendulum is four times the radius, $l = 4a$; while the "angle of deflection" is half the angle turned by the rolling circle, $\varphi = u/2$.

Anharmonic Oscillators. Molecular Vibrations and Planetary Motion

The oscillators that we have considered up to now do not describe natural processes very well, unless the oscillations have comparatively small amplitudes. Forces that increase without limit with distance, as assumed in the mathematical expression $F = -ax$ for a quasielastic force, do not exist in nature; on the contrary, forces between particles will always *decrease* for sufficiently great distances.

This will be the case for relative oscillations of particles like the atoms in a molecule, for instance. The forces decrease for large distances; but attractive and repulsive forces may balance each other at a certain given distance, and two atoms may then oscillate toward and away from each other about this state of equilibrium.

Similar considerations hold for the revolutions of the planets around the sun or of an electron around an atomic nucleus. Here the attractive forces are gravitational and electrostatic, respectively; the repulsive forces are represented by the centrifugal forces acting on the planet or electron. If the attraction and repulsion balance exactly, the orbit is a circle; however, if this equilibrium is disturbed, the orbiting body will oscillate about this circle in such a way as to describe an elliptic orbit. This phenomenon may be considered an oscillation in the radial distance from the central body, that is, by means of one single coordinate.

For molecules it is necessary to distinguish between polar and nonpolar bonds. In the former case the attraction is electrostatic and decreases with the square of the distance. The repulsive forces may be of several different types; for simplicity we shall assume here that they behave like centrifugal forces.

The oscillator equation may now be written as

$$m\ddot{r} = -\frac{B}{r^2} + \frac{C}{r^3}, \tag{6.16}$$

or

$$m\ddot{r} = -\frac{dV}{dr}, \quad V = -\frac{B}{r} + \frac{C}{2r^2}. \tag{6.17}$$

Multiplication by $\dot{r}\, dt = dr$ and integration leads to

$$\frac{1}{2} m\dot{r}^2 = -\frac{A}{2} - V, \tag{6.18}$$

where $-A/2$ is a constant of integration, equal to the total energy E. The function $V(r)$ is the potential energy.

240 / Mechanics and Statistics

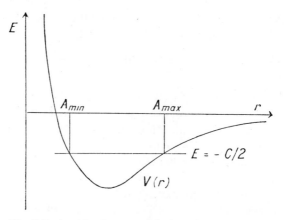

Fig. 1-8. A molecular potential: the anharmonic oscillator.

Figure 1-8 illustrates the case of such an oscillator. It oscillates about the equilibrium position r_0 at the bottom of the curve, with maximum and minimum deflection at $r_{\max}$ and $r_{\min}$.

Therefore (6.18) may be written as

$$m\dot{r}^2 = -A + \frac{2B}{r} - \frac{C}{r^2}. \tag{6.19}$$

The maximum and minimum deflections are found by setting $\dot{r} = 0$, obtaining a second-degree equation in r with the following roots

$$r = \frac{B}{A} \pm \left(\frac{B^2}{A^2} - \frac{C}{A}\right)^{1/2} = a \pm c; \tag{6.20}$$

here we have put

$$B = aA, \qquad C = b^2 A, \qquad a^2 - b^2 = c^2. \tag{6.21}$$

Hence

$$\dot{r} = \pm \left(\frac{A}{m}\right)^{1/2} \left(-1 + \frac{2a}{r} - \frac{b^2}{r^2}\right)^{1/2} \tag{6.22}$$

or

$$\left(\frac{A}{m}\right)^{1/2} t = -\int_{a+c}^{r} \frac{r\, dr}{[-(r-a)^2 + c^2]^{1/2}}. \tag{6.23}$$

Substituting

$$r = a + c \cos u, \tag{6.24}$$

we finally obtain

$$\left(\frac{A}{m}\right)^{1/2} t = \int_0^u (a + c \cos u)\, du = au + c \sin u. \tag{6.25}$$

Equations (6.24) and (6.25) show how r oscillates between a maximum and minimum value, lying on opposite sides of the point of equilibrium. The auxiliary quantity u does not, however, increase linearly with t, but is a periodic function of t. As we see, t depends on u in a fairly simple way; while, on the other hand, u is a more complicated function of t.

We might also have chosen another parametric description of r, namely,

$$\frac{a}{b} - \frac{b}{r} = \frac{c}{b} \cos \varphi \tag{6.26}$$

or

$$r = \frac{b^2/a}{1 - (c/a) \cos \varphi}. \tag{6.27}$$

Thus, writing (6.22) in the form

$$\dot{r} = \pm \left(\frac{A}{m}\right)^{1/2} \{(c^2/b^2) - [(a/b) - (b/r)]^2\}^{1/2} = \pm \left(\frac{A}{m}\right)^{1/2} \frac{c}{b} \sin \varphi, \tag{6.28}$$

we should have obtained

$$\left(\frac{A}{m}\right)^{1/2} t = \left(\frac{A}{m}\right)^{1/2} \int \frac{dr}{\dot{r}} = \int_0^\varphi \frac{b^3/a^2 \, d\varphi}{[1 - (c/a) \cos \varphi]^2}. \tag{6.29}$$

The connection between t and the auxiliary angle φ is here much more complicated than the corresponding relation (6.25) connecting t and u; however, the connection between r and φ is just as simple as that between r and u.

We have not, in this section, calculated the proper frequencies for the different amplitudes, or other features characteristic of oscillators. For the sake of comparison we should then have had to place the origin of our coordinate system at the equilibrium point r_0 (Fig. 1-8). Instead we have calculated the energy relatively to that for an infinite distance between the oscillating particles and adapted our exposition to facilitate a subsequent comparison with more complex problems: for example, the Kepler problem. To this we shall return later; at the present stage our intention has mainly been to elucidate the general possibilities and methods of integration for one-dimensional problems.

7. HIGHER-DIMENSIONAL PROBLEMS

Kinetic and Potential Energy for Conservative Forces

We have defined earlier the concepts of force and work, and shown in the example (5.19) how the work performed by a force accumulates in a quantity called *mechanical energy*, which in turn splits up into *kinetic* and *potential energy* (or, as they are sometimes called, the *energy of motion* and *energy of position*, respectively). We shall now investigate these forms of energy more closely.

When we pass from one-dimensional to higher-dimensional motions, the difficulties of integrating the equations of motion increase considerably; the conditions that must be satisfied in order that an analytic integration of these equations be possible become correspondingly restrictive.

In the case of one-dimensional motions we have shown that the equations may always be integrated, provided that the forces occurring depend only on position. For higher-dimensional motion we must add the condition that *the forces be derivable from a potential* or, as it is often put, that the forces be *conservative*. This condition is identically satisfied for purely position-dependent forces in the one-dimensional case, and thus unnecessary to state explicitly in the treatment of such problems: as already shown in (3.4), we may write $F = -dV/dx$, $V = -\int F\, dx$.

The treatment of higher-dimensional motions, driven by purely position-dependent forces, must therefore be tackled anew. We take, for instance, the three-dimensional motion of a single particle in space

$$m\ddot{x} = F_x(x, y, z),$$
$$m\ddot{y} = F_y(x, y, z), \quad (7.1)$$
$$m\ddot{z} = F_z(x, y, z).$$

We multiply these equations in turn by $\dot{x}\, dt = dx$, $\dot{y}\, dt = dy$, $\dot{z}\, dt = dz$, add the resulting equations together, and integrate, thus obtaining

$$[\tfrac{1}{2}m(\dot{x}^2 + \dot{y}^2 + \dot{z}^2)]_{t_0}^{t} = \int_{s_0}^{s}(F_x\, dx + F_y\, dy + F_z\, dz). \quad (7.2)$$

On the right-hand side we have an expression denoting work, namely, that performed by the forces on the particle in the course of its motion from a point s_0 to another point s, during the time interval from t_0 to t. This work is equal to the quantity on the left-hand side,

$$T = \tfrac{1}{2}mv^2, \quad (7.3)$$

which henceforth will be referred to as the *kinetic energy;* more correctly, it denotes the difference in the kinetic energy of the particle between the two times t and t_0. We have here a simple formulation of the *energy law:* that work is converted to energy.

To compute the quantity on the right-hand side of (7.2), the *work*, we must in general know the trajectory of the particle. Let us now assume, however, that the nature of the force under discussion is such that the integral $\int \mathbf{F} \cdot d\mathbf{r}$ from s_0 to s is independent of the trajectory, that is, of the curve along which the integral is to be calculated. $\mathbf{F} \cdot d\mathbf{r}$ must then be a total differential of a function of x, y, z, so that

$$F_x\,dx + F_y\,dy + F_z\,dz = -dV(x, y, z) = -\left(\frac{\partial V}{\partial x}dx + \frac{\partial V}{\partial y}dy + \frac{\partial V}{\partial z}dz\right), \quad (7.4)$$

or

$$F_x = -\frac{\partial V}{\partial x}, \qquad \mathbf{F} = -\nabla V. \tag{7.5}$$

Denoting the kinetic energy at the times t and t_0 by T and T_0, and the values of the function $V(x, y, z)$ at the corresponding points s and s_0 by V and V_0, we obtain from (7.2)

$$T - T_0 = V_0 - V$$

or

$$T + V = T_0 + V_0 = E. \tag{7.6}$$

This being valid for all times t, it follows that the sum on the left-hand side (which we have denoted by E) is time-independent. Henceforth V will be referred to as the potential energy of the system.

The result of these considerations may then be expressed as follows: The system has a *constant total energy E*, equal to the sum of its potential and kinetic energies. This is the law of energy conservation for mechanical systems. The forces studied here, the nature of which (derivability from a potential) made the deduction of this law possible, are therefore generally referred to as conservative.

For systems where this law is not valid, we must look for any possibilities that energy may have been supplied to or removed from the system. For example, energy can be supplied by time-dependent forces, exerted by bodies outside the system; also, frictional forces may remove mechanical energy from a system by converting it to heat energy.

Integration of the Equations of Motion

The assumption previously made, that the forces be conservative, has helped us to make some progress toward the integration of the equations of motion. Equation 7.6 is a so-called *first integral*, which we now write as

$$\tfrac{1}{2}m(\dot{x}^2 + \dot{y}^2 + \dot{z}^2) + V(x, y, z) = E \tag{7.7}$$

or

$$\frac{1}{2m}(p_x^2 + p_y^2 + p_z^2) + V(x, y, z) = E. \tag{7.8}$$

In (7.8) we have replaced the velocity components by the corresponding components of momentum

$$p_x = m\dot{x}, \qquad p_y = m\dot{y}, \qquad p_z = m\dot{z}. \tag{7.9}$$

244 / Mechanics and Statistics

This formulation is not of great importance in the immediately following considerations; however, it plays a dominant role in Hamilton's formulation of mechanics, to be discussed later.

It is immediately clear that (7.7) cannot be treated in the same way as the one-dimensional equation

$$\tfrac{1}{2}m\dot{x}^2 + V(x) = E, \tag{7.10}$$

from which we obtained the integrated equation of motion

$$t = \int_{x_0}^{x} \frac{dx}{[(2/m)(E - V)]^{1/2}}, \tag{7.11}$$

since (7.7) contains three variables x, y, z, while (7.10) contains only one. It is, however, not unreasonable to assume that if we only knew the actual trajectory of the particle, so that its motion might be treated as somewhat one-dimensional along this trajectory, then it should be possible to obtain some equation analogous to (7.11).

Reformulating (7.11) as

$$t = \frac{\partial S}{\partial E}, \quad S = S(x, E) = \int_{x_0}^{x} [2m(E - V)]^{1/2}\, dx, \tag{7.12}$$

this possibility of an analogy becomes even more apparent. For, differentiation of (7.12) with respect to the coordinate x leads to

$$\frac{\partial S}{\partial x} = [2m(E - V)]^{1/2} = m\dot{x} = p_x, \tag{7.13}$$

or

$$S = \int_{x_0}^{x} p_x\, dx. \tag{7.14}$$

Here p_x is the momentum of the particle (mass point), and the integral (7.14) is called an *action integral*, while the function $S(x, E)$ in (7.12) and (7.14) is usually referred to as the *action function*.

We may now define a similar function for higher-dimensional motion,

$$S = \int_{r_0}^{r} (p_x\, dx + p_y\, dy + p_z\, dz). \tag{7.15}$$

If this integral, taken from a fixed point r_0 to a variable point r, is independent of the path of integration, we have

$$p_x = \frac{\partial S}{\partial x}, \quad p_y = \frac{\partial S}{\partial y}, \quad p_z = \frac{\partial S}{\partial z}, \tag{7.16}$$

or, more briefly,

$$\mathbf{p} = \nabla S. \tag{7.17}$$

Elementary Mechanical Problems | 245

This puts the problem, so to speak, on a vector-analytic basis. If we consider, not only one single trajectory for a particle, but many (or, actually, all possible) trajectories, it becomes natural to think of the momentum $\mathbf{p} = m\mathbf{v}$, or the velocity $\mathbf{v}$, as a function of the position coordinates, as a vector field. Equation 7.17 then indicates that this field should be expressible in terms of an auxiliary three-dimensional scalar function, namely, as the gradient of the action function S, which we have assumed to be definable. The problem may then be formulated as follows: under which conditions, and by what methods, may such a function be determined?

It is immediately clear that this function must satisfy (7.7) and (7.8). Substituting (7.16), we find that the latter becomes

$$\frac{1}{2m}\left[\left(\frac{\partial S}{\partial x}\right)^2 + \left(\frac{\partial S}{\partial y}\right)^2 + \left(\frac{\partial S}{\partial z}\right)^2\right] + V(x, y, z) = E. \qquad (7.18)$$

This is the so-called *Hamilton-Jacobi partial differential equation*, of paramount importance in Hamiltonian mechanics.

We shall here content ourselves with demonstrating that, if this equation is solvable, then the problem of motion may also be solved, that is, the integrated equations of motion may be found.

Let us assume that the function $S(x, y, z)$ satisfies (7.18). From (7.15) it follows that it must also depend on three constants of integration; these might, for instance, be the three components p_x, p_y, p_z of the momentum at the initial position $r = r_0$. Hence, by (7.7) and (7.8), the total energy E is also determined. Thus the quantity E, which may in principle take any value in (7.18), does not in this case represent any degree of freedom; conversely, if the total energy of the system is given initially, then only two of the initial components of momentum may be chosen freely. Altogether, we may say that the action function S must depend on the coordinates and on three constants of integration a_1, a_2, a_3, one of which may be the energy, say $a_1 = E$; the solution of (7.18) may then be written as

$$S = S(x, y, z, E, a_2, a_3). \qquad (7.19)$$

We shall now show that the integrated equations may be expressed in terms of the three derivatives

$$\frac{\partial S}{\partial E}, \quad \frac{\partial S}{\partial a_2}, \quad \frac{\partial S}{\partial a_3}.$$

To this end, we differentiate (7.18) with respect to one of the three constants, say a_2. The right-hand side gives $\partial E/\partial a_2 = 0$, since E itself is one of the constants. Utilizing the fact that

$$\frac{1}{m}\frac{\partial S}{\partial x} = \frac{1}{m}p_x = \dot{x}, \qquad (7.20)$$

we obtain, on the left-hand side,

$$\dot{x}\frac{\partial^2 S}{\partial x\,\partial a_2} + \dot{y}\frac{\partial^2 S}{\partial y\,\partial a_2} + \dot{z}\frac{\partial^2 S}{\partial z\,\partial a_2} = \frac{d}{dt}\frac{\partial S}{\partial a_2}$$

or

$$\frac{d}{dt}\frac{\partial S}{\partial a_2} = 0. \tag{7.21}$$

Hence $\partial S/\partial a_2$ is constant, that is, time-independent.

The differentiation of (7.18) with respect to E runs along exactly the same lines, except that the right-hand side becomes $\partial E/\partial E = 1$, so that

$$\frac{d}{dt}\frac{\partial S}{\partial E} = 1. \tag{7.22}$$

It follows from (7.21) and (7.22) that it must be possible to express the integrated equations of motion in the form

$$\frac{\partial S}{\partial E} = t + b_1,$$

$$\frac{\partial S}{\partial a_2} = b_2, \quad \frac{\partial S}{\partial a_3} = b_3, \tag{7.23}$$

where b_1, b_2, b_3 are three new constants of integration. This, together with a_1, a_2, a_3 (or E, a_2, a_3), gives altogether six constants of integration, that is, the complete number associated with three second-order differential equations. We may adjust these constants to accommodate the initial conditions: for example, the position x_0, y_0, z_0, and the velocity $\dot{x}_0$, $\dot{y}_0$, $\dot{z}_0$, at the time $t = 0$ or $t = t_0$.

The last two equations (7.23) each describe a connection between the coordinates x, y, z; that is, each of them defines a *surface*. Hence, when taken together, they define a *curve*, namely, the *curve of intersection between the two surfaces*. The two equations thus determine the trajectory of the particle in space. The first equation (7.23) then describes the motion of this particle along the trajectory; as we have already anticipated, this equation is clearly completely analogous to (7.12) for a one-dimensional motion.

Motion of a Free Particle

The difficult part of the method just described for determining the motion of a particle is, of course, the solving of the partial differential equation (7.18), that is, the determination of the action function. The treatment of this problem, which requires a close analysis of both coordinate transformations and the so-called canonical transformations (of both coordinates and momenta), will be deferred until later. However, we shall attempt to illustrate the problem by considering the simplest possible example: a *free particle* (i.e. one that is not acted on by any forces whatsoever).

Since the field of force $\mathbf{F} = -\nabla V$ vanishes for a free particle, the potential is constant, and we can choose $V = 0$; (7.18) then becomes

$$\left(\frac{\partial S}{\partial x}\right)^2 + \left(\frac{\partial S}{\partial y}\right)^2 + \left(\frac{\partial S}{\partial z}\right)^2 = 2mE. \tag{7.24}$$

Here we may take, for instance,

$$\frac{\partial S}{\partial y} = a_2, \quad \frac{\partial S}{\partial z} = a_3, \quad \frac{\partial S}{\partial x} = (2mE - a_2^2 - a_3^2)^{1/2} = a_1 \tag{7.25}$$

or

$$S = (2mE - a_2^2 - a_3^2)^{1/2} x + a_2 y + a_3 z. \tag{7.26}$$

This gives us the trajectory of the particle as a straight line, namely, the line of intersection between the planes

$$\frac{\partial S}{\partial a_2} = -\frac{a_2}{(2mE - a_2^2 - a_3^2)^{1/2}} x + y = b_2,$$

$$\frac{\partial S}{\partial a_3} = -\frac{a_3}{(2mE - a_2^2 - a_3^2)^{1/2}} x + z = b_3. \tag{7.27}$$

The time dependence is described by

$$\frac{\partial S}{\partial E} = \frac{m}{(2mE - a_2^2 - a_3^2)^{1/2}} x = t + b_1. \tag{7.28}$$

Combining (7.27) and (7.28), we finally obtain a parametric description of the motion

$$m(x - x_0) = a_1 t, \quad m(y - y_0) = a_2 t, \quad m(z - z_0) = a_2 t, \tag{7.29}$$

where the "parameter" is the time t.

8. LAWS OF LINEAR AND ANGULAR MOMENTUM FOR PARTICLE SYSTEMS OR SOLID BODIES

Definitions of Center of Mass, Angular Momentum, and Moment of Inertia

We consider a system of mass points $m_1, m_2, \cdots, m_i, \cdots$, and define m as their sum:

$$m = m_1 + m_2 + \cdots + m_i + \cdots = \sum_i m_i. \tag{8.1}$$

Denoting the position vectors of these mass points by $\mathbf{r}_1, \mathbf{r}_2, \cdots, \mathbf{r}_i, \cdots$, we may define a certain point (or position vector) $\mathbf{r}$ in space by

$$m\mathbf{r} = \sum_i m_i \mathbf{r}_i; \tag{8.2}$$

we shall refer to this point as the *center of mass* of the system. This definition applies whether the mass points move relative to each other or are rigidly interconnected (e.g. parts of a solid body).

Differentiating (8.2) with respect to time, we obtain

$$m\dot{\mathbf{r}} = \sum_i m_i \dot{\mathbf{r}}_i = \sum_i \mathbf{p}_i = \mathbf{p}, \tag{8.3}$$

where $\dot{\mathbf{r}}$ is the vectorial velocity of the center of mass; since $m\dot{\mathbf{r}}$ denotes the total momentum of the system, the total mass m is thus a measure of the inertia of the system as a whole with respect to translational motion.

We may also define an analogous measure of the inertia of the system with respect to rotation about an axis. First, however, it will be convenient to define a quantity called the moment of momentum or *angular momentum* of the system with respect to a fixed point or a fixed axis in space.

Taking the point in question to be the origin, we define the angular momentum of the system with respect to this point as the vectorial quantity

$$\mathbf{M} = \sum_i \mathbf{r}_i \times \mathbf{p}_i = \sum_i \mathbf{r}_i \times m_i \dot{\mathbf{r}}_i. \tag{8.4}$$

The three vector components M_x, M_y, M_z, where

$$M_z = \sum_i m_i(x_i \dot{y}_i - y_i \dot{x}_i), \quad \text{etc.,} \tag{8.5}$$

will be referred to as the angular momentum with respect to the three coordinate axes.

If we assume that the mass points are rigidly interconnected, forming a solid body, then the rotation of this body about an axis, say the z-axis, will be describable in terms of an angle φ. The distance of each point from this axis is $\rho_i = (x_i^2 + y_i^2)^{1/2}$; thus we have

$$x_i = \rho_i \cos \varphi, \quad y_i = \rho_i \sin \varphi, \tag{8.6}$$

where ρ_i may be considered as a constant under the rotation. Substituting in (8.5), we obtain

$$M_z = \sum_i m_i \rho_i^2 \dot{\varphi} = I_z \dot{\varphi}, \tag{8.7}$$

where

$$I_z = \sum_i m_i \rho_i^2 \tag{8.8}$$

is called the *moment of inertia* of the body or system with respect to the given axis. If the system consists of mass points in motion relative to each other, we may regard the quantity (8.8) as a kind of instantaneous moment of inertia with respect to the said axis.

The Center of Mass Law

We now set up the equations of motion for each of the mass points m_i and add them together, subdividing the forces acting on the mass points into *external* and *internal* forces $\mathbf{F}_i$ and $\mathbf{F}'_i$; thus we obtain

$$\frac{d}{dt}\sum_i (m_i \dot{\mathbf{r}}_i) = \sum_i \mathbf{F}_i + \sum_i \mathbf{F}'_i. \tag{8.9}$$

By internal forces we mean those exerted by other particles in the system. We designate the force exerted by the particle k upon the particle i by $\mathbf{F}'_{ik}$, and correspondingly that exerted by i upon k by $\mathbf{F}'_{ki}$; it then follows from Newton's third law (the action-reaction principle) that these two forces are equal in magnitude and opposite in direction,

$$\mathbf{F}'_{ki} = -\mathbf{F}'_{ik}, \tag{8.10}$$

acting along the straight line connecting the particles (so-called *central forces*). This leads to

$$\mathbf{F}'_i = \sum_k \mathbf{F}'_{ik}, \tag{8.11}$$

and

$$\sum_i \mathbf{F}'_i = \sum_{ik} \mathbf{F}'_{ik} = \sum_{i<k}(\mathbf{F}'_{ik} + \mathbf{F}'_{ki}) = 0. \tag{8.12}$$

Hence the last sum on the right-hand side of (8.9) is zero, whence this equation may be written

$$\frac{d}{dt}(m\dot{\mathbf{r}}) = \mathbf{F}, \qquad \mathbf{F} = \sum_i \mathbf{F}_i, \tag{8.13}$$

using (8.3).

This result may be expressed in words as follows:

The center of mass for a material system moves as if its total mass were concentrated, and all external forces were acting, in this point.

This is the *center of mass law* or *momentum law*, which may also be expressed by saying that the change (per unit time) of the total momentum of the system is equal to the vector sum of the forces acting on the system, and occurs along the direction of this resultant force. In other words, Newton's second law is just as valid for the system as a whole as for single mass points. Thus we have removed that weak point in Newton's original formulation of the laws of motion, namely, that they were expressly formulated to hold for "bodies," and not for material points.

The Law of Areas

We now multiply the equations of motion for each mass point,

$$\frac{d}{dt}(m_i \dot{\mathbf{r}}_i) = \mathbf{F}_i + \mathbf{F}'_i, \tag{8.14}$$

vectorially by the position vector $\mathbf{r}_i$, taken from an arbitrary point chosen as the origin, and add the resulting equations. Utilizing the fact that $\mathbf{F}'_{ki} + \mathbf{F}'_{ik} = 0$, we find that

$$\sum_i \mathbf{r}_i \times \mathbf{F}'_i = \sum_{i<k} (\mathbf{r}_i - \mathbf{r}_k) \times \mathbf{F}'_{ik} = 0, \tag{8.15}$$

since $\mathbf{r}_i - \mathbf{r}_k = \mathbf{r}_{ik}$ and the central force $\mathbf{F}'_{ik}$ are parallel. Further, on account of $\dot{\mathbf{r}}_i \times \dot{\mathbf{r}}_i = 0$, we obtain

$$\frac{d}{dt}(\mathbf{r}_i \times m_i \dot{\mathbf{r}}_i) = \mathbf{r}_i \times \frac{d}{dt}(m_i \dot{\mathbf{r}}_i), \tag{8.16}$$

and thus, finally,

$$\dot{\mathbf{M}} = \mathbf{G}, \qquad \mathbf{G} = \sum_i \mathbf{r}_i \times \mathbf{F}_i, \tag{8.17}$$

where $\mathbf{M}$ defined in (8.4) denotes the angular momentum of the system with respect to a fixed point; $\mathbf{G}$ is the sum of all *external* torques, that is, the moments of all external forces, about this point. The corresponding sum $\mathbf{G}'$ for the internal forces is by (8.15) always zero.

Equation 8.17 is the *law of torques* for a system of particles. If the total moment of all external forces (the total external torque) is zero, or if there are no such forces acting (but possibly internal forces), then the total angular momentum is constant. In this case, the law is sometimes called the *law of areas*; this expression has its origin in Kepler's second law, which states that the radius vector from the sun to a planet sweeps over equal areas in equal intervals of time.

We shall now demonstrate an important extension of the law of torques: that it is valid for the center of mass, even if this point has an arbitrary motion. Thus, denoting the position vector, angular momentum, and torque with respect to the center of mass by dashed symbols, we obtain

$$\mathbf{G} = \mathbf{r} \times \sum_i \mathbf{F}_i + \sum_i \mathbf{r}'_i \times \mathbf{F}_i = \mathbf{r} \times \mathbf{F} + \mathbf{G}', \tag{8.18}$$

$$\mathbf{M} = \sum_i (\mathbf{r} + \mathbf{r}'_i) \times m_i(\dot{\mathbf{r}} + \dot{\mathbf{r}}'_i)$$

$$= \mathbf{r} \times m\dot{\mathbf{r}} + \sum_i \mathbf{r}'_i \times m_i \dot{\mathbf{r}}'_i = \mathbf{r} \times m\dot{\mathbf{r}} + \mathbf{M}'. \tag{8.19}$$

This is a consequence of the fact that, by definition (8.2) of the center of mass, we have

$$\sum_i m_i \mathbf{r}'_i = 0, \qquad \sum_i m_i \dot{\mathbf{r}}'_i = 0, \qquad (8.20)$$

so that two of the four product sums in (8.19) vanish. Hence, substituting (8.18) and (8.19) in (8.17) and utilizing (8.13), we find that the law of torques is also valid if all the quantities are referred to the moving center of mass; in other words,

$$\dot{\mathbf{M}}' = \mathbf{G}'. \qquad (8.21)$$

Work and Kinetic Energy Referred to the Center of Mass

By considerations similar to the above, we deduce that

$$dA = \sum_i \mathbf{F}_i (d\mathbf{r} + d\mathbf{r}'_i) = \mathbf{F}\, d\mathbf{r} + dA', \qquad dA' = \sum_i \mathbf{F}_i\, d\mathbf{r}'_i, \qquad (8.22)$$

$$T = \sum_i \tfrac{1}{2} m_i (\dot{\mathbf{r}} + \dot{\mathbf{r}}'_i)^2 = \tfrac{1}{2} m \dot{\mathbf{r}}^2 + T', \qquad T' = \sum_i \tfrac{1}{2} m_i \dot{\mathbf{r}}'^2_i. \qquad (8.23)$$

It is evident that both the work performed by the external forces, and the kinetic energy of the system, has split neatly into two parts. The first part corresponds to the work performed by the fictitious resultant force on the fictitious mass m concentrated in the center of mass during its motion, and to the kinetic energy of this fictitious mass point, respectively. The second part describes the corresponding quantities for a motion of the system relative to the center of mass.

These results are of paramount importance in the treatment of translational and rotational motions of solid bodies.

Application of the Law of Areas to the Determination of Planetary Orbits

We consider the sun and a planet which attract each other with a force $-MmG/r^2$, corresponding to a potential

$$V = -\frac{MmG}{r}. \qquad (8.24)$$

M and m are the masses of the sun and the planet, which we shall treat as mass points, G is the gravitational constant, and r the distance. For simplicity we assume M to be so large, relative to m, that the center of mass lies in the center of the sun; r is then the distance of the planet from this fixed point.

We now introduce plane polar coordinates r and φ, where φ is the polar angle of the planetary orbit, and write the sum [constant by (7.6)] of potential and kinetic energy as

$$\tfrac{1}{2} m (\dot{r}^2 + r^2 \dot{\varphi}^2) + V = E, \qquad (8.25)$$

the radial and transversal components of $\mathbf{V}$ being $\dot{r}$ and $r\dot{\varphi}$.

252 / Mechanics and Statistics

On the other hand, the angular momentum is constant, by the law of areas

$$r \cdot mr\dot{\varphi} = mr^2\dot{\varphi} = \text{const} = k. \tag{8.26}$$

Hence, putting

$$E = -\tfrac{1}{2}A, \qquad MmG = B, \qquad \frac{k^2}{m} = C, \tag{8.27}$$

we obtain

$$m\dot{r}^2 = -A + \frac{2B}{r} - \frac{C}{r^2} = A\left(-1 + \frac{2a}{r} - \frac{b^2}{r^2}\right). \tag{8.28}$$

Compare (6.19).

From (6.26) and (6.27) it follows that

$$\dot{\varphi} = \left(\frac{C}{m}\right)^{1/2} \cdot \frac{1}{r^2}, \tag{8.29}$$

and this, using (6.19) and (6.22), leads to

$$\frac{d\varphi}{dr} = \frac{\dot{\varphi}}{\dot{r}} = -\frac{b/r^2}{(-1 + 2a/r - b^2/r^2)^{1/2}},$$

or

$$d\varphi = -\frac{d(-b/r)}{[c^2/b^2 - (a/b - b/r)^2]^{1/2}} = d \arccos \frac{b}{c}\left(\frac{a}{b} - \frac{b}{r}\right). \tag{8.30}$$

Thus we obtain the same expression as (6.24) or (6.25), but φ now denotes the polar angle of the orbit, which is an ellipse.

It is evident from (6.23) and (6.21) that the period of oscillation (or, equivalently, the period of revolution for the planet) is

$$T = 2\pi a\left(\frac{m}{A}\right)^{1/2} = 2\pi a^{3/2}\left(\frac{m}{B}\right)^{1/2}, \tag{8.31}$$

where B is a constant, by (8.27). This is *Kepler's third law:*

The square of the period of revolution for any planet is proportional to the cube of the major axis of its orbit, the factor of proportionality being the same for all planets.

In addition, the total energy becomes, by (8.27) and (6.21),

$$E = -\frac{MmG}{2a}. \tag{8.32}$$

This is half of the potential energy of the system when the planet is at a distance from the sun equal to the major semiaxis, that is, $r = a$.

9. RELATIVE MOTION

Coordinate Systems with Nonuniform Motions

The equations of motion which we have deduced in the preceding sections, using Newton's second law, presuppose that the motion is referred to a Galilean coordinate system.

We may, however, also describe the motion with reference to other systems; and in practice we nearly always do just that, since the earth does not constitute a Galilean system, both because of its revolution around the sun and its diurnal rotation. For motions referred to such a nonuniformly moving coordinate system, other equations of motion apply; and these may be found by means of transformations from one system to another. (See Fig. 1-9.)

Let us assume that the motion of a mass point, referred to a Galilean system, is given by the vector

$$\mathbf{r}_0 = x_0 \mathbf{i}_0 + y_0 \mathbf{j}_0 + z_0 \mathbf{k}_0. \tag{9.1}$$

A nonuniformly moving system may be defined by a vector

$$\boldsymbol{\rho} = \xi \mathbf{i}_0 + \eta \mathbf{j}_0 + \zeta \mathbf{k}_0, \tag{9.2}$$

and three unit vectors $\mathbf{i}, \mathbf{j}, \mathbf{k}$, describing the motion of the origin and the orientations of the coordinate axes, respectively. The instantaneous change of these orientations may, as we shall soon see, be described by means of a (constant or variable) angular velocity vector $\boldsymbol{\omega}$.

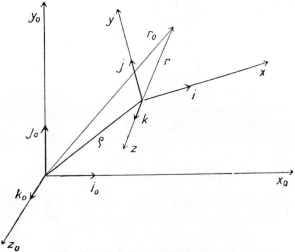

Fig. 1-9. Relative coordinate systems.

The position of a mass point may now be written as

$$\mathbf{r}_0 = \mathbf{\rho} + \mathbf{r},$$
$$\mathbf{r} = x\mathbf{i} + y\mathbf{j} + z\mathbf{k}, \tag{9.3}$$

where x, y, z are the coordinates of the point in the moving system.

Relative Velocity

We now find, for the velocity $\mathbf{v}_0$ relative to the Galilean system

$$\mathbf{v}_0 = \dot{\mathbf{r}}_0 = \dot{\mathbf{\rho}} + \dot{\mathbf{r}},$$
$$\dot{\mathbf{r}} = \mathbf{v} + x\dot{\mathbf{i}} + y\dot{\mathbf{j}} + z\dot{\mathbf{k}}, \tag{9.4}$$
$$\mathbf{v} = \dot{x}\mathbf{i} + \dot{y}\mathbf{j} + \dot{z}\mathbf{k}.$$

Here $\mathbf{v}$ is called the *relative velocity;* because of the rotational motion of the system, this will generally differ from $\dot{\mathbf{r}}$. However, we may write

$$\mathbf{v} = \frac{d\mathbf{r}}{dt}. \tag{9.5}$$

by adopting the convention that the operator d/dt in the following considerations denote differentiation with respect to the moving system: that is, differentiation of the coordinates x, y, z, but not the orientations $\mathbf{i}, \mathbf{j}, \mathbf{k}$, of the coordinate axes.

In order to obtain a more convenient expression for $\dot{\mathbf{r}}$, we now eliminate the quantities $\dot{\mathbf{i}}, \dot{\mathbf{j}}, \dot{\mathbf{k}}$. Since $\mathbf{i}^2 = 1$, $\mathbf{i}\dot{\mathbf{i}} = 0$; hence we may write

$$\dot{\mathbf{i}} = \omega_z \mathbf{j} - \omega_y' \mathbf{k},$$
$$\dot{\mathbf{j}} = \omega_x \mathbf{k} - \omega_z' \mathbf{i}, \tag{9.6}$$
$$\dot{\mathbf{k}} = \omega_y \mathbf{i} - \omega_x' \mathbf{j}.$$

Regarding the unknown coefficients in these expressions as the components of two vectors $\boldsymbol{\omega}$ and $\boldsymbol{\omega}'$, we shall now show that $\boldsymbol{\omega} = \boldsymbol{\omega}'$, and that $\boldsymbol{\omega}$ is the angular velocity vector of the system.

From $\mathbf{i} = \mathbf{j} \times \mathbf{k}$ it follows, by substitution of the last two equations of (9.6) that

$$\dot{\mathbf{i}} = \dot{\mathbf{j}} \times \mathbf{k} + \mathbf{j} \times \dot{\mathbf{k}} = \omega_z' \mathbf{j} - \omega_y \mathbf{k}, \tag{9.7}$$

hence $\boldsymbol{\omega}' = \boldsymbol{\omega}$, by comparison with the first equation (9.6). Using this result, we now easily find that the last part of the expression for $\dot{\mathbf{r}}$ in (9.4) may be written

$$x\dot{\mathbf{i}} + y\dot{\mathbf{j}} + z\dot{\mathbf{k}} = \mathbf{i}(\omega_y z - \omega_z y) + \cdots = \boldsymbol{\omega} \times \mathbf{r}, \tag{9.8}$$

so that

$$\dot{\mathbf{r}} = \mathbf{v} + (\boldsymbol{\omega} \times \mathbf{r}) = \left(\frac{d}{dt} + \boldsymbol{\omega} \times\right)\mathbf{r}. \tag{9.9}$$

These considerations and operations, performed above with respect to the position vector **r**, may clearly be applied to any vector; hence the operator relation (9.9) is generally valid, connecting the time derivatives of an arbitrary vector with respect to a Galilean and a rotating system. In particular, we note that

$$\dot{\boldsymbol{\omega}} = \frac{d\boldsymbol{\omega}}{dt} + (\boldsymbol{\omega} \times \boldsymbol{\omega}) = \frac{d\boldsymbol{\omega}}{dt}. \tag{9.10}$$

that is, the angular acceleration $\dot{\boldsymbol{\omega}}$ or $d\boldsymbol{\omega}/dt$ is the same in both systems.

Relative Acceleration

In order to formulate the new equations of motion, it is necessary to find transformed expressions for the acceleration. Writing (9.4) as

$$\mathbf{v}_0 = \dot{\boldsymbol{\rho}} + \mathbf{v} + (\boldsymbol{\omega} \times \mathbf{r}). \tag{9.11}$$

and differentiating according to (9.9), we then obtain

$$\dot{\mathbf{v}}_0 = \ddot{\boldsymbol{\rho}} + \frac{d\mathbf{v}}{dt} + 2(\boldsymbol{\omega} \times \mathbf{v}) + (\dot{\boldsymbol{\omega}} \times \mathbf{r}) + \boldsymbol{\omega} \times (\boldsymbol{\omega} \times \mathbf{r}). \tag{9.12}$$

Several new terms appear here. The first, $\ddot{\boldsymbol{\rho}}$, represents the linear acceleration of the system (or, more accurately, of its origin). The term $\mathbf{a} = d\mathbf{v}/dt$ is the *relative acceleration* of the mass point. The term $2(\boldsymbol{\omega} \times \mathbf{v})$ is called the *Coriolis acceleration*, after the French mathematician G. Coriolis (1792–1843). The next to the last term represents the effect of any variable rotation of the coordinate system; if the angular velocity is constant, this term vanishes. This may, in practice, be taken to hold for the earth. Actually, the diurnal rotation of the earth is not quite uniform (thus giving rise to the phenomena of precession and nutation, to be discussed later), since the axis of rotation traces out a circle on the celestial sphere in the course of some 26,000 years. However, this motion is so slow (and, consequently, $\dot{\boldsymbol{\omega}}$ so small) that its dynamical effects on the motion of bodies relatively to the earth are imperceptible.

The last term $\boldsymbol{\omega} \times (\boldsymbol{\omega} \times \mathbf{r})$ represents the *centripetal acceleration*. It is perpendicular to the axis of rotation and directed inward; its magnitude is $\omega^2 r \sin \vartheta$, where ϑ is the spherical polar angle ($\vartheta = 0, \pi$ at the poles, and $\vartheta = \pi/2$ at the equator). It may also be expressed by $\omega^2 \rho$, where ρ is the distance from the axis.

Equations of Motion in Relative Coordinates

It is a distinctive feature of the formulation of equations in relative coordinates, that terms which in reality represent a true acceleration are interpreted as forces; these are generally referred to as *kinetic reaction forces*, or simply as *inertial forces*.

Assuming that a particle of mass m is acted on by a force $\mathbf{F}$, and using (9.12), we may rewrite the equation $m\dot{\mathbf{v}}_0 = \mathbf{F}$ as

$$m\frac{d\mathbf{v}}{dt} = \mathbf{F} - m\ddot{\boldsymbol{\rho}} - 2m(\boldsymbol{\omega} \times \mathbf{v}) - m(\dot{\boldsymbol{\omega}} \times \mathbf{r}) - m\boldsymbol{\omega} \times (\boldsymbol{\omega} \times \mathbf{r}). \quad (9.13)$$

The first term $-m\ddot{\boldsymbol{\rho}}$ will usually be imperceptible for motions of bodies relative to the earth. This is due to the fact that the total force $\mathbf{F}$ must contain the attraction of the sun on the body, and this attraction will be very closely equal to $m\ddot{\boldsymbol{\rho}}$, where $\ddot{\boldsymbol{\rho}}$ is the acceleration of the center of mass of the earth. On the side of the earth facing the sun, however, the attraction will be somewhat greater, and on the opposite side somewhat less. This is the primary cause of the tides of the oceans; however, these *tidal motions* are enormously complicated by the effects of the Coriolis force (presently to be discussed), and also by the form of the coastal shores and the varying depths of the ocean bottoms.

The second last term of (9.13) may be ignored, since the rotation of the earth is practically uniform, as already mentioned. The last term, called the *centrifugal force*, is directed perpendicularly outward from the axis of rotation. On the earth this force is greatest at the equator, where it partly counteracts the force of gravity. Generally, the centrifugal force will manifest itself through variations in the gravity field, since it will influence the measured values of the acceleration of gravity at different places on the surface of the Earth.

The middle term $-2m(\boldsymbol{\omega} \times \mathbf{v})$ is called the *Coriolis force;* its horizontal component, in particular, plays a very prominent role for motions in the atmosphere and oceans. This component is determined by the vertical component of the rotation vector, which points upward in the Northern Hemisphere and downward in the Southern; it is proportional to the sine of the latitude α, or equal to $\omega \cos \vartheta$, if we denote the angular velocity of rotation by ω and the polar angle by ϑ. ($\cos \vartheta = \sin \alpha$ is positive in the Northern Hemisphere, and negative in the Southern.) The Coriolis force is thus zero at the equator, and attains its largest values at the poles. It readily follow from (9.13) that it acts perpendicularly to the direction of motion, deviating it to the right in the Northern Hemisphere, and to the left in the Southern. It is this force that (in the Northern Hemisphere) deflects the warm southerly ocean currents eastward, and the cold northerly ones westward, and determines the characteristics of cyclones and anticyclones: anticlockwise o clockwise rotations, respectively.

Free Fall Relative to the Rotating Earth

Within a limited region of the earth, and up to a limited height, the acceleration of gravity may be considered to be constant; we now assume that it contains all the terms on the right-hand side of (9.13), except the Coriolis force. The equation of motion (9.13) then simplifies to read

$$m \frac{d\mathbf{v}}{dt} = \mathbf{F} - 2m(\boldsymbol{\omega} \times \mathbf{v}). \tag{9.14}$$

Putting $\mathbf{F} = -m g \mathbf{k}$, where $\mathbf{k}$ denotes the positive z-axis (directed upward), this may be written

$$\frac{d\mathbf{v}}{dt} = -g\mathbf{k} - (2\boldsymbol{\omega} \times \mathbf{v}), \tag{9.15}$$

or with components, letting the x-axis point eastward and the y-axis northward,

$$\frac{dv_x}{dt} = -2\omega_y v_z + 2\omega_z v_y,$$

$$\frac{dv_y}{dt} = -2\omega_z v_x, \tag{9.15a}$$

$$\frac{dv_z}{dt} = -g + 2\omega_y v_x,$$

since $\omega_x = 0$.

We shall now treat the problem of a vertical fall from a small height; to simplify matters, we assume that $v_x = v_y = 0$ when $t = 0$. Since ω_x and ω_y are very small relatively to g, we may then put, as a first approximation,

$$v_z = v_0 - gt. \tag{9.16}$$

Similarly, we may neglect the term containing v_y in the first equation (9.15a) in comparison to that containing v_z, and integrate to obtain

$$v_x = -2\omega_y(v_0 t - \tfrac{1}{2}gt^2). \tag{9.16a}$$

Substitution of this in the second equation (9.15a) and integration lead to

$$v_y = 4\omega_y \omega_z(\tfrac{1}{2}v_0 t^2 - \tfrac{1}{6}gt^3). \tag{9.16b}$$

The corresponding positions are

$$z = h_0 + v_0 t - \tfrac{1}{2}gt^2,$$
$$x = -2\omega_y(\tfrac{1}{2}v_0 t^2 - \tfrac{1}{6}gt^3) \tag{9.16c}$$
$$y = 4\omega_y \omega_z(\tfrac{1}{6}v_0 t^3 - \tfrac{1}{24}gt^4).$$

It follows from (9.16c) that a body that is falling (or thrown vertically upward) suffers a small deflection in the east-west (E-W) direction, and an even smaller deflection in the north-south (N-S) direction.

To calculate these deflections we must know the rotational velocity ω of the earth; this may be roughly computed by reckoning a year to be equal to 365 solar days and 366 sidereal days, thus obtaining

$$\omega = \frac{2\pi}{24 \cdot 3600} \cdot \frac{366}{365} = 0.729 \cdot 10^{-4}. \tag{9.17}$$

We then have $\omega_y = \omega \sin \vartheta$, $\omega_z = \omega \cos \vartheta$, where ϑ is the polar angle.

We shall first calculate the E-W deflection for a given time of falling $t = T$, when $v_0 = 0$; we then calculate the same deflection, assuming that the body is thrown vertically upward with an initial velocity $v_0 = gT$, so that it takes the time T to travel both upward and downward. In the first case we obtain

$$x_1 = \tfrac{1}{3}\omega g \sin \vartheta \, T^3, \tag{9.17a}$$

and in the last case

$$x_2 = -\tfrac{2}{3}\omega g \sin \vartheta \, T^3,$$
$$x_3 = -\tfrac{4}{3}\omega g \sin \vartheta \, T^3, \tag{9.17b}$$

at the inflection point ($t = T$) and the moment when the body has fallen back to its initial position ($t = 2T$), respectively. The corresponding values for y are

$$y_1 = -\tfrac{1}{6}\omega^2 g \sin \vartheta \cos \vartheta \, T^4,$$
$$y_2 = \tfrac{1}{2}\omega^2 g \sin \vartheta \cos \vartheta \, T^4, \tag{9.17c}$$
$$y_3 = \tfrac{8}{3}\omega^2 g \sin \vartheta \cos \vartheta \, T^4.$$

As an example, let us take $T = 10$, whence the height of fall is roughly (assuming $g = 10$ m/sec^2) 500 m. This then gives us the values

$$x_1 = 24.3 \text{ cm}, \qquad x_2 = -48.6 \text{ cm}, \qquad x_3 = -72.9 \text{ cm}, \tag{9.17d}$$

for (respectively) the *easterly deflection when the body is falling with zero initial velocity*, and the *westerly deflection when it is thrown upward*.

At the equator the N-S deflection is zero; but at higher latitudes (where the E-W deflection is reduced by the factor $\sin \vartheta$) it appears. Taking $\vartheta = 45°$, $\cos \vartheta \sin \vartheta = \tfrac{1}{2}$, the N-S deflection is smaller than the equatorial E-W deflection by a factor of the order $\tfrac{1}{2}\omega T = 0.36 \cdot 10^{-3}$, and thus imperceptible. To obtain deflections of a few centimeters in magnitude we should have to increase the time of falling about threefold, and thus the height of fall about tenfold. If, for instance, the latter were 5000 m, this would give us

$$y_1 = -0.45 \text{ cm}, \qquad y_2 = 1.35 \text{ cm}, \qquad y_3 = 7.2 \text{ cm}. \tag{9.17e}$$

Hence a fall with *zero initial velocity* is subject to a *southerly deflection*, and one with *nonzero initial upward velocity* to a *northerly deflection*.

Finally we remark that the set of equations (9.15a) may also be solved exactly; the first and second of these equations then leads to

$$\frac{d^2v_x}{dt^2} + 4\omega^2 v_x = 2\omega_y g. \tag{9.18}$$

The exact solution thus contains sine and cosine terms with the frequency 2ω, corresponding to a semidiurnal period of oscillation; the approximate solutions above are only power expansions of these terms, where higher-order terms are ignored. These slow oscillations have, of course, no influence on rapid falling motions; however, they are of decisive importance for such phenomena as tidal motions and cyclone formation.

The Foucault Pendulum

We conclude this chapter with an example, describing a very simple application of the theory of relative motion. In 1851, Foucault (1819–1868) performed his famous pendulum experiment in the Panthéon in Paris: an experiment that may be said to demonstrate the rotation of the earth by mechanical means. A pendulum that initially oscillates in a given vertical plane, and is not disturbed too much, will be observed to suffer a deflection: in the course of a sidereal day the plane of oscillation will have turned through an angle $2\pi \sin \alpha$, where α is the latitude at which the experiment is performed.

We assume that the proper frequency of the pendulum is ω_0; the equations of motion for the horizontal coordinates (for small amplitudes relatively to the length of the pendulum, and neglecting the Coriolis force) are then

$$\ddot{x} + \omega_0^2 x = 0,$$
$$\ddot{y} + \omega_0^2 y = 0. \tag{9.19}$$

We now add the horizontal components of the Coriolis force $-2m(\boldsymbol{\omega} \times \mathbf{v})$, and divide by m to obtain $2\boldsymbol{\omega} \times \mathbf{v}$ on the left-hand side of the equations. Considering the pendulum motion to be purely horizontal, we need only take into account the vertical component $\omega_z = \omega \sin \alpha$ of the rotation of the earth. Thus the set of equations (9.19) may be replaced by the set

$$\ddot{x} + 2\omega \sin \alpha \dot{y} + \omega_0^2 x = 0,$$
$$\ddot{y} - 2\omega \sin \alpha \dot{x} + \omega_0^2 y = 0. \tag{9.20}$$

At this stage we introduce a new system of coordinates x', y', rotating clockwise with the angular velocity $\Omega = \omega \sin \alpha$,

$$x' = x \cos \Omega t + y \sin \Omega t,$$
$$y' = -x \sin \Omega t + y \cos \Omega t. \tag{9.21}$$

Substituting this in (9.20), we obtain

$$\ddot{x}' + (\omega_0^2 + \Omega^2)x' = (\ddot{x} + \omega_0^2 x + 2\Omega \dot{y}) \cos \Omega t + (\ddot{y} + \omega_0^2 y - 2\Omega \dot{x}) \sin \Omega t$$

and a corresponding equation for y'. By comparison with (9.20), we may then write

$$\ddot{x}' + \omega_0'^2 x' = 0, \qquad \ddot{y}' + \omega_0'^2 y' = 0,$$
$$\omega_0' = \sqrt{\omega_0^2 + \Omega^2}, \qquad \Omega = \omega \sin \alpha. \tag{9.22}$$

This means that in a coordinate system rotating with the angular velocity $\omega \sin \alpha$, the pendulum will oscillate as if the earth did not rotate, except for the very slight change of frequency from ω_0 to ω_0'. In the course of a sidereal day, then, the oscillation plane of the pendulum will have turned through the angle $2\pi \sin \alpha$. If the pendulum does not oscillate in a vertical plane but moves in an ellipse, then the axes of this ellipse will turn with the same angular velocity, as given by (9.22).

The last equation (9.22) may be written

$$\omega_0^2 = \omega_0'^2 - \Omega^2. \tag{9.23}$$

Here ω_0' is the natural frequency of the pendulum, that is, the frequency it exhibits for oscillations in a Galilean system. In a rotating system the frequency ω_0 (due to the quasielastic force driving the pendulum) is smaller, even becoming zero if $\Omega = \omega_0'$; this is because the quasielastic force $-\omega_0'^2 r'$, due to the gravity field, is reduced by the centrifugal force $\Omega^2 r'$. When $\Omega = \omega_0'$, these forces (being equal in magnitude and oppositely directed) cancel each other. Thus it will be observed, in the *rotating* system (where $\omega_0 = 0$, and the quasielastic force $-\omega_0^2 r$ consequently vanishes), that the pendulum may stand still in any position. In the *fixed* (Galilean) system, on the other hand, the pendulum weight will then describe a circle where the quasielastic and centrifugal force balance each other all the time.

CHAPTER 2

Lagrangian and Hamiltonian Mechanics

10. THE PRINCIPLE OF VIRTUAL WORK AND D'ALEMBERT'S PRINCIPLE

Degrees of Freedom for a System of Mass Points

The state of a mechanical system at any instant of time may be given by the values of certain parameters, or positional coordinates, and their time derivatives at this instant. Each of these parameters is said to represent a degree of freedom, and the number of degrees of freedom for the system is thus equal to the number of positional parameters necessary for its complete description.

For a system of particles moving relative to each other these parameters may be chosen simply as the Cartesian coordinates of the particles; their time-derivatives are then the Cartesian components of the particle velocities. The number of degrees of freedom for such a system of N particles is thus $3N$.

The number of degrees of freedom may, however, be decreased by various types of so-called *constraints*. Two particles may, for instance, be rigidly interconnected, so that their mutual separation does not change with time; this is, to a good approximation, the case for a strongly bound diatomic molecule. A particle or body may also be constrained by external forces to move along a given trajectory, or on a given surface; as examples of the latter we may consider a ball rolling on a flat tabletop, or in a curved bowl. We then speak of *constraining forces*.

A constraint is often expressible by an equation

$$f(q_1, q_2, \cdots, q_n) = \text{const}, \tag{10.1}$$

where $q_1, q_2, \cdots q_n$ are the positional coordinates of the system under discussion. Such constraints are called *holonomic*. In mechanics we also study constraints of a more general type, as

$$\sum_{i=1}^{n} g_i(q_1, q_2, \cdots, q_n) \, dq_i = 0, \tag{10.2}$$

where the left-hand side is not a total differential, so that $g_i = \partial f / \partial q_i$, and so forth. Such constraints, which are called *nonholonomic*, will not be discussed further here.

We shall now construct a model of a rigid body from a set of mass points. (Our knowledge of the composition of matter, more specifically the lattice arrangement of atoms in crystals, tells us that such a model is not only a formal device, but corresponds very closely to reality. It is the "solid" body, regarded as a real continuum, that is fictitious.)

Let us consider two particles, assuming that their mutual distance is constant:

$$r = [(x_1 - x_2)^2 + (y_1 - y_2)^2 + (z_1 - z_2)^2]^{1/2} = \text{const} = r_0. \quad (10.3)$$

Keeping x_1, y_1, z_1 as positional coordinates, we now replace x_2, y_2, z_2 by the relative coordinates

$$x = x_2 - x_1, \quad y = y_2 - y_1, \quad z = z_2 - z_1.$$

Introducing spherical polar coordinates r, ϑ, φ,

$$x = r \sin \vartheta \cos \varphi, \quad y = r \sin \vartheta \sin \varphi, \quad z = r \cos \vartheta, \quad (10.4)$$

and observing that the distance $r = r_0$ does not vary, we then see that the number of degrees of freedom of the system is reduced from six to five.

Considering three rigidly interconnected particles, that is assuming the mutual distances r_{12}, r_{13}, r_{23} to be constant, we obtain only one additional degree of freedom, that is, six altogether. The new positional coordinate may be taken to be the angle of inclination between the plane of the three particles and the line r_{12}. The three angles thus introduced correspond to the *Eulerian angles* for a rigid body. Adding more particles, all rigidly interconnected, to the system does not further increase the number of degrees of freedom. For the fourth particle we have the constraining conditions $r_{14} = \text{const}$, $r_{24} = \text{const}$, $r_{34} = \text{const}$, which "consumes" all the three degrees of freedom of this particle. For the fifth particle we obtain, at first sight, four constraints; however, these are not mutually independent, as the value of r_{45} is determined by those of r_{15}, r_{25}, r_{35}. Similarly each new particle introduces only three independent additional constraints on the system. Thus it is sufficient, for the complete description of the system, to state the (by assumption constant) distance of any particle from three of the other particles. This, again, corresponds closely with reality, as the position of an atom in a crystal lattice is determined mainly by the forces exerted on it by the neighboring atoms.

The number of degrees of freedom of a single particle may be similarly reduced. Let us assume that it is constrained to move on a surface in space, described by

$$f_1(x, y, z) = \text{const} = a_1. \quad (10.5)$$

Lagrangian and Hamiltonian Mechanics / 263

This means that during the motion of the particle it is continually being subjected to forces acting normally to the surface, and thus kept in place. For a ball rolling on a tabletop, it is the reaction force from the table (due to the gravitational force pulling the ball downward) which thus constrains the motion.

Assume that another constraint on the motion

$$f_2(x, y, z) = a_2, \qquad (10.6)$$

is present; this means that the particle must move on another surface as well. These surfaces intersect along a curve, which then becomes the trajectory of the particle. Such a curve may be given parametrically,

$$x = x(s), \quad y = y(s), \quad z = z(s), \qquad (10.7)$$

where s is an arbitrary parameter, for example, the arc length of the trajectory or any function of this arc length; s is then the positional coordinate of the one-dimensional motion of the particle.

If only one constraint obtains, as given by (10.5), we may introduce a network of lines on the surface $f_1 = a_1$, corresponding to different values of *two* parameters ξ and η; $x = x(\xi, \eta)$, and so forth, then determine the new positional coordinates ξ and η.

Virtual Displacement and Virtual Work

Several of the laws of mechanics may be deduced in an easy way, and many computations considerably simplified, by the application of certain very general principles. One of these is the principle of *virtual work*, often also referred to as the *principle of virtual displacements*, which apparently was first employed (albeit in a rather intuitive, nonquantitative way) by Galilei. The further development and quantitative formulation of this principle, and its application to statics, is to a large extent due to Simon Stevin (1548–1620) and the brothers James (1654–1705) and John Bernoulli (1667–1748). Somewhat later, d'Alembert applied the principle to dynamics.

The principle of virtual work was first applied to *statics*, that is, the theory of equilibrium states of bodies, or equilibria of forces. The lever principle of Archimedes may easily be deduced from this principle, since the displacements of the end points of a two-armed lever are proportional to the lengths of the arms. The work (force times distance moved) performed by two forces in equilibrium will then be the same if the forces are inversely proportional to the arms. In the same way one may easily calculate the gain in expended force obtained by the use of a pulley: It is compensated by the "loss of distance moved" by the object upon which the work is performed.

Virtual displacements of a mechanical system are, by definition, arbitrary displacements of the components of the system satisfying the constraints, that is, arbitrary variations $\delta q_1, \delta q_2, \cdots \delta q_n$ of the positional coordinates $q_1, q_2, \cdots q_n$ of a system possessing n degrees of freedom. The word "virtual" is used to signify that the displacements are *arbitrary*, in the sense that they need not correspond to any actual motion executed by the system.

Let us consider a system of freely moving mass points m_i, $i = 1, 2, \cdots f$. On each of these points there acts a force $\mathbf{F}_i$, which is the resultant of all external (and internal) forces acting on the particle. The condition for equilibrium is that all these resultant forces are zero,

$$\mathbf{F}_1 = 0, \quad \mathbf{F}_2 = 0, \cdots, \quad \mathbf{F}_f = 0. \tag{10.8}$$

Hence

$$\delta A = \sum_{i=1}^{f} \mathbf{F}_i \, \delta \mathbf{r}_i = 0 \tag{10.9}$$

for arbitrary displacements $\delta \mathbf{r}_i$ of the mass points. Conversely, (10.8) follows from (10.9) if the $\delta \mathbf{r}_i$ are assumed to be completely arbitrary: one may, for instance, assume $\delta \mathbf{r}_1 \neq 0$, $\delta \mathbf{r}_2 = \delta \mathbf{r}_3 = \cdots = \delta \mathbf{r}_f = 0$, and in this way obtain $\mathbf{F}_1 = 0$. Equation 10.9 thus asserts that the virtual work δA is zero for virtual displacements of a mechanical point system. From such systems one may pass to any mechanical system, for example, rigid bodies, by suitably allowing for the constraints and the corresponding reduction of the degrees of freedom.

Let us consider a single particle m; it then follows from (10.9) that

$$F_x \, \delta x + F_y \, \delta y + F_z \, \delta z = 0, \tag{10.10}$$

where F is the resultant of all forces acting on m. This equation is valid even if the number of degrees of freedom is reduced, for instance, by the constraint

$$f(x, y, z) = \text{const.} \tag{10.11}$$

We must, however, then be careful: in addition to measurable physical forces (gravitational, electrostatic, hydrostatic, etc.), we must also include in (10.10) the *constraining forces*; the latter have, one might say, a geometric origin, being deducible from the constraint (10.11). Such forces are always directed perpendicularly to the constraining surface, automatically compensating any resultant of the external physical forces in this direction. Separating out the constraining forces, and labeling them as $\mathbf{F}'$, we may write (10.10) as

$$\delta A = (F_x + F'_x) \, \delta x + (F_y + F'_y) \, \delta y + (F_z + F'_z) \, \delta z = 0, \tag{10.12}$$

whence, of course,

$$F_x + F'_x = 0, \quad \text{etc.} \tag{10.13}$$

Assuming that the constraining forces $\mathbf{F}'$ are perpendicular to the surface $f(x, y, z) = \text{const}$, and that the virtual displacements occur only along the surface, it follows that

$$F'_x \, \delta x + F'_y \, \delta y + F'_z \, \delta z = 0, \tag{10.14}$$

so that

$$\delta A = F_x \, \delta x + F_y \, \delta y + F_z \, \delta z = 0 \tag{10.15}$$

as before. In this case, however, the displacements are not wholly arbitrary, being connected by the relation

$$\delta f = \frac{\partial f}{\partial x} \, \delta x + \frac{\partial f}{\partial y} \, \delta y + \frac{\partial f}{\partial z} \, \delta z = 0. \tag{10.16}$$

In order to deal with equations like (10.15) and (10.16), it is often of great advantage to employ a very useful procedure known as the method of *Lagrange multipliers*. This method consists of introducing one or more new variables; in this case we define one such multiplier λ, by which we multiply (10.16), and add to (10.15). The equation

$$\delta A + \lambda \delta f = \left(F_x + \lambda \frac{\partial f}{\partial x} \right) \delta x + \left(F_y + \lambda \frac{\partial f}{\partial y} \right) \delta y + \left(F_z + \lambda \frac{\partial f}{\partial z} \right) \delta z = 0$$

$$\tag{10.17}$$

must then still be valid, subject to the already specified condition that the virtual displacements do not violate the constraints on the system. The essential point is now that it is possible to choose λ such that (10.17) becomes valid for arbitrary displacements δx, δy, δz.

As a consequence of this, we have

$$F_x + \lambda \frac{\partial f}{\partial x} = 0 \quad \text{or} \quad F_x = -\lambda \frac{\partial f}{\partial x}, \quad \text{etc.} \tag{10.18}$$

conversely, forces of this type will strike an equilibrium for arbitrary values of λ. Thus, the condition for equilibrium is that the resultant **F** of the external forces is everywhere perpendicular to the surface f. This is, as already mentioned, the case for the constraining forces; it follows from (10.13) that these are given by

$$\mathbf{F}' = -\mathbf{F} = \lambda \, \nabla f. \tag{10.19}$$

Generalized Coordinates

In order to facilitate the transition to a more general formalism, we shall henceforth designate the Cartesian coordinates of a set of particles by one set of coordinates $x_1, x_2, \cdots, x_f$, where $f = 3N$ for N particles. If the masses m_i of the particles are needed, we shall of course assume $m_1 = m_2 = m_3$; moreover, x_1, x_2, x_3 correspond to the usual x, y, z. The principle of virtual work may then for static equilibrium be stated thus:

$$\delta A = \sum_k F_k \, \delta x_k = 0. \tag{10.20}$$

If the number of degrees of freedom is reduced by constraints from f to n, the state of the system may be described by n new coordinates $q_1, q_2, \cdots, q_n$. Taking $x_1, x_2, \cdots, x_n$ to be such coordinates, we may then express the remaining coordinates $x_{n+1}, \cdots, x_f$ as functions of $x_1, \cdots x_n$, by means of the $(f-n)$ constraints of the type $\Phi(x, y, z)$. Furthermore, we may introduce arbitrary coordinate transformations

$$\begin{aligned} x_1 &= x_1(q_1, \cdots, q_n), \\ x_n &= x_n(q_1, \cdots, q_n), \end{aligned} \tag{10.21}$$

which then lead to

$$x_{n+1} = x_{n+1}(q_1, \cdots, q_n),$$

$$\vdots \tag{10.21a}$$

$$x_f = x_f(q_1, \cdots, q_n),$$

or, in general,

$$x_k = x_k(q_i), \quad k = 1, \cdots, f, \quad i = 1, \cdots, n. \tag{10.21b}$$

The parameters, or positional coordinates q_i will henceforth be referred to as *generalized coordinates*.

The principle of virtual work (10.20) may now be written

$$\begin{aligned} \delta A &= \sum_i \Phi_i \, \delta q_i = 0, \\ \Phi_i &= \sum_k F_k \frac{\partial x_k}{\partial q_i}. \end{aligned} \tag{10.22}$$

The quantities Φ_i will be referred to as *generalized forces*. Since the coordinates q_i generally do not correspond to lengths (they may be angles or other measurable quantities), the Φ_i will not in general have the dimension of a force; however, the product $\Phi_i \, \delta q_i$ will always have the dimension of work.

If, in particular, the forces are conservative, that is, deducible from a potential

$$F_k = -\frac{\partial V}{\partial x_k}, \tag{10.23}$$

it follows from (10.22) that

$$\Phi_i = -\frac{\partial V}{\partial q_i}. \tag{10.24}$$

when V is expressed as a function of the generalized coordinates.

D'Alemberts' Principle

The equilibrium condition $F_k = 0$, and so forth, in statics corresponds to the equations of motion

$$F_k - m_k \ddot{x}_k = 0 \qquad (10.25)$$

in dynamics. The two conditions may be considered formally identical if we regard the quantities $-m_k \ddot{x}_k$ as forces; these, representing the inertia of the particles, will be referred to as *inertial forces*. Similarly to the static case we may now put

$$\sum_k (F_k - m_k \ddot{x}_k) \delta x_k = 0; \qquad (10.26)$$

conversely, (10.25) follows from (10.26) if the displacements δx_k are assumed to be arbitrary, (10.26) expresses the so-called *d'Alembert's principle*, which is a generalization to dynamics of the principle of virtual work in statics.

The "virtual" character of the displacements δx_k are made even more evident in dynamics, since we here have to deal with actual motions, where one may define differentials $dx_k = \dot{x}_k \, dt$, corresponding to "real" displacements during a time interval dt.

In those cases where constraints necessitate the introduction of new coordinates, or if other considerations make it desirable to pass over to non-Cartesian coordinates, the form (10.26) of d'Alembert's principle is no longer practical to use.

For general coordinate transformations, as in (10.21), we have

$$\delta x_k = \sum_i \frac{\partial x_k}{\partial q_i} \delta q_i, \qquad (10.27)$$

regardless of whether the number of coordinates q_i is less than that of the coordinates x_k. Substituting (10.27) in (10.26), we obtain

$$\sum_i (\Phi_i + \Phi'_i) \delta q_i = 0; \qquad (10.28)$$

here the Φ_i are the generalized forces corresponding to the coordinates q_i [compare (22)], while the Φ'_i are the corresponding inertial forces. The work performed by the external forces is

$$\delta A = \sum_i \Phi_i \delta q_i. \qquad (10.29)$$

Similarly,

$$\delta A' = \sum_i \Phi'_i \delta q_i \qquad (10.30)$$

is the "work" performed by the inertial forces, and the sum of these is zero,

$$\delta A + \delta A' = 0. \qquad (10.31)$$

268 / Mechanics and Statistics

This may also be expressed by saying that the external work is expended in combating the inertial forces.

These forces may, by (10.26) and (10.27), be written as

$$\Phi'_i = -\sum_k m_k \ddot{x}_k \frac{\partial x_k}{\partial q_i}, \tag{10.32}$$

we shall examine this expression in the next section.

11. LAGRANGE'S EQUATIONS IN GENERALIZED COORDINATES

Inertial Forces Expressed in Terms of Kinetic Energy

In order to reformulate the generalized inertial forces Φ_i occurring in (10.28) and (10.32), we shall make use of the following important relations

$$\frac{\partial x_k}{\partial q_i} = \frac{\partial \dot{x}_k}{\partial \dot{q}_i}, \qquad \frac{d}{dt}\frac{\partial x_k}{\partial q_i} = \frac{\partial \dot{x}_k}{\partial q_i}, \tag{11.1}$$

valid for our general coordinate transformations. The first follows immediately from

$$\dot{x}_k = \sum_i \frac{\partial x_k}{\partial q_i} \dot{q}_i, \tag{11.2}$$

if we regard the q_i and $\dot{q}_i$ as $2n$ independent variables. The second is obtained from (11.2) by differentiating with respect to q_i,

$$\frac{\partial \dot{x}_k}{\partial q_i} = \frac{\partial}{\partial q_i}\sum_j \dot{q}_j \frac{\partial x_k}{\partial q_j} = \sum_i \dot{q}_j \frac{\partial^2 x_k}{\partial q_j \partial q_i} = \frac{d}{dt}\frac{\partial x_k}{\partial q_i}. \tag{11.3}$$

Using (11.1), we now write

$$m_k \ddot{x}_k \frac{\partial x_k}{\partial q_i} = \frac{d}{dt} m_k \dot{x}_k \frac{\partial \dot{x}_k}{\partial \dot{q}_i} - m_k \dot{x}_k \frac{\partial \dot{x}_k}{\partial q_i}. \tag{11.4}$$

Introducing the kinetic energy of the system,

$$T = \sum_k \frac{1}{2} m_k \dot{x}_k^2 = \frac{1}{2}\sum_{ij} T_{ij} \dot{q}_i \dot{q}_j, \tag{11.5}$$

$$T_{ij} = \sum_k m_k \frac{\partial x_k}{\partial q_i}\frac{\partial x_k}{\partial q_j}, \tag{11.5a}$$

the interial force Φ'_i may then by (11.4) be expressed as

$$\Phi'_i = -\left(\frac{d}{dt}\frac{\partial T}{\partial \dot{q}_i} - \frac{\partial T}{\partial q_i}\right). \tag{11.6}$$

This equation furnishes, for one thing, a more convenient way to calculate the inertial force than that given by (10.32), namely, by means of the kinetic energy T. Moreover, it leads straight away to the correct number of equations of motion, corresponding to the number of coordinates $q_1, \cdots q_n$. These equations may now, by (10.28), be written as

$$\frac{d}{dt}\frac{\partial T}{\partial \dot{q}_i} - \frac{\partial T}{\partial q_i} = \Phi_i, \qquad (11.7)$$

which will be referred to as the general *Lagrange's equations*. If, in particular, the generalized forces are deducible from a potential V, as assumed in (10.24), (11.7) takes the form

$$\frac{d}{dt}\frac{\partial T}{\partial \dot{q}_i} - \frac{\partial T}{\partial q_i} = -\frac{\partial V}{\partial q_i}. \qquad (11.8)$$

Furthermore, since V depends only on the coordinates q_i, and not on their derivatives $\dot{q}_i$ (which we shall refer to as *generalized velocities*), these equations may be written as

$$\frac{d}{dt}\frac{\partial L}{\partial \dot{q}_i} - \frac{\partial L}{\partial q_i} = 0, \qquad i = 1, \cdots, n, \qquad (11.9)$$

where the function

$$L = T - V, \qquad (11.10)$$

the difference between the kinetic and the potential energies, is generally called the *Lagrangian*.

As we have made no specific assumptions concerning the nature of the coordinates q_i, it follows that Lagrange's equations assume the general form (11.7), or (11.8) and (11.9), for all possible choices of coordinates. This may also be shown directly by transforming these equations to another set of coordinates $Q_i = Q_1, Q_2, \cdots, Q_n$. The procedure is similar to that employed in (11.1) and (11.6), utilizing relations analogous to (11.1).

12. THE PRINCIPLES OF FERMAT, MAUPERTUIS, AND HAMILTON

Fermat's Principle of Shortest Light-Time or Shortest Optical Light-Path

Let us assume that we have an isotropic medium with a variable index of refraction $n = n(x, y, z)$. By the wave theory of light this means that the velocity of light in the medium is c/n, c being the velocity of light in empty space. The time taken by a ray of light to travel between two spatial points

r_0 and r_1 is then

$$t = \int_{r_0}^{r_1} \frac{n}{c}\, ds, \tag{12.1}$$

where ds is an element of the trajectory described by the light ray. Multiplying by the frequency ν of the light, and putting $\lambda_0 = c/\nu$, $\lambda = c/n\nu$, we obtain

$$\nu t = \int_{r_0}^{r_1} \frac{n}{\lambda_0}\, ds = \int_{r_0}^{r_1} \frac{ds}{\lambda}, \tag{12.2}$$

which expresses the distance traveled in terms of wavelengths, that is, the optical light-path.

Fermat's principle now says that a ray of light always follows the trajectory of *shortest light-traveling time* between any two points on it, or (equivalently) that of the shortest *optical light-path*. Conversely, one may deduce the trajectory of the ray of light from this principle. We shall not discuss this any further here, as several examples were given in the section on variational principles in Part I, but shall confine ourselves here to comparing this well-known "minimum principle" with the following integral principles.

Maupertuis's Principle of Least Action

The integral principles which are presently to be compared with Fermat's principle are all equivalent versions of one basic principle of physics, usually referred to as the *principle of least action*. Before proceeding, we should therefore first discuss the physical properties of the concept of *action*. The dimension of this quantity is energy × time, or (equivalently) momentum × distance. (This may be compared with, for instance, the concept of *power*: i.e. energy output per time unit, which has the dimension energy/time.) The prominent role played by the concept of action in mechanics is perhaps, at first sight, somewhat surprising. However, it is of basic importance even in classical mechanics; and the discovery of the elementary quantum of action $h = 6.525 \cdot 10^{-27}$ erg/sec, and the subsequent development of quantum theory, has even more clearly displayed the fundamental physical significance of this concept.

Maupertuis published his principle in 1747. It was later given a more precise mathematical formulation by Euler and Lagrange; however, the principle appears to have been thought of even earlier, by Leibniz (1646–1716).

Maupertuis' formulation was as follows: "Among all possible motions Nature chooses that which achieves its goal by means of the least action." The motion "chosen" by Nature is thus more "efficient" or "expedient" than any other motion we might imagine executed by the system; the principle is, in other words, of a *teleological* character. (It is perhaps not very surprising that Maupertuis, in accordance with the rationalistic spiritual climate of his time, also endowed his principle with theological implications.)

We may express Maupertuis' principle mathematically as follows, considering for simplicity a single particle of mass m. Letting ds be an element of a possible light-ray trajectory between two spatial points a and b, we define an integral

$$S = \int_a^b mv\, ds, \tag{12.3}$$

which denotes the *total action* along the trajectory. The trajectory actually pursued by the particle is that for which S is a minimum. If the dependence of the velocity v on the coordinates x, y, z is known, this would be formally equivalent to Fermat's principle in geometrical optics, which we have already stated in (12.1) and (12.2).

We shall not discuss further how in general to deduce equations of motion from this minimum condition on S, other than to refer to the well-known results of geometrical optics: A ray of light is always perpendicular to a family of surfaces $S(x, y, z) = $ const, and the function S determining these surfaces is the solution of the partial differential equation

$$\nabla^2 S = n^2(x, y, z). \tag{12.4}$$

Here S is called the *wave front* or *eikonal*.

We may regard mv in (12.3) as a function of the coordinates x, y, z, if we take the forces to be conservative, and utilize the energy equation

$$\tfrac{1}{2}mv^2 + V(x, y, z) = E, \quad \text{i.e.}$$
$$mv = [2m(E - V)]^{1/2}. \tag{12.5}$$

We must therefore assume that the motion of the particle by (12.3) is always directed perpendicularly to a family of surfaces $S = $ const, determined by the differential equation

$$\nabla^2 S = 2m[E - V(x, y, z)]. \tag{12.6}$$

The function $S(x, y, z)$ is called the action function; since $\mathbf{p} = m\mathbf{v}$ must always be perpendicular to the surfaces $S = $ const, we obtain the relation

$$\mathbf{p} = \nabla S. \tag{12.7}$$

The differential equation (12.6) then results from the substitution of (12.7) in the energy equation

$$\tfrac{1}{2}m\mathbf{p}^2 + V = E. \tag{12.8}$$

We shall now illustrate Fermat's (and thus, to a certain extent, also Maupertuis') principle by the following simple considerations. Let us assume that we have two optical media, with different constant indices of refraction n_1 and n_2, which are separated by a plane surface. This problem was treated in Part I in the section on variational principles, from which the law of refraction has been deduced directly. Here we shall, instead, consider the differential equation (12.4) for the wave front.

We need not solve this equation. It is enough to note that the increase of S in the direction perpendicular to the wave fronts is proportional to n and that the distance between two surfaces $S = a$ and $S = a + 1$ is $1/n$. This gives us the following picture. On the one side of the bordering plane we draw plane surfaces with a given normal direction and separation $1/n_1$; on the other side we draw plane surfaces with separation $1/n_2$. The normal direction of these surfaces is then determined by the requirement that they "join" the surfaces across the bordering plane continuously. Denoting the angles between the surfaces and the bordering plane by α_1 and α_2, we thus obtain

$$\frac{\sin \alpha_1}{\sin \alpha_2} = \frac{1/n_1}{1/n_2} = \frac{n_2}{n_1}. \qquad (12.9)$$

Since α_1 and α_2 are the angles of incidence and refraction, this is just the well-known law of Snellius for optical refraction.

If we assume that the index of refraction varies continuously, instead of being constant in each medium and changing discontinuously across a bordering surface, we may still learn something from Fig. 2-1. It is evident that light has a tendency to deviate toward regions of higher index of refraction. In mechanics such an index is, by (12.5) and (12.6), to be replaced by $[2m(E - V)]^{1/2}$. This means that the trajectory of a particle or body will

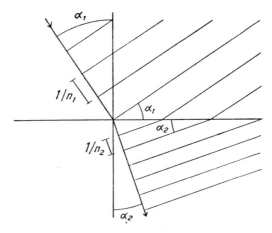

Fig. 2-1. Deduction of the refraction law by means of $\nabla^2 S = n^2$.

deviate toward regions of lower potential V; which, of course, also follows from the fact that the deviation occurs along the mechanical force $\mathbf{F} = -\nabla V$. Substituting $ds = v\, dt$ in (12.3), we obtain

$$S = \int_{t_0}^{t_1} mv^2\, dt = \int_{t_0}^{t_1} 2T\, dt, \qquad (12.10)$$

where $T = \frac{1}{2}mv^2$ is the kinetic energy; this is the form corresponding to the considerations of Leibniz on this subject. Further, putting $T = E - V$, we arrive at

$$S = E(t_1 - t_0) + \int_{t_0}^{t_1} (T - V)\, dt, \qquad (12.11)$$

a formulation that corresponds to the principle to be discussed next, *Hamilton's principle*.

Hamilton's Principle

This is actually only another version of the principle of least action, and thus an integral principle, in contrast to d'Alembert's principle, for example. In Hamilton's principle, however, the integral principles of mechanics attain their most perfect form; and it is therefore only natural that we discuss it in somewhat greater detail.

This principle is often introduced as a postulate, justified by the fact that the equations of motion for physical systems are deducible from it. We shall here proceed somewhat differently, and show that it may be deduced from differential principles, like that of d'Alembert.

We take d'Alembert's equation in the form of (10.28), with external and internal forces Φ_i and Φ'_i, and express the latter by (11.6); multiplication by dt and integration between two instants of time t_0 and t_1 of the motion then yields

$$\int_{t_0}^{t_1} \sum_i \left[\Phi_i - \left(\frac{d}{dt} \frac{\partial T}{\partial \dot{q}_i} - \frac{\partial T}{\partial q_i} \right) \right] \delta q_i\, dt = 0. \qquad (12.12)$$

We must now sharpen our notion of the displacements δq_i, which in d'Alembert's principle are taken to be quite arbitrary quantities. In (12.12) it is natural to assume that the δq_i are continuous functions of the time t; thus we write

$$\delta q_i = \varepsilon_i \eta_i(t), \qquad (12.13)$$

where the continuous function $\eta_i(t)$ is quite arbitrary, both in magnitude and shape. The quantity ε_i is taken to be independent of t and infinitesimal, since we shall require that *displacements* δq_i, as well as *differentials* dq_i, be of an infinitesimal order of magnitude.

The restrictions on the variation of the δq_i (i.e. the requirements of continuity and infinitesimality) does not impair the validity of d'Alembert's principle, which is assumed to hold for arbitrary displacements δq_i. Conversely, the restrictions do not prevent us from deducing the equations of motion from d'Alembert's principle.

By partial integration of (12.12) and the substitution of (10.29), we obtain

$$\int_{t_0}^{t_1}\left[\delta A + \sum_i \left(\frac{\partial T}{\partial \dot{q}_i}\frac{d}{dt}(\delta q_i) + \frac{\partial T}{\partial q_i}\delta q_i\right)\right] dt - \sum_i \left[\frac{\partial T}{\partial \dot{q}_i}\delta q_i\right]_{t_0}^{t_1} = 0. \quad (12.14)$$

Utilizing the definition (12.13) of the displacements δq_i, we may now introduce the coordinates of the "imagined" trajectory

$$\bar{q}_i = q_i + \delta q_i$$
$$\bar{q}_i - q_i = \varepsilon_i \eta_i(t) = \delta q_i \quad (12.15)$$
$$\delta \dot{q}_i = \dot{\bar{q}}_i - \dot{q}_i = \varepsilon_i \dot{\eta}_i(t) = \frac{d}{dt}(\delta q_i),$$

where the $\delta \dot{q}_i$ are the variations of the velocities $\dot{q}_i$. Since the kinetic energy T is a function of coordinates and velocities alone, the last term of the integrand in (12.14) may be written δT, if we replace $d/dt(\delta q_i)$ by $\delta \dot{q}_i$. Finally we make the assumption, which is crucial for the formulation of Hamilton's principle, that all variations

$$\delta q_i = \bar{q}_i - q_i \quad (12.16)$$

are zero at the two instants of time t_0 and t_1, that is, at the beginning and end of the trajectory interval discussed. Equation 12.14 then takes the form

$$\int_{t_0}^{t_1} (\delta A + \delta T)\, dt = 0. \quad (12.17)$$

If, in particular, the forces are conservative, that is, if the generalized forces defined in (10.23) and (10.24) are deducible from a potential V: $\Phi_i = -\partial V/\partial q_i$ we clearly have $\delta A = -\delta V$. In this case the variational symbol may be placed outside the integral sign, thus

$$\delta \int_{t_0}^{t_1} (T - V) = 0, \quad (12.18)$$

or

$$\int_{t_0}^{t_1} (T - V)\, dt = \text{extr.} \quad (12.19)$$

This is the simplest formulation of Hamilton's principle. Expressed in words, it postulates that the equations of motion for a mechanical system

Lagrangian and Hamiltonian Mechanics / 275

are deducible from the integral principle

$$\int_{t_0}^{t_1} L(q_i, \dot{q}_i, t)\, dt = \text{extr.} \quad (12.20)$$

where the *Lagrangian function* L is given, for time-independent conservative forces, by

$$L = T - V. \quad (12.21)$$

For more general forces, this function may not have the simple form (12.21), and may, for instance, be explicitly dependent on time. In any case the boundary conditions are taken to be

$$\delta q(t_0) = \delta q(t_1) = 0. \quad (12.22)$$

The problem of whether the equations of motion for nonconservative systems are deducible from this principle will need a separate investigation.

13. OTHER DEDUCTIONS AND APPLICATIONS OF LAGRANGE'S EQUATIONS

Deduction of Lagrange's Equations from Hamilton's Principle

The equations of motion deduced from Hamilton's principle are

$$\frac{d}{dt}\frac{\partial L}{\partial \dot{q}_i} - \frac{\partial L}{\partial q_i} = 0, \quad i = 1, 2, \cdots, n. \quad (13.1)$$

These are found by varying q_i and $\dot{q}_i$ in (12.20), writing $\delta \dot{q}_i = d(\delta q_i)/dt$ in accordance with (12.15), and integrating the resulting expression in parts; utilizing the boundary condition (12.22), we then obtain

$$\int_{t_0}^{t_1} \frac{\partial L}{\partial \dot{q}_i} \delta \dot{q}_i\, dt = -\int_{t_0}^{t_1} \left(\frac{d}{dt}\frac{\partial L}{\partial \dot{q}_i} \delta q_i\right) dt. \quad (13.2)$$

Thus the variation of (12.20) leads to

$$\int_{t_0}^{t_1} \sum_i \left(\frac{\partial L}{\partial q_i} - \frac{d}{dt}\frac{\partial L}{\partial \dot{q}_i}\right) \delta q_i\, dt = 0. \quad (13.3)$$

The only way to satisfy (13.3) for all imaginable variations $\delta q_i = \bar{q}_i(t) - q_i(t)$ of the coordinates, or (equivalently) for all "neighboring" virtual trajectories $\bar{q}_i(t)$, is to put the expression in braces equal to zero, that is, to assume the validity of the equations (13.1).

Direct Deduction from the Equations of Motion

A direct deduction of Lagrange's equations from a set of equations in Cartesian coordinates is not difficult, if we note the simplifications that the relations (11.1) make possible. Since the argument runs entirely along the same lines as that of Section 11, we shall only sketch it very briefly.

First we define generalized coordinates by a set of equations

$$x_k = x_k(q_i),$$
$$k = 1, \cdots, f, \quad i = 1, \cdots, n. \tag{13.4}$$

These apply both for $f = n$ (no constraints), and $f > n$ (number of degrees of freedom reduced from f to n by $f - n$ constraints).

Second, we write the equations of motion as follows

$$\frac{d}{dt}\frac{\partial T}{\partial \dot{x}_k} - F_k = 0, \quad T = \sum_k \frac{1}{2} m_k \dot{x}_k^2. \tag{13.5}$$

After the transformation (13.4), the kinetic energy will be expressed as a function of the generalized coordinates and velocities q_i and $\dot{q}_i$, since

$$\dot{x}_k = \sum_i \frac{\partial x_k}{\partial q_i} \dot{q}_i; \tag{13.6}$$

it will, moreover, be a quadratic form in these velocities, as shown in (11.5).

We now differentiate the expression

$$\frac{\partial T}{\partial \dot{q}_i} = \sum_k \frac{\partial T}{\partial \dot{x}_k} \cdot \frac{\partial \dot{x}_k}{\partial \dot{q}_i} = \sum_k \frac{\partial T}{\partial \dot{x}_k} \cdot \frac{\partial x_k}{\partial q_i}, \tag{13.7}$$

with respect to time, and substitute (11.1), to obtain

$$\frac{d}{dt}\frac{\partial T}{\partial \dot{q}_i} = \sum_k \left(\frac{d}{dt}\frac{\partial T}{\partial \dot{x}_k} \cdot \frac{\partial x_k}{\partial q_i} + \frac{\partial T}{\partial \dot{x}_k} \cdot \frac{\partial \dot{x}_k}{\partial q_i} \right). \tag{13.8}$$

However,

$$\frac{\partial T}{\partial q_i} = \sum_k \frac{\partial T}{\partial \dot{x}_k} \cdot \frac{\partial \dot{x}_k}{\partial q_i}, \tag{13.9}$$

whence, by subtraction,

$$\frac{d}{dt}\frac{\partial T}{\partial \dot{q}_i} - \frac{\partial T}{\partial q_i} = \sum_k \frac{d}{dt}\frac{\partial T}{\partial \dot{x}_k} \cdot \frac{\partial x_k}{\partial q_i}. \tag{13.10}$$

Introducing the definition (10.22) of the generalized forces,

$$\Phi_i = \sum_k F_k \frac{\partial x_k}{\partial q_i}, \tag{13.11}$$

we then have

$$\frac{d}{dt}\frac{\partial T}{\partial \dot{q}_i} - \frac{\partial T}{\partial q_i} - \Phi_i = \sum_k \left(\frac{d}{dt}\frac{\partial T}{\partial \dot{x}_k} - F_k \right) \frac{\partial x_k}{\partial q_i} = 0. \tag{13.12}$$

using (13.5). When the forces are conservative,

$$\Phi_i = -\frac{\partial V}{\partial q_i}, \qquad (13.13)$$

as in (10.23) and (10.24), we may substitute $L = T - V$ in (13.12), obtaining

$$\frac{d}{dt}\frac{\partial L}{\partial \dot{q}_i} - \frac{\partial L}{\partial q_i} = 0. \qquad (13.14)$$

Formulation of Lagrange's Equations for Nonconservative Forces

We now consider the question of whether the Φ_i may in general be expressed, either as in (13.13), or as

$$\Phi_i = \frac{\partial M}{\partial q_i} - \frac{d}{dt}\frac{\partial M}{\partial \dot{q}_i}. \qquad (13.15)$$

If so, we must have

$$F_k = -\frac{\partial V}{\partial x_k} \quad \text{or} \quad F_k = \frac{\partial M}{\partial x_k} - \frac{d}{dt}\frac{\partial M}{\partial \dot{x}_k} \qquad (13.16)$$

Evidently several types of nonconservative forces may be expressed in the latter way. As a first example, assume that the forces are explicitly dependent on time,

$$F_k = F_k(t). \qquad (13.17)$$

In this case we obtain $F_k = -\partial V/\partial x_k$ by putting

$$V = -\sum_k x_k F_k(t). \qquad (13.18)$$

Central forces originating in a body outside the system may also be deduced from a potential; the latter will then, however, in general be time-dependent. Consider, for instance, a celestial body moving along a known trajectory. This body will then exert a known time-dependent force on another celestial body, say, a planet. If the force between two such bodies is given as

$$\mathbf{F}_k = -\nabla_k V(r_{kl}), \qquad (13.19)$$

where l refers to the coordinates of the other body, then this equation will still be valid if x_l, y_l, z_l are given functions of time, so that the force $\mathbf{F}_k$ is time-dependent.

Several types of forces are nonconservative because they are not central; an important example is that of electromagnetic forces. An electron moving in a magnetic field is acted on by a force (the Lorentz force) which is perpendicular both to the direction of motion and the field. However, the

278 / Mechanics and Statistics

electromagnetic forces

$$\mathbf{F} = e\mathbf{E} + \frac{e}{c}(\mathbf{v} \times \mathbf{H}) \tag{13.20}$$

may easily be deduced from the function

$$M = \frac{e}{c}(\dot{x}A_x + \dot{y}A_y + \dot{z}A_z) - e\varphi, \tag{13.21}$$

where $\mathbf{A}$ and φ, are, respectively, the magnetic vector potential and electrostatic scalar potential, defined by the equations

$$\mathbf{H} = \operatorname{curl} \mathbf{A},$$

$$\mathbf{E} = -\nabla\varphi - \frac{1}{c}\frac{\partial \mathbf{A}}{\partial t}. \tag{13.22}$$

We have

$$\frac{\partial M}{\partial x} - \frac{d}{dt}\frac{\partial M}{\partial \dot{x}} = -e\frac{\partial \varphi}{\partial x} + \frac{e}{c}\left(\dot{x}\frac{\partial A_x}{\partial x} + \dot{y}\frac{\partial A_y}{\partial x} + \dot{z}\frac{\partial A_z}{\partial x} - \frac{dA_x}{dt}\right). \tag{13.23}$$

Furthermore, since the particle coordinates are functions of t, we may write

$$\frac{dA_x}{dt} = \dot{x}\frac{\partial A_x}{\partial x} + \dot{y}\frac{\partial A_x}{\partial y} + \dot{z}\frac{\partial A_x}{\partial z} + \frac{\partial A_x}{\partial t}. \tag{13.23a}$$

and substitute this in (13.23a) to obtain

$$-e\frac{\partial \varphi}{\partial x} - \frac{e}{c}\frac{\partial A_x}{\partial t} + \frac{e}{c}\left[\dot{y}\left(\frac{\partial A_y}{\partial x} - \frac{\partial A_x}{\partial y}\right) - \left(\frac{\partial A_x}{\partial z} - \frac{\partial A_z}{\partial x}\right)\right]$$

$$= eE_x + \frac{e}{c}(\mathbf{v} \times \mathbf{H})_x. \tag{13.23b}$$

Lagrange's equations for a charged particle in an electromagnetic field may thus be written

$$\frac{d}{dt}\frac{\partial L}{\partial \dot{x}} - \frac{\partial L}{\partial x} = 0, \quad L = T + M - V, \tag{13.24}$$

where V is any potential of nonelectromagnetic origin.

We have thus shown, by an example, that Lagrange's equations may sometimes be written in a form corresponding to Hamilton's principle, even though the external forces depend on the velocity components. In the case of motion of a particle in an electromagnetic field, however, this is because the Lorentz force is perpendicular to the direction of motion, hence performs no work. In general, Lagrange's equations may not be written in the form (13.24) if the forces are velocity-dependent; we must then content ourselves with stating the general form Φ_i for those forces that are not deducible from a Lagrangian, thus writing

$$\frac{d}{dt}\frac{\partial L}{\partial \dot{q}_i} - \frac{\partial L}{\partial q_i} = \Phi_i. \tag{13.25}$$

One example of this difficulty is the friction term $2k\dot{x}$ in the equation for damped oscillations

$$\ddot{x} + 2k\dot{x} + \omega_0^2 x = 0. \tag{13.26}$$

There exists no auxiliary function M such that this term may be expressed as in (13.15); the function $M = x\dot{x}$, for instance, only yields $\Phi_i = F_x = 0$.

Lagrangian Formulation of the Relativistic Equation

Finally, we shall show how Lagrange's equations must be modified in order to allow for the relativistic dependence of mass on velocity.

The Newtonian equations of motion must now be written

$$\frac{d}{dt}(m\dot{x}) = F_x, \tag{13.27}$$

where

$$m = \frac{m_0}{(1 - v^2/c^2)^{1/2}} \tag{13.28}$$

is the *variable mass* and m_0 the *rest mass* of the particle. If the force is deducible from a potential, $\mathbf{F} = -\nabla V$, it is evident that (13.27) may be written in Lagrangian form by means of the function

$$L = T' - V, \tag{13.29}$$

provided that we can find a function T' such that

$$m\dot{x} = \frac{\partial T'}{\partial \dot{x}}. \tag{13.30}$$

The function

$$T' = m_0 c^2 \left[1 - \left(1 - \frac{v^2}{c^2}\right)^{1/2}\right], \tag{13.31}$$

$$v^2 = \dot{x}^2 + \dot{y}^2 + \dot{z}^2.$$

satisfies this requirement.

We shall show later that (13.31) differs from the function T which, according to the principle of conservation of energy, we should have to denote as the kinetic energy. This latter function is

$$T = m_0 c^2 \left[\frac{1}{(1 - v^2/c^2)^{1/2}} - 1 \right], \tag{13.32}$$

and the two become approximately identical only for very small values of v/c,

$$T' \approx T \approx \tfrac{1}{2} m_0 v^2, \tag{13.33}$$

corresponding to the assumptions of nonrelativistic mechanics.

Incidentally, we might just as well use the functions

$$T' = -m_0 c^2 \left(1 - \frac{v^2}{c^2}\right)^{1/2}, \quad T = m_0 c^2 \Big/ \left(1 - \frac{v^2}{c^2}\right)^{1/2}, \tag{13.34}$$

because of the fact, brought to light by the theory of relativity, that any particle of rest mass m_0 has a rest energy

$$E = m_0 c^2. \tag{13.35}$$

This has been strikingly confirmed by the mutual annihilation of particles such as *electrons* and *positrons*, with the total conversion of their energy to electromagnetic radiation.

14. PARTIAL INTEGRATION OF LAGRANGE'S EQUATIONS

Generalized Momenta

The components of the momentum or quantity of motion, $\mathbf{p} = m\mathbf{v}$, may be expressed in Cartesian coordinates as follows:

$$p_x = \frac{\partial T}{\partial \dot{x}} = \frac{\partial L}{\partial \dot{x}}, \tag{14.1}$$

using the kinetic energy $T = \tfrac{1}{2} m v^2$, or the Lagrangian $L = T - V$. It is then natural to define generalized momenta, corresponding to a set of generalized coordinates q_i, by the equations

$$p_i = \frac{\partial L}{\partial \dot{q}_i}, \quad i = 1, \cdots, n. \tag{14.2}$$

One of the advantages of this definition is that we may now write Lagrange's equations as

$$\dot{p}_i = \frac{\partial L}{\partial q_i}, \tag{14.3}$$

that is, as a set of first-order simultaneous differential equations. However, the number of variables has now increased from n to $2n$, since the momenta

p_i are to be regarded as new independent variables. We therefore need another set of n differential equations; such a set is (14.2), regarded as differential equations in the p_i.

We need hardly stress that these momenta p_i are not momenta in the usual physical sense. Their dimension depends on that of the corresponding coordinates q_i, in such a way that $p_i \, dq_i$ always has the dimension of action; if, say, q_i represents an angle of rotation about an axis, then p_i is the angular momentum with respect to this axis (as already mentioned, angular momentum has the dimension of action).

It should, moreover, be clearly understood that the quantities p_i may not always be thus identified with either linear or angular momenta. In electrodynamics, where noncentral forces occur, we have, by (13.21) and (13.24),

$$p_x = \frac{\partial T}{\partial \dot{x}} + \frac{\partial M}{\partial \dot{x}} = m\dot{x} + \frac{e}{c} A_x. \tag{14.4}$$

The generalized angular momenta will then contain similar additional terms; this means that the usual angular momenta will not be constant, so that the law of areas will no longer be valid (at least not in its original formulation). On the other hand, the generalized momenta may be constants of the motion.

It is, in particular, to be noted that $\dot{p}_i$ and $\partial L/\partial q_i$ in (14.3) are not to be identified with the inertial and external forces $-\Phi'_i$ and Φ_i. For, since $L = T - V$, we have

$$\dot{p}_i = \frac{d}{dt} \frac{\partial T}{\partial \dot{q}_i}, \qquad -\Phi'_i = \frac{d}{dt} \frac{\partial T}{\partial \dot{q}_i} - \frac{\partial T}{\partial q_i},$$
$$\frac{\partial L}{\partial q_i} = \frac{\partial T}{\partial q_i} - \frac{\partial V}{\partial q_i}, \qquad \Phi_i = -\frac{\partial V}{\partial q_i}. \tag{14.5}$$

Lagrange's Equations and the Energy Function

Lagrange's equations are not in general solvable by elementary methods of integration. An important step toward a solution is, of course, to choose the most suitable set of generalized coordinates; but even when this is done, elementary integration may not be possible.

It is therefore important to note that in certain cases it is always possible to obtain at least a first integral of these equations, namely, when the forces are conservative. Actually, the general condition for this possibility of partial integration is somewhat less restrictive, requiring only that the Lagrangian be explicitly independent of time; it is therefore also possible to integrate partially the equations of motion for systems where electromagnetic fields occur. The present considerations will be made in such a way as to facilitate the transition—even for systems with a time-dependent Lagrangian—to Hamilton's formulation of mechanics.

Multiplying (14.2) by $\ddot{q}_i$ and (14.3) by $\dot{q}_i$, we obtain

$$\sum_i \left(p_i \ddot{q}_i + \dot{p}_i \dot{q}_i - \frac{\partial L}{\partial q_i} \dot{q}_i - \frac{\partial L}{\partial \dot{q}_i} \ddot{q}_i \right) - \frac{\partial L}{\partial t} = -\frac{\partial L}{\partial t}. \tag{14.6}$$

Introducing the function

$$H = \sum_i p_i \dot{q}_i - L, \tag{14.7}$$

we may write (14.6) as

$$\frac{dH}{dt} = -\frac{\partial L}{\partial t}. \tag{14.8}$$

If L is explicitly independent of t, that is, depends only indirectly on time through the coordinates and velocities q_i and $\dot{q}_i$, so that $\partial L/\partial t = 0$, then H must be constant. Thus we write

$$\sum_i p_i \dot{q}_i - L = E, \tag{14.9}$$

where E is a constant of integration. For conservative forces E corresponds to that quantity which we have defined earlier as the total energy: the sum of the kinetic and potential energies. Hence we shall refer to H as the *energy function;* and we now proceed to prove the assertion stated above.

Let us, as in (11.5), write

$$T = \sum_{ij} \frac{1}{2} T_{ij} \dot{q}_i \dot{q}_j. \tag{14.10}$$

It follows from this that

$$\sum_i p_i \dot{q}_i = \sum_i \frac{\partial T}{\partial \dot{q}_i} \dot{q}_i = \sum_{ij} T_{ij} \dot{q}_i \dot{q}_j = 2T, \tag{14.11}$$

since every term in the sum (14.10) occurs twice.

Since we have assumed $L = T - V$, it follows from (14.7) and (14.9) that

$$H = 2T - (T - V) = T + V = E. \tag{14.12}$$

Kinetic Energy in Relativistic Mechanics

We shall utilize the energy integral (14.9) to demonstrate the connection between T and T' postulated in (13.31) to (13.34). Taking T' to be given by (13.34), we then have

$$\dot{x} \frac{\partial T'}{\partial \dot{x}} + \dot{y} \frac{\partial T'}{\partial \dot{y}} + \dot{z} \frac{\partial T'}{\partial \dot{z}} = \frac{m_0 v^2}{(1 - v^2/c^2)^{1/2}}$$

$$= \frac{m_0 c^2}{(1 - v^2/c^2)^{1/2}} - m_0 c^2 \left(1 - \frac{v^2}{c^2} \right)^{1/2} = T + T'. \tag{14.13}$$

Since $L = T - V$, this leads to
$$H = T + T' - (T' - V) = T + V, \tag{14.14}$$
if we put
$$T = \frac{m_0 c^2}{(1 - v^2/c^2)^{1/2}}. \tag{14.15}$$

Kinetic Energy for Electromagnetic Forces

In the consideration of motion in an electromagnetic field we must introduce the Lagrangian
$$L = T + M - V,$$
$$M = \frac{e}{c}(\dot{x}A_x + \dot{y}A_y + \dot{z}A_z) - e\varphi. \tag{14.16}$$

The generalized momenta are then given by (14.4), so that
$$\dot{x}p_x + \dot{y}p_y + \dot{z}p_z = m\dot{r}^2 + \frac{e}{c}\dot{\mathbf{r}}\mathbf{A} = 2T + M + e\varphi. \tag{14.17}$$
Hence it follows by (14.7) that
$$H = T + (e\varphi + V). \tag{14.18}$$

It is evident from (14.18) that the kinetic energy is the same as the function T in the Lagrangian, but different from the function $T' = T + M$, which may be taken (formally, at least) to replace T in L. The potential energy V is replaced in (14.18) by $V + e\varphi$ in a very natural way, by the addition of the electric potential energy.

15. HAMILTON'S EQUATIONS

Energy as a Function of Coordinates and Momenta

The Hamiltonian formulation of the equations of motion may be defined as that which consistently describes the state of a system by two sets of variables q_i and p_i, $i = 1, 2, \cdots, n$, that is by $2n$ independently variable coordinates and momenta.

As a matter of fact, we have already written down a version of these equations, in (14.2) and (14.3); their true Hamiltonian form, however, is obtained by solving (14.2) with respect to the $\dot{q}_k$,
$$\dot{q}_k = \dot{q}_k(q_i, p_i) \tag{15.1}$$
and substituting these functions in (14.3), thus obtaining
$$\dot{p}_k = \dot{p}_k(q_i, p_i). \tag{15.2}$$

284 / Mechanics and Statistics

These are purely formal ways of writing Hamilton's equations. The advantage of this procedure is that the right-hand sides of (15.1) and (15.2) may be expressed by means of a function $H(q_k, p_k, t)$, called the *Hamiltonian* or *energy function*. Actually the latter is identical with the energy H of (14.7), but there is a formal difference: the $\dot{q}_i$ are assumed to be eliminated by means of (15.1), so that the Hamiltonian becomes a function of coordinates and momenta. It then turns out that the equations of motion attain the following simple form

$$\dot{q}_k = \frac{\partial H}{\partial p_k}, \qquad \dot{p}_k = -\frac{\partial H}{\partial q_k}, \qquad k = 1, \cdots, n. \tag{15.3}$$

Let us first of all examine the Hamiltonian function in some typical cases. For a single particle and conservative forces we have

$$H = T + V, \qquad T = \tfrac{1}{2}m(\dot{x}^2 + \dot{y}^2 + \dot{z}^2). \tag{15.4}$$

Eliminating the velocity by means of the relation

$$p_x = \frac{\partial T}{\partial \dot{x}} = m\dot{x}, \qquad \text{etc.}, \tag{15.5}$$

we obtain

$$H = \frac{1}{2m}(p_x^2 + p_y^2 + p_z^2) + V. \tag{15.6}$$

In relativistic mechanics we also have $H = T + V$ by (14.12), but

$$T = \frac{m_0 c^2}{(1 - v^2/c^2)^{1/2}}, \qquad p_x = \frac{\partial T'}{\partial \dot{x}} = \frac{m_0 \dot{x}}{(1 - v^2/c^2)^{1/2}}. \tag{15.7}$$

It follows from these equations that

$$p_x^2 + p_y^2 + p_z^2 + m_0^2 c^2 = \frac{m_0^2 c^2}{1 - v^2/c^2}, \tag{15.8}$$

whence we obtain

$$T = c(p_x^2 + p_y^2 + p_z^2 + m_0^2 c^2)^{1/2}. \tag{15.9}$$

This illustrates the fact that the form of the expression for kinetic energy, or the energy function, may change radically with the replacement of velocities by momenta.

If we have an electromagnetic field, the velocities occurring in the kinetic energy must be eliminated by means of the equations

$$m\dot{x} = p_x - \frac{e}{c} A_x, \tag{15.10}$$

which follow from (14.4). Thus the kinetic energy, expressed in terms of generalized momenta, achieves the form

$$T = \frac{1}{2m}\left(\mathbf{p} - \frac{e}{c}\mathbf{A}\right)^2. \tag{15.11}$$

In the relativistic case we obtain as the final expression for the Hamiltonian in an electromagnetic field

$$H = c\left[\left(\mathbf{p} - \frac{e}{c}\mathbf{A}\right)^2 + m_0^2 c^2\right]^{1/2} + (e\varphi + V). \tag{15.12}$$

For an electron in an atomic field of force, we must replace e by $-e$ in this expression; the symbol e then denotes the *magnitude* of the elementary quantum of electric charge.

Deduction of Hamilton's Equations from Lagrange's Equations

Hamilton's equations may be deduced from Lagrange's in the following simple way. We put, as in (14.7),

$$H(q_k, p_k, t) = \sum_i p_i \dot{q}_i - L(q_i, \dot{q}_i, t), \tag{15.13}$$

regarding the velocities $\dot{q}_i$ as functions of the q_k and p_k.

We now perform a partial differentiation of (15.13) with respect to p_k, and also with respect to q_k, and utilize the fact that

$$\begin{aligned}\sum_i \left(p_i - \frac{\partial L}{\partial \dot{q}_i}\right)\frac{\partial \dot{q}_i}{\partial p_k} &= 0, \\ \sum_i \left(p_i - \frac{\partial L}{\partial \dot{q}_i}\right)\frac{\partial \dot{q}_i}{\partial q_k} &= 0, \end{aligned} \tag{15.14}$$

which follows from the general definition (14.2) of the momenta. Furthermore we have $\dot{p}_k = \partial L/\partial q_k$ by (14.3), so that we finally arrive at

$$\begin{aligned}\frac{\partial H}{\partial p_k} &= \dot{q}_k, \\ \frac{\partial H}{\partial q_k} &= -\dot{p}_k, \end{aligned} \tag{15.15}$$

which is the form of Hamilton's equation already stated in (15.3).

Multiplying these equations by $\dot{p}_k$ and $\dot{q}_k$, and adding, we obtain

$$\dot{q}_k \frac{\partial H}{\partial q_k} + \dot{p}_k \frac{\partial H}{\partial p_k} = 0. \tag{15.16}$$

Since
$$\sum_k \left(\dot{q}_k \frac{\partial H}{\partial q_k} + \dot{p}_k \frac{\partial H}{\partial p_k} \right) = \frac{dH}{dt} - \frac{\partial H}{\partial t}, \quad (15.17)$$
we thus, by summation over k in (15.16), obtain the relation
$$\frac{dH}{dt} = \frac{\partial H}{\partial t}. \quad (15.18)$$

If both the Lagrangian L and the Hamiltonian H are explicitly independent of time, that is, in accordance with (14.8)
$$\frac{\partial H}{\partial t} = -\frac{\partial L}{\partial t} = 0, \quad (15.18a)$$
we arrive back at the earlier result
$$H(q_k, p_k) = E, \quad (15.19)$$
where E is a constant, representing the total energy.

The Connection between the Canonical Equations and Hamilton's Principle

Hamilton's equations of motion are often referred to as *canonical equations*. As would be expected, these equations may also be deduced directly from a variational principle; we shall now demonstrate briefly that this is the principle of Hamilton previously discussed.

According to (14.7) we may put
$$L = \sum_k p_k \dot{q}_k - H \quad (15.20)$$
and express Hamilton's principle as follows
$$\int_{t_0}^{t_1} \left[\sum_k p_k \dot{q}_k - H(q_k, p_k, t) \right] dt = \text{extr.} \quad (15.21)$$
If we here regard the momenta p_k as functions of the coordinates and velocities defined by means of the Lagrangian L, that is, $p_k = \partial L/\partial \dot{q}_k$, there is no essential difference between (15.21) and our previous formulation of Hamilton's principle. It is, however, very natural to assume that this principle is valid even if we take the q_k and p_k as the new $2n$ variables, to be varied independently of each other. We must then postulate the requirement $\delta q_k(t_0) = \delta q_k(t_1) = 0$, that is, the boundary conditions already laid down for the variation, to be valid still. On the other hand, there is no reason for assuming similar boundary conditions for the p_k. Since a function with a given value at a point may have very different derivatives, the requirement

$\delta q_k = 0$ imposes no limitations on $\delta \dot{q}_k$, and thus none on δp_k either. Variation of the momenta in (15.21) leads immediately to the first equation of Hamilton

$$\dot{q}_k - \frac{\partial H}{\partial p_k} = 0. \tag{15.22}$$

To obtain the second, we must first integrate in parts the first term of the integrand; the subsequent variation of q_k then gives us

$$\dot{p}_k + \frac{\partial H}{\partial q_k} = 0. \tag{15.23}$$

Equation (15.21) is not the only form of Hamilton's principle that leads to the canonical equations. The integral principle

$$\int_{t_0}^{t_1} \left[-\sum_k q_k \dot{p}_k - H \right] dt = \text{extr.} \tag{15.24}$$

will also serve, if we take as new boundary conditions $\delta p_k(t_0) = \delta p_k(t_1) = 0$. This is easily shown in a way analogous to the above.

The alternative formulation (15.24) will be of importance for our subsequent consideration of a certain class of simultaneous transformation of both coordinates and momenta: the so-called canonical transformations.

16. GENERAL SOLUTION METHODS FOR THE CANONICAL EQUATIONS

Cyclic Variables and Constants of Motion

In practice, the equations of motion for a mechanical system are only rarely solvable by elementary methods of integration; and, correspondingly, there exists no general method of solving the canonical equations either. However, in certain cases a partial integration is always possible: if so-called *cyclic variables* occur.

A cyclic variable is generally defined as one on which the Hamiltonian does not depend explicitly. Assuming q_n to be a cyclic coordinate, we may write

$$H = H(q_1, \cdots, q_{n-1}, p_1, \cdots, p_n). \tag{16.1}$$

Thus we have, for the corresponding momentum,

$$\dot{p}_n = \frac{\partial H}{\partial q_n} = 0, \qquad p_n = \text{const.}, \tag{16.2}$$

p_n is then called a *constant* of the motion.

This furnishes another first integral, in addition to the energy equation. For every cyclic coordinate, we may thus deduce one first integral; hence it is clearly of advantage to choose coordinates in such a way that as many as possible of them are cyclic. However, it is not in general possible to obtain more than a limited number of cyclic variables by means of coordinate transformations. On the other hand, one may employ a more general class of transformations of both coordinates and momenta, the so-called *canonical transformations*, to obtain a complete set of cyclic variables.

Such cyclic variables are not, however, always pure coordinates of position. In general, they are what is usually termed *canonically* conjugate to certain other quantities, which happen to be constants of the motion. An instructive example is the case of a time-independent Hamiltonian: the energy E is then a constant, and the time t canonically conjugate to it.

One instance of a cyclic variable is the polar angle ψ for the motion of a particle in a central field $V(r)$. We know that this motion must be in a plane which we take to be the xy-plane,

$$x = r \cos \psi, \quad y = r \sin \psi. \tag{16.3}$$

Thus

$$T = \tfrac{1}{2}m(\dot{x}^2 + \dot{y}^2) = \tfrac{1}{2}m(\dot{r}^2 + r^2\dot{\psi}^2), \tag{16.4}$$

which, since

$$p_r = m\dot{r}, \quad p_\psi = mr^2\dot{\psi}, \tag{16.5}$$

leads to

$$H = \frac{1}{2m}\left(p_r^2 + \frac{1}{r^2}p_\psi^2\right) + V(r). \tag{16.6}$$

Hence ψ is cyclic, and $p_\psi = $ const.

Introducing spherical polar coordinates r, ϑ, φ,

$$x = r \sin\vartheta \cos\varphi, \quad y = r \sin\vartheta \sin\varphi, \quad z = r \cos\vartheta \tag{16.7}$$

we obtain

$$T = \tfrac{1}{2}m[\dot{r}^2 + r^2(\dot{\vartheta}^2 + \sin^2\vartheta\,\dot{\varphi}^2)],$$

$$H = \frac{1}{2m}\left[p_r^2 + \frac{1}{r^2}\left(p_\vartheta^2 + \frac{1}{\sin^2\vartheta}p_\varphi^2\right)\right] + V(r). \tag{16.8}$$

Clearly, the azimuthal angle φ is cyclic, so that

$$p_\varphi = \frac{\partial T}{\partial \dot{\varphi}} = mr^2 \sin^2\vartheta\,\dot{\varphi} = \text{const.} \tag{16.9}$$

On the other hand, $p_\vartheta = mr^2\dot\vartheta$ is not constant. However, it follows from (16.6) that

$$p_\psi^2 = p_\vartheta^2 = \frac{1}{\sin^2\vartheta}p_\varphi^2, \tag{16.10}$$

so that the square of the total angular momentum p_ψ, is constant; ψ and p_ψ are then as in (3.6), the polar angle of the plane of motion and the corresponding canonical momentum variable: that is, the total angular momentum. In general,

$$p_\varphi = p_\psi \cos \alpha \tag{16.11}$$

where α is the angle between the plane of motion and the xy-plane, or between the trajectory normal and the z-axis; p_φ is then the z-component of the constant angular momentum p_ψ.

It should also be noted that

$$\int_a^b \sum_i p_i \, dq_i = \int_{t_0}^{t_1} 2T \, dt \tag{16.12}$$

by (14.11), that is, that the integral on the left-hand side, taken between two points a, b on the trajectory, does not depend on the coordinates chosen. A consequence of this is that

$$\int p_\psi \, d\psi = \int p_\vartheta \, d\vartheta + \int p_\varphi \, d\varphi, \tag{16.13}$$

a relation of some importance for the calculation of the so-called *phase integrals* or *action variables*.

Having shown already that the angular momentum is a constant for motion in a central field, we shall now demonstrate the same by means of Hamilton's equations. Thus, putting

$$M_z = xp_y - yp_x, \tag{16.14}$$

we obtain

$$\dot{M}_z = \dot{x}p_y - \dot{y}p_x + x\dot{p}_y - y\dot{p}_x$$

$$= p_y \frac{\partial H}{\partial p_x} - p_x \frac{\partial H}{\partial p_y} - \left(x \frac{\partial H}{\partial y} - y \frac{\partial H}{\partial x} \right) = 0, \tag{16.14a}$$

since H only depends on the expressions $p^2 = p_x^2 + p_y^2 + p_z^2$ and $r^2 = x^2 + y^2 + z^2$.

The three components of angular momentum M_x, M_y, and M_z are constants of motion, each corresponding to a canonically conjugate cyclic variable; namely, the angular position relative to the appropriate coordinate axis; it may then seem strange that we do not here immediately obtain three cyclic variables. By geometric considerations, however, we realize that the three angles mentioned above are not independent of each other. The maximum number of cyclic coordinates resulting from a constant angular momentum therefore reduces to 2.

The Hamilton-Jacobi Differential Equation and the Action Function

We have previously discussed the possibility for the momentum vector of a single particle to be representable as a sort of irrotational vector field $\mathbf{p} = \nabla S$, where the coordinate-dependent quantity S is generally termed the action function. This means that an infinite family of possible particle trajectories is perpendicular to the surfaces $S = $ const, which may be expressed as follows:

$$p_k = \frac{\partial S}{\partial q_k}. \tag{16.15}$$

For conservative forces and constant energy such a function must clearly exist, since it only has to satisfy the requirement

$$H\left(q_k, \frac{\partial S}{\partial q_k}\right) = E \tag{16.16}$$

and thus becomes a function of the coordinates.

Equation 16.16 is the Hamilton-Jacobi differential equation for the action function S; in general this equation will be of the first order, but quadratic in S. A solution will then depend on the n coordinates q_k, and also on n constants of integration a_i, $i = 1, 2, \cdots, n$.

Integration of the Equations of Motion

If the Hamilton-Jacobi equation had always been solvable in a reasonably simple way, the further integration of the equations of motion would (at least in principle) be straightforward; this, unfortunately, is not the case. However, we shall take the liberty of assuming that there exist solutions

$$S = S(q_k, a_i), \tag{16.17}$$

which, when substituted in (16.16), lead to

$$E = E(a_1, a_2, \cdots, a_n). \tag{16.18}$$

Once such a complete solution is known, the set of equations

$$\frac{\partial S}{\partial a_i} = \omega_i t + b_i, \qquad \omega_i = \frac{\partial E}{\partial a_i} \tag{16.19}$$

would give us the integrated equations of motion; we note that the correct number, $2n$, of constants of integration occur:

$$a_1, \cdots, a_n, b_1, \cdots, b_n.$$

The justification of this assumption may be demonstrated directly by differentiating (16.19) with respect to t. On the left-hand side we then obtain

$$\frac{d}{dt}\frac{\partial S}{\partial a_i} = \sum_k \dot{q}_k \frac{\partial^2 S}{\partial a_i \partial q_k} = \sum_k \frac{\partial H}{\partial p_k}\frac{\partial p_k}{\partial a_i} = \frac{\partial H}{\partial a_i} = \frac{\partial E}{\partial a_i} = \omega_i. \quad (16.20)$$

Defining
$$\varphi_i = \omega_i t + b_i, \quad (16.21)$$
it follows from (16.20) that
$$\dot{\varphi}_i = \frac{\partial H}{\partial a_i}. \quad (16.22)$$

Since H is independent of t, and thus also of φ_i, and since the a_i are constants, we may put

$$\dot{a}_i = -\frac{\partial H}{\partial \varphi_i}, \quad (16.23)$$

both sides of this expression being zero. This implies that we may regard the a_i and φ_i as canonically conjugate quantities (respectively, constants of motion and cyclic variables), satisfying the canonical equations (16.22) and (16.23). We shall presently show how such quantities result from the already-discussed canonical transformations.

17. CANONICAL TRANSFORMATIONS

Coordinate-Momentum Space

The n positional coordinates $q_1, \cdots, q_n$ of a mechanical system with n degrees of freedom may be regarded as points in an abstract space of n dimensions; to describe the state of a system it is, in addition, necessary to introduce the n momenta $p_1, \cdots, p_n$. These two sets of variables may thus together be taken to constitute a $2n$-dimensional *phase space* or *coordinate-momentum space;* this concept is of fundamental importance in several theoretical disciplines of physics, notably statistical mechanics and quantum theory.

Consider now the class of all possible transformations (of coordinates and momenta) in this space. Clearly this class is much more general than that of the coordinate transformation hitherto employed. In fact, it is *too general;* we shall, in the following, restrict ourselves to the consideration of a certain subclass, variously termed *contact transformations* or *canonical transformations*.

We shall define these transformations by requiring that they leave the generalized element of volume

$$d\tau = dq_1 \cdots dq_n\, dp_1 \cdots dp_n \quad (17.1)$$

invariant, and that they satisfy a new set of canonical equations (hence the name canonical transformations). Thus labeling the new coordinates and momenta $Q_1, \cdots Q_n, P_1, \cdots P_n$, we must have

$$\dot{Q}_k = \frac{\partial \bar{H}}{\partial P_k}, \qquad \dot{P}_k = -\frac{\partial \bar{H}}{\partial Q_k}, \qquad (17.2)$$

where $\bar{H}(Q_k, P_k, t)$ is the new Hamiltonian, possibly different from the original $H(q_k, p_k, t)$.

The Canonical Transformation Function

Substituting $p_k = Q_k, q_k = -P_k$ in the canonical equations, we obtain the form

$$\dot{Q}_k = \frac{\partial H}{\partial P_k}, \qquad \dot{P}_k = -\frac{\partial H}{\partial Q_k}. \qquad (17.3)$$

This implies that, formally, coordinates and momenta may exchange roles in the canonical equations; this particular type of canonical transformation is then nothing but a change of names, without any real physical significance. Of the four different possibilities in this respect we shall choose only one, as follows.

We utilize the two different formulations (15.21) and (15.24) of Hamilton's principle and put the sets of equations (15.15) and (15.2) equivalent to the variational principles

$$\delta \int_{t_0}^{t_1} \left(\sum_k p_k \dot{q}_k - H \right) dt = 0,$$
$$\delta \int_{t_0}^{t_1} \left(-\sum_k Q_k \dot{P}_k - \bar{H} \right) dt = 0. \qquad (17.4)$$

Here the coordinates q_k and the momenta P_k, respectively, are taken to have fixed values at the times $t = t_0$ and $t = t_1$. This implies that the difference between the two integrands in (17.4) must be a total derivative with respect to t,

$$\sum_k (p_k \dot{q}_k + Q_k \dot{P}_k) - H + \bar{H} = \frac{dF}{dt}, \qquad (17.5)$$

where F is a function of q_k and P_k with fixed boundary values

$$F = F(q_k, P_k, t). \qquad (17.6)$$

Hence the transformation equations

$$p_k = \frac{\partial F}{\partial q_k}, \qquad Q_k = \frac{\partial F}{\partial P_k}, \qquad \bar{H} = H + \frac{\partial F}{\partial t} \qquad (17.7)$$

follow immediately.

The said function is called the generating function of the transformation, or simply the transformation function. It clearly gives rise to an infinite number of different possible transformations; as an example, the identity transformation is characterized by the generating function

$$F = \sum_k q_k P_k. \tag{17.8}$$

Another possible choice is

$$F = P_x r \cos \varphi + P_y r \sin \varphi, \tag{17.9}$$

where r and φ are plane polar coordinates; this leads to

$$X = \frac{\partial F}{\partial P_x} = r \cos \varphi, \quad p_r = \frac{\partial F}{\partial r} = P_x \cos \varphi + P_y \sin \varphi,$$

$$Y = \frac{\partial F}{\partial P_y} = r \sin \varphi, \quad p_\varphi = \frac{\partial F}{\partial \varphi} = xP_y - yP_x, \tag{17.10}$$

from which it follows that (17.9) specifies the transition from Cartesian to polar coordinates in a plane.

In the case of a time-independent Hamiltonian, we may restrict ourselves to a correspondingly time-independent transformation function

$$F(q_k, P_k, t) = S(q_k, P_k). \tag{17.11}$$

This turns out to be identical with the previously discussed action function S, since the transformation equations evidently are

$$p_k = \frac{\partial S}{\partial q_k}, \quad Q_k = \frac{\partial S}{\partial P_k}, \quad \bar{H} = H, \tag{17.12}$$

and the new canonical equations

$$\dot{Q}_k = \frac{\partial \bar{H}}{\partial P_k}, \quad \dot{P}_k = -\frac{\partial \bar{H}}{\partial Q_k}. \tag{17.13}$$

If we now require that $p_k = \partial S/\partial q_k$ satisfy the energy equation

$$H\left(q_k, \frac{\partial S}{\partial q_k}\right) = E, \tag{17.14}$$

S becomes a function of the coordinates and a set of constants of integration $a_1, \cdots, a_n$, which we identify with the new momenta $P_1, \cdots, P_n$. This means that the energy E, and thus also the new Hamiltonian $\bar{H}$, becomes a function of these momenta alone,

$$\bar{H} = E(P_1, \cdots, P_n). \tag{17.15}$$

Hence, by (17.13), $\dot{P}_k = 0$, $P_k = \text{const}$, as already assumed; furthermore

$$\dot{Q}_k = \frac{\partial E}{\partial P_k} = \text{const} = \omega_k, \quad Q_k = \omega_k t + b_k. \tag{17.16}$$

The new canonical variables Q_k are thus identical with the quantities φ_k in (16.21). The advantage of these canonical transformations, where the action S acts as the generating function, is that a complete set of generalized cyclic variables is obtained; the latter are then canonically conjugate to a complete set of constants of motion.

18. SOLUTION OF THE PROBLEM OF MOTION BY SEPARATION OF THE VARIABLES

The only case where an algorithm for solving the Hamilton-Jacobi equation exists is when this equation is separable in the variables, that is, when a set of coordinates $q_1, \cdots, q_n$ exists, in which the solution has the form

$$S = S_1(q_1) + S_2(q_2) + \cdots + S_n(q_n) = \sum_k S_k(q_k). \tag{18.1}$$

This leads to ordinary first-order differential equations for the functions $S_k(q_k)$:

$$\left(\frac{dS_k}{dq_k}\right)^2 = F(q_k) \tag{18.2}$$

or

$$\frac{dS_k}{dq_k} = f(q_k). \tag{18.2a}$$

Hence S_k may be found merely by integrating the function $f(q_k)$.

Each $S_k(q_k)$ will, of course, depend on one or several (or even all) of the constants of integration, so that we must put

$$S_k = S_k(q_k, a_i), \quad i = 1, 2, \cdots, n. \tag{18.3}$$

The integrated equations of motion are thus

$$\sum_k \frac{\partial S_k(q_k, a_i)}{\partial a_i} = \omega_i t + b_i. \tag{18.4}$$

Equations (16.6) and (16.8) are examples of Hamiltonians for which the corresponding Hamilton-Jacobi equations are separable. The possibility of separation ensues from the choice of variables, the radius vector r entering into the potential energy $V(r)$, and one or two angular coordinates ψ or ϑ and φ, which (together with the r-coordinate) constitute an orthogonal curvilinear coordinate system.

Lagrangian and Hamiltonian Mechanics / 295

From (16.8) we obtain the Hamilton-Jacobi equation

$$\left(\frac{\partial S}{\partial r}\right)^2 + \frac{1}{r^2}\left[\left(\frac{\partial S}{\partial \vartheta}\right)^2 + \frac{1}{\sin^2 \vartheta}\left(\frac{\partial S}{\partial \varphi}\right)^2\right] = 2m(E - V), \qquad (18.5)$$

whence, by separation of the variables,

$$S = S_r(r) + S_\vartheta(\vartheta) + S_\varphi(\varphi),$$

$$\frac{dS_\varphi}{d\varphi} = a_3,$$

$$\left(\frac{dS_\vartheta}{d\vartheta}\right)^2 + \frac{a_3{}^2}{\sin^2 \vartheta} = a_2{}^2, \qquad (18.6)$$

$$\left(\frac{dS_r}{dr}\right)^2 + \frac{a_2{}^2}{r^2} = 2m(E - V).$$

We shall not at this point discuss these equations any further, as they will be treated in more detail in the chapter on Bohr's atomic theory in Part IV. Instead we shall examine Eq. (16.6) somewhat more closely, because from a purely mechanical point of view it contains all the essential features of the problem.

Orbits and Orbital Motion

Before proceeding, we shall first indicate how the treatment of the problem may be simplified by a suitable choice of the constants of motion (actually: of integration) $a_1, a_2, \cdots, a_n$. Since the energy is a function of these quantities $E = E(a_1, \cdots, a_n)$, we are at liberty to choose it as equal to one of them. Thus, putting $a_1 = E$, for example, we obtain E independent of $a_2, \cdots, a_n$. The integrated equations of motion (16.9) then take the form

$$\frac{\partial S}{\partial a_1} = t + b_1, \qquad \frac{\partial S}{\partial a_i} = b_i, \qquad i = 2, \cdots, n. \qquad (18.7)$$

The last $n - 1$ of these equations are time-independent, and determine the coordinates q_i as functions of each other, or of one coordinate q_1, say. In other words, they determine the "trajectory" of the system in n-dimensional coordinate space; however, we must utilize the first of the equations (18.7) to obtain the motion of the trajectory point as a function of t.

We shall now illustrate all this by demonstrating how the plane trajectory of a particle in a central field may be deduced from (16.6). This equation leads to

$$p_\psi = \frac{dS_\psi}{d\psi} = a_2, \quad S_\psi = a_2\psi,$$

$$p_r = \frac{dS_r}{dr} = \pm[2m(E - V) - a_2^2/r^2]^{\frac{1}{2}}, \qquad (18.8)$$

$$S = \pm \int_{r_0}^{r} \left[2m(E - V) - \frac{a_2^2}{r^2}\right]^{\frac{1}{2}} dr + a_2\psi.$$

The double sign occurs because the square root changes sign at inflection points of the trajectory. For increasing/decreasing r the root is to be taken as positive/negative, respectively; thus, if r_0 corresponds to the maximum value of r or to a trajectory point where r is decreasing, the minus sign applies.

We may now determine the trajectory by putting $\partial S/\partial a_2 = b_2 = 0$, $r_0 = r_{\max}$, which means taking $\psi = 0$ at $r = r_{\max}$; this leads to the equation

$$\psi = -\int_{r_{\max}}^{r} \frac{a_2/r^2}{[2m(E - V) - a_2^2/r^2]^{\frac{1}{2}}} dr. \qquad (18.9)$$

for the trajectory.

The same result may evidently be obtained from the energy equation, expressed in terms of the velocities

$$\tfrac{1}{2}m(\dot{r}^2 + r^2\dot{\varphi}^2) + V = E \qquad (18.10)$$

and the law of areas

$$mr^2\dot{\varphi} = \text{const} = a_2, \qquad (18.11)$$

that is, relations that may be deduced from Lagrange's equations. The latter may be written

$$m\dot{r} = -\left[2m(E - V) - \frac{a_2^2}{r^2}\right]^{\frac{1}{2}},$$

$$m\dot{\varphi} = a_2/r^2. \qquad (18.12)$$

whence, dividing one equation by the other, we obtain

$$\frac{\dot{\varphi}}{\dot{r}} = \frac{d\varphi}{dr} = -\frac{a_2/r^2}{[2m(E - V) - a_2^2/r^2]^{\frac{1}{2}}}. \qquad (18.13)$$

If $V(r)$ denotes a Coulomb electrostatic potential, or (as is the case for the planetary orbits of astronomy) a gravitational potential of the form $V = -\text{const}/r$, the trajectory becomes elliptic or hyperbolic, according as the energy is negative or positive.

We shall not consider any further the specification of the constants in (18.9) and (18.13), as an analogous equation is given in (8.30). This leads to the well-known elliptic orbit.

$$r = \frac{b^2/a}{1 - c/a \cos \psi}, \qquad (18.14)$$

where r and φ are plane polar coordinates.

The motion of the mass point along this orbit is determined, as a function of the time, by the equation

$$\frac{\partial S}{\partial E} = t - t_0, \qquad (18.15)$$

where S is the action function of (18.8). Further specialized treatment of this problem will be deferred to the consideration of atomic theory in Part IV.

CHAPTER 3

Dynamics of Continuously Extended Bodies

19. MOTION OF SOLID BODIES

Moment of Inertia and Products of Inertia for a Body

We have deduced, in Chapter 1, Section 8, several results that will enable us to simplify considerably the treatment of the dynamics of rigid bodies. In particular, the center of mass law, the law of torques, and the results obtained for work and kinetic energy referred to the center of mass, will prove convenient in this respect.

We have seen that the inertia of a mechanical system with respect to rotational motion about an axis is represented by a scalar quantity, namely, the moment of inertia. More generally, the moment of inertia may be referred to three axes within the body, or to an arbitrary axis given as a function of its direction angles; it is then a symmetric second-rank tensor. The diagonal elements of this tensor are called *moments of inertia* proper, and the off-diagonal elements are called *products of inertia*. Being symmetric, the tensor of inertia may be reduced to diagonal form by a suitable choice of orthogonal coordinate axes; these are then referred to as the principal axes of the body.

The definition (8.8) of the moment of inertia has the disadvantage that it becomes time-dependent if the z-axis considered is taken to be fixed in space. We shall henceforth define the moments of inertia of a body with respect to an axis fixed in the body; it is then clear that we must utilize the previously given considerations of relative motion.

For simplicity we first assume that the body rotates about a fixed internal point; the linear velocity of a mass point m_i in the body is then

$$\mathbf{v}_i = \boldsymbol{\omega} \times \mathbf{r}_i \tag{19.1}$$

and the kinetic energy

$$T = \sum_i \frac{1}{2} m_i \mathbf{v}_i^2 = \sum_i \frac{1}{2} m_i [\omega^2 \mathbf{r}_i^2 - (\boldsymbol{\omega} \mathbf{r}_i)^2]. \tag{19.2}$$

This may be written as

$$2T = I_{xx}\omega_x^2 + I_{xy}\omega_x\omega_y + I_{xz}\omega_x\omega_z + I_{yx}\omega_y\omega_x + I_{yy}\omega_y^2 + I_{yz}\omega_y\omega_z$$
$$+ I_{zx}\omega_z\omega_x + I_{zy}\omega_z\omega_y + I_{zz}\omega_z^2 = \boldsymbol{\omega} I \boldsymbol{\omega}, \quad (19.3)$$

where

$$I_{xx} = \sum_i m_i(y_i^2 + z_i^2), \quad I_{xy} = -\sum_i m_i x_i y_i, \quad \text{etc.}, \quad (19.4)$$

are the components of the inertia tensor I.

Similarly we find the angular momentum vector

$$\mathbf{M} = \sum_i m_i \mathbf{r}_i \times (\boldsymbol{\omega} \times \mathbf{r}_i)$$
$$= \sum_i m_i \{\boldsymbol{\omega}(\mathbf{r}_i)^2 - \mathbf{r}_i(\boldsymbol{\omega}\mathbf{r}_i)\} = I\boldsymbol{\omega}. \quad (19.5)$$

Denoting the inertia tensor with respect to the center of mass by I', we then easily obtain, by referring I and I' to parallel coordinate axes,

$$I = I' + I_0,$$
$$I_{0xx} = m(y^2 + z^2), \quad I_{0xy} = -mxy, \quad \text{etc.}, \quad (19.6)$$

where I_0 is the inertia tensor cooresponding to the concentration of the total mass m of the body in the center of mass $\mathbf{r} = x\mathbf{i} + y\mathbf{j} + z\mathbf{k}$.

When the principal axes of the body are chosen as coordinate axes, the inertia tensor becomes diagonal, and the kinetic energy is given by

$$T = \tfrac{1}{2}(I_1\omega_x^2 + I_2\omega_y^2 + I_3\omega_z^2). \quad (19.7)$$

In general none of the principal axes will then pass through the center of mass.

It may be shown that the products of inertia I_{xy} vanish if one of the principal axes passes through the center of mass. Thus, if the inertia tensor I' has been transformed to principal axes, the diagonal form of the inertia tensor I will in general be maintained if and only if the point of reference lies on one of these axes.

Equations of Motion for Rigid Bodies

The law of torques is often very useful in the treatment of the motion of rigid bodies. For instance, if the moment of forces about an axis through a fixed point in the body (e.g. the center of mass) is zero, the angular momentum about this axis will be constant.

Such direct applications of the law of torques are sometimes convenient; however, they often put a great strain on one's attentiveness and imagination. We shall, therefore, not make use of such elementary considerations here,

Dynamics of Continuously Extended Bodies | 301

but instead immediately deduce Lagrange's equations for the motion. These are

$$\frac{d}{dt}\frac{\partial T}{\partial \dot{q}_i} = \Phi_i. \tag{19.8}$$

As usual, the Φ_i are generalized forces, defined by the work

$$dA = \sum_i \Phi_i \, dq_i \tag{19.9}$$

performed by infinitesimal displacements of the positional coordinates.

If

$$dA = -dV, \tag{19.10}$$

(19.8) may be written

$$\frac{d}{dt}\frac{\partial L}{\partial \dot{q}_i} - \frac{\partial L}{\partial q_i} = 0, \quad L = T - V. \tag{19.11}$$

Eulerian Angles for Rotational Motion

In the description of the rotational motion of a rigid body, it is frequently advantageous to employ the Eulerian angles ϑ, φ, and ψ, as illustrated in Fig. 3.1. X, Y, and Z are axes fixed in the rotating body, conveniently chosen as the principal axes of the body so that the inertia tensor only has the three diagonal components (i.e. moments of inertia) I_1, I_2, I_3. The angle ϑ is the angle between the Z-axis and the Z_0-axis of a fixed Galilean coordinate frame X_0, Y_0, Z_0; φ is the angle between X_0 and the intersection line S of the X_0Y_0- and the XY-plane; and ψ is the angle between S and X.

The three angular velocities $\dot{\vartheta}$, $\dot{\varphi}$, and $\dot{\psi}$ each contribute to the instantaneous rotational vector. The component $\dot{\vartheta}$ is directed along S, $\dot{\varphi}$ along Z_0, and

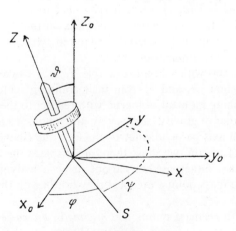

Fig. 3-1. Eulerian angles.

$\dot{\psi}$ along Z. Thus we obtain the components

$$\omega_x = \dot{\vartheta} \cos \psi + \dot{\varphi} \sin \vartheta \sin \psi,$$
$$\omega_y = -\dot{\vartheta} \sin \psi + \dot{\varphi} \sin \vartheta \cos \psi, \qquad (19.12)$$
$$\omega_z = \dot{\varphi} \cos \vartheta + \dot{\psi}.$$

referred to the three axes in the body, and the kinetic energy

$$T = \tfrac{1}{2}(I_1 \omega_x^2 + I_2 \omega_y^2 + I_3 \omega_z^2) \qquad (19.13)$$

expressed in terms of the angular velocities (19.12).

Hence it follows, in general, that at most φ may be a cyclic variable: i.e. when the potential energy is independent of φ, so that the torque with respect to the Z_0-axis, $M_{z_0} = -\partial V/\partial \varphi$, is zero. This will, for instance, be the case if the only force acting on the body is gravity, directed along the Z_0-axis, and if the center of mass lies on the Z-axis.

However, the variable ψ may also be cyclic, if it should happen that $I_1 = I_2$. This does not mean that the body must necessarily have rotational symmetry, geometrically speaking, with respect to the Z-axis; such a rotational symmetry may be said to obtain, but then in a purely dynamical sense. Thus a body with a regular polygonal cross section, for instance, will have such a dynamical rotational symmetry; however, bodies of irregular shape may also possess this symmetry.

If $I_1 = I_2$, the kinetic energy takes the form

$$T = \tfrac{1}{2}[I_1(\dot{\vartheta}^2 + \dot{\varphi}^2 \sin^2 \vartheta) + I_3(\dot{\varphi} \cos \vartheta + \dot{\psi})^2]. \qquad (19.14)$$

Here both φ and ψ are cyclic, provided that the potential energy V, or the torque $M_z = P_\psi = -\partial v/\partial \psi$, is zero. If the only force acting on the body is gravity, this implies that the center of mass lies on the Z-axis. Thus a complete dynamical rotational symmetry is required, with respect to both the center of mass and the moment of inertia.

Let us illustrate this with a simple example. First, we take two mass points m_1 and m_2, at distance x_1 and x_2 from their common center of mass; these then have a vanishing moment of inertia with respect to the axis through the points and the center of gravity (the x-axis), and a nonzero moment of inertia with respect to an axis perpendicular to the first one (the y-axis). To obtain equal moments of inertia, we must therefore displace the z-axis away from the center of mass. Otherwise we cannot achieve a dynamical rotational symmetry for two mass points, except if they both lie on the axis of rotation (the z-axis).

We now take three mass points m_1, m_2, m_3 in an xy-plane, with the six coordinates x_1, y_1, x_2, y_2, and so forth. There are then altogether four

Dynamics of Continuously Extended Bodies / 303

conditions to be satisfied: two for the center of mass, and two for the inertia tensor $I_{xx} = I_{yy}$ and $I_{xy} = 0$. The two degrees of freedom left for the mass points are rotation (of the system about the center of mass) and dilation (i.e. an increase or decrease of the distances from the points to the center of gravity).

For an increasing number of mass points we obtain correspondingly increasing possibilities of locating the particles more or less arbitrarily, subject only to the requirement of dynamical rotational symmetry.

20. MOTION OF A BODY AROUND A FIXED POINT OR AROUND THE CENTER OF MASS

The Gyroscope

We shall now apply the theory of rigid body motion to bodies with dynamical rotational symmetry. We may distinguish between two possibilities: (1) the body is constrained to rotate about a fixed point, and (2) it is freely movable. In Case (2), we may take the motion to be compounded by the motion of the center of mass and the rotation of the body about this point, which may then be regarded as fixed. This problem of treating rotational motion about a fixed point, which we shall refer to as the *gyroscope problem*, is fundamental in the theory of rigid body motion.

To attack this problem, it is necessary to know the external torque acting on the rotating body, which we shall call a *gyroscope*. We assume first that it rotates about a fixed point on the rotational axis of symmetry (the Z-axis), at a distance l from the center of mass; compare Fig. 3-1. The torque about the line S is then

$$M_3 = mgl \sin \vartheta, \qquad (20.1)$$

where m is the total mass of the body.

Since

$$M_3 \, d\vartheta = dA = -dV = -d(mgl \cos \vartheta), \qquad (20.2)$$

the potential energy is

$$V = mgl \cos \vartheta. \qquad (20.3)$$

Hence, by means of (19.14), we obtain the Lagrangian

$$L = \tfrac{1}{2}[I_1(\dot\vartheta^2 + \dot\varphi^2 \sin^2 \vartheta) + I_3(\dot\varphi \cos \vartheta + \dot\psi)^2] - mgl \cos \vartheta. \qquad (20.4)$$

Clearly φ and ψ are cyclic variables, and

$$p_\varphi = \frac{\partial L}{\partial \dot\varphi} = I_1 \dot\varphi \sin^2 \vartheta + I_3(\dot\varphi \cos \vartheta + \dot\psi) \cos \vartheta = a_2,$$

$$p_\psi = \frac{\partial L}{\partial \dot\psi} = I_3(\dot\varphi \cos \vartheta + \dot\psi) = a_3, \qquad (20.5)$$

are constants; $p_\varphi = a_2$ and $p_\psi = a_3$ are the constant angular momenta about the fixed spatial axis Z_0 and the rotational axis Z of the body, respectively.

Before discussing the general solution of this problem, we shall first analyze a few simpler motions.

Let us take $a_2 = a_3 = 0$, so that $\dot\varphi = \dot\psi = 0$. It then follows that

$$L = \tfrac{1}{2}I_1\dot\vartheta^2 - mgl\cos\vartheta, \tag{20.6}$$

which is the Lagrangian for the physical pendulum. Assuming that the energy

$$E = T + V = \tfrac{1}{2}I_1\dot\vartheta^2 + mgl\cos\vartheta \tag{20.7}$$

is not sufficiently large to permit complete revolutions of the pendulum, it will thus oscillate about the vertical position $\cos\vartheta = -1$ with the center of mass directly below the fixed point.

We now let the body simultaneously execute a rotational motion in φ or ψ. This motion will then prevent the axis of the body from taking the position $\cos\vartheta = -1$; instead it will oscillate between a maximum and minimum value of $\cos\vartheta$. This results in a combined horizontal rotation, or *precession*, and vertical oscillation, or *nutation*, of the axis.

Let us investigate the possibilities of obtaining a pure precession, that is, a constant value of ϑ. By (20.5) we must have

$$I_1\dot\varphi = \frac{(a_2 - a_3\cos\vartheta)}{\sin^2\vartheta}, \tag{20.8}$$

$$I_3\dot\psi = a_3 - I_3\dot\varphi\cos\vartheta,$$

whence it follows that $\dot\varphi$ and $\dot\psi$ are constants, if ϑ is constant.

The differential equation for the coordinate ϑ is that of Lagrange

$$\frac{d}{dt}\frac{\partial L}{\partial \dot\vartheta} - \frac{\partial L}{\partial \vartheta} = 0$$

or

$$I_1\ddot\vartheta - (I_1 - I_3)\dot\varphi^2\sin\vartheta\cos\vartheta + I_3\dot\varphi\dot\psi\sin\vartheta = mgl\sin\vartheta. \tag{20.9}$$

We shall assume $I_1 > I_3$. Putting $\vartheta = \text{const}$, $\dot\vartheta = \ddot\vartheta = 0$, only two terms in (20.9) survive to balance the torque $mgl\sin\vartheta$; assuming furthermore that $\dot\psi = 0$ so that the gyroscope has no intrinsic rotation about the Z-axis, we obtain

$$-(I_1 - I_3)\dot\varphi^2\sin\vartheta\cos\vartheta = mgl\sin\vartheta. \tag{20.10}$$

Hence it follows that $\cos\vartheta$ must be negative, so that the axis of rotation points downward. The left-hand side of (20.9) is a kind of centrifugal force, a part of which [the term $(ml\dot\varphi^2\sin\vartheta)(-\cos\vartheta)$] arises from the actual centrifugal force acting on the center of mass. The latter, as is easily visualized, stabilizes the gyroscope only when the axis of rotation points downward.

As a further consequence of (20.9), a stable configuration is only possible when $\dot\varphi \neq 0$; the last term will then stabilize the system for a fairly small $\dot\varphi$, provided that $\dot\psi$ is then fairly large. Conversely, stability results for a large $\dot\varphi$ and small $\dot\psi$ if the term

$$-(I_1 - I_3)\dot\varphi^2 \sin \vartheta \cos \vartheta$$

is positive; this, in turn, implies that $\cos \vartheta$ is negative for $I_1 > I_3$. If, on the other hand, $I_3 > I_1$, a stable configuration may be attained even for positive values of $\cos \vartheta$.

To sum up, stability may be obtained in two ways: through an intrinsic rotation of the gyroscope about its axis of symmetry Z, and by rotation about the fixed axis Z_0. The latter is difficult to realize in practice, and (as already mentioned) may only occur when the gyroscope axis points *downward*, if $I_1 > I_3$.

Hence we shall, primarily, consider the case of fairly large $\dot\psi$ and fairly small φ, in which the main stabilizing factor is the last term of (20.9); it then follows that $\dot\varphi$ and $\dot\psi$ must have the same sign.

Having thus discussed pure precessional motion, we now proceed to the more general case of gyroscope motion with nutation. Instead of the Lagrangian (20.4), we introduce the Hamiltonian $T + V$; by (20.5) or (20.8); this may be written

$$\frac{1}{2I_1}\left[I_1^2\dot\vartheta^2 + \frac{(a_2 - a_3 \cos \vartheta)^2}{\sin^2 \vartheta}\right] + \frac{1}{2I_3} a_3^2 + mgl \cos \vartheta = E. \quad (20.11)$$

Putting
$$\xi = \cos \vartheta, \quad \dot\xi = -\sin \vartheta \dot\vartheta, \quad (20.12)$$

this leads to
$$I_2 \dot\xi = \sqrt{f(\xi)},$$

$$f(\xi) = (a - b\xi)(1 - \xi^2) - (a_2 - a_3\xi)^2, \quad (20.13)$$

$$a = 2I_1(E - a_3^2/2I_3), \quad b = 2I_1 mgl.$$

The exact solution of this equation is

$$t - t_0 = \int_{\xi_0}^{\xi} \frac{I_1 \, d\xi}{(f(\xi))^{1/2}}, \quad (20.14)$$

which is an elliptic integral, since $f(\xi)$ is a third degree expression in ξ.

We need not solve this integral, in order to obtain a qualitative survey of the different types of motion described by (20.13). Motion is possible only when $f(\xi)$ is positive. Since $f(\pm\infty) = \pm\infty$, and $f(-1)$ and $f(1)$ are both negative, the function $f(\xi)$ can have only two roots ξ_1 and ξ_2 in the interval

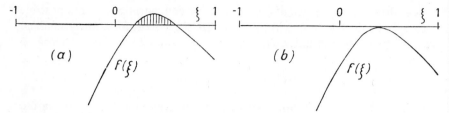

Fig. 3-2. The function $f(\xi)$ in (20.13), describing a precession (a) *with*, and (b) *without*, a nutation.

$-1 < \xi < 1$; between these roots, that is, in the interval $\xi_1 < \xi < \xi_2$, $f(\xi)$ must then be positive (compare Fig. 3-2).

We now assume that a_3 is so large that the intrinsic rotational velocity $\dot{\psi}$ is always positive; by (20.8), $\dot{\varphi}$ will then have the same sign as $a_2 - a_3\xi$. For the upper limit of libration, the maximum value of $\xi = \cos\vartheta$, φ will attain its minimum value. If a_2 is sufficiently small, $\dot{\varphi}$ may be negative at this limit, so that the gyroscope precesses backward at this point. Also, $\dot{\varphi}$ may also be zero at the upper limit of libration, so that the curve drawn by the Z-axis on a sphere (centered in the fixed point constraining the motion of the gyroscope) exhibits a cusp at this point. Finally, for large values of a_2 and a rapid precession, the latter occurs in the forward direction all the time. These three forms of combined precession and nutation are demonstrated in Fig. 3-3.

Lastly, we shall examine more closely the condition for a purely precessional motion corresponding to Fig. 12b, namely, that $f(\xi) = 0$ has a double root. Then this must also be a root of the equation $f'(\xi) = 0$; hence the two equations may be combined to obtain a new equation having the same root. This new equation may be chosen so as to eliminate, for instance, the constant a, which contains the energy.

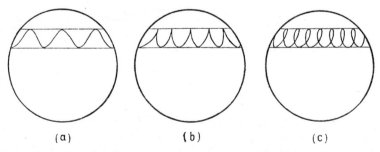

Fig. 3-3. Precession with nutation of spin axis: (a) slow spin with rapid precession; (b) mean spin; and (c) rapid spin with slow precession and retrograde motion at upper libration limit.

Thus utilizing the expression

$$I_1 \dot{\varphi} = \frac{a_2 - a_3 \xi}{1 - \xi^2}, \tag{20.15}$$

which corresponds to (20.8), we may write

$$(I_1 \dot{\varphi})^2 - \frac{a_3}{\xi}(I_1 \dot{\varphi}) - \frac{b}{2\xi} = 0 \tag{20.16}$$

or

$$I_1 \dot{\varphi} = \frac{a_3}{2\xi}\left[1 \pm \left(1 - \frac{2b\xi}{a_3^2}\right)^{\!\!1/2}\right]. \tag{20.16a}$$

This yields one *large* and one *small* value of $\dot{\varphi}$. Taking the *negative* sign in (20.16a), and expanding the square root, we obtain approximately

$$2 I_1 a_3 \dot{\varphi} \approx b = 2 I_1 mgl, \tag{20.17}$$

or by (20.5),

$$I_3 (\dot{\varphi} \cos \vartheta + \dot{\psi}) \dot{\varphi} \approx mgl. \tag{20.17a}$$

This corresponds to a stabilization of the motion by means of the last term in (20.9), that is, by a large intrinsic rotation $\dot{\psi}$. Similarly the *positive* sign in (20.16a), and the large value of $\dot{\varphi}$ relatively to $\dot{\psi}$ thus obtained, corresponds to a stabilization by rotation about the fixed Z_0-axis.

It is immediately clear that the latter is difficult to realize in practice, and hence of small importance, compared with the case of a rapid intrinsic rotation about an axis combined with a slow precession. However, we have included it in our considerations, due to the fact that it shows up somewhat unexpectedly in many treatments of the gyroscope problem; for this reason it frequently attains a certain aura of mystery. The nature of this kind of motion as shown by (9.10) ought now to be clear, and need not concern us further. It is a phenomenon analogous to a circular pendulum, the weight of which travels in a circle; the centrifugal and gravitational forces on the weight are then in mutual equilibrium.

Precession of the Equinoxes

As an example of the rotation of a body around the center of mass, we shall consider the precession and nutation of the earth due to the attraction of the moon and the sun; more specifically, we shall show how the oblateness of the earth makes it possible for the moon and sun to exert a torque on it.

The treatment of the motion of the earth under the influence of this torque may be carried out largely along the same lines as in the previous section; the most troublesome part of the problem is to calculate the torque.

It is, however, clear that this torque must be proportional to $I_3 - I_1$, the difference between the moments of inertia of the earth with respect to the polar axis and an axis passing through the center of mass in the equatorial plane. This difference is equivalent to a configuration of two negative masses ($-m$) on the polar axis at a distance l from the center of the earth, and a mass $2m$ in the center; the magnitudes of m and l are then given by

$$I_3 - I_1 = 2ml^2. \tag{20.18}$$

We now calculate, say, the potential energy of the sun with respect to these (fictitious) mass points. Letting M be the mass of the sun, γ the gravitational constant, and Θ the angle between the polar axis of the earth and the direction toward the sun, we obtain for the mutual potential energy of the sun and the mass points

$$\frac{2mM\gamma l^2}{R^3} P_2(\cos \Theta)$$

or

$$V = CP_2(\cos \Theta), \qquad C = \gamma M(I_3 - I_1)/R^3 \tag{20.19}$$

by (20.18).

The angle Θ changes during the annual revolution of the sun on the celestial sphere, as shown in Fig. 3-4.

Here Z_0 is a fixed axis perpendicular to the (circular) trajectory of the sun in the ecliptic, Z the polar axis, and X an axis passing through the equinoctial points. The angular distance of the sun from the autumnal equinox X is $\Omega t - \varphi$, where Ω is the angular velocity of the sun on the celestial sphere, and φ is the angular distance of X from a fixed position X_0 on the ecliptic.

Simple geometric considerations now lead to

$$\cos \Theta = -\sin \vartheta \sin (\Omega t - \varphi), \tag{20.20}$$

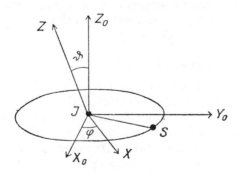

Fig. 3-4. Precession of the equinoxes.

so that, by (20.19)

$$V = \frac{3}{2} C \left(\sin^2 \vartheta \sin^2 (\Omega t - \varphi) - \frac{1}{3} \right) \tag{20.21}$$

or

$$-\frac{\partial V}{\partial \vartheta} = -\frac{3}{2} C \sin \vartheta \cos \vartheta (1 - \cos 2(\Omega t - \varphi)),$$

$$-\frac{\partial V}{\partial \varphi} = \frac{3}{2} C \sin^2 \vartheta \sin 2(\Omega t - \varphi). \tag{20.22}$$

These torques are to balance the kinetic reaction forces

$$\frac{d}{dt} \frac{\partial T}{\partial \dot\vartheta} - \frac{\partial T}{\partial \vartheta} \quad \text{and} \quad \frac{d}{dt} \frac{\partial T}{\partial \dot\varphi} - \frac{\partial T}{\partial \varphi}. \tag{20.23}$$

In the expression (20.4) for the kinetic energy, or L, the dominant term is the last one, which leads to the cross-term $I_3 \dot\varphi \dot\psi \cos \vartheta$. With $\dot\psi = \omega$, this becomes approximately

$$\frac{d}{dt} \frac{\partial T}{\partial \dot\vartheta} - \frac{\partial T}{\partial \vartheta} = I_3 \dot\varphi \omega \sin \vartheta,$$

$$\frac{d}{dt} \frac{\partial T}{\partial \dot\varphi} - \frac{\partial T}{\partial \varphi} = -I_3 \dot\vartheta \omega \sin \vartheta. \tag{20.24}$$

Combining (20.23) and (20.24), the equations of motion then take the form

$$\dot\varphi = -\frac{3}{2} \frac{C}{I_3 \omega} \cos \vartheta (1 - \cos 2(\Omega t - \varphi)),$$

$$\dot\vartheta = -\frac{3}{2} \frac{C}{I_3 \omega} \sin \vartheta \sin 2(\Omega t - \varphi), \tag{20.25}$$

with approximate solutions

$$\varphi - \varphi_0 = -\frac{3}{2} \frac{C}{I_3 \omega} \cos \vartheta \cdot t,$$

$$\vartheta - \vartheta_1 = \frac{3}{2} \frac{C \sin \vartheta}{I_3 \omega \cdot 2\Omega} \cos 2(\Omega t - \varphi). \tag{20.26}$$

The nutational frequency is twice as large as that of the celestial revolution of the sun (or moon), and the nutation has the character of a forced oscillation. Measuring the amplitudes of the nutation, we may find the constant C, or

$$\frac{3}{2} \frac{C}{I_3 \omega} = \frac{3}{2} \frac{\gamma M}{\omega R^3} \frac{I_3 - I_1}{I_3} \tag{20.27}$$

and thus the relative ratio between the two principal moments of inertia I_1 and I_3 of the earth. This, in turn, enables us to calculate $\varphi - \varphi_0$, or the angular velocity $\dot\varphi$ for the equinoctial precession.

The nutational amplitude is about 4", and the period of revolution of the equinoxes about 25,800 years. In the course of this period the polar axis of the earth, or the celestial north pole, describes a circle on the celestial sphere with the radius 23.5°.

The Gyrocompass

A fly wheel with a large moment of inertia I_3 about the axis of rotation and a large intrinsic rotational velocity $\dot\psi$ is supported as shown in Fig. 3-5; it is thus constrained to precess about a vertical axis perpendicular to its own axis of intrinsic rotation. The support is so arranged that the rotational axis of the wheel points horizontally in a longitudinal direction; hence this axis is inclined at the angle α to the polar axis of the earth, where α is the latitude of the geographical location of the wheel.

The rotation of the earth induces in the fly wheel, or gyrocompass, a precession about the north-south axis with the angular velocity $\omega \cos \alpha$, where ω is the rotational velocity of the earth; this corresponds to an angular velocity $\omega \sin \alpha$ about the vertical axis.

Denoting the angle of horizontal westerly deflection by φ, we may write the three components of the rotational precession velocity as

$$\omega_1 = \dot\varphi + \omega \sin \alpha, \qquad \omega_2 = \omega \cos \alpha \sin \varphi,$$
$$\omega_3 = \dot\psi + \omega \cos \alpha \cos \varphi. \tag{20.28}$$

The kinetic energy is approximately

$$T = \tfrac{1}{2} I_1 \omega_1^2 + \tfrac{1}{2} I_3 \omega_3^2, \tag{20.29}$$

if we assume $\dot\psi \gg \omega$ and neglect the term $\tfrac{1}{2} I_1 \omega_2^2$.

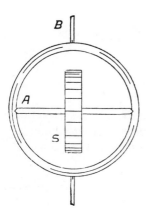

Fig. 3-5. The gyrocompass.

Since no external torque is acting, Lagrange's equation for the angle φ becomes

$$\frac{d}{dt}\frac{\partial T}{\partial \dot{\varphi}} - \frac{\partial T}{\partial \varphi} = I_1\ddot{\varphi} + I_3\omega_3\omega \cos \alpha \sin \varphi = 0, \qquad (20.30)$$

or approximately,

$$\ddot{\varphi} + \omega_0^2\varphi = 0, \qquad \omega_0^2 = \dot{\psi}\Omega \cos \alpha \frac{I_3}{I_1}. \qquad (20.31)$$

Hence the gyrocompass will precess about the meridian with the angular frequency ω_0. It may therefore replace the magnetic compass and has the advantageous property, relative to the latter, of being independent of the magnetic field of the earth.

An interesting feature of the gyrocompass, when employed for navigational purposes on a ship, is that its precession is not quite independent of the course and velocity of the ship. If the latter were able to circumnavigate the earth westward in one day, the gyrocompass would not work (i.e. not precess). If the time of such a circumnavigation were n days, the quasielastic force and the frequency ω_0 would decrease relatively as the ratios $(n-1)/n$ and $(2n-1)/2n$, respectively.

Still more interesting is the fact that a gyrocompass onboard a ship on a northerly or southerly course would exhibit a deviation; the reason for this is that the gyroscope will then precess with angular velocity $\dot{\alpha}$ around the horizontal east-west axis. If the earth were not rotating, the gyrocompass would then point westward for a positive α; because of this rotation, however, it suffers a westerly deviation, corresponding to

$$\alpha \approx \tan \alpha = \frac{\dot{\alpha}}{\Omega \cos \alpha}. \qquad (20.32)$$

For a ship at a latitude of $60°$, with a sailing velocity of 25 km/hr, or 600 km/day, this would result in $\varphi \approx 18°$.

21. INTERNAL MOTION IN SOLID BODIES

Displacement and Strain

In our treatment of the motion of bodies we have so far neglected internal changes of the body. We shall now discuss this topic in some detail, taking advantage of the fact that we have already considered the motion of a body as a whole, by means of the center of mass coordinates and Eulerian angles. Deferring the treatment of fluid motion to a later chapter, we may thus, in our discussion of internal motion, take the body as a whole to be at rest. We thus consider elastically deformable solid bodies, where each infinitesimal mass element may be characterized by a vector $\mathbf{r} = \mathbf{i}x + \mathbf{j}y + \mathbf{k}z$, that is, by three Cartesian coordinates.

If such a body is deformed by static or time-dependent forces, the mass elements will in general be displaced from their equilibrium positions; their new position will then be given by a vector

$$\mathbf{r} + \mathbf{u} = \mathbf{i}(x + u_x) + \mathbf{j}(y + u_y) + \mathbf{k}(z + u_z). \tag{21.1}$$

The vector $\mathbf{u}$ thus denotes what we call the *displacement* of each mass element. It will generally be different for different parts of the body, so that $\mathbf{u}$ becomes a function of the coordinates x, y, z; the latter are then to be considered as *independent* variables, while $\mathbf{u}$ is the *dependent* variable. For nonstatic displacements $\mathbf{u}$ will, in addition, be time-dependent; thus we have, in general,

$$\mathbf{u} = \mathbf{u}(x, y, z, t). \tag{21.2}$$

However, the vector $\mathbf{u}$ is not, in itself, of primary interest. If, for instance, $\mathbf{u}$ is the same for all points of the body, we would only obtain a trivial total displacement of the body as a whole, corresponding to a displacement $\mathbf{u}$ of the center of mass. No internal change would ensure; hence a uniform displacement $\mathbf{u}$ is, as far as internal motion is concerned, equivalent to the displacement $\mathbf{u} = 0$.

Only *nonuniform displacements* bring about internal changes in the body; in this case, a relative displacement of two mass points $\mathbf{r}$ and $\mathbf{r} + d\mathbf{r}$ ensues, given by

$$d\mathbf{u} = \frac{\partial \mathbf{u}}{\partial x} dx + \frac{\partial \mathbf{u}}{\partial y} dy + \frac{\partial \mathbf{u}}{\partial z} dz. \tag{21.3}$$

This relative displacement, which for static forces would correspond to a change of shape, or even of volume, of the body is generally referred to as the *strain*.

The Strain Tensor

In (21.3) there are altogether nine quantities characterizing the relative displacement of the mass points, namely three partial derivatives, with respect to x, y, z, of each of the three displacement components.

The nine quantities

$$\frac{\partial u_x}{\partial x} \quad \frac{\partial u_y}{\partial x} \quad \frac{\partial u_z}{\partial x}$$

$$\frac{\partial u_x}{\partial y} \quad \frac{\partial u_y}{\partial y} \quad \frac{\partial u_z}{\partial y}$$

$$\frac{\partial u_x}{\partial z} \quad \frac{\partial u_y}{\partial z} \quad \frac{\partial u_z}{\partial z}$$

together constitute a tensor called the relative displacement tensor and denoted by grad **u**; the top line then contains the components of the vector $\partial \mathbf{u}/\partial x$, etc. Equation (21-3) may thus be written

$$d\mathbf{u} = d\mathbf{r}\,\text{grad}\,\mathbf{u} = \left(dx\frac{\partial}{\partial x} + dy\frac{\partial}{\partial y} + dz\frac{\partial}{\partial z}\right)\mathbf{u}, \tag{21.4}$$

or

$$du_x = d\mathbf{r}\,\nabla u_x, \quad \text{etc.} \tag{21.5}$$

It should be noted that not all such tensors will result in a deformation or strain. Assume, for instance, that the only nonvanishing components of this tensor are

$$\frac{\partial u_z}{\partial y} = -\frac{\partial u_y}{\partial z} = \Phi_x,$$

$$\frac{\partial u_x}{\partial z} = -\frac{\partial u_z}{\partial x} = \Phi_y, \tag{21.6}$$

$$\frac{\partial u_y}{\partial x} = -\frac{\partial u_x}{\partial y} = \Phi_z,$$

Then, by (21.3), we may write

$$d\mathbf{u} = \mathbf{i}(\Phi_y\,dz - \Phi_z\,dy) + \cdots = \boldsymbol{\Phi} \times d\mathbf{r}, \tag{21.7}$$

regarding Φ_x, Φ_y, Φ_z as the components of a vector $\boldsymbol{\Phi}$. The relative displacement $d\mathbf{u}$ thus corresponds to an infinitesimal rotation about an axis, the positive direction of which is determined by the vector $\boldsymbol{\Phi}$; the angular magnitude of this rotation is $\frac{1}{2}|\text{curl}\,\mathbf{u}|$, the magnitude of $\boldsymbol{\Phi}$.

If $\boldsymbol{\Phi}$ is the same for all parts of the body, we only obtain a net rotation of the latter, and hence no strain results. The point is now that a corresponding change in an infinitesimal element of volume will not deform it, and thus not give rise to any elastic forces. Hence, if the components of the symmetric tensor

$$u_{xx} = \frac{\partial u_x}{\partial x}, \quad u_{xy} = \frac{1}{2}\left(\frac{\partial u_y}{\partial x} + \frac{\partial u_x}{\partial y}\right), \quad \text{etc.} \tag{21.8}$$

are zero, no strain will occur; we therefore call the symmetric tensor u in (21.8) the *strain tensor*.

Introducing, in addition, the antisymmetric tensor

$$u'_{xy} = \frac{1}{2}\left(\frac{\partial u_y}{\partial x} - \frac{\partial u_x}{\partial y}\right) = \frac{1}{2}\text{curl}_z\,\mathbf{u}, \tag{21.9}$$

or

$$u' = \begin{Bmatrix} 0 & \frac{1}{2}\text{curl}_z\,\mathbf{u} & -\frac{1}{2}\text{curl}_y\,\mathbf{u} \\ -\frac{1}{2}\text{curl}_z\,\mathbf{u} & 0 & \frac{1}{2}\text{curl}_x\,\mathbf{u} \\ \frac{1}{2}\text{curl}_y\,\mathbf{u} & -\frac{1}{2}\text{curl}_x\,\mathbf{u} & 0 \end{Bmatrix}, \tag{21.9a}$$

the relative displacement tensor may be written

$$\text{grad } \mathbf{u} = u + u',$$

$$u_{xx} = \frac{\partial u_x}{\partial x}, \quad u_{xy} = \frac{1}{2}\left(\frac{\partial u_y}{\partial x} + \frac{\partial u_x}{\partial y}\right), \cdots$$

$$u'_{xx} = 0, \quad u'_{xy} = \frac{1}{2}\left(\frac{\partial u_y}{\partial x} - \frac{\partial u_x}{\partial y}\right), \cdots \quad (21.10)$$

The important point to note is then that the elastic forces depend only on the symmetric strain tensor u; this will be discussed in more detail later. For irrotational, or curl-free, motions the antisymmetric tensor vanishes; this, however, must not lead us to believe that a rotational, or divergence-free, motion conversely results only in an antisymmetric tensor, and hence no elastic forces. The three conditions (21.6), which we write

$$\frac{\partial u_z}{\partial y} + \frac{\partial u_y}{\partial z} = 0, \quad (21.11)$$

are quite different from the condition for divergence-free motion

$$\text{div } u = \frac{\partial u_x}{\partial x} + \frac{\partial u_y}{\partial y} + \frac{\partial u_z}{\partial z} = 0. \quad (21.12)$$

Thus a deformable solid body will exhibit both *longitudinal* (irrotational) and *transversal* (rotational) waves.

Examples of Strain. One-, Two-, and Three-Dimensional Continua

In addition to ordinary solid bodies, we may also consider simpler continua of one or two dimensions, typically exemplified by stretched vibrating strings and membranes, respectively. Other examples that may be idealized as one-dimensional continua, are, an organ pipe with a fairly narrow column of air, an oscillating metal spring, etc. A wavy liquid surface may also, in a certain sense, be considered as a two-dimensional continuum.

These examples show that vibrations in one- and two-dimensional continua may be both transversal and longitudinal; however, only one of these types of oscillation, and one dependent variable, will occur in each particular case. For the vibrating string or membrane this dependent variable is the deflection u perpendicular to the direction of the string/membrane when at rest. Denoting the coordinates of the string in this direction by x, u becomes a function of x and t,

$$u = u(x, t), \quad (21.13)$$

or similarly, introducing plane Cartesian coordinates for the membrane,

$$u = u(x, y, t), \quad (21.14)$$

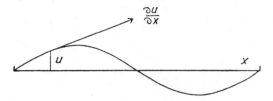

Fig. 3-6. Deformation of a vibrating string.

For the string we have only one component of strain $\partial u/\partial x$, corresponding to (say) $\partial u_z/\partial x$ for a three-dimensional body; for the membrane two components $\partial u/\partial x$ and $\partial u/\partial y$ occur, corresponding to, say, $\partial u_z/\partial x$ and $\partial u_z/\partial y$. (See Fig. 3-6.)

When treating the strain of a solid body, we shall consider each component

$$u_{xx} = \frac{\partial u_x}{\partial x}, \qquad u_{xy} = \frac{1}{2}\left(\frac{\partial u_y}{\partial x} + \frac{\partial u_x}{\partial y}\right), \qquad \text{etc.}$$

separately; this enables us to use plane figures for illustration.

The component u_{xx} is easily interpretable, representing simply an elongation of the body in the x-direction with no transverse contraction. This is not to be confused with the usual stretching of a body by an external force, which is in general accompanied by such a contraction. We here touch upon the question of how the strain depends on external forces; at the moment however, we shall restrict ourselves to the discussion of how the elastic forces depend on the strain.

We may, of course, immediately illustrate simultaneously the effect of the two components of strain u_{xx} and u_{yy}. Assuming that u_{xx} is positive and represents an elongation, while u_{yy} is negative and corresponds to a contraction, we obtain Fig. 3-7.

We consider next the component O_{xy}. If

$$\frac{\partial u_y}{\partial x} = \frac{\partial u_x}{\partial y} \quad \text{or} \quad \frac{\partial u_y}{\partial x} - \frac{\partial u_x}{\partial y} = 0, \qquad (21.15)$$

no rotation of the body results. If the strain is uniform throughout the body, the point x, y will receive a displacement, relatively to the center of mass, with the components

$$\frac{\partial u_x}{\partial y} y, \qquad \frac{\partial u_y}{\partial x} x,$$

as illustrated in Fig. 3-8.

This type of deformation corresponds to an elongation in one diagonal direction, and a contraction of the same magnitude in the other diagonal

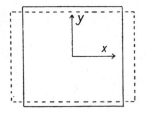

Fig. 3-7. Elongation and contraction of a body along two principal directions: $U_{xx} > 0$, $U_{yy} < 0$.

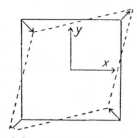

Fig. 3-8. A double shear, corresponding to $U_{xy} = U_{yx}$. This gives rise to equal elongations and contractions in each of the two diagonal directions.

direction; no change of volume then ensues. Hence we may rotate the coordinate axes 45° to obtain a strain tensor with only diagonal components

$$u_{xx} = \frac{\partial u_x}{\partial x} \quad \text{and} \quad u_{yy} = \frac{\partial u_y}{\partial y}$$

of the same magnitude, but of opposite signs; in other words,

$$\frac{\partial u_x}{\partial x} + \frac{\partial u_y}{\partial y} = 0,$$

implying a divergence-free displacement **u** in the xy-directions.

If, in addition to the component u_{xy}, the diagonal terms u_{xx} and u_{yy} are different from zero, a suitable rotation of the coordinate axes would then also result in the vanishing of u_{xy}; however, the angle of rotation would then in general be another angle. Similarly, one may bring about the vanishing of all off-diagonal components of the three-dimensional strain tensor by a suitable rotation of the spatial coordinate axes. This is referred to as transforming the tensor to its principal axes, and any strain may thus be regarded as an elongation (or contraction) along three mutually perpendicular spatial directions.

22. HAMILTON'S PRINCIPLE FOR CONTINUOUS ELASTIC MEDIA

Equations of Motion and Potential Energy for a Vibrating String

The equations of motion for a vibrating string are easily found by considering an element dx of the string, with the mass $m = \rho\, dx$; its possibilities of motion are then restricted to deflections $u(x, t)$ perpendicular to the string.

Dynamics of Continuously Extended Bodies / 317

In a stretched string there is a tension p, equal to the force with which we must hold the string back if it were cut. For an ordinary string this tension is constant, that is, independent of x. Formally, we might of course also consider strings of both variable tensions, $p = p(x)$, and nonuniform mass densities, $\rho = \rho(x)$; in the present considerations, however, we shall primarily discuss homogeneous material bodies where such complications do not occur.

We must now find the net force, acting on the element $p\, dx$. Assuming that the element extends from x to $x + dx$, the tension will have a component

$$p(x)\frac{\partial u(x)}{\partial x}$$

directed toward the equilibrium position of the element, and a corresponding oppositely directed component in $x + dx$. The resultant force, directed away from the equilibrium position, is then

$$dx\, \frac{\partial}{\partial x}\left(p\, \frac{\partial u}{\partial x}\right),$$

so that the equation of motion for the string may be written

$$\rho\, \frac{\partial^2 u}{\partial t^2} = \frac{\partial}{\partial x}\left(p\, \frac{\partial u}{\partial x}\right) \tag{22.1}$$

or

$$\rho\, \frac{\partial^2 u}{\partial t^2} = p\, \frac{\partial^2 u}{\partial x^2} \tag{22.1a}$$

if p is assumed to be constant.

Next we find the potential energy of such a string, relative to that of its equilibrium position, by subjecting the deflection to an arbitrary displacement $\delta u(x)$; the change in potential energy for one single element is then

$$-\delta u\, dx\, \frac{\partial}{\partial x}\left(p\, \frac{\partial u}{\partial x}\right) = -\delta u\, d\!\left(p\, \frac{\partial u}{\partial x}\right), \tag{22.2}$$

and for the whole string

$$\delta V = -\int \delta u\, d\!\left(p\, \frac{\partial u}{\partial x}\right) = \int p\, \frac{\partial u}{\partial x}\, \frac{\partial (\delta x)}{\partial x}\, dx$$

$$= \int p\, \frac{\partial u}{\partial x}\, \delta\, \frac{\partial u}{\partial x}\, dx = \delta \int \frac{1}{2} p\!\left(\frac{\partial u}{\partial x}\right)^{\!2} dx \tag{22.3}$$

if δu is zero at the end points of the string. Thus the potential energy of the string is

$$V = \int \frac{1}{2} p\!\left(\frac{\partial u}{\partial x}\right)^{\!2} dx, \tag{22.4}$$

318 / Mechanics and Statistics

that is, it is, for each element, a quadratic function of the component of strain $\partial u/\partial x$.

The kinetic energy is given by a similar integral

$$T = \int \frac{1}{2} \rho \left(\frac{\partial u}{\partial t}\right)^2 dx. \tag{22.5}$$

It may be shown that the equation of motion (22.1) for the vibrating string is equivalent to the usual formulation of Hamilton's principle

$$\delta \int_{t_0}^{t_1} (T - V) \, dt = 0. \tag{22.6}$$

Actually this integral principle differs from the previously formulated ones only in that it contains a double integral in x and t, while, in point mechanics, we may express T and V by summation over a finite set of coordinates. The proof of the equivalence of (22.1) and (22.6) follows from our considerations (22.2) to (22.4), taken in the reverse order.

Potential Energy and Stress for a Vibrating Membrane

The equation of motion for a vibrating membrane may be deduced similarly to that of a vibrating string. However, we may easily imagine the membrane to be stretched in such a way that the tension is different along two principal directions; by tension we then mean the force we should have to exert, per unit length of a section, in order to hold the membrane in place if it were cut along this section. Denoting the tensions along the two principal directions by p_1 and p_2, the equations of motion may be written

$$\rho \frac{\partial^2 u}{\partial t^2} = p_1 \frac{\partial^2 u}{\partial x^2} + p_2 \frac{\partial^2 u}{\partial y^2}, \tag{22.7}$$

where ρ is the density of mass per unit area.

The potential energy of the membrane is

$$V = \int U \, dx \, dy, \qquad U = \frac{1}{2} p_1 \left(\frac{\partial u}{\partial x}\right)^2 + \frac{1}{2} p_2 \left(\frac{\partial u}{\partial y}\right)^2, \tag{22.8}$$

and the potential energy density U thus once more turns out to be quadratic in the components of strain.

However, this simple form of the potential function only comes about because we have oriented the coordinate axes along the directions of the principal tensions p_1 and p_2; this form is then invariant only if $p_1 = p_2$, that is, if the membrane is dynamically isotropic. In general, for arbitrarily chosen coordinate axes, we would have

$$U = \frac{1}{2} p_{11} \left(\frac{\partial u}{\partial x}\right)^2 + p_{12} \frac{\partial u}{\partial x} \frac{\partial u}{\partial y} + \frac{1}{2} p_{22} \left(\frac{\partial u}{\partial y}\right)^2, \quad (22.9)$$

that is, a quadratic in the components of strain with both square and product terms. As for the vibrating string, the equation of motion for the membrane may be deduced from an integral principle of the form (22.6), with

$$T = \int \frac{1}{2} \rho \left(\frac{\partial u}{\partial t}\right)^2 dx\, dy, \quad (22.10)$$

$$T = \int \Theta\, dx\, dy, \quad \Theta = \frac{1}{2} \rho \left(\frac{\partial u}{\partial t}\right)^2.$$

Introducing

$$L = \Theta - U, \quad (22.11)$$

(22.7) may be written

$$\frac{\partial}{\partial t} \frac{\partial L}{\partial(\partial u/\partial t)} + \frac{\partial}{\partial x} \frac{\partial L}{\partial(\partial u/\partial x)} + \frac{\partial}{\partial y} \frac{\partial L}{\partial(\partial u/\partial y)} = 0. \quad (22.12)$$

Let us examine a little more closely the last two terms of this equation:

$$\frac{\partial U}{\partial(\partial u/\partial x)} = p_{11} \frac{\partial u}{\partial x} + p_{21} \frac{\partial u}{\partial y},$$

$$\frac{\partial U}{\partial(\partial u/\partial y)} = p_{12} \frac{\partial u}{\partial x} + p_{22} \frac{\partial u}{\partial y}, \quad (22.13)$$

by (22.9); the expressions (22.13) then represent the forces acting perpendicularly to the membrane at the edge of a section perpendicular to, respectively, the x- and y-axis. These quantities are usually referred to in the theory of elasticity as *stresses* or *surface forces* (here "line forces" along the section); denoting them by S_1 and S_2, respectively, the force per unit area acting perpendicularly to the membrane becomes

$$F = \frac{\partial S_1}{\partial x} + \frac{\partial S_2}{\partial x} = p_{11} \frac{\partial^2 u}{\partial x^2} + 2p_{12} \frac{\partial^2 u}{\partial x\, \partial y} + p_{22} \frac{\partial^2 u}{\partial y^2}, \quad (22.13a)$$

or, if we orient the axes along the directions of the principal tensions, the same as (22.7).

We note that stresses are not the same as tensions, that is, the forces stretching the membrane or string. Stresses are the forces acting in points of the string or along sectional lines of the membrane (per unit length), which drive the string or membrane back toward the position of equilibrium. These forces are linear functions of the components of strain, while the differential

320 / Mechanics and Statistics

potential energy is correspondingly quadratic. The tensions, for example, p_1 and p_2, occur as coefficients in the expressions for potential energy and stresses, and are often referred to as coefficients of elasticity.

For solid (i.e. three-dimensional) bodies it is natural to expect the potential energy to be expressible as a quadratic function of the components of strain $u_{xx}, u_{xy}, \cdots$, and the stresses or surface forces in such a medium similarly as linear functions of these components. We shall now demonstrate the results to which this assumption leads.

Hamilton's Principle for Three-Dimensional Elastic Bodies

In order to avoid rather complicated expressions, we now introduce a somewhat different notation for the six components of strain of an elastic body, writing

$$u_1 = u_{xx} = \frac{\partial u_x}{\partial x}, \qquad u_2 = u_{yy}, \qquad u_3 = u_{zz}, \qquad (22.14a)$$

but

$$u_4 = 2u_{yz} = \frac{\partial u_z}{\partial y} + \frac{\partial u_y}{\partial z},$$

$$u_5 = 2u_{zx} = \frac{\partial u_x}{\partial z} + \frac{\partial u_z}{\partial x}, \qquad (22.14b)$$

$$u_6 = 2u_{xy} = \frac{\partial u_y}{\partial x} + \frac{\partial u_x}{\partial y}.$$

A factor 2 has been introduced in (22.14b), for two reasons: this is in accordance with common notational practice in elasticity theory, where symbols like

$$z_y + y_z = \frac{\partial u_z}{\partial y} + \frac{\partial u_y}{\partial z} = u_4, \qquad \text{etc.,} \qquad (22.14c)$$

are frequently employed; and second, differential operator identities like

$$\frac{\partial}{\partial\left(\frac{\partial u_z}{\partial y}\right)} = \frac{\partial}{\partial\left(\frac{\partial u_y}{\partial z}\right)} = \frac{\partial}{\partial u_4}, \qquad \text{etc.,} \qquad (22.14d)$$

are simplified (if we had put, say, $u_y = u_{yz}$ a factor $\frac{1}{2}$ would have shown up in (22.14d)). In the theory of elasticity the quantities (22.14b) are often referred to as *components of deformation*.

As previously announced we now assume that the equations of internal motion in an elastic body are deducible from Hamilton's principle

$$\delta \int_{t_0}^{t_1} (T - V)\, dt = 0,$$

$$T = \int \Theta \, dv = \int \frac{1}{2} \rho \dot{\mathbf{u}}^2 \, dv, \qquad V = \int U \, dv, \tag{22.15}$$

where $dv = dx\, dy\, dz$, $\dot{\mathbf{u}} = \partial \mathbf{u}/\partial t$, and where the potential energy per unit volume is a quadratic function of the components of strain; this function we write as

$$U = \tfrac{1}{2} c_{11} u_1{}^2 + c_{12} u_1 u_2 + \cdots,$$

$$= \frac{1}{2} \sum_{i,k} c_{ik} u_i u_k. \tag{22.16}$$

that is, as a quadratic form with $\tfrac{1}{2} \cdot 6F = 21$ coefficients c_{ik}, the so-called coefficients of elasticity.

The deduction of the Euler equations corresponding to the variational principle (22.15) is somewhat involved, and will not be carried through in detail here. It is in principle very simple: as in (22.12), we differentiate the function $\Theta - U$ with respect to $\partial u_x/\partial t$, $\partial u_x/\partial x$, and so forth, and then once more with respect to t, x, and so on. As a final result, we obtain for the force per unit volume of the body

$$F_x = \frac{\partial S_{xx}}{\partial x} + \frac{\partial S_{xy}}{\partial y} + \frac{\partial S_{xz}}{\partial z} = \operatorname{div} \mathbf{S}_x \tag{22.17}$$

$$\mathbf{S}_x = S_{xx} \mathbf{i} + S_{xy} \mathbf{j} + S_{xz} \mathbf{k},$$

where

$$S_{xx} = \frac{\partial U}{\partial u_1}, \qquad S_{xy} = \frac{\partial U}{\partial u_6}, \qquad S_{xz} = \frac{\partial U}{\partial u_5}. \tag{22.18}$$

These elastic forces F_x or $\mathbf{S}_x$ are, by (22.16) and (22.18), linear functions of the components of strain $u_1, \cdots, u_6$; they are thus proportional to the strain, in accordance with the fundamental law of elasticity first proposed by Robert Hooke (1635–1703).

Utilizing the form of F_x in (22.17) and the Gauss theorem, the volume forces hitherto envisaged are easily replaced by surface forces. Thus we obtain

$$\int F_x \, dv = \int \mathbf{n} \mathbf{S}_x \, df, \tag{22.19}$$

by integration over a closed spatial volume; here df denotes an element, with unit normal vector $\mathbf{n}$, of the bounding surface.

Similar considerations for the other components F_y and F_z lead to

$$\int \mathbf{F}\, dv = \int \mathbf{n} S\, df, \tag{22.20}$$

where S is the so-called stress tensor

$$S = \mathbf{i} \cdot \mathbf{S}_x + \mathbf{j} \cdot \mathbf{S}_y + \mathbf{k} \cdot \mathbf{S}_z$$

$$= \begin{pmatrix} S_{xx} & S_{xy} & S_{xz} \\ S_{yx} & S_{yy} & S_{yz} \\ S_{zx} & S_{zy} & S_{zz} \end{pmatrix}. \tag{22.21}$$

This tensor is symmetric, since

$$S_{yx} = \frac{\partial U}{\partial u_6} = S_{xy}. \tag{22.22}$$

We have expressed the elastic forces as surface forces, that is, as contact forces acting from one part of the body on another contiguous part through a (fictitious) bordering surface. The force $\mathbf{F}$ of the previous considerations is the resultant of the force per unit volume, acting on an infinitesimal element of volume; introducing the symbol Div for the divergence of a tensor, this force may then by (22.17) be written as

$$\mathbf{F} = \operatorname{Div} S \tag{22.23}$$

and the equations of internal motion for a solid body correspondingly as

$$\rho \ddot{\mathbf{u}} = \operatorname{Div} S. \tag{22.24}$$

23. STRESSES AND COEFFICIENTS OF ELASTICITY

Surface Forces of Stresses

A common method of introducing the concept of stresses is illustrated in Fig. 3-9, where we have drawn an infinitesimal tetrahedral body, with three of its surfaces lying in the three coordinate planes. We assume that the forces (per unit area) $\mathbf{S}_x$, $\mathbf{S}_y$, $\mathbf{S}_z$ on these surfaces are known and wish to find the corresponding force $\mathbf{S}_n$ on the fourth surface. Denoting the area of the latter by σ and the components of its normal unit vector $\mathbf{n}$ by n_x, n_y, n_z, we find that the other three surfaces then have the areas $n_x\sigma, n_y\sigma, n_z\sigma$.

If we now imagine the body shrinking continually to zero, these surface forces will decrease as the second power of the linear dimensions of the body, while the volume, and hence also the mass, will correspondingly decrease as the third power. For infinitesimal bodies, it is therefore necessary that the

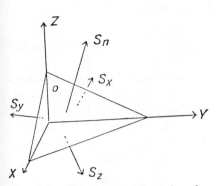

Fig. 3-9. The stress tensor and surface forces.

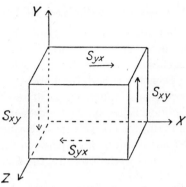

Fig. 3-10. An illustration of the symmetry of surface forces.

four surface forces be in exact equilibrium to prevent the body from receiving an infinite acceleration, by Newton's second law; we must, in other words, have

$$\mathbf{S}_n = n_x \mathbf{S}_x + n_y \mathbf{S}_y + n_z \mathbf{S}_z. \tag{23.1}$$

We now consider the components S_{xy} and S_{yx} of the first two vectors of (23.1), as illustrated in Fig. 3-10. The forces on each of the two surfaces, perpendicular to the x- and y-axis, respectively, will result in a torque proportional to the third power of the linear dimensions of the body. But the moment of inertia decreases correspondingly as the fifth power; hence the two couples of forces must, for infinitesimal bodies, be in exact equilibrium.

As the two torques are given by $S_{xy}\, dv$ and $S_{yx}\, dv$, where dv is the volume of the infinitesimal body, this implies that

$$S_{xy} = S_{yx} \tag{23.2}$$

as in our previous considerations.

We now assume that these surface forces are linear functions of the components of strain $u_1, \cdots, u_6$ in (22.14),

$$S_{xx} = c_{11}u_1 + c_{12}u_2 + c_{13}u_3 + c_{14}u_4 + c_{15}u_5 + c_{16}u_6,$$
$$\vdots \tag{23.3}$$
$$S_{xy} = c_{61}u_1 + c_{62}u_2 + c_{63}u_3 + c_{64}u_4 + c_{65}u_5 + c_{66}u_6.$$

This set of equations may also be deduced from (22.1b) and (22.18). There, however, we obtained symmetric coefficients $c_{12} = c_{21}$, $c_{61} = c_{16}$, and so on, right away; while, in the present section, this assumption had to be justified by other considerations.

Number of Coefficients of Elasticity for a Solid Body

We shall now show how the number of the independent coefficients of elasticity decreases with increasing symmetry of the body, and finally reaches a minimum for isotropic bodies.

Even for a body of known symmetry, there are two theories of elastic coefficients, which we shall refer to as the *multiconstant* and *rariconstant* theories. Equation 23.3 represents the former with the maximum number of independent coefficients of elasticity, 21. The rariconstant theory says that certain relations hold between these coefficients; these are generally termed the Cauchy relations after the French mathematician Augustin Louis Cauchy (1789–1857).

The Cauchy relations are

$$c_{23} = c_{44}, \quad c_{56} = c_{14},$$
$$c_{31} = c'_{55}, \quad c_{64} = c_{25}, \qquad (23.4)$$
$$c_{12} = c_{66}, \quad c_{45} = c_{36},$$

and may be explained as follows.

Let us assume that the forces acting between two particles k and k' are purely central, that is, expressible in terms of a potential $\varphi_{kk'}(r)$, or $\varphi(r)$ for short; we introduce, in addition, the displacements $\mathbf{u}_k$ and $\mathbf{u}_{k'}$, and the relative displacement $\mathbf{u}_{kk'} = \mathbf{u}_k - \mathbf{u}_{k'}$. For small displacements we may take the strain to be homogeneous, that is, $u_{xx}, u_{xy}, \cdots$ or $u_1, u_2, \cdots$ to be constant; thus we may write

$$u_{kk'x} = u_{xx}x_{kk'} + u_{xy}y_{kk'} + u_{xz}z_{kk'}, \quad \text{etc.} \qquad (23.5)$$

We now expand $\varphi_{kk'}(r)$, where $r = |\mathbf{r}_{kk'} + \mathbf{u}_{kk'}|$, in powers of the components of $\mathbf{u}_{kk'}$. This results in a zero-order term representing the energy of cohesion which is not of interest for the present considerations, and a first-order term which must vanish for small displacements from the natural equilibrium state of the body. The term of interest is that of second order; and we note that

$$\frac{\partial^2 \varphi}{\partial x^2} = P(r) + Q(r)x^2,$$
$$\frac{\partial^2 \varphi}{\partial x \, \partial y} = Q(r)xy, \qquad (23.6)$$
$$P(r) = \frac{1}{r}\frac{d\varphi}{dr}, \quad Q(r) = \frac{1}{r}\frac{d}{dr}\left(\frac{1}{r}\frac{d\varphi}{dr}\right).$$

It is easily shown that the term $P(r)$ of the first equation may be neglected in our present considerations, because of the equilibrium conditions already mentioned. Thus the second-order potential energy is

$$U = \frac{1}{2} \sum_{xy} \sum_{x'y'} C_{xyx'y'} u_{xy} u_{x'y'}, \tag{23.7}$$

where

$$C_{xyx'y'} = \sum_{kk'} x_{kk'} y_{kk'} Q_{kk'}(r_{kk'}) x'_{kk'} y'_{kk'}, \tag{23.8}$$

and the summation is to be taken independently over the three coordinates x, y, z.

The quantities $C_{xyx'y'}$ are essentially the same as our coefficients of elasticity $c_{11}, c_{12}, \cdots$. When $x \neq y$ or $x' \neq y'$, we may replace u_{xy} by $2u_{xy}$, or $u_{x'y'}$ by $2u_{x'y'}$ if we sum only once over the pair of indices x, y or x', y'. Clearly, these quantities do not change on permutation of the four coordinates x, y, x', y'; hence

$$C_{yyzz} = C_{yzyz}, \quad C_{zxzy} = C_{xxyz}. \tag{23.9}$$

By (22.14a) and (22.14b), this corresponds to the two first equations (23.4).

Hence we have, apparently, proved the rariconstant theory for particles interacting by means of central forces. However, the Cauchy relations are in general not exactly valid. Classical elasticity theory attempted to explain this by the hypothesis that the particles of a body are, in reality, extended molecules that may subject each other to internal torques; correspondingly modern theory assumes the occurrence of internal displacements of atoms in a crystal lattice. These displacements are then more complex than those envisaged in (23.5), that is, a homogeneous strain throughout the body; as a consequence of this, the symmetry properties exhibited by (23.8), and hence also the Cauchy relations, are no longer valid.

Reduction of the Number of Independent Coefficients of Elasticity for Higher Lattice Symmetries

The results hitherto deduced hold for bodies of arbitrarily low symmetry. The lowest degree of symmetry exhibited by crystalline bodies is, as is well known, that of triclinic crystals. In monoclinic crystals one of the crystal axes is perpendicular to the other two; a simplification of the elastic equations will therefore ensue if one coordinate axis, say the z-axis, is oriented along this particular crystal axis.

For monoclinic crystals we assume that the potential energy is the same for two strains, one of which may be derived from the other by reflection across the xy-plane. Such a "mirror image" of a given strain results by the change of sign in u_{yz} and u_{zx}, that is, u_4 and u_5 in (22.14b). Hence the eight coefficients of elasticity

$$c_{14} \quad c_{24} \quad c_{34} \quad c_{64}$$
$$c_{15} \quad c_{25} \quad c_{35} \quad c_{65}$$

326 / Mechanics and Statistics

all vanish. By the multiconstant theory 13 coefficients then remain, and by the rariconstant theory nine, since two of the Cauchy relations (23.4) now vanish, so that the difference becomes four.

For orthorhombic crystals all axes are mutually perpendicular, so that the potential energy is invariant with respect to reflection across all three coordinate planes. As a consequence of this, not only all of the nine coefficients corresponding to combinations of 1, 2, 3, with 4, 5, 6, vanish, but also the three coefficients c_{56}, c_{64}, c_{45}. Only three of the Cauchy relations then remain, whence the multiconstant and rariconstant theories yield nine, respectively six, coefficients of elasticity.

For uniaxial, tetragonal, trigonal, hexagonal, and rhombohedral crystals, the potential energy must be invariant with respect to rotations of the coordinate system about the axis of symmetry, say the z-axis. Hence we may exchange x with y and y with $\pm x$, thus obtaining

$$c_{11} = c_{22}, \quad c_{13} = c_{23}, \quad c_{44} = c_{55}. \qquad (23.10)$$

Only two Cauchy relations then survive, and the two theories yield, respectively, six and four independent coefficients. (See Fig. 3-11.)

Finally, there is the regular or cubic crystal lattice which has the highest degree of symmetry and occurs in bodies with isotropic or direction-independent elastic properties. Here the potential energy will be invariant with respect to any permutation of coordinate axes, so that we obtain

$$c_{11} = c_{22} = c_{33}, \quad c_{12} = c_{13} = c_{33}, \quad c_{44} = c_{55} = c_{66} \qquad (23.11)$$

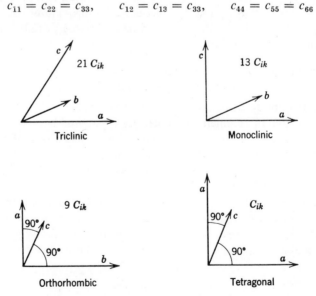

Fig. 3-11. The number of coefficients of elasticity for different crystal symmetries.

and only one Cauchy relation

$$c_{12} = c_{66}. \tag{23.12}$$

Thus we obtain for isotropic bodies only three or two independent coefficients of elasticity by the multiconstant or rariconstant theory, respectively.

A further decrease of the number of coefficients results for uniaxial and regular crystals if we assume that the coefficients are invariant with respect to arbitrary rotations of the coordinate system. We note that the strain

$$\frac{\partial u_x}{\partial x} = -\frac{\partial u_y}{\partial y} = 0, \quad \frac{\partial u_x}{\partial y} = \frac{\partial u_y}{\partial x} = \varepsilon \tag{23.13}$$

is identical with the strain

$$\frac{\partial u_x}{\partial x} = -\frac{\partial u_y}{\partial y} = \varepsilon, \quad \frac{\partial u_x}{\partial y} = \frac{\partial u_y}{\partial x} = 0 \tag{23.14}$$

in a system rotated 45° relatively to the first one. Comparing the expressions for the potential energy in the two cases and assuming that the coefficients of elasticity are invariant, we obtain

$$c_{11} - c_{12} = 2c_{66}, \tag{23.15}$$

so that the Cauchy relation $c_{66} = c_{12}$ leads to

$$c_{12} = \tfrac{1}{3}c_{11}, \tag{23.16}$$

a relation that was deduced by S. Poisson (1781–1840) from his molecular theory of elasticity (the rariconstant theory).

Equation 23.16 does not usually hold, which demonstrates the fact that the rariconstant theory is not generally valid. However, we shall in the following assume that (23.15) is correct; this will be so, at least, for isotropic media composed of disordered microscopic crystals.

24. WAVES IN ISOTROPIC ELASTIC MEDIA

Deduction of the Elastic Force Density

It is evident from the preceding discussion that elastic waves in a more or less anisotropic crystal become very complicated. The treatment of waves in a uniaxial crystal is not especially difficult, but of more practical interest in optics, namely, for optically uniaxial crystals. Here we shall restrict ourselves to the consideration of wave propagation in *isotropic* elastic media.

We may then write

$$S_{xx} = c_{11}\frac{\partial u_z}{\partial x} + c_{12}\left(\frac{\partial u_y}{\partial y} + \frac{\partial u_z}{\partial z}\right)$$

$$= c_{12}\,\text{div }\mathbf{u} + (c_{11} - c_{12})\frac{\partial u_x}{\partial x},$$

$$S_{xy} = c_{66}\left(\frac{\partial u_y}{\partial x} + \frac{\partial u_x}{\partial y}\right) \tag{24.1}$$

$$= (c_{11} - c_{12})\frac{1}{2}\left(\frac{\partial u_y}{\partial x} + \frac{\partial u_x}{\partial y}\right),$$

$$S_{xz} = (c_{11} - c_{12})\frac{1}{2}\left(\frac{\partial u_x}{\partial z} + \frac{\partial u_z}{\partial x}\right),$$

or in a more compact notation,

$$S = Ic_{12}\,\text{div }\mathbf{u} + (c_{11} - c_{12})u, \tag{24.2}$$

where $I = \mathbf{i}\cdot\mathbf{i} + \mathbf{j}\cdot\mathbf{j} + \mathbf{k}\cdot\mathbf{k}$ is the unit tensor.

It is easily seen, using the components of the strain tensor u in (24.1), that

$$\text{Div } u = \nabla\,\text{div }\mathbf{u} - \tfrac{1}{2}\,\text{curl curl }\mathbf{u}, \tag{24.3}$$

and since

$$\text{Div } I\,\text{div }\mathbf{u} = \nabla\,\text{div }\mathbf{u}, \tag{24.4}$$

we obtain the force per unit volume

$$\mathbf{F} = c_{11}\,\nabla\,\text{div }\mathbf{u} - c_{66}\,\text{curl curl }\mathbf{u}. \tag{24.5}$$

The equation of motion for the isotropic medium is therefore

$$\rho\ddot{\mathbf{u}} = c_{11}\,\nabla\,\text{div }\mathbf{u} - c_{66}\,\text{curl curl }\mathbf{u}. \tag{24.6}$$

Transversal and Longitudinal Waves

We shall now decompose the displacement field $\mathbf{u}$ into an irrotational and a rotational part (as is, indeed, possible for any vector field);

$$\mathbf{u} = -\nabla\Phi + \text{curl }\mathbf{A} \tag{24.7}$$

and obtain the two independent equations

$$\rho\,\nabla\ddot{\Phi} = c_{11}\,\nabla\nabla^2\Phi,$$
$$\rho\,\text{curl }\ddot{\mathbf{A}} = -\text{curl}\,(c_{66}\,\text{curl curl }\mathbf{A}). \tag{24.8}$$

By introducing the additional auxiliary requirement

$$\text{div }\mathbf{A} = 0, \tag{24.9}$$

these equations are then satisfied putting

$$\rho\ddot{\Phi} = c_{11} \nabla^2\Phi,$$
$$\rho\ddot{\mathbf{A}} = c_{66} \nabla^2\mathbf{A}. \qquad (24.10)$$

Solutions of (24.10) are plane waves

$$\Phi = f(\mathbf{sr} - v_1 t), \quad v_1 = \left(\frac{c_{11}}{\rho}\right)^{1/2} \qquad (24.11)$$

and

$$\mathbf{A} = \mathbf{A}_0 g(\mathbf{sr} - v_2 t), \quad v_2 = \left(\frac{c_{66}}{\rho}\right)^{1/2}, \qquad (24.12)$$

where $\mathbf{s}$ is a unit vector pointing along the direction of wave propagation, and $\mathbf{A}_0$ a constant vector perpendicular to $\mathbf{s}$,

$$\mathbf{A}_0 \mathbf{s} = 0. \qquad (24.13)$$

It follows from

$$\mathbf{u}_1 = -\nabla\Phi = -\mathbf{s} f'(\mathbf{sr} - v_1 t) \qquad (24.14)$$

and

$$\mathbf{u}_2 = \operatorname{curl} \mathbf{A} = (\mathbf{s} \times \mathbf{A}_0) g'(\mathbf{sr} - v_2 t) \qquad (24.15)$$

that the *irrotational* (curl-free) waves (24.14) are *longitudinal*, and the *rotational* (divergence-free) waves (24.15) *transversal*. The two types of waves have different velocities of propagation, the largest being that of the longitudinal waves; both these velocities are nondispersive, that is, independent of the frequency or wavelength.

Longitudinal Waves in Fluid Media

The coefficient of elasticity $c_{66} \equiv \mu$ is often referred to as the *modulus of rigidity*. We now introduce another quantity k, which we shall call the *modulus of compression*, defined as the reciprocal of the compressibility $\kappa = 1/k$.

The compressibility is defined as the ratio between the relative decrease of volume $-\operatorname{div} \mathbf{u}$ and the increase of pressure dp. For constant normal surface forces p is equal to the stress, with the opposite sign,

$$S_{xx} = S_{yy} = S_{zz} = -p. \qquad (24.16)$$

Assuming that we have an isotropic expansion or compression of a body,

$$\frac{\partial u_x}{\partial x} = \frac{\partial y_y}{\partial y} = \frac{\partial u_z}{\partial z}, \quad \frac{\partial u_y}{\partial z} = 0, \qquad (24.17)$$

we obtain

$$S_{xx} = S = c_{11}\frac{\partial u_x}{\partial x} + c_{12}\left(\frac{\partial u_y}{\partial y} + \frac{\partial u_z}{\partial z}\right) = \frac{1}{3}(c_{11} + 2c_{12})\operatorname{div} \mathbf{u}. \qquad (24.18)$$

Hence by (23.16) we find

$$k = \tfrac{1}{3}(c_{11} + 2c_{12}) = c_{11} - \tfrac{4}{3}\mu, \qquad c_{11} = k + \tfrac{4}{3}\mu, \qquad (24.19)$$

thus expressing the linear coefficient of elasticity c_{11} in terms of the moduli of compression and rigidity k and μ. For fluid bodies, that is, $\mu = 0$, we have $c_{11} = k$.

Applying this to an ideal gas,

$$pV = RT, \qquad -\frac{dV}{V} = \frac{dp}{p},$$

$$\kappa = \frac{1}{p}, \qquad k = p, \qquad (24.20)$$

we obtain the velocity of wave propagation

$$v = \left(\frac{p}{\rho}\right)^{1/2} = \sqrt{pV} = \sqrt{RT}, \qquad (24.21)$$

V being the volume of a gram of the gas and R the gas constant.

At the temperature of $0°C$, and atmospheric pressure, $p = 1.033 \text{ kg/cm}^2 = 1.014 \cdot 10^6 \text{ dyn/cm}^2$, and $\rho = 0.0013 \text{ g/cm}^3$, this yields

$$v = 2.79 \cdot 10^4 \text{ cm/sec} = 279 \text{ m/sec}. \qquad (24.22)$$

This value is too low; the propagation velocity of sound in air is considerably higher. The reason for this is that the air is heated during compressions and cooled during expansions, and these variations of temperature do not have even nearly enough time to smooth out. We have, in other words, an adiabatic compression and expansion; consequently we should have employed the adiabatic equation of state

$$pV^\gamma = \text{const}, \qquad -\frac{dV}{V} = \frac{dp}{\gamma p},$$

$$k = \gamma p, \qquad v = \left(\frac{\gamma p}{\rho}\right)^{1/2}. \qquad (24.23)$$

For air, consisting of diatomic nitrogen and oxygen molecules, the adiabatic constant is

$$\gamma = \tfrac{7}{5} = 1.4. \qquad (24.14)$$

this then leads to the velocity

$$v = 3.30 \cdot 10^4 \text{ cm/sec} = 330 \text{ m/sec} \qquad (24.25)$$

which is very close to that actually measured.

The velocity (24.22) was calculated by Newton, and the correct value (24.25) somewhat later by Laplace.

CHAPTER 4

Hydrodynamics

25. KINEMATICS OF FLUIDS

Instantaneous Coordinates and Identification Coordinates

In our study of the motion of fluids we shall encounter many new and curious features; it is therefore advantageous to consider first, in some detail, the kinematics of fluids.

In the case of strains or oscillations in solid bodies, a particle could be identified simply by its rest coordinates x, y, z or position vector **r**; the instantaneous coordinates were then given by a vector **u**, denoting the displacement from the point of equilibrium.

Such a mode of description is, in general, much too restrictive in the case of fluid motion. We shall—as for solid bodies—use the designation (*fluid*) *particle* for a small *element of volume* of the fluid with its *center of mass* in x, y, z. As a rule, such a particle will be carried about from one position to another in the fluid; hence it is better to denote its instantaneous position by x, y, z. We may, however, also identify the particles by means of their coordinates x_0, y_0, z_0 at a given time, say, $t = t_0$. The coordinates of a particle are, of course, functions of time alone, on the other hand, the coordinates of the totality of particles are, as a whole, functions both of t and the initial coordinates, that is,

$$x = x(x_0, y_0, z_0, t), \quad \text{etc.} \tag{25.1}$$

Velocity Field

The velocity components of a single particle are found as usual by differentiating the coordinates with respect to time,

$$v_x = \frac{dx}{dt}. \tag{25.2}$$

The velocities of all particles taken together, however, constitute a velocity field **v** which depends both on time and the initial coordinates x_0, y_0, z_0

identifying each particle. The most comprehensive way of writing these velocity components is therefore

$$v_x = \left(\frac{\partial x}{\partial t}\right)_{x_0, y_0, z_0 = \text{const}} = \frac{Dx}{Dt}. \tag{25.3}$$

The quantity on the extreme right in (25.3) is called the *individual derivative* of x with respect to time. Other scalar and vectorial quantities also have their individual derivatives, defined analogously to (25.3). In general it will cause no confusion if we write d/dt instead of D/Dt, and refer to these quantities as *absolute* or *total derivatives*.

The velocity components are often, in hydrodynamics, denoted by u, v, w; we shall employ the usual vector symbols v_x, v_y, v_z. The velocity field is generally taken to be a function of the time t and the instantaneous coordinates x, y, z of the particle, that is,

$$\mathbf{v} = \mathbf{v}(x, y, z, t). \tag{25.4}$$

For given values of x, y, z, $\mathbf{v}$ denotes the velocities of different fluid particles, namely, those passing through the fixed point x, y, z in the course of time; at a given value of t, (25.4) provides a "snapshot" of the velocity field at all spatial points.

The first three of the derivatives

$$\frac{\partial \mathbf{v}}{\partial x}, \frac{\partial \mathbf{v}}{\partial y}, \frac{\partial \mathbf{v}}{\partial z}, \frac{\partial \mathbf{v}}{\partial t},$$

denote the local change of the velocity field along the coordinate axes, while the last one describes the change in time of the velocity at the fixed point x, y, z. The latter quantity, $\partial \mathbf{v}/\partial t$, is therefore called the *local* time derivative of the velocity; taking x, y, z, t to be independent variables, it may also be referred to as the *partial* time derivative of the velocity.

Acceleration and Total Derivative in Hydrodynamics

The local derivative of $\mathbf{v}$ is not to be confused with the acceleration of the fluid particle at x, y, z. To find the latter, we must regard $\mathbf{v}$ as a function of t and a set of identification coordinates x_0, y_0, z_0, as previously mentioned, thus expressing the dependence of x, y, z on t as in (25.1). Thus we obtain

$$\dot{\mathbf{v}} = \frac{D\mathbf{v}}{Dt} = \frac{\partial \mathbf{v}}{\partial x}\frac{Dx}{Dt} + \frac{\partial \mathbf{v}}{\partial y}\frac{Dy}{Dt} + \frac{\partial \mathbf{v}}{\partial z}\frac{Dz}{Dt} + \frac{\partial \mathbf{v}}{\partial t}$$

$$= \frac{\partial \mathbf{v}}{\partial x}\dot{x} + \frac{\partial \mathbf{v}}{\partial y}\dot{y} + \frac{\partial \mathbf{v}}{\partial z}\dot{z} + \frac{\partial \mathbf{v}}{\partial t}. \tag{25.5}$$

Hydrodynamics / 333

Corresponding expressions ensue for the individual derivative of any scalar or vectorial field. Regarding the coordinates as functions of time in accordance with (25.1), we may also (as already mentioned) replace the name *individual derivative* by *total* or *absolute derivative*, and use the common notation **v** or $d\mathbf{v}/dt$. This is easily shown by defining the total time derivative of a function $\psi(x, y, z, t)$ by

$$\dot{\psi} = \lim_{\Delta t \to 0} \frac{1}{\Delta t} [\psi(x + \dot{x}\Delta t, y + \dot{y}\Delta t, z + \dot{z}\Delta t, t + \Delta t) - \psi(x, y, z, t)] \quad (25.6)$$

and expanding (25.6) in a Taylor series in Δt; this leads to an equation of the form

$$\dot{\psi} = \dot{x}\frac{\partial \psi}{\partial x} + \dot{y}\frac{\partial \psi}{\partial y} + \dot{z}\frac{\partial \psi}{\partial z} + \frac{\partial \psi}{\partial t} \quad (25.7)$$

for both scalar and vectorial functions ψ. Hence we may define individual or total time differentiation by the operator

$$\frac{D}{Dt} = \frac{d}{dt} = v_x \frac{\partial}{\partial x} + v_y \frac{\partial}{\partial y} + v_z \frac{\partial}{\partial z} + \frac{\partial}{\partial t} = \mathbf{v}\nabla + \frac{\partial}{\partial t}. \quad (25.8)$$

It is often convenient to reformulate the expression (25.5) for the acceleration by means of the relation

$$\mathbf{v} \times (\text{curl } \mathbf{v}) = \mathbf{v} \times (\nabla \times \mathbf{v}) = \tfrac{1}{2} \nabla(\mathbf{v}\mathbf{v}) - (\mathbf{v}\nabla)\mathbf{v}, \quad (25.9)$$

obtained by writing the curl operator as $\nabla\times$ and employing the well-known identity for triple vector products. (The factor $\tfrac{1}{2}$ occurs because the operator ∇ in (25.9) is assumed to act on both factors of the product **vv**, instead of on only the postfactor.)

To aid in the visualization of Eq. (25.9), we calculate one of the components $(\mathbf{v}\nabla)\mathbf{v}$:

$$\begin{aligned}
(\mathbf{v}\nabla)v_x &= v_x \frac{\partial v_x}{\partial x} + v_y \frac{\partial v_x}{\partial y} + v_z \frac{\partial v_x}{\partial z} \\
&= v_x \frac{\partial v_x}{\partial x} + v_y \frac{\partial v_y}{\partial x} + v_z \frac{\partial v_z}{\partial x} \\
&\quad - v_y \left(\frac{\partial v_y}{\partial x} - \frac{\partial v_x}{\partial y}\right) + v_z \left(\frac{\partial v_x}{\partial z} - \frac{\partial v_z}{\partial x}\right) \\
&= \frac{\partial}{\partial x}\left(\frac{1}{2}v^2\right) - (\mathbf{v} \times \text{curl } \mathbf{v})_x.
\end{aligned} \quad (25.10)$$

334 / Mechanics and Statistics

Thus we have, in general

$$\dot{\mathbf{v}} = \frac{\partial \mathbf{v}}{\partial t} + (\mathbf{v}\,\mathbf{\nabla})\mathbf{v} = \frac{\partial \mathbf{v}}{\partial t} + \mathbf{\nabla}\left(\frac{1}{2}v^2\right) - \mathbf{v} \times \mathrm{curl}\,\mathbf{v}. \quad (25.11)$$

The product terms (of the velocity components and their derivatives) which occur in this expression for the acceleration result in serious complications of the hydrodynamic equations of motion. We have already seen the enormous significance that the property of linearity or nonlinearity has for the treatment of ordinary differential equations; this significance is even greater for the simultaneous partial differential equations of hydrodynamics. As a matter of fact the hydrodynamic equations of motion are, in general, not solvable by analytic methods of integration, this being possible only in very special cases.

However, (25.11) may often be replaced, to a very good approximation, by a set of linear equations. This will, for instance, be the case when the velocities are so small that the nonlinear terms of (25.11) may be neglected; we then speak of *slow motions*. For waves and oscillations this may also be expressed by saying that the deflections have *small amplitudes:* more precisely, small relative to the wavelength.

Equation of Continuity for Compressible and Incompressible Fluids

In addition to the purely kinematic complications just discussed, other difficulties appear in hydrodynamics. In particular, the treatment of inhomogeneous fluids is problematic. The inhomogeneity is a "historical" element, that is, it must be specified in some initial state of the fluid. In sea water, for instance, it results mainly from variations of salinity in different locations. Variations of temperature with time also contribute to the inhomogeneity and complicate matters even more, being usually more rapid than secular variations of salinity. As a rule the only way to attack problems involving inhomogeneous fluids is to regard them as incompressible; even then, the problems are generally far from easy to solve.

An important relation for the treatment of density variations in fluids is the hydrodynamic *equation of continuity* for mass, by which the local variation of mass density with time is determined by the mass flux div $(\rho \mathbf{v})$ out of the corresponding element of volume. Thus

$$\mathrm{div}\,\rho\mathbf{v} + \frac{\partial \rho}{\partial t} = 0 \quad (25.12)$$

where ρ is the density. Since

$$\mathrm{div}\,\rho\mathbf{v} = \mathbf{v}\,\mathbf{\nabla}\rho + \rho\,\mathrm{div}\,\mathbf{v} \quad (25.13a)$$

and

$$\frac{d\rho}{dt} = \mathbf{v}\,\nabla\rho + \frac{\partial \rho}{\partial t}, \qquad (25.13b)$$

the equation of continuity may also be written

$$\rho\,\mathrm{div}\,\mathbf{v} + \frac{d\rho}{dt} = 0. \qquad (25.14)$$

For incompressible fluids each particle has a constant mass density, that is, $\dot{\rho} = 0$; hence by (25.14)

$$\mathrm{div}\,\mathbf{v} = 0, \qquad (25.15)$$

the physical meaning of which is easily visualized.

By (25.13b) the local variation of density with time in an incompressible fluid is expressible in terms of the instantaneous density ρ and instantaneous velocity field $\mathbf{v}$, thus

$$\frac{\partial \rho}{\partial t} = -\mathbf{v}\,\nabla\rho. \qquad (25.16)$$

The physical significance of (25.16) is also evident: after an infinitesimal time interval dt the spatial point $\mathbf{r}$ will be occupied by the fluid particle previously situated at the point $\mathbf{r} - \mathbf{v}\,dt$, and the density of this particle is then $\rho - dt\,\mathbf{v}\,\nabla\rho$.

As already shown, the individual derivative of the pressure or any other field quantity satisfies the relation

$$\frac{dp}{dt} = \mathbf{v}\,\nabla p + \frac{\partial p}{\partial t}\,; \qquad (25.17)$$

if this derivative vanishes, equations corresponding to (25.16) ensue

$$\frac{\partial p}{\partial t} = -\mathbf{v}\,\nabla p, \quad \text{etc.} \qquad (25.18)$$

This equation may be used to specify, say, boundary conditions for waves on a free surface of water. In this case the pressure at the actual surface, that is, acting on individual particles at the surface, is constant and equal to the atmospheric pressure. However, the local pressure at the (imaginary) surface of equilibrium will vary, as shown by (25.18).

Conversely, a fluid particle will experience an individual change of pressure

$$\dot{p} = \mathbf{v}\,\nabla p, \qquad (25.19)$$

even in the case of stationary flow and pressure ($\partial p/\partial t = 0$).

Velocity Potential and Laplace's Equation for Incompressible Fluids

The equation of continuity div $\mathbf{v} = 0$ for an incompressible fluid leads to particularly simple expressions if we introduce the so-called velocity potential ψ, defined by

$$\mathbf{v} = -\nabla \psi. \tag{25.20}$$

Taking the divergence of (25.20), we then find that ψ must satisfy Laplace's equation

$$\nabla^2 \psi = 0. \tag{25.21}$$

It is to be noted that this equation is of a purely kinematical nature, expressing the constraint on the motion resulting from the incompressibility of the fluid. (A similar, though not quite so simple, constraint results also for compressible fluids.) This constraint has the consequence that Laplace's kinematic equation assumes, in a certain sense, the role of an equation of motion; for this reason, (25.21) becomes almost as important in certain branches of hydrodynamics as it is in electrostatics.

Kinematic Boundary Conditions

The issue of the so-called boundary conditions is very important in hydrodynamics; according to their physical nature these may be classified as *kinematic* or *dynamic*. An example of the latter is given by (25.18): on a wavy water surface the force of gravity will tend to pull down the wave crests and elevate the troughs. Equation 25.18, which expresses this tendency, does not have the form of an equation of motion but it may be deduced from such equations.

Of primary importance, however, are the kinematic boundary conditions. In general, a fluid must be confined in some way by a container if it is not assumed to be of infinite extension. At the walls of this container (or, in general, at fixed boundary surfaces) the normal component of velocity must necessarily vanish,

$$v_n = 0. \tag{25.22}$$

In terms of a velocity potential this may be equivalently written as

$$\frac{\partial \psi}{\partial n} = 0. \tag{25.23}$$

For idealized frictionless fluids only this boundary condition is usually taken to apply. For real fluids exhibiting internal friction (so-called *viscous* fluids), on the other hand, one must assume that the layers of fluid contiguous to the fixed boundary surface adhere to it, due to cohesive forces between the molecules of the fluid and those of the container walls. This means that the fluid cannot glide along this surface, whence we obtain a new boundary condition $v_t = 0$, where v_t is the tangential velocity. Thus the complete boundary condition for a real fluid at the surface of a fixed solid body is simply

$$\mathbf{v} = 0. \tag{25.24}$$

26. THE HYDRODYNAMIC EQUATIONS OF MOTION

Equations of Motion for Frictionless Fluids

Elastically speaking, fluids may be characterized as having no rigidity or elasticity of shape; the internal stress therefore become very simple. From our previous discussion of the theory of elasticity, it follows that the force per unit volume is equal to $-\nabla p$, where p is the hydrostatic pressure. This may also easily be shown by considering the resultant force on an infinitesimal parallelepiped $dx\,dy\,dz$. The pressure on the faces x and $x + dx$ gives rise to the forces

$$p\,dy\,dz \quad \text{and} \quad \left(p + \frac{\partial p}{\partial x}dx\right)dy\,dz,$$

respectively; the difference, taken along the positive x-axis, is then

$$-\frac{\partial p}{\partial x}dx\,dy\,dz.$$

The external forces acting on the fluid are, in most cases, of gravitational origin; such forces will usually be deducible to a very good approximation from a potential. In the present section we shall consider only forces of this kind and assume that they are proportional to the weight or density of the fluid. Thus we express the external forces by

$$\mathbf{F} = -\rho\,\nabla\Phi \tag{26.1}$$

where Φ is a given potential function. Denoting the resultant force density by $\rho\dot{\mathbf{v}}$, where $\dot{\mathbf{v}}$ is the acceleration, we then obtain the vectorial equation of motion

$$\dot{\mathbf{v}} = -\frac{1}{\rho}\nabla p - \nabla\Phi, \tag{26.2}$$

having divided by ρ. If the fluid is homogeneous, so that $\rho = \rho(p)$ is a function of hydrostatic pressure alone, this equation further reduces to

$$\dot{\mathbf{v}} = -\nabla(P + \Phi),$$
$$P = \int \frac{dp}{\rho(p)}. \tag{26.3}$$

Equation (26.2) is the Eulerian formulation of the hydrodynamic equations of motion. A different formulation of these equations has been given by Lagrange; in this book, however, only the Eulerian form will be considered.

It follows from (25.11) that (26.2) may be written

$$\frac{\partial \mathbf{v}}{\partial t} - \mathbf{v} \times \operatorname{curl} \mathbf{v} = -\frac{1}{\rho} \nabla \left[p - \left(\frac{1}{2} v^2 + \Phi \right) \right]. \qquad (26.4)$$

Equations of Motion for Fluids with Internal Friction

If the internal friction of the fluid is to be taken into account, new unpleasant complications appear, in addition to those arising from the nonlinear terms of (4). As we have seen, the stress tensor of a body is always symmetric with respect to the components of strain. For fluid motion the layers of fluid may suffer a relative displacement, and a permanent strain result, without any accompanying elastic forces. On the other hand, frictional forces arise during the deformation; these are proportional to the *rate of strain*.

A pure rotation of a fluid particle can, of course, not give rise to any frictional forces; thus the internal friction is independent of the rotational velocity

$$\boldsymbol{\omega} = \tfrac{1}{2} \operatorname{curl} \mathbf{v},$$

of, say, the antisymmetric tensor component

$$2\omega_z = \frac{\partial v_y}{\partial x} - \frac{\partial v_x}{\partial y}. \qquad (26.5)$$

Hence we must assume that the instantaneous internal stress due to friction depend only on the symmetric tensor components of the rate of strain. Denoting the stress tensor by S', we may then write

$$S'_{xy} = S'_{yx} = \eta \left(\frac{\partial v_y}{\partial x} + \frac{\partial v_x}{\partial y} \right), \qquad (26.6)$$

where η is called the *coefficient of internal friction* or the *viscosity*.

This definition of the tensor component S'_{xy} is fairly straightforward. However, there remains one difficulty, which is not of great practical importance, but should nevertheless be considered for the sake of formal completeness. This difficulty concerns the augmentation of (26.6) to a complete set of tensorial equations.

In order to obtain a close analogy with the theory of elasticity, we write

$$\eta = c'_{66}. \qquad (26.7)$$

Introducing two additional constants c'_{11} and c'_{12}, we may then write

$$S'_{xx} = c'_{11} \frac{\partial v_x}{\partial x} + c'_{12} \left(\frac{\partial v_y}{\partial y} + \frac{\partial v_z}{\partial z} \right) \qquad (26.8)$$

corresponding to (24.1). As yet, nothing has been assumed about the connection between these coefficients, except that they must satisfy the relation

$$c'_{11} - c'_{12} = c'_{66}, \qquad (26.9)$$

corresponding to (23.15), in order to be invariant with respect to rotations of coordinate axes.

One of these coefficients is still indeterminate. A natural way of determining it is to assume that a purely isotropic expansion or compression of the fluid does not give rise to frictional forces. Thus utilizing (24.18) and (24.19), and introducing the coefficient

$$k' = \tfrac{1}{3}(c'_{11} + 2c'_{12}) \qquad (26.10)$$

analogous to the modulus of compression k, all frictional coefficients are uniquely determined by putting $k' = 0$; we then obtain

$$c'_{11} = \tfrac{4}{3}\eta, \qquad c'_{12} = -\tfrac{2}{3}\eta, \qquad c'_{66} = \eta. \qquad (26.11)$$

The tensor equation for the frictional force may then, analogous to (24.2), be written

$$S' = -\tfrac{2}{3}\eta I \,\text{div}\, \mathbf{v} + 2\eta v_s, \qquad (26.12)$$

where I is the unit tensor and v_s the rate of strain tensor,

$$I = \mathbf{i}\cdot\mathbf{i} + \mathbf{j}\cdot\mathbf{j} + \mathbf{k}\cdot\mathbf{k},$$

$$\begin{aligned}v_s &= \mathbf{i}\cdot\mathbf{i}\frac{\partial v_x}{\partial x} + \mathbf{i}\cdot\mathbf{j}\cdot\frac{1}{2}\!\left(\frac{\partial v_y}{\partial x} + \frac{\partial v_x}{\partial y}\right) + \cdots \\ &\quad + \mathbf{j}\cdot\mathbf{i}\cdot\frac{1}{2}\!\left(\frac{\partial v_y}{\partial x} + \frac{\partial v_x}{\partial y}\right) + \mathbf{j}\cdot\mathbf{j}\frac{\partial v_y}{\partial y} + \cdots \\ &\quad + \cdots.\end{aligned} \qquad (26.13)$$

Since it is the tangential rate of strain of the fluid that primarily produces the friction, and not the compression, it is evident that the term in (26.12) containing div $\mathbf{v}$ is of little practical importance.

The frictional force $\mathbf{F'} = -\text{div}\, S'$ thus appears as the divergence of (26.12); the calculation runs along exactly the same lines as that leading to (24.5) for the elastic force density for an isotropic solid body and will not be repeated here. We obtain as the final result

$$\mathbf{F'} = \tfrac{4}{3}\eta\,\boldsymbol{\nabla}\,\text{div}\,\mathbf{v} - \eta\,\text{curl}\,\text{curl}\,\mathbf{v}. \qquad (26.14)$$

The hydrodynamic equations of motion (26.2) thus receive an additional term,

$$\dot{\mathbf{v}} = -\frac{1}{\rho}\boldsymbol{\nabla} p - \boldsymbol{\nabla}\Phi + \mathbf{F'}. \qquad (26.15)$$

For incompressible fluids the first term in the expression (26.14) for $\mathbf{F'}$ vanishes.

27. HYDRODYNAMICS OF IDEAL FLUIDS

Definition of Ideal Fluids

We shall, in general, define an ideal fluid as one not possessing internal friction. It is also possible to idealize such fluids further by assuming that they are homogeneous, or even incompressible, or by taking the density to be a function of, say, the pressure alone. The crucial step in the transition from real to ideal fluids is, however, the neglect of friction.

The advantage of such an idealization is not that we thereby may separate out certain fluids as "almost ideal," that is, having very small coefficients of internal friction. The difference between ideal and nonideal fluid motion does not arise from the nature of different fluids but from *different forms of motion*. The utility of the concept of an ideal fluid lies in the fact that we often have occasion to study fluid motions where friction plays a subordinate role; this is, for instance, the case for slow or large scale motions. As another example, it follows from (26.14) that the second term of the frictional force vanishes for irrotational motion; the first term is, as already pointed out, generally very small, and even zero for incompressible fluids. For nonideal fluids this favorable state of affairs no longer obtains; the reason for this is, from a mathematical point of view, that the frictional forces at the fixed boundary surfaces confining the fluid give rise to rotational motions (eddies).

Stationary Flow in Ideal Fluids

We first discuss the case of an ideal fluid where the density is a function of the pressure alone, that is,

$$\rho = \rho(p) \quad \text{and} \quad \frac{1}{\rho} \nabla p = \nabla P, \tag{27.1}$$

as in (26.3). The equation of motion (26.4) may then be written

$$\frac{\partial \mathbf{v}}{\partial t} - \mathbf{v} \times \text{curl } \mathbf{v} = -\nabla \left(\frac{1}{2} v^2 + P + \Phi \right). \tag{27.2}$$

We now assume that at a certain instant of time the motion is irrotational everywhere, curl $\mathbf{v} = 0$; the second term on the left-hand side is then a constant (possibly even vanishing) vector. Taking the curl of (27.7) and observing that the operators curl and $\partial/\partial t$ commute, we obtain

$$\text{curl } \frac{\partial \mathbf{v}}{\partial t} = \frac{\partial}{\partial t} \text{curl } \mathbf{v} = 0. \tag{27.3}$$

Thus curl $\mathbf{v}$ does not change with time—it remains constantly zero. This is a consequence of the fact that the right-hand side of (27.2) is expressible as a gradient, or (physically speaking) that the forces present do not generate rotational motions.

Since the motion is irrotational, we may write

$$\mathbf{v} = -\nabla\psi \tag{27.4}$$

which leads to

$$\nabla\left(-\frac{\partial\psi}{\partial t} + \frac{1}{2}v^2 + P + \Phi\right) = 0 \tag{27.4a}$$

or

$$\frac{1}{2}v^2 + P + \Phi = \frac{\partial\psi}{\partial t} + f(t), \tag{27.4b}$$

where $f(t)$ is an arbitrary function of t. For *stationary* motion this reduces to

$$\tfrac{1}{2}v^2 + P + \Phi = C. \tag{27.5}$$

For an incompressible fluid an even simpler equation results, namely,

$$\frac{1}{2}v^2 + \frac{1}{\rho}p + \Phi = C, \tag{27.6}$$

which is known as *Bernoulli's theorem*. It was stated by Daniel Bernoulli (1700–1782) in his book *Hydrodynamica* (1738).

Torricelli's Law

From (27.6) we may also deduce the well-known *Torricelli's law* for the efflux velocity from an orifice, in a reservoir of water with the depth h. We assume that the orifice is narrow relative to the dimensions of the reservoir, so that the velocity of the water at the surface may be set equal to zero; the velocity at the orifice is denoted by v. The (atmospheric) pressure p acting on the water surface will be assumed to be practically equal to that acting at the orifice. Taking the acceleration of gravity g to be constant, the gravitational potential may be written

$$\Phi = \Phi_0 + gz, \tag{27.7}$$

where z is the altitude. Putting $z = 0$ at the orifice and $z = h$ at the water surface, it follows from (27.6) that

$$\frac{1}{2}v^2 + \frac{1}{\rho}p + \Phi_0 = C = \frac{1}{\rho}p + \Phi_0 + gh \tag{27.8}$$

or

$$\tfrac{1}{2}v^2 = gh, \quad v = \sqrt{2gh}. \tag{27.9}$$

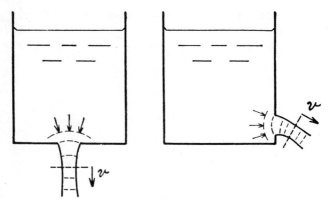

Fig. 4-1. Efflux from a container, with contraction of the jet of water (*vena contracta*).

Thus the efflux velocity of the water is that of a body having fallen freely from the altitude h. (See Fig. 4.1.)

Vena Contracta

An interesting phenomenon connected with such an efflux is the subsequent cross-sectional contraction of the jet of water; the point of the jet at which this contraction is complete is called, characteristically, the *vena contracta*. This phenomenon has no connection with the physical properties of the fluid, such as molecular cohesion, surface forces, and so forth; the latter only appear later, in the well-known partitioning of the jet into separate drops. The appearance of the vena contracta may at least approximately be deduced from purely mechanical principles: the laws of energy and momentum.

The contraction of the jet is, of course, accompanied by an increase of the velocity; the latter is fairly small just inside of the orifice, to which the water flows from all directions, and increases to a (uniform) maximum velocity v at the vena contracta.

The excess of pressure at the bottom of the reservoir, above the atmospheric pressure, is $\rho g h$. If the orifice cross section is f, this excess pressure will impart to the jet a momentum per unit time equal to $\rho g h f$. However, the amount of water issuing through the orifice per unit time is $v f'$, if we denote the minimum jet cross section by f'. Since each unit volume possesses the momentum ρv, this leads to the equation

$$\rho v^2 f' = \rho g h f$$

or

$$\frac{v^2 f'}{f} = gh. \qquad (27.10)$$

Comparing this with (27.9), we find that we must put

$$f' = \tfrac{1}{2}f. \tag{27.11}$$

Hence it follows from the law of conservation of momentum that the excess pressure in the reservoir at the orifice is not sufficient to maintain a jet of the same cross section as the orifice. The jet is thus contracted to approximately half of its original (orifice) cross section. As is natural, the fluid needs some room in which to reorientate its direction of flow from the lateral (toward the orifice) to the rectilinear (at the vena contracta). A more accurate analysis leads, as might be expected, to a factor slightly different from (and in excess of) the $\tfrac{1}{2}$ appearing in (27.11).

It is interesting to note that Bernoulli's theorem also applies, with certain limitations, to rotational fluid motion. Multiplying (27.2) by $\mathbf{v}$, we obtain

$$\frac{\partial}{\partial t}\left(\frac{1}{2}v^2\right) = -(\mathbf{v}\,\mathbf{\nabla})\left(\frac{1}{2}v^2 + P + \Phi\right), \tag{27.12}$$

so that

$$(\mathbf{v}\,\mathbf{\nabla})(\tfrac{1}{2}v^2 + P + \Phi) = 0 \tag{27.13}$$

for stationary flow, where

$$\frac{\partial}{\partial t} = 0 \quad \text{and} \quad \frac{d}{dt} = (\mathbf{v}\,\mathbf{\nabla}). \tag{27.14}$$

Hence it follows from (27.13) that

$$\tfrac{1}{2}v^2 + P + \Phi = \text{const} \tag{27.15}$$

along a streamline, $(\mathbf{v}\,\nabla) = 0$, or (equivalently) for an individual fluid particle $d/dt = 0$.

28. ROTATIONAL FLOW IN IDEAL FLUIDS

We have hitherto considered irrotational fluid motions only. When we now commence our study of rotational flow, new difficulties arise. In restricting our treatment to ideal fluids we remove ourselves, in a certain sense, further from reality, since it is indeed for rotational motion that frictional effects primarily manifest themselves. However, a discussion of such motions in ideal (frictionless) fluids is necessary as a first step toward the understanding of more complex fluid motions. In the present considerations we shall not proceed any further in this direction; the influence of friction, and in particular the damping of the rotational motion, constitutes a problem requiring separate treatment.

Kelvin's Circulation Theorem

Let us consider a loop, that is, a closed curve, which either remains at rest or follows the motion of the fluid. We denote the positional vector of points on the loop, and vectorial line elements along it, by **s** and *d***s**, respectively. For a loop at rest we may regard **s** as a function of one parameter ξ only, $\mathbf{s} = \mathbf{s}(\xi)$; while, for a loop participating in the fluid motion, **s** will also depend on time, $\mathbf{s} = \mathbf{s}(\xi, t)$.

The *circulation* round the loop is defined as the line integral

$$C = \oint \mathbf{v} \, d\mathbf{s} \tag{28.1}$$

taken once around the closed loop. Assuming that the density of our ideal fluid is a function of the pressure, we may then, as in (26.3), write

$$\frac{d\mathbf{v}}{dt} = -\nabla(P + \Phi). \tag{28.2}$$

For a loop fixed in space, this leads to

$$\frac{dC}{dt} = \oint \frac{d\mathbf{v}}{dt} d\mathbf{s} = -\int d(P + \Phi) = 0, \tag{28.3}$$

that is, the circulation along a fixed loop does not change with time.

If, on the other hand, the loop moves with the fluid, an additional term

$$\mathbf{v} \frac{d}{dt} d\mathbf{s} = \mathbf{v} \frac{d}{dt}\left(\frac{\partial \mathbf{s}}{\partial \xi} d\xi\right) = \mathbf{v} \frac{\partial^2 \mathbf{s}}{\partial \xi \, \partial t} d\xi = \mathbf{v} d\left(\frac{\partial \mathbf{s}}{\partial t}\right)$$

appears in the integrand. According to our assumptions, however, $d\mathbf{s}/dt = \mathbf{v}$, so that the additional term is $\mathbf{v} \cdot d\mathbf{v} = d(\frac{1}{2}v^2)$. Thus the circulation theorem for a moving loop becomes

$$\frac{dC}{dt} = \oint d\left(\frac{1}{2}v^2 - P - \Phi\right) = 0, \tag{28.4}$$

since the integrand is a complete differential.

This circulation theorem is due to Lord Kelvin. A special case has already been employed: namely, that when the circulation along an arbitrary closed curve is zero at one time, that is, curl $\mathbf{v} = 0$, then this state of affairs will be maintained.

An extension of this result is the circulation laws of V. Bjerknes for inhomogeneous fluids. Here we obtain

$$\frac{dC}{dt} = -\oint \frac{1}{\rho} \nabla p \, d\mathbf{s} = -\oint \frac{dp}{\rho}. \tag{28.5}$$

Defining the circulation of momentum for an incompressible fluid as

$$C' = \oint \rho \mathbf{v}\, ds, \tag{28.6}$$

we then easily find

$$\frac{dC'}{dt} = -\oint \rho \frac{d\mathbf{v}}{dt} ds = -\oint \rho \nabla \Phi\, ds = \oint \mathbf{F}\, ds, \tag{28.7}$$

since $dp/dt = 0$.

Vortices and Vortex Filaments

Applying Stokes' theorem

$$\oint \mathbf{v}\, ds = \int \operatorname{curl} \mathbf{v}\, d\mathbf{f} \tag{28.8}$$

to an infinitesimal circle perpendicular to the vector curl $\mathbf{v}$, we obtain

$$v \cdot 2\pi r = |\operatorname{curl} \mathbf{v}|\, \pi r^2, \qquad \omega = \frac{v}{r} = \frac{1}{2} |\operatorname{curl} \mathbf{v}|. \tag{28.9}$$

This equation is valid if we neglect translational motion; v is the rotational velocity of the fluid around the center of the circle. The angular velocity ω, regarded as a vector, may then be written

$$\boldsymbol{\omega} = \tfrac{1}{2} \operatorname{curl} \mathbf{v}, \tag{28.10}$$

and is usually referred to as the *vorticity* of the motion. This is analogous to the equation $\boldsymbol{\Omega} = \tfrac{1}{2} \operatorname{curl} \mathbf{u}$ for the angular rotation of an element of volume in the course of an arbitrary displacement $\mathbf{u}$.

The simultaneous differential equations

$$\frac{dx}{\omega_x} = \frac{dy}{\omega_y} = \frac{dz}{\omega_z} \tag{28.11}$$

now define, at any given moment t, curves pointing along the vector $\boldsymbol{\omega}$ or curl $\mathbf{v}$; these curves are called vortex lines.

If vortex lines are drawn through a loop they enclose a *vortex* or *vortex filament*. The circulation along the loop is equal to the surface integral of curl $\mathbf{v}$. If the vortex is enclosed by another loop, this surface integral must be the same, as an immediate consequence of the application of the Gauss theorem to a closed surface consisting of the two end surfaces of the vortex filament and its enclosing cylindrical surface. Since div curl $\mathbf{v} = 0$, the volume integral over this filament section vanishes, as does the surface integral over the cylindrical boundary surface, since curl $\mathbf{v}$ is directed along the latter. The integrals over the end surfaces must therefore be equal; thus the strength of the vortex, defined as the circulation along any loop enclosing it, is an invariant quantity. Hence a vortex filament cannot begin or end at an interior point of the fluid, but only at the fluid boundaries; the only other possibility is that it constitutes, in itself, a closed loop. This is a law of general validity, that is, not restricted to ideal fluids.

Helmholtz' Law of Conservation of Vortex Strength

We now deduce the famous law, due to Helmholtz (1821–1894), of the *conservation of vortex strength*. Let us consider two surfaces S_1 and S_2, which are generated by vortex lines and consequently intersect each other along such a line ω. The circulation along a loop arbitrarily drawn in any of these surfaces then vanishes; if the surfaces participate in the motion of the fluid, this circulation will remain zero. Hence the two surfaces must continue to be generated by vortex lines; for, if such lines were to cut through the surface in any region, then the circulation around a loop in the surface enclosing this region and the corresponding surface integral of curl $\mathbf{v}$ would not vanish. But the intersection line ω between S_1 and S_2 consists continually of the same particles; any "string" of particles that, at one time, constitute a vortex line will continue to do so. In other words: *vortex lines are conserved during the motion of the fluid*. This applies even to the lines bounding a vortex filament; thus *the filament will permanently consist of the same fluid particles*. Finally a loop enclosing the vortex filament, and moving with the fluid, will continually enclose this filament; hence, since the circulation C is constant by Kelvin's theorem, *the vortex strength is conserved*, that is, does not change with time. (See Fig. 4-2.)

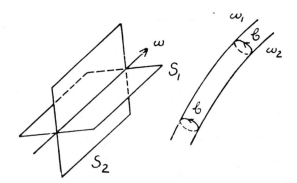

Fig. 4-2

Deduction of the Velocity Field when the Vorticity is Known

For rotational motion, say, of an incompressible fluid, we may always write

$$\mathbf{v} = \text{curl } \mathbf{A}, \quad \text{div } \mathbf{A} = 0. \tag{28.12}$$

Assuming the vorticity to be known, and stationary,

$$\text{curl } \mathbf{v} = \mathbf{c}(x, y, z) = 2\boldsymbol{\omega}, \tag{28.13}$$

we obtain the vectorial Poisson equation

$$\nabla^2 \mathbf{A} = -\mathbf{c}. \tag{28.14}$$

by means of (28.12) and (28.13). For the solution of this equation, we refer the reader to the chapters on vector analysis and electrostatics in Parts 1 and 3, respectively. The solution may be written

$$\mathbf{A}(x, y, z) = \frac{1}{4\pi} \int \frac{\mathbf{c}(\xi, \eta, \zeta)}{r} d\xi \, d\eta \, d\zeta, \tag{28.15}$$

$$r^2 = (x - \xi)^2 + (y - \eta)^2 + (z - \zeta)^2,$$

where r is the distance between the point x, y, z at which the vector potential $\mathbf{A}$ is to be calculated and the auxiliary point of integration ξ, η, ζ.

The Rectilinear Homogeneous Vortex Filament

In a small spatial region containing a bounded vortex filament, the latter may be approximated by the rectilinear filament

$$\mathbf{A} = \mathbf{k}\Psi, \quad \text{div } \mathbf{A} = \frac{\partial \Psi}{\partial z} = 0, \quad \Psi = \Psi(x, y). \tag{28.16}$$

Assuming that the filament cross section is circular and that the vorticity magnitude $|\text{curl } \mathbf{v}| = 2\omega$ is constant inside the filament and vanishes outside it, we then obtain

$$\frac{\partial^2 \Psi}{\partial x^2} + \frac{\partial^2 \Psi}{\partial y^2} = \begin{cases} -2\omega, & r < a, \\ 0, & r > a, \end{cases} \tag{28.17}$$

by (27.14) and (27.16).

Here we have $r = \sqrt{x^2 + y^2}$. Introducing polar coordinates r, φ in the xy-plane, and assuming that the vortex filament has axial symmetry, $\Psi = \Psi(r)$, (28.17) then becomes

$$\frac{1}{r} \frac{d}{dr} r \frac{d\Psi}{dr} = \begin{cases} -2\omega, \\ 0. \end{cases} \tag{28.18}$$

The solutions of (28.18) are subject to certain requirements. Both $\Psi'(r)$ and $\Psi''(r)$ must be continuous at $r = a$, and finite both for $r = 0$ and $r \to \infty$. The homogeneous equation has the solutions $\Psi = $ const, and $\Psi = \log r$; the former is then chosen so as to ensure the continuity of Ψ at $r = a$. Since $\Psi = -\tfrac{1}{2}\omega r^2$ is a solution of the inhomogeneous equation for $r < a$, it may then be added to the homogeneous solution without any extra constant of integration. The logarithmic solution is not admissible for $r < a$, since both Ψ and Ψ' then become infinite when $r \to 0$. Hence the solutions that satisfy the requirement of continuity may be written

$$\Psi = -\frac{1}{2}\omega r^2, \quad r < a, \quad \Psi = -\omega a^2\left(\frac{1}{2} + \log\frac{r}{a}\right), \quad r > a. \quad (28.19)$$

Utilizing

$$v_x = \frac{\partial \Psi}{\partial y}, \quad v_y = -\frac{\partial \Psi}{\partial x}, \quad (28.20)$$

we find

$$v_r = \frac{x}{r}v_x + \frac{y}{r}v_y = \frac{1}{r}\left(x\frac{\partial \Psi}{\partial y} - y\frac{\partial \Psi}{\partial x}\right) = \frac{1}{r}\frac{\partial \Psi}{\partial \varphi} = 0,$$

$$v_\varphi = \frac{x}{r}v_y - \frac{y}{r}v_x = -\frac{1}{r}\left(x\frac{\partial \Psi}{\partial x} + y\frac{\partial \Psi}{\partial y}\right) = -\frac{\partial \Psi}{\partial r}. \quad (28.21)$$

Thus the radial velocity vanishes, as would be expected. The tangential velocity is

$$v_\varphi = \omega r, \quad r < a, \quad v_\varphi = \frac{\omega a^2}{r}, \quad r > a. \quad (28.22)$$

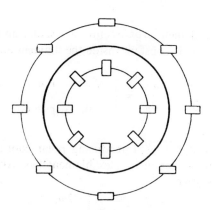

Fig. 4-3

Hence we see, in this case, that the vortex filament itself rotates as a solid body with angular velocity ω. Outside the filament the rotational velocity decreases with increasing radius. The difference between rotational and irrotational fluid motion may be visualized by illustrating the motion of a small element of fluid inside and outside the vortex, as shown in Fig. 4-3.

The result (28.22) may, of course, be arrived at in a much simpler way by purely intuitive geometric considerations; however, our discussion demonstrates the general mathematical methods that must be used for more complicated problems.

29. FLOW IN AN INCOMPRESSIBLE IDEAL FLUID

Two-Dimensional Flow and Conformal Mappings

For an incompressible fluid we must always have

$$\text{div } \mathbf{v} = 0. \tag{29.1}$$

Furthermore, if the motion is assumed to be irrotational, we may write

$$\mathbf{v} = -\nabla \psi, \tag{29.2}$$

implying that the velocity potential must satisfy Laplace's equation

$$\nabla^2 \psi = 0. \tag{29.3}$$

This is another example of the fact, which has already been mentioned previously, that for certain hydrodynamic problems the solution of Laplace's equation is more important than the discussion of the equations of motion. For the presently considered potential flow of an incompressible fluid, the problem is solved once we have solved Laplace's equation with appropriate boundary conditions, which in this case primarily is $\partial \psi / \partial n = 0$ on the fixed surfaces containing the fluid. In addition we may, of course, have to allow for specific dynamic boundary conditions.

This apparently complete elimination of the equations of motion may be explained as follows. Fluid particles which change their velocity are subject to an acceleration, due to forces, in particular pressure forces. If the fluid is assumed to have a very small compressibility, any accumulation of fluid in a region would be accompanied by enormous pressure gradients, which would very rapidly "smooth out" the fluid density, or at least counteract any further compression. If the fluid is perfectly incompressible, such accumulations are impossible; this means that the necessary counterforces automatically adjust themselves, so to speak, in such a way as to effect the accelerations required (to prevent compression). The situation is similar to that of a material body or mass point moving along a frictionless surface; in both cases, constraints lead to a decrease in the number of degrees of freedom.

Several problems of hydrodynamics may profitably be studied in two dimensions. Consider, for instance, the airfoils or wings of an airplane; here the flow in a plane perpendicular to the wing is of primary interest, while lateral flows at the wing tips are not so important.

We shall, therefore, first discuss two-dimensional flow in an xy-plane, and assume that the velocity in the z-direction is zero. By (29.2) and (29.3) we then have

$$v_x = -\frac{\partial \psi}{\partial x}, \quad v_y = -\frac{\partial \psi}{\partial y},$$

$$\frac{\partial^2 \psi}{\partial x^2} + \frac{\partial^2 \psi}{\partial y^2} = 0.$$

(29.4)

An interesting method of solving (29.4) results from the theory of functions. We introduce a complex variable

$$z = x + iy,$$
(29.5)

not to be confused with the third Cartesian coordinate z, which we have suppressed. It then follows from the theory of analytic functions,

$$f(z) = P(x, y) + iQ(x, y),$$
(29.6)

that the components P and Q both satisfy Laplace's equation (29.4).

We now transform to arbitrary new coordinates ξ, η which define another complex variable

$$\zeta = \xi + i\eta.$$
(29.7)

The transformation is said to be a *conformal mapping* if it may be written

$$x + iy = z = g(\zeta) = p(\xi, \eta) + iq(\xi, \eta),$$
(29.8)

where $g(\zeta)$ is an analytic function. This means that if two curves intersect in the xy-plane at a certain angle, the transformed curves will intersect at the same angle in the $\xi\eta$-plane; the mapping thus *preserves angles*.

In this way we may easily obtain new solutions from a given solution of the problem of flow by choosing different analytic functions $g(\zeta)$ for the transformation. Using (29.6), for instance, which is a solution of Laplace's equation, we find a new solution

$$F(\zeta) = f[g(\zeta)],$$

valid in a $\xi\eta$-plane, or

$$F(z) = f[g(z)]$$
(29.9)

in the actual xy-plane.

Thus we may, to mention one particular example, obtain solutions valid for obstacles of elliptic cross section, once we know the solution for an

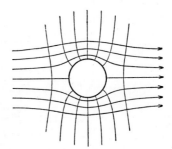

Fig. 4-4. Flow around a cylinder or sphere.

obstacle of circular cross section. Even problems of apparently much greater complexity may be solved by this general method; we shall not, however, pursue such considerations any further.

Instead we now consider a particular example of a solution of (29.9); as illustrated in Fig. 4-4, we envisage a fixed cylindrical obstacle, with circular cross section, placed perpendicularly to the direction of flow. The unperturbed flow—that which would obtain if the obstacle were absent—is assumed to be linear and homogeneous, with the constant velocity v_0 in the x-direction; it may then be expressed by the function

$$\psi_0 = -v_0 x \qquad (29.10)$$

which is immediately seen to satisfy Laplace's equation. We may also write

$$\psi_0 = -v_0 r \cos \varphi \qquad (29.11)$$

in plane polar coordinates. However, we may easily find another solution that is more suitable for the present purpose, namely,

$$\psi_1 = \frac{c}{r} \cos \varphi = c \frac{x}{r^2}; \qquad (29.12)$$

by inspection this function is immediately seen to satisfy Laplace's equation, either in the form (29.4) or in polar coordinates

$$\frac{1}{r} \frac{\partial}{\partial r}\left(r \frac{\partial \psi}{\partial r}\right) + \frac{1}{r^2} \frac{\partial^2 \psi}{\partial \varphi^2} = 0. \qquad (29.13)$$

The formulation (29.12) does not apply at $r = 0$; this, however, is of no consequence, since we are only interested in the flow outside the fixed obstacle, that is, for $r \geq a$, where a is the radius of the cylinder. We easily check that the boundary condition $\partial \psi/\partial n = \partial \psi/\partial r = 0$ for $r = a$ is met by setting $c = -a^2$, and adding the solutions (29.11) and (29.12), thus obtaining

$$\psi = \psi_0 + \psi_1 = -v_0\left(r + \frac{a^2}{r}\right) \cos \varphi \qquad (29.14)$$

or

$$v_r = -\frac{\partial \psi}{\partial r} = v_0\left(1 - \frac{a^2}{r^2}\right)\cos\varphi,$$

$$v_\varphi = -\frac{1}{r}\frac{\partial \psi}{\partial \varphi} = -v_0\left(1 + \frac{a^2}{r^2}\right)\sin\varphi,$$

$$v_x = v_r \cos\varphi - v_\varphi \sin\varphi = v_0\left(1 - \frac{a^2}{r^2}\cos 2\varphi\right), \quad (29.15)$$

$$v_y = v_r \sin\varphi + v_\varphi \cos\varphi = -v_0 \frac{a^2}{r^2}\sin 2\varphi.$$

By (29.15) the velocity v_x in the direction of flow has its minimum value just in front of and behind the cylinder at $\varphi = 0$ and $\varphi = \pi$, and its maximum value just at the top and bottom of the cylinder, at $\varphi = \pm\pi/2$. Compare Fig. 4-4.

Flow around a Sphere

We shall now consider a problem related to the preceding one, the flow of an incompressible ideal fluid around a sphere, that is, a three-dimensional problem. Let us, for the moment, assume that it is the sphere that moves through the fluid with constant velocity. This problem attracted a considerable amount of attention at one time, due to the fact that the sphere is found to meet with no resistance from the fluid. However, this is the case only if the motion of the sphere is uniform. If the velocity is to be increased, we must perform work: not only that corresponding to the increase in the kinetic energy of the sphere, but also some additional work, which then appears as increased kinetic energy of the fluid. This may be said to manifest itself as an apparent increase in the mass of the sphere; this increase then, rather remarkably, turns out to be exactly equal to half the mass of the displaced fluid.

To begin with, we assume that the situation is analogous to that of the cylindrical obstacle previously treated: that the sphere is at rest, and the fluid executing an unperturbed flow in the z-direction, with a constant velocity $v_z = v_0$; the x- and y-axes are then perpendicular to the direction of flow, and the radius of sphere is a, so that we only need solutions for $r > a$.

The solutions of Laplace's equation which we are seeking must be symmetrical with respect to the z-axis. The first is

$$\psi_0 = -v_0 z$$

which determines the unperturbed flow

$$\mathbf{v}_0 = -\nabla\psi_0 = \mathbf{k}v_0;$$

the second is

$$v_1 = -\frac{Cz}{r^3} = C\frac{\partial}{\partial z}\left(\frac{1}{r}\right).$$

That this is a solution follows immediately from the equation $\nabla^2(1/r) = 0$, since $(\partial/\partial z)\nabla^2 = \nabla^2(\partial/\partial z)$.

The correct choice of the constant C then leads to

$$\psi = -v_0 z\left(1 + \frac{a^3}{2r^3}\right) = -v_0\left(r + \frac{a^3}{2r^2}\right)\cos\vartheta, \qquad (29.16)$$

where ϑ is the polar angle, that is, the angle between the z-axis and the radius vector from the center of the sphere. The velocity components v_r and v_ϑ are found by applying the operators $-(\partial/\partial r)$ and $-(\partial/r\,\partial\vartheta)$ to ψ; thus

$$v_r = v_0\left(1 - \frac{a^3}{r^3}\right)\cos\vartheta, \qquad v_\vartheta = -v_0\left(1 + \frac{a^3}{2r^3}\right)\sin\vartheta. \qquad (29.17)$$

We now imagine the fluid to be at rest and the sphere to be moving through it with the velocity v_0 in the negative z-direction. The first terms in (29.17), due to the unperturbed flow, then vanish, and we obtain

$$v_r = -v_0\frac{a^3}{r^3}\cos\vartheta, \qquad v_\vartheta = -v_0\frac{a^3}{2r^3}\sin\vartheta,$$

$$v^2 = v_r^2 + v_\vartheta^2 = \frac{v_0^2 a^6}{4r^6}(4\cos^2\vartheta + \sin^2\vartheta). \qquad (29.18)$$

In order to find the fluid pressure and its effect on the motion, we utilize Bernoulli's theorem which leads to a decrease of pressure:

$$\Delta p = -\tfrac{1}{2}\rho v^2. \qquad (29.19)$$

Since, by (29.18), v^2 is the same at corresponding points on the front and rear sides of the sphere, that is, for ϑ and $\pi - \vartheta$, the resultant force on the sphere due to pressure vanishes. Thus the sphere meets with no resistance.

We now calculate the kinetic energy of the fluid. Since $\overline{\cos^2\vartheta} = \tfrac{1}{3}$, $\overline{\sin^2\vartheta} = \tfrac{2}{3}$, and $d\tau = 4\pi r^2\,dr$ is the volume of a spherical shell of thickness dr, we obtain

$$T = \rho v_0^2 \pi a^3 \cdot \int_a^\infty \frac{a^3\,dr}{r^4} = \rho v_0^2 \frac{\pi}{3} a^3$$

or

$$T = \frac{1}{2}M'v_0^2, \qquad M' = \frac{1}{2}M, \qquad M = \frac{4\pi}{3}a^3\rho, \qquad (29.20)$$

where M is the mass of the fluid displaced by the sphere.

This kinetic energy of the fluid flow, due to the sphere, increases as the second power of the velocity; it, as well as the kinetic energy of the sphere, comes into play when the velocity of the sphere is to be increased. The net result is an apparent increase $M' = \frac{1}{2}M$ in the mass of the sphere. In reality, of course, the increased resistance arises because the pressure field is no longer spherically symmetric for accelerated motions of the sphere.

30. WAVES IN IDEAL FLUIDS

Gravity Waves

We shall now study some simple forms of wave motion, beginning with waves that are driven by gravity, and hence rather naturally referred to as gravity waves.

We assume that the fluid possesses a natural state of rest, and that all velocities are so small that their squares may be neglected in comparison with the velocities themselves. In this way we may replace the acceleration $\dot{\mathbf{v}}$ by the partial derivative $\partial \mathbf{v}/\partial t$. Further, we shall describe the motion by means of the displacement $\mathbf{u}$ of the fluid particles from their equilibrium position, similar to the previous treatment of elastic continua. The equations of motion may then be written

$$\frac{\partial^2 \mathbf{u}}{\partial t^2} = -\frac{1}{\rho} \nabla(p + P) - \nabla \Phi, \tag{30.1}$$

where P is the hydrostatic pressure obtaining in the state of rest, p the pressure *increment*, and $\partial \mathbf{u}/\partial t = \mathbf{v}$.

In the equilibrium state we have $\mathbf{u} = 0$, $p = 0$, that is,

$$\nabla P = -\rho \nabla \Phi. \tag{30.2}$$

If the fluid is homogeneous and the density uniform, this leads to

$$P = \rho(\Phi_0 - \Phi). \tag{30.3}$$

For ordinary fields of gravity, we may put $\Phi = gz$, where z is the altitude; neglecting the pressure at the fluid surface $z = z_0$, we then obtain

$$P = \rho g(z_0 - z). \tag{30.4}$$

If the density is nonuniform, say, increasing with depth so as to result in stable layers of fluid, the pressure is

$$P = g \int_z^{z_0} \rho(z) \, dz. \tag{30.5}$$

Stabilizing forces will then arise, driving the fluid particles back to the layers where they belong, if they (the particles) are raised or lowered by the wave motion.

These possibilities are, of course, of considerable interest; we shall not, however, consider them here, but regard the fluid as homogeneous. The forces driving the waves to be studied arise chiefly from the raising or lowering of the surface of the fluid: because of the discontinuity of the mass distribution across this surface, these forces are much greater than a relative raising or lowering of fluid particles of slightly different densities. Hence we utilize (30.1) and (30.2) to obtain

$$\frac{\partial^2 \mathbf{u}}{\partial t^2} = -\frac{1}{\rho} \nabla p. \qquad (30.6)$$

For the pressure, on the other hand, we must have

$$\frac{dp}{dt} = \frac{\partial p}{\partial t} + \mathbf{v} \nabla (P + p). \qquad (30.7)$$

From the preceding considerations we may here neglect $\mathbf{v} \cdot \nabla p$ as small of the second order. At the surface the pressure is constant, that is, $dp/dt = 0$; (30.7) then becomes

$$\frac{\partial p}{\partial t} = -\mathbf{v} \nabla P \qquad (30.8)$$

which may be integrated to give

$$p = -\mathbf{u} \nabla P = \rho g u_z. \qquad (30.9)$$

The physical significance of this result is easily visualized. The increment of pressure at the fluid layer of equilibrium altitude $z = z_0$ (and, approximately, also in the immediately contiguous layers below) equals the product of the displacement u_z of the actual surface and the specific weight ρg of the fluid.

We now assume that the displacement is expressible in terms of a scalar function ψ, which we might call the "displacement potential," in analogy to the previously considered velocity potential; thus we put

$$\mathbf{u} = -\nabla \psi \qquad (30.10)$$

and substitute this into (30.6) to obtain

$$p = \rho \frac{\partial^2 \psi}{\partial t^2}. \qquad (30.11)$$

The combination of (30.9) and (30.10) then leads to

$$g \frac{\partial \psi}{\partial z} = -\frac{\partial^2 \psi}{\partial t^2}. \qquad (30.12)$$

This relation, deduced from the equations of motion and hence of a *dynamic* character, here appears as a *boundary condition for the free surface;* the fundamental equation to be solved for the fluid motion, however, is of a purely *kinematic* nature, namely,

$$\nabla^2 \psi = 0 \qquad (30.13)$$

which is obtained by (30.10) by putting div $\mathbf{u} = 0$. In addition to the dynamic boundary condition (30.12), we also have the kinematic condition $u_n = 0$, or

$$\frac{\partial \psi}{\partial n} = 0 \qquad (30.14)$$

at the fixed boundary surfaces containing the fluid.

The only containing surface that we shall take into account is a plane bottom at the constant depth h; furthermore we shall discuss only propagating waves. Putting $z = 0$ for the surface and $z = -h$ for the bottom, we consider plane waves propagating in the x-direction, that is, waves that do not depend on the y-coordinate. Equation 30.13 may then be written

$$\frac{\partial^2 \psi}{\partial x^2} + \frac{\partial^2 \psi}{\partial z^2} = 0 \qquad (30.15)$$

and solved by separation of the variables.

Let v be the velocity of wave propagation and λ the wavelength, and define $k = 2\pi/\lambda$. The function

$$\psi = -\left(\frac{a}{k}\right) \sin k(x - vt) \cosh k(z + h) \qquad (30.16)$$

is then a solution of (30.15), satisfying the boundary condition (30.14) at the bottom, $z = -h$. The dynamic condition (30.12) for the surface $z = 0$ results in the relation

$$gk \sinh kh = k^2 v^2 \cosh kh$$

or

$$v^2 = \left(\frac{g}{k}\right) \tanh kh \qquad (30.17)$$

between k, h, and v.

We shall discuss two extreme cases: $kh \ll 1$ or $h \ll \lambda$, and $kh \gg 1$, $h \gg \lambda$. The former corresponds to, for instance, tidal waves. We may here replace the hyperbolic tangent by its argument, obtaining

$$v^2 = gh. \qquad (30.18)$$

The latter case, where the depth is very large compared to the wavelength, corresponds to ordinary ocean waves in deep water. Here we may put

$\tan kh = 1$, which leads to

$$v^2 = \frac{g}{k} = \frac{g\lambda}{2\pi}. \tag{30.19}$$

Hence the velocity of propagation is independent of the depth and proportional to the wavelength. In shallow water, (30.19) will no longer be valid; by (30.17) v must eventually begin to decrease with the depth h. This is in accordance with the fact that sea waves slow down as they near the shoreline.

Waves Due to Molecular Forces

The small ripples often observed on a water surface are not caused by gravity but by the surface tension. Denoting this by T, we find that a bending of the surface results in a force per unit area in, say, the z-direction, equal to

$$T \frac{\partial^2 u_z}{\partial x^2},$$

similar to the case of an oscillating string or membrane. This results in a corresponding decrease of pressure immediately below the surface, so that we obtain the new boundary condition

$$p = \rho g u_z - T \frac{\partial^2 u_z}{\partial x^2} \tag{30.20}$$

instead of (30.9). By (30.11) this may be written

$$-\frac{\partial^2 \psi}{\partial t^2} = g \frac{\partial \psi}{\partial z} - \left(\frac{T}{\rho}\right) \frac{\partial^3 \psi}{\partial z \, \partial x^2}. \tag{30.21}$$

The substitution of the solution (30.16) then gives us

$$v^2 = \left(\frac{g}{k} + \frac{T}{\rho} k\right) \tanh kh, \tag{30.22}$$

or, in deep water, $h \to \infty$,

$$v^2 = \frac{g}{k} + \frac{kT}{\rho}. \tag{30.22a}$$

It follows from this that the velocity of propagation for surface waves does not vanish as $\lambda \to 0$, $k \to \infty$. There exists a minimum velocity for $k^2 = g\rho/T$, given by

$$v^2 = 2\left(\frac{gT}{\rho}\right)^{1/2}. \tag{30.23}$$

The wind velocity must not be less than this minimum in order to produce waves. For greater wind velocities two types of waves result: one with the approximate wavelength

$$\lambda = \frac{2\pi v^2}{g} \qquad (30.24)$$

which for increasing v approaches a pure gravity wave; and another with the approximate wavelength

$$\lambda = \frac{2\pi T}{\rho v^2}. \qquad (30.25)$$

It is this latter type that appears as ripples on a water surface, moving more or less independently of the greater wind or gravity waves.

Pressure and Sound Waves

We shall now consider an irrotational wave motion in a *compressible* ideal fluid; as in the preceding section, we neglect flows and regard the motion as small oscillations about an equilibrium state, in which the equilibrium pressure P is assumed to balance the external forces. The equation of motion for the fluid is then the same as in (30.6),

$$\ddot{\mathbf{u}} = -\frac{1}{\rho}\nabla p, \qquad (30.26)$$

writing, for brevity, $\ddot{\mathbf{u}}$ instead of $\partial^2 \mathbf{u}/\partial t^2$. For the pressure we have the equation

$$p = -k \operatorname{div} \mathbf{u}, \qquad (30.27)$$

where k is the modulus of compression. Hence, by substitution in (30.26), we obtain

$$\ddot{\mathbf{u}} = \frac{k}{\rho}\nabla \operatorname{div} \mathbf{u}, \qquad (30.28)$$

or, utilizing the identity

$$\nabla \operatorname{div} \mathbf{u} - \nabla^2 \mathbf{u} = \operatorname{curl} \operatorname{curl} \mathbf{u} = 0, \qquad (30.29)$$

the well-known form of the wave equation

$$\ddot{\mathbf{u}} = \frac{k}{\rho}\nabla^2 \mathbf{u}. \qquad (30.30)$$

Differentiating with respect to time, we find the same wave equation for the velocity,

$$\ddot{\mathbf{v}} = \frac{k}{\rho}\nabla^2 \mathbf{v}, \qquad (30.31)$$

and even for the velocity potential ψ,

$$\ddot{\psi} = \frac{k}{\rho}\nabla^2 \psi, \qquad \mathbf{v} = -\nabla\psi. \qquad (30.32)$$

For monochromatic waves, where the time dependence is given by a harmonic factor $\sin \omega t$, $\cos \omega t$ or $e^{i\omega t}$, we arrive at the equation

$$\nabla^2 \psi + k'^2 \psi = 0, \qquad k' = \frac{2\pi}{\lambda} = \omega \left(\frac{\rho}{k}\right)^{1/2}. \tag{30.33}$$

for the amplitude of the wave. This is generally to be solved in conjunction with boundary conditions such as $v_n = -(\partial \psi/\partial n) = 0$ on the surfaces of solid bodies, which is often a very difficult problem.

For fluids of infinite spatial extension we obtain solutions in the form of plane longitudinal waves, for example,

$$\psi = a \cos \omega \left(t - \frac{x}{c}\right),$$

$$\mathbf{v} = -\mathbf{i}\left(\frac{a\omega}{c}\right) \sin \omega \left(t - \frac{x}{c}\right), \tag{30.34}$$

where

$$c = \frac{\omega}{k'} = \left(\frac{k}{\rho}\right)^{1/2} \tag{30.35}$$

is the phase velocity of the pressure waves.

We may also find solutions in the form of spherical waves

$$\psi = \frac{a}{r} \cos \omega \left(t - \frac{r}{c}\right),$$

$$\mathbf{v} = \left\{\frac{a\omega}{cr} \sin \omega \left(t - \frac{r}{c}\right) + \frac{a}{r^2} \cos \omega \left(t - \frac{r}{c}\right)\right\} \frac{\mathbf{r}}{r}. \tag{30.36}$$

These waves propagate radially outward, from a point in space, and for great distances from this point the first term of $\mathbf{v}$ in (30.36) dominates, as in the similar case of electromagnetic waves.

The solution (30.36) ceases to apply at $r = 0$. It may, however, be imagined to be generated by a pulsating sphere in the fluid. We may even find more complicated spherical waves, such as

$$\Psi = \frac{\partial \psi}{\partial z} = \frac{a\omega z}{cr^2} \sin \omega \left(t - \frac{r}{c}\right) - \frac{az}{r^3} \sin \omega \left(t - \frac{r}{c}\right), \tag{30.37}$$

which is obtained from ψ in (30.36) by differentiation with respect to z; since $\partial/\partial t$ commutes both with ∇^2 and $\partial^2/\partial t^2$, the derivative of ψ is therefore also a solution. Equation 30.37 may be referred to as a dipole wave; it will correspond approximately to the wave radiated by a sphere executing small harmonic oscillations in the z-direction.

It is evident from these considerations that the radiation of pressure or sound waves from a center of disturbance may be a very complicated phenomenon. The same applies for problems involving reflection and absorption; we shall not discuss such phenomena here, but instead give a brief account of the transport of energy by a pressure wave.

Energy and Energy Transport in a Pressure Wave

We introduce the symbols T and V for the kinetic and potential energy density of a wave. For the former, we may immediately write $T = \frac{1}{2}\rho v^2$. The latter is found by noting that a pressure increment dp will bring about a relative volume decrement dp/k; multiplying this by p and integrating from 0 to p, we obtain an increase $V = p^2/2k$ of the potential energy density.

This may also be found by a somewhat different argument. Let us consider an arbitrary closed surface with the normal velocity component v_n pointing outwards. The pressure acting on the inside of the surface then performs, for each surface element, a work per unit time equal to pv_n; hence the energy flow may be written

$$\mathbf{I} = p\mathbf{v}. \tag{30.38}$$

Thus the energy output per unit volume and time is given by

$$-\text{div } \mathbf{I} = -\mathbf{v}\,\boldsymbol{\nabla} p - p\,\text{div }\mathbf{v} = \rho\mathbf{v}\dot{\mathbf{v}} + \frac{1}{k}p\dot{p} = \dot{T} + \dot{V} \tag{30.39}$$

with the previously given expressions for T and V. We have then, in accordance with (30.27), $\dot{p} = -k \text{ div } \mathbf{v}$.

31. FLUIDS WITH INTERNAL FRICTION

General Considerations

The mathematical treatment of real (i.e. viscous) fluids is far more complicated than that of ideal fluids. We shall therefore simplify our discussion as much as possible, for example, by considering only incompressible homogeneous fluids, where div $\mathbf{v} = 0$ and $\rho = $ const; the essential features of viscous fluids are not much affected by these idealizations.

As already mentioned, the equations of motion for viscous fluids are

$$\rho \frac{d\mathbf{v}}{dt} = -\boldsymbol{\nabla} p - \rho\,\boldsymbol{\nabla}\Phi + \eta(2\,\boldsymbol{\nabla}\,\text{div }\mathbf{v} - \text{curl curl }\mathbf{v}). \tag{31.1}$$

We now assume that the motion is so slow that the individual derivative may be replaced, to a good approximation, by the partial derivative. The simplifications introduced above then lead to

$$\rho \frac{\partial \mathbf{v}}{\partial t} = -\nabla(p + \rho\Phi) + \eta \nabla^2 \mathbf{v} \tag{31.2}$$

on reformulation of the expression curl curl **v**.

The problem is now to find, in a reasonably simple way, solutions of this equation. It is immediately evident that we may obtain a separate Laplace equation for the scalar function $p + \rho\Phi$,

$$\nabla^2(p + \rho\Phi) = 0 \tag{31.3}$$

by taking the divergence of (31.2).

Equation (31.3) is, as a rule, not difficult to solve, provided that the surfaces containing the fluid are relatively simple. However, unique solutions $p + \rho\Phi$ result only if, in addition, the velocity field **v** is calculated and adjusted to the given boundary conditions.

The reason for this complication is that these boundary conditions differ from those valid for ideal fluids. In both cases the condition $\mathbf{v}_n = 0$ at the surface of solid bodies must be satisfied; for viscous fluids, however, the additional boundary condition $\mathbf{v}_t = 0$ must be introduced as well. Due to cohesive forces between the solid body and the fluid molecules, the latter will adhere to the body; hence the motion is characterized by layers of fluid sliding along each other. The force arising as a result of this sliding is determined by the *coefficient of internal friction* or *viscosity* appearing in (31.1).

We have previously demonstrated that in ideal fluids the circulation along a loop, which is fixed or participating in the fluid motion, will be constant; this implies that if the motion is irrotational at one time, this state of affairs will be maintained. For viscous fluids this is no longer the case, because of the frictional term $\eta \nabla^2 \mathbf{v}$. In other words, we may no longer express the velocity in terms of a potential, $\mathbf{v} = -\nabla\psi$; if we try to do this, we will find that the resulting solutions cannot satisfy the boundary conditions.

Since div **v** = 0 for incompressible fluids, it is natural to put

$$\mathbf{v} = \text{curl } \mathbf{A}. \tag{31.4}$$

Furthermore **A** may be assumed to have a vanishing divergence; hence it may, in many cases, be convenient to write

$$\mathbf{A} = \text{curl } \mathbf{B}, \quad \mathbf{v} = \text{curl curl } \mathbf{B} = \nabla \text{ div } \mathbf{B} - \nabla^2 \mathbf{B}. \tag{31.5}$$

We may now simplify matters by noting that $\nabla(p + \rho\Phi)$ is a Laplacian vector, by (31.3); thus it is expressible as the curl of a vector potential $\mathbf{A}_1$, or as the curl curl of a vector $\mathbf{B}_1$,

$$-\nabla(p + \rho\Phi) = \text{curl } \mathbf{A}_1 = \text{curl curl } \mathbf{B}_1 = \nabla \text{ div } \mathbf{B}_1 - \nabla^2 \mathbf{B}_1. \tag{31.6}$$

Substituting (31.4), (31.5), and (31.6) in (31.2), we find that the latter will be satisfied by solutions of the equations

$$\rho \frac{\partial \mathbf{A}}{\partial t} = \eta \, \nabla^2 \mathbf{A} + \mathbf{A}_1 \tag{31.7}$$

or

$$\rho \frac{\partial \mathbf{B}}{\partial t} = \eta \, \nabla^2 \mathbf{B} + \mathbf{B}_1. \tag{31.8}$$

We shall not discuss here the general question of whether the solutions of (31.7) and (31.8) constitute a complete set of solutions of (31.2), but shall only consider this problem for some particular cases.

A further simplification results from the assumption that the flow is stationary. Apart from (31.3), which we may then regard as solved, we obtain the equations

$$\eta \, \nabla^2 \mathbf{v} - \nabla(p + \rho \Phi) = 0, \tag{31.9}$$

$$\eta \, \nabla^2 \mathbf{A} + \mathbf{A}_1 = 0, \tag{31.10}$$

$$\eta \, \nabla^2 \mathbf{B} + \mathbf{B}_1 = 0, \tag{31.11}$$

according to which of the above described procedures we choose to follow. We shall now illustrate each of these three procedures by examples.

A Fluid between Two Rotating Cylinders

Let us consider the situation illustrated in Fig. 4-5, that is, a viscous fluid between a fixed outer cylinder with radius b and a rotating inner cylinder with radius a and angular velocity ω_0. The potential Φ may be set equal to zero, since gravitational forces are of no essential importance for this problem. We observe that the pressure must be independent of the azimuthal angle φ, due to the rotational symmetry; nor can it depend on r, except for sufficiently large centrifugal forces (of the order of v^2), which we exclude from consideration by assuming that the motion is slow. Hence we may take the

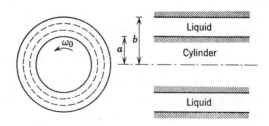

Fig. 4-5. Fluid between two rotating cylinders, to determine the viscosity.

pressure to be constant, that is, $\nabla p = 0$ or $\nabla(p + p\Phi) = 0$. We must therefore solve the equation $\nabla^2 \mathbf{v} = 0$, and then adjust the solutions to the boundary conditions $v_\varphi = \omega_0 a$ for $r = a$ and $v_\varphi = 0$ for $r = b$.

Let us assume that $v_r = 0$, $v_x = -v \sin \varphi$, $v_y = v \cos \varphi$. The equation $\nabla^2 \mathbf{v} = 0$ may thus be written

$$\frac{1}{r}\frac{d}{dr} r \frac{dv}{dr} - \frac{v}{r^2} = 0, \qquad r^2 = x^2 + y^2, \tag{31.12}$$

with the solutions

$$v = Ar + \frac{B}{r}. \tag{31.13}$$

The adjustment of the constants then leads to

$$v = \omega_0 a \frac{b/r - r/b}{b/a - a/b} = \omega r. \tag{31.14}$$

The drag of the rotating cylinder on the fluid may be measured by the torque that must be exerted on the cylinder in order to maintain its rotation. We must then note that when the rotation is uniform, the fluid exhibits no sliding motions. Hence the force per unit area is not $\eta \, dv/dr$, as it would be in a Cartesian coordinate system, but

$$\left(\eta r \frac{d\omega}{dr}\right)_{r=a} = -\eta \omega_0 \cdot \frac{2b^2}{b^2 - a^2}. \tag{31.15}$$

The total torque necessary to counteract these forces is then obtained by multiplication by the surface $2\pi a l$ and the radius a, thus

$$M = 4\pi \eta \omega_0 l \frac{a^2 b^2}{b^2 - a^2}. \tag{31.16}$$

If we wish to make use of (31.10), we must note that the equation

$$-\nabla(p + \rho\Phi) = \operatorname{curl} \mathbf{A}_1 = 0 \tag{31.17}$$

has both $\mathbf{A}_1 = 0$ and $\mathbf{A}_1 = \text{const}$ as solutions. For the present rotational motion it seems natural to take the vector potential to be directed along the z-axis; we must then assume that this applies also for $\mathbf{A}$. Hence we put $\mathbf{A}_1 = C\mathbf{k}$ and $\mathbf{A} = \psi(x, y)\mathbf{k}$, since the velocity field must be independent of z. Thus we obtain

$$\eta \nabla^2 \psi + c = 0. \tag{31.18}$$

As a consequence of the rotational symmetry ψ must be a function of $r = \sqrt{x^2 + y^2}$ only, whence (31.18) may be written

$$\frac{1}{r}\frac{d}{dr} r \frac{d\psi}{dr} = -\frac{c}{\eta} \tag{31.19}$$

with the solutions

$$\psi = -\frac{C}{4\eta} r^2 + D + E \log r. \qquad (31.20)$$

This yields the azimuthal velocity

$$v_\varphi = v = -\frac{d\psi}{dr} = \frac{C}{2\eta} r - \frac{E}{r} \qquad (31.21)$$

in accordance with (31.13).

Poiseuille's Law. Flow of a Fluid through a Pipe

We shall consider an example that is perhaps simpler than the preceding one, namely, the well-known law of Poiseuille describing the flow through a circular pipe.

We choose the axis of the pipe as the z-axis. Furthermore, we put $\Phi = 0$, since the difference in the gravitational field at the two ends of a vertical or sloping pipe may be assumed equivalent to a difference in pressure at these ends. (See Fig. 4-6.)

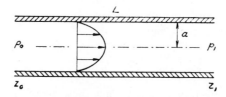

Fig. 4-6. Flow through a pipe.

It seems natural to expect the pressure to decrease uniformly along the axial direction, independently of the distance from the axis; as a solution of the equation $\nabla^2 p = 0$ we may then take

$$p = p_0 - \frac{p_0 - p_1}{L} z, \qquad \nabla p = -\frac{p_0 - p_1}{L} \mathbf{k}. \qquad (31.22)$$

This may now be substituted in (31.9) to yield

$$\nabla^2 \mathbf{v} = -\frac{p_0 - p_1}{\eta L} \mathbf{k}. \qquad (31.23)$$

We must assume **v** to be directed along the z-axis, so that the layers of fluid slide on each other; such a state of affairs is often called a *laminar flow*. Since div $\mathbf{v} = \partial v_z/\partial z = 0$, **v** must be independent of z, so that $\mathbf{v} = v(x, y)\mathbf{k}$. From the relation $\nabla^2(Ax^2 + 2Bxy + Cy^2) = 2(A + C)$, it then follows that

$$v = D - (Ax^2 + 2Bxy + Cy^2) \qquad (31.24)$$

Hydrodynamics | 365

is an admissible solution. The condition $v = 0$ may thus easily be met on an elliptical cylinder surface. For a pipe of circular cross section and radius a, we must put $A = C$ and $B = 0$; the adjustment of A and D then yields

$$v = \frac{p_0 - p}{4\eta L}(a^2 - r^2). \tag{31.25}$$

The amount of fluid passing through a cross section of the pipe per unit time is therefore given by

$$Q = 2\pi \int_0^a vr\, dr = \frac{\pi(p_0 - p)a^4}{8\eta L}, \tag{31.26}$$

which is *Poiseuille's law*.

Motion of a Sphere through a Viscous Fluid

Most expositions of this problem are rather complicated; we shall, however, give it a fairly brief and simple treatment by means of (31.11).

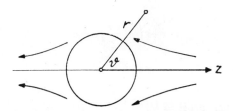

Fig. 4-7. Sphere in a viscous fluid.

We employ the usual artifice of regarding the sphere as fixed in space, and the fluid as having an unperturbed velocity $-v_0\mathbf{k}$ in the opposite direction to that of the envisaged uniform motion of the sphere through the fluid. In this way we obtain an artificial *stationary* field of flow, as demonstrated in Fig. 4-7. Here we have drawn the z-axis, the radius vector r, the polar angle ϑ, and a few streamlines. As a consequence of the rotational symmetry about the z-axis, all scalar quantities must be independent of the azimuthal angle. We once again put $\Phi = 0$. As solutions of the equation $\nabla^2 p = 0$ we may choose between quantities such as

$$P_0 = 1, \quad \frac{1}{r}P_0 = \frac{1}{r}, \quad rP_1 = r\cos\vartheta, \quad \frac{1}{r^2}P_1 = \frac{\cos\vartheta}{r^2}, \cdots.$$

The first three are, for various reasons, without interest; we therefore try the last one, describing a perturbation pressure which decreases rapidly outward from the sphere. Thus we write

$$p = C\frac{\cos\vartheta}{r^2} = C\frac{z}{r^3} = -C\frac{\partial}{\partial z}\left(\frac{1}{r}\right) = -C\operatorname{div}\frac{\mathbf{k}}{r}. \qquad (31.27)$$

Since $\nabla^2(1/r) = 0$, we have

$$-\nabla p = C\left(\nabla\operatorname{div}\frac{\mathbf{k}}{r} - \nabla^2\frac{\mathbf{k}}{r}\right) = C\operatorname{curl}\operatorname{curl}\frac{\mathbf{k}}{r}. \qquad (31.28)$$

With

$$\mathbf{v} = -v_0\mathbf{k} + \operatorname{curl}\operatorname{curl}\mathbf{B} \qquad (31.29)$$

we then obtain

$$\nabla^2\mathbf{B} = -\frac{1}{\eta}\mathbf{B}_1 = -\frac{C}{\eta}\frac{\mathbf{k}}{r}. \qquad (31.30)$$

corresponding to (31.11). Since $\nabla^2 r = 2/r$, we can easily find a particular solution of this equation, to which solutions of the homogeneous equation may be added. We put $C = c\eta a$, where a is the radius of the sphere, and assume that the solution

$$\mathbf{B} = \mathbf{k}\left(-\frac{ca}{2}r + d\frac{a^3}{r}\right) \qquad (31.31)$$

is sufficiently general. This leads to

$$\operatorname{curl}\operatorname{curl}\mathbf{B} = \nabla\left(-\frac{ca}{2}\frac{z}{r} - da^3\frac{z}{r^3}\right) + ca\frac{\mathbf{k}}{r}. \qquad (31.32)$$

The r- and ϑ-components of the velocity are then, by (31.24),

$$v_r = -\cos\vartheta\left[v_0 - c\frac{a}{r} - 2d\left(\frac{a}{r}\right)^3\right],$$

$$v_\vartheta = \sin\vartheta\left[v_0 - \frac{c}{2}\frac{a}{r} + d\left(\frac{a}{r}\right)^3\right]. \qquad (31.33)$$

The boundary conditions $v_r = 0$, $v_\vartheta = 0$ for $r = a$ may, as is readily seen, be satisfied by putting $c = \frac{3}{2}v_0$, $d = -\frac{1}{4}v_0$, whence we obtain

$$v_r = -v_0\cos\vartheta\left[1 - \frac{3}{2}\frac{a}{r} + \frac{1}{2}\left(\frac{a}{r}\right)^3\right],$$

$$v_\vartheta = v_0\sin\vartheta\left[1 - \frac{3}{4}\frac{a}{r} - \frac{1}{4}\left(\frac{a}{r}\right)^3\right]. \qquad (31.34)$$

Since $C = c\eta a = \frac{3}{2}\eta v_0 a$, the pressure corresponding to this solution is

$$p = \frac{3}{2}\eta v_0 a \frac{\cos \vartheta}{r^2} \qquad (31.35)$$

by (31.27).

Due to the difference in pressure acting on the right- and left-hand sides of the sphere in Fig. 4-7, a resultant force on the latter arises in the direction of the fluid flow. Since the pressure acts perpendicular to the surface, we obtain the z-component by multiplication by $\cos \vartheta$; the resultant force is then

$$\frac{3}{2}\eta v_0 a \cdot 2\pi \int_0^\pi \cos^2 \vartheta \sin \vartheta \, d\vartheta = 2\pi \eta v_0 a. \qquad (31.36)$$

However, the fluid flow also gives rise to a force that acts in the same direction; per unit area, this force is

$$\left(\eta \frac{\partial v_\vartheta}{\partial r}\right)_{r=a} = v_0 \sin \vartheta \cdot \frac{3}{2} \cdot \frac{1}{a}. \qquad (31.37)$$

As it is tangential to the sphere, we must multiply by $\sin \vartheta$ in order to obtain the z-component; integration then leads to the resultant force

$$\frac{3}{2}\eta v_0 a \cdot 2\pi \int_0^\pi \sin^2 \vartheta \sin \vartheta \, d\vartheta = 4\pi \eta v_0 a. \qquad (31.38)$$

The total resistance experienced by a sphere of radius a, moving through a fluid of viscosity η with the velocity v_0, is thus

$$K = 6\pi \eta v_0 a. \qquad (31.39)$$

This is the well-known *Stokes' law*.

CHAPTER 5

Thermodynamics

Classical thermodynamics originates in the work of the Frenchman Sadi Carnot (1796–1832). In a famous article of 1824, "Réflexions sur la Puissance Motrice du Feu $\cdots$," he formulates, for the first time, the principle later to become known as the *law of entropy* or the *second law of thermodynamics*. This principle expresses the assumption that heat must be transported from a warmer to a colder place in order to be able to perform mechanical work. This idea was, at the time, compatible with the still widely held notion that heat was a special kind of substance, *caloric*, one of the last of the many *imponderabilia* (i.e. weightless substances) of ancient times.

Other ideas concerning the nature of heat had already been proposed, by Humphrey Davy (1778–1829) and notably by Benjamin Thompson, later to become Earl Rumford (1753–1814). Both Davy and Earl Rumford based their views on very comprehensive and convincing experimental evidence.

However, a clear formulation of the principle known today as the law of conservation of energy, of which the *first law of thermodynamics* appears as a special case, was not put forward until 1842 by Robert Julius Mayer (1814–1878). We shall discuss this later, but mention at the present stage that Carnot actually seems to have had a clear idea of this principle: In an unpublished paper, which did not become known until after his death, he even calculated the mechanical equivalent of heat, in a way similar to that employed later by Mayer.

To sum up, then, we may state that classical thermodynamics was founded on two main principles: the laws of *energy* and *entropy*, or the first and second law. In later times a third main principle has been added, the nature of which may be completely understood only on the basis of modern atomic physics. This principle is due to Walter Nernst (1864–1941) and is usually referred to as the *third law of thermodynamics;* it states that the so-called entropy is zero for all substances at the temperature of absolute zero, about $-273°C$.

32. TEMPERATURE AND TEMPERATURE SCALES

Measurement and Definition of Temperatures

Irrespective of their notions of the true nature of heat, everyone seems to have regarded it as a quantity that can be measured, so that one may speak of *quantities of heat*. A body becomes warmer or cooler, according as it is supplied with or deprived of a quantity of heat; the amount of heating or cooling is then measured by the so-called *temperature*.

For normal temperatures we may as a rule immediately decide which of two bodies is the warmer; on the other hand, it is difficult to determine by how much it is warmer. Since, however, most bodies expand when heated, this provides a handy method of measuring differences of temperature and defining temperature scales. The most common instrument used for measuring everyday temperatures is the mercury thermometer. The scale of any thermometer may be one of several, for example, Fahrenheit, Reaumur, or centigrade. The centigrade scale has its zero point at the temperature of melting ice under a pressure of one atmosphere, and the scale point 100 at the temperature of boiling water under the same pressure (actually at the temperature of the steam, since the water may be superheated).

The mercury thermometer has, however, inherent upper and lower limitations. For one thing, mercury freezes at about $-39°C$; at this and lower temperatures such thermometers are often replaced by alcohol thermometers. For increasing temperatures, all liquid-column thermometers rapidly become useless. The most sensible way of obtaining measurements on an extended range of high temperatures seems to be that of using gas thermometers, since gases remain gases irrespective of how much they are heated. For low temperatures, it is clearly advantageous to use gases with a low boiling point (i.e. temperature of condensation); such are the "permanent" gases of earlier times: air, oxygen, nitrogen, hydrogen, carbon dioxide, and so forth. The most "permanent" of these, and hence the most useful for scientific purposes, is hydrogen; while helium, and possibly also other noble gases, such as argon or neon, would be even better.

It should be noted that these substances define individual temperature scales. The expansion of volume, for instance, is not proportional for all substances, not even for the permanent gases. Since it is often more convenient to measure a pressure than a volume, gas thermometers are often constructed as pressure thermometers, maintaining a constant volume for the gas. However, each gas defines its own individual scale of temperature in this case also.

We shall discuss later how one may establish one special temperature scale, which has particular advantages from a scientific point of view. This is the so-called *absolute* or *thermodynamic* scale, also frequently referred to as the Kelvin scale. Its zero point is the previously mentioned absolute zero temperature; otherwise, it may be described as a modified centigrade scale, since the interval 0°C–100°C still contains one hundred degrees. Kelvin scale temperatures are written °K; the melting point of ice is then about 273°K.

Absolute Zero and Absolute Temperature

The thermodynamic temperature scale corresponds to that defined by a thermometer filled with an imagined idealized substance, which is called an *ideal* gas. The notion of such a substance derives, to a high degree of precision, from the observed properties of real gases; hydrogen, in particular, obeys approximately the laws for an ideal gas over a very large range of temperature. A hydrogen thermometer, calibrated in centigrade degrees, will provide an operational definition of temperature that is valid down to very low temperatures, somewhere between $-200°C$ and $-300°C$. Since the thermal coefficient of expansion of hydrogen (as of the other permanent gases), calculated in the centigrade scale, is very close to $1/273$ at $0°C$, an extension of this temperature scale downward will end at about $-273°C$. Here the volume, or rather the pressure, of the gas is zero. This is the so-called *absolute zero point*, the value of which will vary with the different gases used in our thermometers; taking it as the zero point of our temperature scale, we obtain the so-called *absolute temperature*, commonly denoted by T. The gas then satisfies approximately the equation

$$PV = RT, \qquad (32.1)$$

where P is the pressure of the gas, V the volume, and R a constant depending on the amount of gas present. For gas thermometers in which the volume is kept constant while the pressure varies (pressure thermometers), (32.1) then defines the temperature in terms of the observed pressure. This equation is also, to a large extent, valid for many gases in which the pressure and volume both vary at a constant temperature; we then usually write the equation

$$PV = \text{const} \qquad (32.2)$$

and call it *Boyle's* or *Mariotte's law*.

For other gases than that used in our thermometer to define the temperature scale, the relation (32.1) will only be approximately valid; the deviations will, however, in general be relatively small. The idealized substance that we have called an ideal gas, however, satisfies (32.1) and (32.2) exactly for all values of P and V, as long as T is defined by means of this same ideal gas. Another criterion of an ideal gas, that the so-called internal energy is to be a function of the temperature alone, will be discussed later.

In the thermodynamic temperature scale (32.1) is thus, strictly speaking, valid only for ideal gases; the behavior of real gases will exhibit large or small deviations from that described by this equation. Such deviations become large when a gas is cooled or compressed so much that it approaches the limit of condensation.

The (absolute) zero point of the rational thermodynamic temperature scale lies at a temperature of $-273.16°C$, a value that has been uniquely determined by theoretical considerations. Thermal measurements of different kinds have verified this value to a high degree of precision. We shall discuss later how one may employ a gas thermometer to calculate the corrections necessary to arrive at the thermodynamic temperature scale.

33. THE EQUATION OF STATE

States

As is well-known, a body may appear in three different states of aggregation or phases: the solid, the liquid, or the gaseous phase. Here we shall discuss primarily the latter two of these phases, and in particular, the gaseous phase.

These different states of aggregation may, under certain physical conditions, coexist in the same system: gas and liquid together, say, or even all three phases together, as for melting ice. In this case, the description of the state of the system must include certain additional parameters specifying how much of the substance is in this or that phase. Otherwise the state of a given homogeneous amount of matter—as examples of which we shall chiefly consider gases—is determined by two state variables, the temperature and pressure.

State Variables

A homogeneous substance may be characterized by several different physical quantities. Some of these, such as the pressure P or temperature T, are *intensive*, that is, properties describing small parts of the substance. Others, such as volume V, internal energy U, entropy S, are *extensive*, that is, characteristic of the whole amount of the substance. Quantities of the kind mentioned here are called state variables. We also have other thermodynamic quantities, such as the external work W performed by a gas, or a quantity of heat Q supplied to a system; these are of an essentially different nature, referring not only to the instantaneous state of the system (as do the state variables) but also to its past history. We shall discuss this difference in more detail later.

It now turns out that the state of a homogeneous substance, a gas, is uniquely determined by the values of two state variables only; these will be called the independent state variables. Any other state variables are then given as unique functions of these two quantities and may consequently be denoted as dependent state variables. We may freely choose any two state variables as independent. Most commonly, two of the three quantities pressure P, temperature T, and volume V are so chosen; there are then the three possible combinations (P, V), (P, T), and (V, T). As a matter of fact, the great number of possible equivalent formulations of the equations of thermodynamics is rather a nuisance. In addition to the three mentioned, the entropy S, the internal energy U, and three other so-called thermodynamic potentials frequently appear. Since thermodynamic equations are habitually expressed in terms of partial derivatives of one state variable with respect to another, keeping a third such variable constant, it is evident that a multitude of different versions of such equations results. We shall not make any attempt to include all these possibilities in a systematic exposition. Rather, we shall endeavor to give a presentation of the bare essentials necessary to understand the basic principles of thermodynamics. To this end, we now commence by discussing the equation of state.

Connection between Pressure, Temperature, and Volume

Since the state of a substance is uniquely defined by two variables alone, a relation between the three fundamental quantities P, V, and T must exist,

$$F(P, V, T) = 0. \tag{33.1}$$

This is often written in a more explicit form, as $P = P(V, T)$, $V = V(P, T)$, or perhaps most commonly as

$$f(P, V) = T. \tag{33.2}$$

In these expressions T is not, of course, necessarily defined according to the absolute thermodynamic temperature scale; for instance, the scales of any gas thermometer, for example, a hydrogen thermometer, will serve equally well to establish equations of state. However, it is desirable to avoid different equations of state for one and the same substance by adopting the most rational temperature scale, the Kelvin scale; and we shall, in the following, as a rule assume tacitly that T is measured in this scale, even though we still have not formulated the precise definition of the latter.

For an ideal gas (and, approximately, for permanent gases) we then obtain an equation of state in the form (33.2), namely,

$$PV = RT, \tag{33.3}$$

as already noted. Another such equation is the well-known van der Waals equation

$$\left(P + \frac{a}{V^2}\right)(V - b) = RT \tag{33.4}$$

which has particular importance for gases approaching condensation; from it one may deduce important characteristic quantities, like the *critical pressure*, *critical temperature*, and *critical volume*.

State Diagrams

If we by means of equations of state or other equations wish to represent one quantity as a function of two others, we may use so-called state diagrams. The dependent variable is then measured along the ordinate, and one of the independent variables along the abscissa. Keeping the other independent variable constant, the states resulting from variation of the abscissa variable will be represented as points on a curve. Another fixed value of the variable that was kept constant results in another curve. Such curves are usually denoted by the prefix "iso-" or "is-." For instance, if the temperature is the quantity kept constant for each curve, they are called *isothermals;* while curves connecting points that represent states with the same entropy are called *isentropics* or *adiabatics*.

A state diagram that appears very often is the so-called PV-diagram (Fig. 5-1). T_1 and T_2 here denote two different temperatures; if the substance is an ideal gas, the two corresponding isothermals are then ordinary hyperbolas. Also, S_1 and S_2 denote two values of the so-called entropy of the substance, to be defined later, and the two corresponding isentropics resemble hyperbolas, being given by

$$PV^\gamma = \text{const}, \quad \gamma > 1. \tag{33.5}$$

Several other types of state diagram are also fairly commonly used, for example, the TS-diagram. One of the advantages of any such state diagram, the curves of which may be plotted experimentally, is that it enables one to

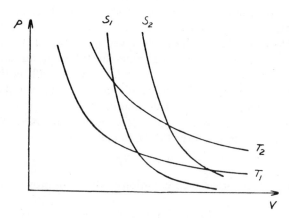

Fig. 5-1. PV-diagram with isothermals and isotropics.

read off directly other important physical quantities. From the *TS*-diagram, for instance, one may read off the quantities of heat supplied to a system; the *PV*-diagram, correspondingly yields the amount of work performed by a gas during expansion, or in the course of a cyclic process. The quadrilateral plane figure bounded by the two isothermals and adiabatics in Fig. 5-1 is often used to describe a so-called Carnot cycle; it then shows both what kind of periodic change the substance has experienced and the work that has simultaneously been performed, the latter being equal to the area of the quadrilateral.

34. THE FIRST LAW OF THERMODYNAMICS

Heat and Energy. The Law of Energy Conservation

The three great names connected with the law of energy conservation are Robert Mayer (whom we have already mentioned), J. P. Joule (1818–1889), and H. Helmholtz (1821–1894). Of these, Mayer was, primarily, the philosopher, the man of ideas and imagination. His arguments were, in particular, based on the calculation of the mechanical equivalent of heat by means of the measured values of the specific heats of different gases at constant volumes or pressures. Joule, who was not a scholar, but an autodidactic, was the great experimenter. In several important fields, notably thermodynamics and electrostatics, he performed a series of extremely (according to the standard of his time) accurate measurements, which have been of fundamental importance in the recognition and understanding of the energy law. Finally, Helmholtz was the scholarly theorist, with the ability to generalize the law and apply it to all branches of natural science, in his monumental treatise *Über die Erhaltung der Kraft* (1847).

However, Sadi Carnot was the predecessor of all these. In the previously mentioned unpublished manuscript found after his death in 1832, we read:

"La chaleur n'est autre chose que la puissance motrice qui a changé de forme. Partout où il y a destruction de puissance motrice, il y a, en même temps, production de chaleur en quantité précisément proportionelle à la quantité de puissance motrice détruite. Réciproquement, où il y a destruction de la chaleur, il y a production de puissance motrice.

Un peut donc poser en thèse générale que la puissance motrice est en quantité invariable dans la nature, qu'elle n'est jamais, a proprement parler, ni produite, ni détruite."

This formulation of the energy law, and the corresponding hypothesis concerning the nature of heat, is as clear as anyone could wish; and it is certainly more positive in its form than the famous statement: "*I do not know what heat is, except it be* MOTION," attributed to Earl Rumford.

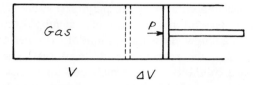

Fig. 5-2. Calculation of the mechanical equivalent of heat.

The argument by which Robert Mayer calculated the mechanical equivalent of heat is illustrated in Fig. 5-2.

Imagine that a gas is heated from the temperature T_1 to T_2, subject to two different conditions: first, being kept at a constant volume, so that the pressure increases; and second, expanding at a constant pressure. A greater amount of heat is absorbed in the latter case than in the former. Denoting the two quantities of heat, measured in calories, by Q_1 and Q_2, respectively, we have

$$Q_2 - Q_1 = (C_P - C_V)(T_2 - T_1), \qquad (34.1)$$

where C_P and C_V are the specific heat capacities of the gas (which was air, in Mayer's discussion) at constant pressure and volume, respectively; the values of these capacities were known to Mayer from experiments conducted earlier by other scientists.

The difference between Q_1 and Q_2 was now assumed by Mayer to be due to the fact that the excess heat, absorbed during the expansion of the gas at constant pressure, had been converted to mechanical work. This work is then given by

$$W = P \, \Delta V, \qquad (34.2)$$

where P is the pressure and ΔV the increase of volume, and may be calculated by means of the known coefficient of thermal expansion for air. Mayer put these two quantities, that is, the excess heat and the work performed, proportional,

$$J(Q_2 - Q_1) = W. \qquad (34.3)$$

The factor of proportionality J is called the mechanical equivalent of heat; its magnitude is

$$J = 427 \text{ kgm/kcal}, \qquad (34.3a)$$

or

$$J = 4.185 \text{ joule/cal}$$

$$\approx \frac{1}{0.24} \text{ joule/cal.} \qquad (34.3b)$$

Internal Energy of a Gas

When a gas is heated it will absorb heat energy. We shall, in the following, measure both mechanical and heat energy in the same units, that is, joules or ergs, in accordance with (34.3b), and may thus omit the heat equivalent factor J.

The heat energy absorbed altogether by the gas, when it is heated from the absolute zero to its present temperature, is called its internal energy, and denoted by U. The energy law, or the first law of thermodynamics, may then be expressed mathematically by

$$\delta Q = dU + P\, dV. \qquad (34.4)$$

Here δQ is the quantity of heat supplied, dU the increase of internal energy, and $P\, dV$ the work performed by the gas, appearing as an increase in the mechanical energy of the surroundings. All of these three quantities are taken to be infinitesimal; this is necessary to ensure that P (or any other functions appearing in this, or similar, equations) be uniquely defined, $P = P(V, T)$. The notation δQ for the infinitesimal quantity of heat expresses the fact, previously mentioned, that the amount of heat supplied is not a state variable, in contradistinction to P, V, and U.

Ideal Gases

We now, first of all, define an ideal gas as one that obeys the Boyle-Mariotte law exactly,

$$PV = \text{const}, \qquad (34.5)$$

and has an internal energy depending only on the temperature,

$$U = U(T). \qquad (34.6)$$

If we take, say, V and T as independent variables, we then obtain

$$dU = \left(\frac{\partial U}{\partial T}\right)_V dT + \left(\frac{\partial U}{\partial V}\right)_T dV = \frac{dU}{dT} dT$$

or

$$\left(\frac{\partial U}{\partial T}\right)_V = \frac{dU}{dT}, \qquad \left(\frac{\partial U}{\partial V}\right)_T = 0. \qquad (34.7)$$

The general energy law (34.4) for gases, written as

$$\delta Q = \left(\frac{\partial U}{\partial T}\right)_V dT + \left[\left(\frac{\partial U}{\partial V}\right)_T + P\right] dV \qquad (34.8)$$

then takes the form

$$\delta Q = \frac{dU}{dT} dT + P\, dV. \qquad (34.9)$$

for an ideal gas.

We have not yet defined the absolute thermodynamic temperature scale, and may thus take T to be measured on the scale of any gas thermometer.

If this gas is assumed to be ideal, the temperature may then clearly be defined in such a way that (34.5) appears as the equation of state

$$PV = RT. \tag{34.10}$$

Consider now a gas expanding at a constant pressure,

$$P\,dV = R\,dT. \tag{34.11}$$

If we substitute this in (34.9), and write

$$C_V(T) = \frac{dU}{dT}, \tag{34.12}$$

where C_V is the *heat capacity*, or *specific heat* at constant volume, we obtain

$$\delta Q = (C_V + R)\,dT = C_P\,dT, \tag{34.13}$$

where C_P is the specific heat at constant pressure.

Hence it follows that

$$C_P - C_V = R, \tag{34.14}$$

where R is the so-called gas constant for the amount of gas considered. Usually, this constant is referred to a certain quantity of the gas, for example, a gram or (more commonly) a gram molecule; the latter choice leads, as we shall see, to a universal gas constant, valid for all gases.

Hitherto we have taken T to be measured on the individual temperature scale defined by the gas itself. However, it is a consequence of the second law of thermodynamics (presently to be discussed) that all ideal gases define the same temperature scale. Hence the gas constant R becomes universal; furthermore, C_P and C_V will turn out to be constants (i.e. independent of the temperature) for ideal gases. This means that the internal energy is proportional to the absolute thermodynamic temperature

$$U = C_V T. \tag{34.15}$$

Entropy of an Ideal Gas

Equation (34.9) may, for an ideal gas, be written

$$\delta Q = C_V\,dT + RT\,\frac{dV}{V}. \tag{34.16}$$

If $\delta Q = 0$, we then obtain

$$dS = C_V\,\frac{dT}{T} + R\,\frac{dV}{V} = 0 \tag{34.17}$$

or

$$S = \int \frac{C_V}{T}\,dT + R\log V = \text{const.} \tag{34.18}$$

The quantity S, which henceforth will be referred to as the *entropy* of the gas, thus becomes a function of V and T, and consequently a state variable. It follows from (34.16) and (34.17) that

$$dS = \frac{\delta Q}{T}. \tag{34.19}$$

Hence the entropy does not change when the gas is subjected to a slow or *reversible*, process without any influx of heat. Such processes are, as previously mentioned, generally called *adiabatic;* a more logical designation would be *isentropic* processes.

This new state variable, the entropy S, is as yet only defined by means of its differential dS, which is equal to the quantity of heat δQ absorbed, divided by the absolute temperature T; it thus contains an arbitrary additive constant. The physical significance of the entropy will be considered in more detail later, when we discuss the second law.

If we may assume that C_V is constant*, the expression for the entropy becomes, by (34.13):

$$S = C_V \log T + R \log V + S_0, \tag{34.20}$$

where S_0 is an arbitrary constant.

Writing

$$\frac{C_P}{C_V} = \gamma, \tag{34.21}$$

and utilizing (34.14) and (34.20), we obtain the following relation connecting V and T for isentropic processes

$$TV^{\gamma-1} = \text{const.} \tag{34.22}$$

Using the equation of state, we thus finally arrive at a relation connecting the pressure and volume, namely,

$$PV^\gamma = \text{const.} \tag{34.23}$$

35. THE SECOND LAW OF THERMODYNAMICS

Reversible and Irreversible Processes

In the following considerations, it will be necessary to distinguish clearly between *reversible* and *irreversible* processes. A reversible process, thermodynamically speaking, is one during which the thermodynamic equilibrium

---

* This assumption may be justified by arguments derived from kinetic gas theory. For an ideal gas, we may here deduce the relation $PV = \frac{2}{3}U$, where U is the total translational kinetic energy of the gas molecules; compare (38.5). However, this results from considerations using purely mechanical, and not thermodynamical, principles.

is never at any time noticeably disturbed. Mathematically, this may be expressed by saying that the process takes place at an infinitesimal rate. In practice, however, processes that are fairly rapid from the mechanical point of view may well be approximately reversible. This is because a gas may often be able to adjust itself very rapidly to a thermodynamic equilibrium with the moving mechanical parts of a heat engine, since molecular velocities are generally very high.

Another important condition that a reversible process must satisfy is that there must be no loss of heat from the system considered to the surroundings, since the conduction of heat from a warmer to a cooler place is a quite irreversible phenomenon. Thus, when heat is supplied to the working substance of a heat engine, one must take care to ensure that the heat reservoir and the substance have the same temperature, if the process is to be reversible.

Perpetual Motion Machines of the Second Kind

The two laws of thermodynamics may be formulated together, as the postulate that it is impossible to construct a so-called *perpetuum mobile*, or perpetual motion machine. We must then distinguish between two different kinds of such machines, usually referred to as perpetual motion machines of the first and second kind. The former may be described briefly as a machine that is able to perform continual work on its surroundings without being continually supplied with energy. Such a process is incompatible with the law of conservation of energy, or the first law of thermodynamics.

A perpetual motion machine of the second kind, however, is not incompatible with the energy law. We might imagine that the amount of heat present in, say, the oceans were directly converted into mechanical work, by means of some ingenious process; this is then possible, provided that we have a cold reservoir to assist in the conversion. For instance, work may be extracted from the vast quantities of heat in the equatorial regions if we use the polar regions as a cold reservoir. Indeed, it is this difference in temperature that gives rise to the great ocean currents; if the kinetic energy of the latter were converted to mechanics work, the source of energy would be equatorial heat.

However, a perpetual motion machine of the second kind is defined as an isolated system that yields mechanical work without a permanent net transport of heat from a warmer to a colder place; an equivalent definition is, then, a machine that is able to transport heat from a colder to a warmer place without being supplied with energy. We shall assume that the construction of such machines is impossible, and refer to this postulate as the second law of thermodynamics; later, we shall deduce certain consequences of the said assumption, which in turn lead to other equivalent formulations of the second law.

Efficiency of a Heat Engine

The main components of a heat engine are: a working substance, a warm reservoir and a cold reservoir, as well as moving mechanical parts that are able to perform work. This is done by the expansion of the working substance, due to heat energy flowing into it from the warm reservoir (which, in turn, receives its heat energy from the combustion of, say, coal or oil). Let us assume that the warm reservoir thus supplies the working substance with a quantity of heat Q that enables the substance to perform a mechanical amount of work W; the efficiency of the heat engine is then, by definition,

$$E = \frac{W}{Q}. \tag{35.1}$$

The efficiency depends on many circumstances: the loss of heat because of poor insulation, the amount of mechanical work being converted to heat by friction and thus lost, and so forth. Primarily, however, it depends on the temperature of the cold reservoir; the lower this temperature is, the higher the efficiency. For instance, steam engine design was greatly improved when the steam was made to condense in a reservoir filled with cold water, instead of in open air, as had previously been the practice. On the other hand, it is not sufficient to have a cold reservoir at a low temperature; one must also ensure that the working substance may be utilized right down to this temperature. In a steam engine, for instance, a cold reservoir with a temperature below 0°C would be of very little help theoretically, and presumably impossible to use in practice.

Efficiency of Reversible Processes

We shall now show that our postulate—that no perpetual motion machine of the second kind may exist—leads to the result that all reversible heat engines working between the same two temperatures T_2 and T_1 have exactly the *same* efficiency, while, on the other hand, the efficiency of any irreversible engine is less.

If this theorem is correct, the efficiency of a reversible heat engine must be

$$E = \frac{T_2 - T_1}{T_2}, \tag{35.2}$$

where T_2 and T_1 are the absolute temperatures of the warm and cold reservoirs, respectively, *measured on the individual temperature scale defined by any ideal gas*. This, in turn, implies that the temperature scale of all ideal gases are equal, provided that they agree at two fixed points.

For an ideal gas, the efficiency (35.2) may easily be deduced by the consideration of a Carnot cycle between two isothermals and two adiabatics in a *PV*-diagram. (See Fig. 5-3.)

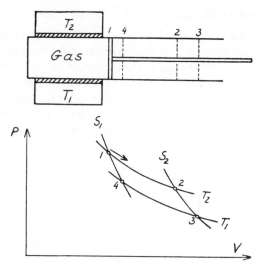

Fig. 5-3. Calculation of the efficiency, in the case where the working substance is an ideal gas.

The gas is assumed to be contained in a cylinder with a movable piston and perfectly insulating walls; the latter may, however, be put into heat-conductive connection with the warm or cold reservoir whenever this is desired. We begin by letting the gas be so connected with the reservoir T_2, and move the piston outward so that the gas expands isothermally and reversibly from state 1 to state 2 (see Fig. 5-3). The cylinder is then isolated from T_2, whereupon the gas expands adiabatically and reversibly along the adiabatic S_2 to state 3, where its temperature has dropped to T_1. At this stage it is connected with the reservoir T_1 and compressed isothermally to state 4, where it intersects the adiabatic S_1. Finally, it is isolated from T_1 and further compressed adiabatically until it attains state 1, thus completing the cycle.

We now calculate the quantities of heat Q_2 and Q_1, absorbed and emitted by the gas at the temperatures T_2 and T_1, and also the total work W performed. At the first step during the isothermal expansion the internal energy does not change; hence the quantity of heat absorbed is simply equal to the work performed

$$Q_2 = \int_{V_1}^{V_2} P \, dV = RT_2 \int_{V_1}^{V_2} \frac{dV}{V} = RT_2 \log \frac{V_2}{V_1}. \tag{35.3}$$

Similarly, the quantity of heat emitted to the cold reservoir during the isothermal compression is

$$Q_1 = RT_1 \log \frac{V_3}{V_4}. \tag{35.4}$$

No more calculations are now necessary, since the total work performed during the cycle must be

$$W = Q_2 - Q_1. \tag{35.5}$$

as a consequence of the first law. Hence the amounts of mechanical work performed by and on the gas during the adiabatic expansion and contraction compensate each other exactly.

However, it follows from the adiabatic equation of state (33.21) that

$$T_2 V_2^{\gamma-1} = T_1 V_3^{\gamma-1}, \qquad T_2 V_1^{\gamma-1} = T_1 V_4^{\gamma-1}, \tag{35.6}$$

or, by division,

$$\frac{V_2}{V_1} = \frac{V_3}{V_4}. \tag{35.7}$$

The logarithms (35.3) and (35.4) are thus equal, so that we obtain

$$\frac{Q_2}{Q_1} = \frac{T_2}{T_1}; \tag{35.8}$$

in other words, the ratios of the two quantities of heat and the corresponding absolute temperatures are equal. Hence we arrive at the efficiency

$$E = \frac{Q_2 - Q_1}{Q_2} = \frac{T_2 - T_1}{T_2}, \tag{35.9}$$

as already announced in (35.2).

Comparison of the Efficiencies of Two Heat Engines

We now consider two reversible heat engines of similar construction to that in Fig. 5-3, connected in a way enabling the one to drive the other; the engines are assumed to have common warm and cold reservoirs. We let both engines run in such a way that the state of the working substance moves in a PV-diagram between two adiabatics with a suitably chosen separation, so that the amounts of work performed on the one engine and by the other during a Carnot cycle are equal:

$$W = Q_2 - Q_1, \qquad W' = Q_2' - Q_1', \qquad W = W'. \tag{35.10}$$

The particular way in which the two engines are connected is immaterial, and will not be considered in detail. The piston movement of the one may, but need not, be directly transmitted to the other piston. In principle we may imagine that the work performed is communicated to a fly wheel with a large moment of inertia and a correspondingly small rotational velocity which may be adjusted to the motions of the pistons. In this fashion the energy input and output of the engines may be stored or tapped as desired, and the engines may consequently be allowed to run through their cycles in any mutually independent ways.

We now assume that the energy of the fly wheel is, on the average, constant in time—or, alternatively, the same at the end of each completed cycle—so that the amounts of work W and W' are performed at the expense of heat energy according to (35.10); this may also be written

$$Q_1 - Q_1' = Q_2 - Q_2'. \tag{35.11}$$

The left-hand side then corresponds to the quantity of heat supplied to the cold reservoir T_1 during a complete cycle, while the right-hand side is that correspondingly extracted from the warm reservoir T_2. These differences may, by our postulate concerning the impossibility of a perpetual motion machine of the second kind, be positive or zero, but not negative, that is,

$$Q_2 \geqq Q_2'. \tag{35.12}$$

Since the engine processes are reversible, however we may imagine that they exchange roles, so that the first is now driven by the second, instead of vice versa. This means that the quantity of heat Q_2' formerly emitted at T_2 is now absorbed, and conversely for Q_1' at T_1; similarly Q_2 and Q_1 assume the roles hitherto played by Q_2' and Q_1'. Hence (35.12) must be replaced by

$$Q_2' \geqq Q_2. \tag{35.13}$$

The only solution compatible with both (35.12) and (35.13) is then

$$Q_2' = Q_2, \quad Q_1' = Q_1, \tag{35.14}$$

whence, by (35.10),

$$E = \frac{W}{Q_2} = \frac{W'}{Q_2'} = E'. \tag{35.15}$$

Thus the efficiencies of the two engines are equal, being that of a single reversible engine with an ideal gas as the working substance, such as that used to calculate the efficiency (35.9).

However, if the first engine is not reversible, the pure inequality (35.12) may hold, so that

$$E = \frac{W}{Q_2} \leqq \frac{W'}{Q_2'} = E'. \tag{35.16}$$

In other words, the efficiency of any heat engine may equal, but never exceed, that of a reversible heat engine.

Entropy as a State Variable

We now define the entropy change of a substance for an infinitesimal reversible process as

$$dS = \frac{\delta Q}{T}. \tag{35.17}$$

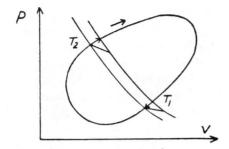

Fig. 5-4. Calculation of the change of entropy during a cyclic process.

This result will now be used to calculate the total entropy change S during an arbitrary cyclic process, as demonstrated in the PV-diagram of Fig. 5-4.

We subdivide the area inside the closed curve in the diagram by infinitesimally separated adiabatics; through the points where the latter intersect the curve we draw infinitesimal isothermals, corresponding at the upper and lower point of the same adiabatic to the temperatures T_2 and T_1, respectively. The amounts of heat absorbed or emitted along the infinitesimal segments of the curve between the adiabatics will be called δQ_2 and δQ_1; and the heat quantities emitted or absorbed along the corresponding segments of the isothermals we call $\delta Q_2'$ and $\delta Q_1'$. Since no heat is absorbed or emitted along an adiabatic, the difference between $\delta Q_2'$ and δQ_2 is given by the area of the upper infinitesimal triangle, representing a certain amount of work. However, this area is quadratically infinitesimal compared with the linear separation between the adiabatics; hence we may put $\delta Q_2 = \delta Q_2'$ and $\delta Q_1 = \delta Q_1'$. Thus by our definition of absolute temperatures and (35.8), we must have

$$\frac{\delta Q_2}{T_2} = \frac{\delta Q_1}{T_1} \tag{35.18}$$

or

$$\sum \left(\frac{\delta Q_2}{T_2} - \frac{\delta Q_1}{T_1} \right) = 0. \tag{35.19}$$

Algebraically, taking emitted quantities of heat to be negative, this may be expressed by the integral

$$\oint dS = \oint \frac{\delta Q}{T} = 0. \tag{35.20}$$

This implies that the integral

$$S_2 - S_1 = \int_1^2 dS \tag{35.21}$$

is independent of the path of integration, that is, that S is a genuine state variable, being a function of P and V only.

The Second Law and Entropy

We shall now demonstrate how the second law of thermodynamics may be expressed in terms of the entropy. We consider an isolated system, which may consist of two or several subsystems of different temperatures. If these subsystems are not insulated from each other, heat would flow from parts of higher temperatures T_2 to parts of lower temperatures T_1, and the entropy of the system would increase, since

$$\frac{\delta Q}{T_1} - \frac{\delta Q}{T_2} > 0.$$

Thus the total entropy of the system would tend to increase. However, processes other than heat flow may occur; the different subsystems may perform work on each other. Formally we may regard this as happening by means of heat engines built into the system and taken to be part of it. We know that such engines may well transport heat from a higher to a lower temperature, if the process runs irreversibly; on the other hand, they may not transport heat from a colder to a warmer place. Thus we have for the whole system at any stage of its development,

$$dS = \sum \frac{\delta Q}{T} \geqq 0. \tag{35.22}$$

The second law may consequently be expressed by saying that the entropy of an isolated system never decreases, but tends to increase steadily. A stable thermodynamic equilibrium is reached when the entropy has attained its maximum value, so that

$$dS = 0, \quad d^2S \leqq 0. \tag{35.23}$$

Clausius, who introduced the concept of entropy into thermodynamics, has expressed this in somewhat more grandiose terms, regarding the whole universe as an isolated system, "Die Energie der Welt ist konstant. Die Entropie der Welt strebt einem Maximum zu."

Application of the Entropy Law to Evaporation and Melting

We shall demonstrate a simple application of the entropy law. Let us consider an evaporating liquid, that is, a system with two phases, with the pressure P, the temperature T, and the specific volumes of the vapor and liquid v_1 and v_2. The functions T, v_1, and v_2 are then functions of the pressure alone, and T is constant during the evaporation; in a PV-diagram the isothermals are then horizontal.

Consider now a gram of the vapor, running through a cyclic process as shown in Fig. 5-5. The term AB is an isothermal compression and liquefaction of the saturated vapor from the volume v_2 to v_1; BC is an isentropic

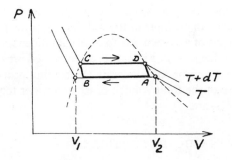

Fig. 5-5. Deduction of the Clausius-Clapeyron formula.

compression of the liquid to the pressure $P + dP$, with practically no change of volume. Also, CD is an isothermal expansion and evaporation of the liquid at the higher pressure $P + dP$ and the higher temperature $T + dT$. Finally, DA is an isentropic expansion of the vapor to the lower pressure and temperature P and T.

We denote the latent heat of vaporization of the liquid at the temperature T by L; at the temperature $T + dT$ it must then be $L + dW$, where dW is the work performed during the cycle, that is, the area of the plane figure $ABCD$:

$$dW = (v_2 - v_1)\, dP.$$

Since the entropy must be the same before and after the cycle, we have

$$\frac{L}{T} = \frac{L + dW}{T + dT}$$

or

$$L\, dT = T\, dW = T(v_2 - v_1)\, dP,$$

whence follows the *Clausius-Clapeyron equation*

$$\frac{dP}{dT} = \frac{L}{T(v_2 - v_1)}. \tag{35.25}$$

A similar equation may be obtained for the phenomenon of melting. Letting M be the latent heat of fusion, and v_2 and v_3 the specific volumes of the liquid and the solidified substance, we obtain

$$\frac{dT}{dP} = \frac{T(v_2 - v_3)}{M}. \tag{35.26}$$

for the inverse equation. From this equation it is evident that the melting point temperature increases with pressure if the specific volume of the liquid is greater than that of the solidified substance. The lowering of the melting point of ice with increasing pressure is due to the fact that ice is lighter than water. This furnishes an example of the well-known principle of Le Chatelier, namely, that changes of a system due to external influences tend to oppose or counteract these influences. For instance, an increased pressure will lower the melting point of ice and hence accelerate the melting process; however, this leads to a decrease of the total volume of ice and water together which tends to decrease the pressure.

36. THERMODYNAMIC POTENTIALS

Entropy and Free Energy. Gibbs' Potential and Enthalpy

We have seen that the entropy of an isolated system tends to increase toward a maximum, that is, $dS \geqq 0$; an isolated system is one into which there is no influx of heat or work. For a gas this means that $dU = dV = 0$. On the other hand, we have

$$T\,dS - dU - P\,dV = 0 \qquad (36.1)$$

for a state of thermodynamic equilibrium. If such an equilibrium is not present, we must assume that (36.1) is to be replaced by the inequality

$$T\,dS - dU - P\,dV \geqq 0, \qquad (36.1a)$$

since this yields $dS \geqq 0$ for constant energy and volume.

This inequality may be reformulated in many ways, for instance,

$$d(U - TS) + S\,dT + P\,dV \leqq 0. \qquad (36.2)$$

This form is convenient if T and V are to be kept constant, as when the system is not heat-insulated, but kept at a constant temperature by means of a heat reservoir. The function

$$F = U - TS \qquad (36.3)$$

will then evidently decrease steadily and attain a minimum value for constant T and V. This function F, being that part of the internal energy that may be converted to mechanical work, is called the *free energy*.

The function

$$G = U - TS + PV \qquad (36.4)$$

is called the thermodynamic potential, or Gibbs' potential. Utilizing (36.4), (36.2) may be written

$$dG + S\,dT - V\,dP \leqq 0. \qquad (36.5)$$

This function thus decreases steadily toward a minimum value for the system when T and P are kept constant.

Finally, we consider the function
$$H = U + PV \qquad (36.6)$$
called the *heat function* or *enthalpy*. It may be utilized to reformulate (36.2) and (36.5) as
$$dH - T\,dS - V\,dP \leq 0. \qquad (36.7)$$
Thus the enthalpy decreases toward a minimum value for a constant entropy and pressure of the system.

Maxwell's Thermodynamic Relations

Let us rewrite the equations (36.1), (36.2), (36.4), and (36.7) for the thermodynamic potentials in a state of equilibrium:
$$\begin{aligned} dU &= T\,dS - P\,dV, \\ dF &= -S\,dT - P\,dV, \\ dG &= -S\,dT + V\,dP, \\ dH &= T\,dS + V\,dP. \end{aligned} \qquad (36.8)$$

As a consequence of the first of these equations
$$T = \left(\frac{\partial U}{\partial S}\right)_V, \qquad -P = \left(\frac{\partial U}{\partial V}\right)_S, \qquad (36.9)$$
so that, comparing the second-order derivatives $\partial^2 U/\partial S\,\partial V$, we obtain
$$\left(\frac{\partial T}{\partial V}\right)_S = -\left(\frac{\partial P}{\partial S}\right)_V. \qquad (36.10)$$
This is one of the four well-known thermodynamic relations named after Maxwell; the remaining three are
$$\left(\frac{\partial S}{\partial V}\right)_T = \left(\frac{\partial P}{\partial T}\right)_V, \qquad (36.11)$$
$$\left(\frac{\partial S}{\partial P}\right)_T = -\left(\frac{\partial V}{\partial T}\right)_P, \qquad (36.12)$$
$$\left(\frac{\partial T}{\partial P}\right)_S = \left(\frac{\partial V}{\partial S}\right)_P. \qquad (36.13)$$

The Joule-Kelvin Experiment

Let us consider a pipe with a porous plug—made of, say, cotton wool or silk—which allows the gas to seep through slowly if the pressures P_1 and P_2 on the two sides are different, $P_1 > P_2$. The pipe is provided with piston on

Fig. 5-6. The Joule-Kelvin experiment.

both sides of the plug; (see Fig. 5-6). The piston on the left is moved inward, compressing the gas before it at the pressure P_1, while the piston on the right recedes, acted on by a smaller counterpressure P_2. Since no work is done on the porous plug, the energy equation yields

$$P_1 V_1 - P_2 V_2 = U_2 - U_1 \qquad (36.14)$$

or

$$U_1 + P_1 V_1 = U_2 + P_2 V_2. \qquad (36.15)$$

Thus the enthalpy or heat function

$$H = U + PV \qquad (36.16)$$

is constant, and we now ask: Does this result in any change of temperature? If H is a function of temperature alone, and conversely $T = T(H)$, no such change will occur. However, if the enthalpy also depends on, say, the pressure, the gas seeping through will attain another temperature. The observed differences of pressure and temperature being $P_1 - P_2$ and $T_1 - T_2$, we may write

$$\frac{T_1 - T_2}{P_1 - P_2} = \left(\frac{\partial T}{\partial P}\right)_H = \xi. \qquad (36.17)$$

For an ideal gas this quantity is zero, since U and PV, and thus H, depends only on the temperature; for nonideal gases it will, as a rule, be positive. Exceptional in this respect are hydrogen, for which ξ is negative right down to about $-80°C$, and helium, for which it is negative down to very low absolute temperatures. This effect, the cooling of the substance by expansion against a lower pressure, is extensively utilized in refrigerators. Because of the anomalous properties of hydrogen and helium, the former must be cooled down by means of liquid air, and the latter in turn by liquid hydrogen. In many new plants devoted to the liquefaction of helium, only liquid air is used; the process employed here was first developed by the Russian physicist P. Kapitza.

Determination of Absolute Temperatures by Means of Nonideal Gases

We shall now see how measurement of the quantity ξ in (36.17) and the coefficient of thermal expansion for a nonideal gas may serve to determine the absolute temperature of the gas.

We shall utilize the equation

$$dH = T\,dS + V\,dP, \tag{36.18}$$

but introduce T as an independent variable insted of S, thus obtaining

$$dH = T\left(\frac{\partial S}{\partial T}\right)_P dT + \left[T\left(\frac{\partial S}{\partial P}\right)_T + V\right]dP,$$

or

$$\xi = \left(\frac{\partial T}{\partial P}\right)_H = -\frac{T(\partial S/\partial P)_T + V}{T(\partial S/\partial T)_P}. \tag{36.19}$$

We re-express the derivative in the numerator by means of Maxwell's relation (36.12),

$$\left(\frac{\partial S}{\partial P}\right)_T = -\left(\frac{\partial V}{\partial T}\right)_P = -V\alpha_P, \tag{36.20}$$

where α_P is the coefficient of thermal expansion at constant pressure; the denominator is similarly rewritten by means of the equation

$$\delta Q = T\,dS = T\left(\frac{\partial S}{\partial T}\right)_P dT + T\left(\frac{\partial S}{\partial P}\right)_T dP,$$

that is,

$$T\left(\frac{\partial S}{\partial T}\right)_P = \left(\frac{\partial Q}{\partial T}\right)_P = C_P. \tag{36.21}$$

The substitution of (36.20) and (36.21) in (36.19) then gives

$$\xi C_P = (\alpha_P T - 1)V. \tag{36.22}$$

The quantity $\xi C_P/V$ is small. If it can be assumed to vanish, we would have $T = 1/\alpha_P$; that is, the absolute temperature would be equal to the reciprocal of the coefficient of expansion at this temperature. For $\xi \neq 0$, this no longer holds.

The quantities ξ, C_P, and α_P are, in (36.22) taken to be measured by means of a thermometer with a Kelvin scale, that is, just the scale we wanted to establish by means of a nonideal gas. For increasing temperatures, however, these three quantities become steadily less dependent on the temperature scale; hence we may write, as an approximation,

$$T = \frac{1}{\alpha_P}\left(1 + \xi\frac{C_P}{V}\right). \tag{36.23}$$

From (36.23) we may, for instance, find the centigrade temperature of the absolute zero point; this is done indirectly by determining the absolute temperature T_0 at, say, 0°C. We then measure ξ, C_P, and α_P at one single temperature, for example, 0°C, 100°C, or any intermediate temperature. However, the results of such measurements, performed at various temperatures, will not be in exact agreement.

To set this right, we imagine that the temperature is measured by means of a gas thermometer, using the nonideal gas discussed; in this temperature scale the gas then obeys the equation of state

$$PV = RT'. \qquad (36.24)$$

Thus we have, at constant pressure,

$$\alpha_P = \frac{1}{V}\left(\frac{\partial V}{\partial T'}\right)_P \frac{dT'}{dT} = \frac{1}{T'}\frac{dT'}{dT}, \qquad (36.25)$$

whence by (36.22),

$$\frac{T}{T'}\frac{dT'}{dT} = 1 + \frac{\xi' C_P'}{V}, \qquad (36.26)$$

since, as is easily seen, $\xi' C_P'$ takes the same value for all temperature scales.

The general solution of (36.26) is

$$\log \frac{T}{T_0} = \int_{T_0'}^{T'} \frac{dT'}{T'[1 + \xi' C_P'/V]}. \qquad (36.27)$$

For the determination of T_0, the Kelvin temperature of the melting point of ice, we may then employ

$$\log \frac{T_0 + 100}{T_0} = \int_{T_0'}^{T_0' + 100} \frac{dT'}{T'[1 + \xi' C_P'/V]}. \qquad (36.28)$$

The value found experimentally is $T_0 = 273.16°K$, corresponding to the value $-273.16°C$ for the absolute zero point.

37. EQUATIONS OF STATE FOR NONIDEAL GASES

Measurement of the quantity ξ furnishes us with a possibility of generalizing the equation of state for an ideal gas, thus obtaining an equation of state with one adjustable parameter. Empirically it is found that ξ, for several gases, obeys the law

$$\xi C_P = \frac{a}{T^2} \qquad (37.1)$$

very accurately, over a large range of temperatures. Thus since

$$\alpha_P = \frac{1}{V}\left(\frac{\partial V}{\partial T}\right)_P, \tag{37.2}$$

we obtain from (35.22) the differential equation

$$T\left(\frac{\partial V}{\partial T}\right)_P = V + \frac{a}{T^2}, \tag{37.3}$$

with the solution

$$V = Tf(P) - \frac{a}{3T^2}. \tag{37.4}$$

In order that this be valid at high temperatures, we must put $f(P) = R/P$; in this way we arrive at the nonideal equation of state

$$PV = RT - \frac{aP}{3T^2}. \tag{37.5}$$

This equation, of course, applies only in a limited temperature range, since the right-hand side diverges for $T \to 0$.

The Van Der Waals Equation of State

The nonideal equation of state which is most widely known is that of van der Waals. For real gases at limited ranges of temperature it is not as accurately valid as (37.5), but it has the advantage of being applicable even to gases at the point of condensation. It may be written

$$\left(P + \frac{a}{V^2}\right)(V - b) = RT. \tag{37.6}$$

The first additional term a/V^2 arises from the assumption that cohesive forces act between the gas molecules, and the second term b expresses the assumption that these molecules have a finite volume.

The calculation of the quantity $\xi = (\partial T/\partial P)_H$ is now not so easy. However, it follows from (35.2) or (35.3) that if $\xi = 0$, $T = 1/\alpha_P$, or

$$\frac{V}{T} = V\alpha_P = \left(\frac{\partial V}{\partial T}\right)_P. \tag{37.7}$$

We may then by means of (37.7) and the equation of state determine the so-called inversion temperature, that is, the transition point from positive to negative ξ. For temperatures higher than that of inversion, the gas becomes warmer when it expands without performing any work, as in the Joule-Kelvin experiment; for lower temperatures it becomes correspondingly cooler. Thus, if this effect is to be utilized for refrigeration purposes, gases with low temperatures of inversion must first be cooled down past this point by other means.

We now differentiate the van der Waals equation (37.6) with respect to T, keeping P constant and replacing $\partial V/\partial T$ everywhere by V/T, and obtain

$$-\frac{2a}{V^2 T}(V-b) + \left(P + \frac{a}{V^2}\right)\frac{V}{T} = R,$$

or eliminating P,

$$\frac{VR}{V-b} - R = \frac{2a}{V^2 T}(V-b). \tag{37.8}$$

This yields the inversion temperature

$$T = \frac{2a}{Rb}\frac{(V-b)^2}{V^2}. \tag{37.9}$$

Hence this temperature increases with a, that is, with increasing cohesive forces; on the other hand, an increase of b, that is, of the molecule size or the repulsive forces, leads to a lowering of it.

Critical Constants of a Van Der Waals Gas

Finally, we shall demonstrate how the van der Waals equation may furnish useful information about the phenomenon of condensation, the transition from the gaseous to the liquid phase.

We write the equation of state as

$$V^3 - \left(b + \frac{RT}{P}\right)V^2 + \frac{a}{P}V - \frac{ab}{P} = 0, \tag{37.10}$$

as an ordered third-degree equation in V; it will then for given values of the temperature and pressure have three roots. At high temperatures and not too high pressures two of these roots will be complex conjugates, while the third will be real and positive.

At lower temperatures and higher pressures all the three roots will be real and at a certain temperature and pressure they will coincide; this triple root then denotes the so-called *critical volume* which obtains at the corresponding *critical temperature* and *critical pressure*.

The equation of state may be somewhat simplified by re-expressing the temperature, pressure, and volume in terms of these three critical quantities which of course are functions of the three constants a, b, and R in (37.10)

The condition for coincidence of the three roots is most easily obtained by solving the differentiated equation

$$V^2 - \frac{2}{3}\left(b + \frac{RT}{P}\right)V + \frac{a}{3P} = 0, \tag{37.11}$$

and requiring that its two roots coincide,

$$V_k = \frac{1}{3}\left(b + \frac{RT_k}{P_k}\right), \qquad V_k^2 = \frac{a}{3P_k}. \tag{37.12}$$

In addition, (37.10) yields the relation

$$\frac{ab}{P_k} = V_k^3. \tag{37.13}$$

The relations in (37.12) give us a and b as functions of V_k, T_k; however, these three quantities are not independent, due to (37.13). We find

$$a = 3P_k V_k^2, \qquad b = \tfrac{1}{3}V_k, \qquad P_k V_k = \tfrac{3}{8}RT_k. \tag{37.14}$$

Substituting these expressions in (37.10) and writing

$$P = pP_k, \qquad V = vV_k, \qquad T = tT_k, \tag{37.15}$$

we finally obtain the equation of state

$$\left(p + \frac{3}{v^2}\right)\left(v - \frac{1}{3}\right) = \frac{8}{3}t. \tag{37.16}$$

By means of (37.16) we may now draw a general pv-diagram for a van der Waals gas. For $t = 1$, we have $p \to \infty$ when $v \to \tfrac{1}{3}$, $p = 1$ when $v = 1$, and $p \to 0$ when $v \to \infty$. At this point $v = 1$ the tangent is horizontal, $\partial p/\partial v = 0$; this is known as the critical point. For $t > 1$ we obtain an "ordinary" isothermal which, however, may deviate strongly from a hyperbola in the vicinity of the critical point. For $t < 1$, two horizontal tangents appear, and the curve will have both a maximum and a minimum; between these two points, it will "slope the wrong way," that is, an increase of pressure will correspond to an increase of volume. This part of the curve and also a small section of it protruding beyond the external points, is to be regarded as unphysical; it is, as a rule, replaced by a horizontal straight-line segment, representing the liquid phase.

The three types of isothermals resulting from the van der Waals equation are demonstrated in Fig. 5-7. The dashed curve denotes the transition between the vapor and vapor + liquid states of aggregation, yielding at each point the volume of the pure liquid phase. It follows from certain thermodynamic laws, which we shall not discuss in detail, that the areas of the two

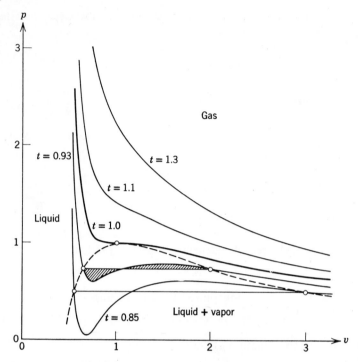

Fig. 5-7. Isothermals and phases, according to the van der Waals equation of state.

shaded—partly unphysical—regions in Fig. 5-7 must be equal; in other words, no net work is performed by an imagined cyclic process following the third-degree curve one way, and the horizontal isothermal back the other way. However, a formal treatment of this problem involves both the solution of a third-degree equation and the calculation of logarithmic integrals, and will therefore not be attempted here.

CHAPTER 6

Kinetic Theory of Gases

PRELIMINARY REMARKS

We have seen how the energy law leads to the conclusion that heat is only a form of motion that is not observable as such under ordinary conditions; the entities which participate in this motion, thus serving to accumulate kinetic energy, are the smallest parts of matter, notably molecules and atoms. Indirectly, however, this heat motion may be observed by various means. An observable phenomenon of primary importance in this respect is that of the so-called Brownian motions of colloidally suspended particles, for example, smoke or dust: such particles receive irregular impacts from the (still smaller) molecules, hence they exhibit a quivering sort of disordered motion.

These correct and fertile notions of the real nature of heat were, of course, soon subjected to theoretical investigations; here the laws of mechanics proved to be readily applicable. However, an additional element proved to be necessary for the construction of such a kinetic theory of heat, namely, the notion of the chaotic form of heat motion.

Thus the kinetic theory of gases takes an intermediate position between classical thermodynamics and the statistical mechanics to be developed later. It makes it possible to visualize directly the laws of thermodynamics; on the other hand, it is not based on such general propositions as statistical mechanics. By attacking the problems more directly and introducing the necessary statistical concepts gradually, it frequently succeeds in giving a more vivid and tangible picture of what is going on than does the more abstract theory of statistical mechanics. On the other hand, it is often more unwieldy in operation than statistical mechanics, where a general basis for the treatment of statistical problems has already been established. Nevertheless, several physical problems connected with the kinetic theory of heat are more easily approached by means of the classical kinetic theory of gases; hence it must, at any rate, be regarded as a very useful preliminary to the study of statistical mechanics.

The first attempts toward a kinetic theory of gases were made by Daniel

Bernoulli (1700–1782). In his previously mentioned book *Hydrodynamica* (1738), he showed that the Boyle-Mariotte law may be explained by means of the notion of moving gas particles. However, no further progress was made in this field until more than a century later; the founders of the first working kinetic theory of gases were, primarily, Clausius (1822–1888) and Maxwell (1831–1879). The further development and formal perfection of the theory now known as *statistical mechanics* owes, in particular, a great deal to Ludwig Boltzmann (1844–1906) and J. W. Gibbs (1839–1903). Maxwell discovered the law of velocity distribution that was later to bear his name; and one of the many great contributions of Boltzmann was the statistical interpretation of the *entropy* concept of classical thermodynamics, which he succeeded in linking with the notion of *thermodynamic probability*. The peak of perfection, with respect to the formulation of classical statistical mechanics, was attained by the work of Gibbs; this form has, moreover, proved to be readily applicable also in quantum statistical mechanics.

38. CONNECTION BETWEEN KINETIC ENERGY AND PRESSURE IN A GAS

Momentum and Pressure for Molecular Collisions

Let us consider a single free molecule in a rectangular box with edges a, b, c. We assume that the molecule has mass m and velocity $\mathbf{v}$, with components v_x, v_y, v_z. In elastic collisions with the walls the normal component will change signs; and the molecule will thus transmit the momentum $2mv_x$, say, to the wall perpendicular to the x-axis, and correspondingly for the other walls. This must be compensated by a counterpressure, which then for all the gas molecules taken together constitutes the hydrostatic pressure of the gas. Taking the counterpressure to be p_x, per molecule, we obtain

$$bcp_x = 2mv_x n_x, \tag{38.1}$$

where n_x is the number of impacts per second,

$$n_x = \frac{v_x}{2a}. \tag{38.2}$$

Substitution in (38.1) then yields

$$P_x = \frac{mv_x^2}{V}, \quad V = abc, \tag{38.3}$$

where V is the volume of the box.

Considering a large number of molecules of the same kind, we may now assume that the averages of the squares of the velocity components are equal, because of the disordered molecular motion obtaining at thermodynamic equilibrium. Furthermore, since a single molecule continually changes its direction of motion due to collisions with other molecules, we may assume that the same equality of averages holds for a single molecule, when the number of collisions is sufficiently high. Thus, since $\overline{v_x^2} = \tfrac{1}{3}v^2$, the contribution from a single molecule is

$$p_x = p_y = p_z = p = \frac{1}{3}\frac{mv^2}{V} = \frac{2}{3}\frac{w}{V}, \qquad (38.4)$$

where w is the kinetic energy of the molecule. Hence, for the whole gas,

$$PV = \frac{2}{3}\sum_i \left(\frac{1}{2}m_i v_i^2\right) = \frac{2}{3}W, \qquad (38.5)$$

where W is the total *translational kinetic energy* of the gas. This is the Boyle-Mariotte law, $PV = $ const, provided that W is a function of the temperature alone.

Equation of State for a Molecular Gas

Since the temperature is a characteristic of the thermal state of a gas (i.e. of its content of heat energy), we may define a temperature scale by putting W equal to an arbitrary function of the temperature. Thus, writing

$$W = \tfrac{3}{2}RT, \qquad (38.6)$$

we obtain

$$PV = RT. \qquad (38.7)$$

Hence the total aggregate of molecules behaves like an ideal gas as a consequence of which the temperature scale defined by (38.6) may be identified with the absolute thermodynamic scale.

If the gas is monatomic, so that the molecules possess no intrinsic rotational energy, the internal gas energy becomes

$$U = \tfrac{3}{2}RT. \qquad (38.8)$$

Furthermore, if we make the reasonable assumption that similar molecules attain, in the long run, the same average kinetic energy due to collisions with each other, it follows that the average translational energy of any molecule is

$$u = \tfrac{3}{2}kT, \qquad R = Nk, \qquad (38.9)$$

where N is the number of molecules in the gas. We shall show later that k, which is usually referred to as *Boltzmann's constant*, is the same for all gases, and thus a universal constant.

Partial Pressures in Mixtures of Gases

Since (38.4) applies to any single molecule, (38.5) will be valid for any aggregate of molecules; this leads to

$$PV = \frac{2}{3} \sum_m W_m, \qquad (38.10)$$

where W_m is the translational energy of the aggregate, which may contain molecules of quite different kinds. Hence, writing $W_m = \frac{3}{2} R_m T$, we obtain

$$PV = RT, \qquad R = \sum_m R_m. \qquad (38.11)$$

Equations (38.10) and (38.11) thus constitute an equivalent formulation of Dalton's law for the partial pressures of gas mixtures.

39. VELOCITY DISTRIBUTION FOR DISORDERED HEAT MOTION

Molecular Collisions

Up to this point we have regarded the molecules of a gas as mass points, and only suggested (somewhat loosely) that the velocity directions of the various molecules may be expected to change all the time due to intermolecular collisions. The next step, as long as we do not wish to consider the forces of molecular interactions in detail, is to represent the molecules by elastic spheres which interact by collisions only. This gives us the opportunity to study the distribution of velocities in an aggregate of molecules in thermodynamic equilibrium, that is, having a state of motion as disordered as possible.

We may then determine the distribution of velocity magnitudes for the molecules of a gas—which may be composed of molecules of one kind only, or of different kinds. First, however, we shall examine in more detail the distribution of *velocity directions*.

Intuitively, we feel that this must be isotropic, since no particular spatial direction has any privileged position, relative to other directions. For a gas in equilibrium, at any rate, the average of the velocity components of similar molecules will surely be zero, since (generally speaking) the same number of molecules must be moving in any one direction and the corresponding opposite direction.

We shall not give any formal proof of the isotropy of velocity direction, as such a proof would seem to be rather superfluous. However, we shall demonstrate, by a simple argument, that one must expect such an isotropic distribution to be reached very quickly.

Kinetic Theory of Gases | 401

We assume the molecules to be elastic spheres with the diameter d, so that the smallest distance between the centers of any two molecules is d. A molecule will then collide with other molecules having their centers inside a cylinder with radius d around the rectilinear trajectory of the first molecule. We shall assume that the probability of finding a molecule with its center anywhere inside this cylinder is uniform, that the distribution of molecules in the space available to them is completely uniform. This does not mean that the molecules are regularly distributed in space, like the atoms of a crystal lattice; on the contrary, they are moving freely, sometimes closely packed, sometimes far away from one another, with continually changing positions.

We now introduce a gross simplification by assuming one molecule to lie permanently at rest in a certain point of space (we may, for instance, regard it as having an infinite mass). We shall investigate the velocity distribution of the molecules which come in from a given direction to collide with the fixed molecule.

We draw a sphere of radius d about the center of the fixed molecule, and consider molecules that travel along a straight line at a distance b (the so-called impact parameter) from the parallel through the center of the sphere. The number of such molecules arriving during a given interval of time may be written $dN = 2\pi nb\, db$, where n is the average number passing through a unit area normal to the direction of motion.

If we denote the scattering angle by ϑ, it is evident from Fig. 6-1 that

$$b = d \cos \frac{\vartheta}{2} \qquad (39.1)$$

from which we obtain

$$dN = 2\pi nb\, db = -\pi n\, d^2 \sin\frac{\vartheta}{2} \cos\frac{\vartheta}{2} d\vartheta = -\pi n\, d^2 \cdot \frac{1}{2} \sin\vartheta\, d\vartheta. \qquad (39.2)$$

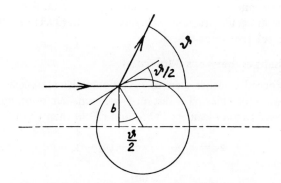

Fig. 6-1. Change of direction by molecular collisions.

It follows from (39.2) that the distribution of scattered molecules is completely isotropic and equal to $\tfrac{1}{4}n\,d^2$ per unit solid angle. Integrating both expressions (2), we find

$$2\pi n \int_0^d b\,db = -\tfrac{1}{2}\pi n\,d^2 \int_\pi^0 \sin\vartheta\,d\vartheta = n\pi\,d^2. \tag{39.3}$$

A more detailed investigation shows that exactly the same result may be deduced for the relative velocity between two molecules. Thus the relative velocity distribution will be practically uniform after, on the average, one collision per molecule.

Let us first consider two molecules with masses m_1 and m_2, colliding centrally with the velocities v_1 and v_2; after collision the velocities are u_1 and u_2. We then have

$$m_1(u_1^2 - v_1^2) = m_2(v_2^2 - u_2^2), \tag{39.4}$$

and

$$m_1(u_1 - v_1) = m_2(v_2 - u_2), \tag{39.5}$$

whence, by division,

$$u_1 + v_1 = v_2 + u_2, \tag{39.6}$$

or

$$u_1 - u_2 = -(v_1 - v_2). \tag{39.7}$$

In other words, the net result of the collision is that the relative velocity changes sign.

We may now, as a next step, imagine that the two molecules also have a relative velocity in the directions of the other two axes. These velocity components do not change at all with the collision. Hence we find, as a general result for collisions between two molecules of arbitrary masses, that the *relative normal component of velocity changes sign*, and nothing else.

We may then utilize Fig. 6-1 once more, referring the motion to a coordinate system in which one of the molecules is permanently at rest. The relative velocity and direction then correspond completely to the true velocity for the molecule previously considered, and (39.1) to (39.3) therefore also apply for relative velocities.

Energy Exchanges between Molecules

We now proceed another step to investigate the statistical exchange of energy between molecules. To this end we consider once more a central collision between two molecules; (39.5) and (39.6) then yield

$$\begin{aligned} u_1 &= v_1 + \frac{2m_2}{m_1 + m_2}(v_2 - v_1) \\ u_2 &= v_2 + \frac{2m_1}{m_1 + m_2}(v_1 - v_2). \end{aligned} \tag{39.8}$$

If, for instance, the masses are equal, it follows that the molecules simply exchange velocities, $u_1 = v_2$, $u_2 = v_1$. On the other hand, if we let $m_2 \to \infty$ and assume $v_2 = 0$, then $u_1 = -v_1$, $u_2 = 0$; that is, the infinitely massive molecule remains at rest, while the velocity of the other molecule changes sign.

Let us now consider two molecules with the relative velocity $\mathbf{V} = \mathbf{v}_1 - \mathbf{v}_2$. The center of the molecule must lie inside a cylinder with the radius d around the molecule m_1, in order that the two molecules collide. However, the probability that the molecule m_2 be in any point inside this cylinder is taken to be constant, and the point of contact at collision may consequently be in any direction from the center of the first molecule; we characterize this direction by a unit vector $\mathbf{s}$. The various possible directions of the latter are not equally probable, since the probability is proportional, not to the elements of area on the sphere but to the projections of these on a plane perpendicular to the direction of the relative velocity. Denoting the component of $\mathbf{s}$ in this direction by s_z, the probability of any given direction is then proportional to s_z.

We now consider a collision such as indicated in Fig. 6-2. Here only the normal component of $\mathbf{V}$ is of significance, that is,

$$\mathbf{s}(\mathbf{V}\mathbf{s}) = \mathbf{s}Vs_z = \mathbf{s}[(\mathbf{v}_1 - \mathbf{v}_2)_z s_z]. \tag{39.9}$$

By (39.8) this is to be multiplied by $2m_2/m_1 + m_2$ in order to find the change of velocity for the first molecule due to the impact, that is,

$$\mathbf{u}_1 = \mathbf{v}_1 - \frac{2m_2}{m_1 + m_2} \mathbf{s}[(\mathbf{v}_1 - \mathbf{v}_2)_z s_z]. \tag{39.10}$$

Let us calculate the average change of kinetic energy for all possible impact normals $\mathbf{s}$, corresponding to the right-hand side of the sphere in Fig. 6-2. Multiplying the equation

$$m_1(\mathbf{u}_1^2 - \mathbf{v}_1^2) = -\frac{4m_1 m_2}{m_1 + m_2}(\mathbf{v}_1 \mathbf{s})[(\mathbf{v}_1 - \mathbf{v}_2)_z s_z]$$
$$+ \frac{4m_1 m_2^2}{(m_1 + m_2)^2}[(\mathbf{v}_1 - \mathbf{v}_2)_z s_z]^2 \tag{39.11}$$

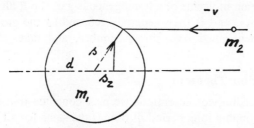

Fig. 6-2. Energy distribution by molecular collision; v = relative velocity.

by s_z, and putting

$$\overline{s_z} = \tfrac{1}{2}, \quad \overline{s_z{}^2} = \tfrac{1}{4}, \quad \overline{s_x s_z{}^2} = \overline{s_y s_z{}^2} = 0,$$

we then obtain

$$\overline{m_1(\mathbf{u}_1{}^2 - \mathbf{v}_1{}^2)} = -\frac{2m_1 m_2}{m_1 + m_2} v_{1z}(v_{1z} - v_{2z}) + \frac{2m_1 m_2{}^2}{(m_1 + m_2)^2}(v_{1z} - v_{2z})^2$$

$$= \frac{2m_1 m_2}{(m_1 + m_2)^2}[m_2 v_{2z}{}^2 - m_1 v_{1z}{}^2 + (m_1 - m_2)v_{1z}v_{2z}]. \quad (39.12)$$

We have assumed, in this discussion, that $v_{1z} > v_{2z}$. If $v_{2z} > v_{1z}$, the relative velocity will have the opposite sign, and corresponding considerations would hold for the left-hand side of the sphere in Fig. 6-2. Hence, if we average all the velocities v_{1z} and v_{2z}, the averages of the products $v_{1z}v_{2z}$ must vanish. Finally, by averaging over all directions of relative velocity, we obtain a factor $\tfrac{1}{3}$ at the right-hand side of (39.12). The final result then becomes, in the notation employed above,

$$\overline{m_1(\mathbf{u}_1{}^2 - \mathbf{v}_1{}^2)} = \frac{2}{3}\frac{m_1 m_2}{(m_1 + m_2)^2}(m_2\overline{\mathbf{v}_2{}^2} - m_1\overline{\mathbf{v}_1{}^2}). \quad (39.13)$$

This clearly implies that the molecules of one kind—in this case, those with mass m_1—continually gains in kinetic energy, as long as the average kinetic energy of the molecules of the other kind is larger, that is, $\overline{m_2 \mathbf{v}_2{}^2} > \overline{m_1 \mathbf{v}_1{}^2}$. A thermodynamic equilibrium is reached when the average energies of both kinds are equal, $\overline{\mathbf{u}_1{}^2} = \overline{\mathbf{v}_1{}^2}$, when

$$\tfrac{1}{2}\overline{m_1 \mathbf{v}_1{}^2} = \tfrac{1}{2}\overline{m_2 \mathbf{v}_2{}^2}. \quad (39.14)$$

This equation is of fundamental importance. It shows that the constant R in (37.8) and (37.9) is the same for all gases with the same number of molecules; in other words, the quantity in (39.9) is a universal constant, which (as previously mentioned) is called Boltzmann's constant. If the N in the same equation is conventionally fixed, to denote (as is usually done) the number of molecules in a gram molecule of a homogeneous gas, the R also has the same value for all gases, and is then generally referred as the *gas constant*. The number N is then called Avogadro's number, and takes the value $N = 0.606 \cdot 10^{24}$.

The Equipartition Theorem

From the preceding considerations we may conclude that the average or mean energy through a long period of time is the same for all molecules of a gas, as long as the latter can be represented as elastic spheres; furthermore this energy is equally distributed among all the three translational degrees of freedom.

This model only fits monatomic gases, such as the noble gases and certain metallic vapors. The great majority of molecules are diatomic and must therefore be represented by, say, elastic ellipsoids of rotation, or by two elastic spheres bound together so as to have a finite, more or less constant, mutual separation (i.e. a "dumbbell"-like structure).

Such a molecule has five degrees of freedom, three translational and two rotational. It now turns out that, in general, the average kinetic energy of any molecule is distributed equally among all its degrees of freedom. Thus a molecule of n degrees of freedom receives an average kinetic energy $\frac{1}{2}kT$ for each, that is, altogether the kinetic energy $\frac{1}{2}nkT$. In addition, some energy appears as potential energy, due to the molecular binding forces. If the binding is purely harmonic, it turns out that the potential energy for the two atoms of a diatomic molecule also contributes an amount $\frac{1}{2}kT$ to the average energy. The total mean energy of a diatomic molecule should then be $\frac{7}{2}kT$, while the rigid-molecule model results in the energy $\frac{5}{2}kT$. From the classical point of view a contradiction appears here, since a gradually hardening molecular binding will all the time give the larger energy of $\frac{7}{2}kT$. The transition from $\frac{7}{2}kT$ to $\frac{5}{2}kT$ is only to be understood by means of quantum theory, and will not be further discussed here.

This result, that the kinetic energy is distributed equally among all the degrees of freedom of the molecule, is called the *equipartition theorem*. We shall not give any complete proof of this theorem here, but content ourselves with some simple and—it is hoped—convincing considerations.

We assume that two elastic spheres have the masses m_1 and m_2, and introduce the center of mass coordinates and the relative coordinates

$$X = \frac{m_1 x_1 + m_2 x_2}{m_1 + m_2}, \qquad x = x_2 - x_1, \qquad \text{etc.} \qquad (39.15)$$

It is easily found that

$$(m_1 + m_2)\dot{X}^2 + \frac{m_1 m_2}{m_1 + m_2} \dot{x}^2 + m_1 \dot{x}_1^2 + m_2 \dot{x}_2^2. \qquad (39.16)$$

The total kinetic energy is thus partitioned into the kinetic energy of the motion of the center of mass, and that of the relative motion.

We know that

$$\overline{m_1 \dot{x}_1^2} = \overline{m_2 \dot{x}_2^2} = kT \qquad (39.17)$$

and assume that $\overline{\dot{x}_1 \dot{x}_2} = 0$, that the motion is completely disordered; this then leads to

$$(m_1 + m_2)\overline{\dot{X}^2} = \frac{\overline{m_1 \cdot m_1 \dot{x}_1^2 + m_2 \cdot m_2 \dot{x}_2^2}}{m_1 + m_2} = kT,$$

$$\frac{m_1 m_2}{m_1 + m_2} \overline{\dot{x}^2} = \frac{\overline{m_2 \cdot m_1 \dot{x}_1^2 + m_1 m_2 \dot{x}_2^2}}{m_1 + m_2} = kT. \quad (39.18)$$

Formally, therefore, the motion of the center of mass may be represented by a free particle with the mass

$$M = m_1 + m_2 \quad (39.19)$$

and the relative motion by one with the mass

$$m = \frac{m_1 m_2}{m_1 + m_2}. \quad (39.20)$$

The relative motion may, furthermore, be resolved into a radial component, describing the variation of the distance r between the particles, and a rotation about the center of mass, determined by the two well-known spherical polar coordinate angles ϑ and φ; thus we obtain

$$x = r \sin \vartheta \cos \varphi, \quad y = r \sin \vartheta \sin \varphi, \quad z = r \cos \vartheta, \quad (39.21)$$

$$m(\dot{x}^2 + \dot{y}^2 + \dot{z}^2) = m(\dot{r}^2 + r^2 \dot{\vartheta}^2 + r^2 \sin^2 \vartheta \dot{\varphi}^2). \quad (39.22)$$

Here we have, for instance,

$$\dot{r} = \dot{x} \sin \vartheta \cos \varphi + \dot{y} \sin \vartheta \sin \varphi + \dot{z} \cos \vartheta. \quad (39.23)$$

Taking the average of $\dot{r}^2$ and keeping the angles ϑ and φ constant, we find

$$\overline{\dot{r}^2} = \overline{\dot{x}^2}(\sin^2 \vartheta \cos^2 \varphi + \sin^2 \vartheta \sin^2 \varphi + \cos^2 \vartheta) = \overline{\dot{x}^2}; \quad (39.24)$$

it has then been assumed that $\overline{\dot{x}\dot{y}} = 0$, and so forth, and $\overline{\dot{x}^2} = \overline{\dot{y}^2} = \overline{\dot{z}^2}$.

Hence the mean kinetic energy is distributed with the same amount to the radial degree of freedom as to each of the three Cartesian degrees of freedom. Twice this amount then remains to be distributed among the two rotational degrees of freedom, and it is easily shown that the latter correspondingly receive equal amounts of average kinetic energy:

$$\overline{m\dot{r}^2} = \overline{mr^2 \dot{\vartheta}^2} = \overline{mr^2 \sin^2 \vartheta \dot{\varphi}^2}. \quad (39.25)$$

The only new feature introduced by these considerations of a diatomic molecule (or of a system of two molecules freely moving relative to each other) is that the motion may be resolved in several ways by introducing different kinds of "orthogonal" coordinates, and that the mean kinetic energy is distributed equally, with the amount $\frac{1}{2}kT$, to each of the corresponding degrees of freedom.

However, if we now assume that the two atoms of our system are rigidly connected, the only difference is that the relative radial motion disappears; only the two rotational degrees of freedom are then left. In this case we must evidently conclude that each of these degrees of freedom receives exactly the same average kinetic energy as before, namely $\frac{1}{2}kT$, while the energy formerly distributed to the radial relative motion now vanishes.

This is the content of the equipartition theorem; that the mean kinetic energy is distributed equally among all the degrees of freedom of the molecule. That it must be so, would appear from the preceding considerations to be evident, even though we have not stated a complete formal proof of this theorem for the general case.

40. MAXWELL'S VELOCITY DISTRIBUTION

Energy Exchanges and the Distribution of Velocities

We have seen that the molecules of a gas in thermodynamic equilibrium are distributed equally among all possible directions of motion, that is, the velocity distribution is *isotropic*. Furthermore, it has been shown by the equipartition theorem that the *mean energy* is equally distributed among all degrees of freedom. This, however, does not mean that the total velocity is the same for all molecules. On the contrary, the velocities of some molecules are large and of others small; but these velocity magnitudes are continually being exchanged among the molecules, so that their mean energy in the long run corresponds to a mean velocity.

We shall now investigate whether it is possible, by combined dynamical and statistical considerations, to determine the mean velocity distribution in an aggregate of molecules.

A simple argument, which is closely similar to the "proof" originally stated by Maxwell himself, but usually regarded as somewhat inadequate, is the following: Let us assume that the number of molecules having an x-component of velocity between v_x and $v_x + dv_x$ is proportional to $f(v_x)\,dv_x$. For an isotropic velocity distribution this would imply that the number of molecules with velocities between $\mathbf{v}$ and $\mathbf{v} + d\mathbf{v}$ is

$$dN = Nf(v_x)f(v_y)f(v_z)\,dv_x\,dv_y\,dv_z \tag{40.1}$$

with the normalization

$$\int_{-\infty}^{+\infty} f(x)\,dx = 1. \tag{40.2}$$

Due to the isotropy, we must now assume that

$$w(v_x, v_y, v_z) = f(v_x)f(v_y)f(v_z) \tag{40.3}$$

depends only on $|\mathbf{v}|$, and not on the components of $\mathbf{v}$, that

$$\begin{aligned} w(x, y, z) = f(x)f(y)f(z) = \text{const} \\ \text{for} \quad x^2 + y^2 + z^2 = \text{const.} \end{aligned} \tag{40.4}$$

Differentiating these expressions and dividing by w, we obtain

$$\begin{aligned} \frac{f'(x)}{f(x)} dx + \frac{f'(y)}{f(y)} dy + \frac{f'(z)}{f(z)} dz = 0, \\ \text{for} \quad x\,dx + y\,dy + z\,dz = 0. \end{aligned} \tag{40.5}$$

Lagrange's method of undetermined multipliers then leads to

$$\left[\frac{f'(x)}{f(x)} - \lambda x\right] dx + \left[\frac{f'(y)}{f(y)} - \lambda y\right] dy + \left[\frac{f'(z)}{f(z)} - \lambda z\right] dz = 0, \tag{40.6}$$

where now dx, dy, dz may vary freely. This implies that

$$f'(x) - \lambda x f(x) = 0, \quad \text{etc.,}$$

$$f(x) = C' e^{-(\lambda/2)x^2}, \tag{40.7}$$

$$w(v_x, v_y, v_z) = C e^{-(\lambda/2)v^2}$$

The normalization (40.2) yields

$$C = \left(\frac{\lambda}{2\pi}\right)^{3/2} \tag{40.8}$$

since, as is well known,

$$\int_{-\infty}^{+\infty} e^{-\alpha x^2} dx = \left(\frac{\pi}{\alpha}\right)^{1/2}.$$

The constant λ in (40.7) is not universal, but depends on the molecular masses and the temperature. To determine this dependence we must calculate the mean kinetic energy

$$\overline{\tfrac{1}{2}m\mathbf{v}^2} = \iiint_{-\infty}^{+\infty} w(v_x, v_y, v_z) \tfrac{1}{2}m\mathbf{v}^2 \, dv_x \, dv_y \, dv_z. \tag{40.9}$$

It is easily seen that this integral may be obtained from the preceding integral of normalization

$$\iiint_{-\infty}^{+\infty} e^{-(\lambda/2)v^2} dv_x \, dv_y \, dv_z = \frac{1}{C} = \left(\frac{2\pi}{\lambda}\right)^{3/2} \tag{40.10}$$

Kinetic Theory of Gases | 409

by differentiating with respect to $-\lambda$ and multiplying by m. We thus arrive at the result

$$\frac{1}{2}m\mathbf{v}^2 = -mC\frac{\partial}{\partial \lambda}\left(\frac{1}{C}\right) = m\frac{1}{C}\frac{\partial C}{\partial \lambda} = m\frac{\partial}{\partial \lambda}\log C = \frac{3}{2}\frac{m}{\lambda}. \quad (40.11)$$

This average kinetic energy is, however, the same as that which we have previously found to equal $\frac{3}{2}kT$. This leads to

$$\lambda = \frac{m}{kT}, \quad (40.12)$$

so that Maxwell's law of velocity distribution should read

$$w(\mathbf{v}) = \left(\frac{m}{2\pi kT}\right)^{3/2} \exp\left(-\frac{m}{2kT}\mathbf{v}^2\right). \quad (40.13)$$

See Fig. 6-3, and also Fig. 6-4.

Introducing the abbreviation $U = \frac{1}{2}m\mathbf{v}^2$, this law may be written, in a simplified version, as

$$w = \text{const } e^{-U/kT}. \quad (40.14)$$

This is Boltzmann's distribution law, which applies even if U denotes the

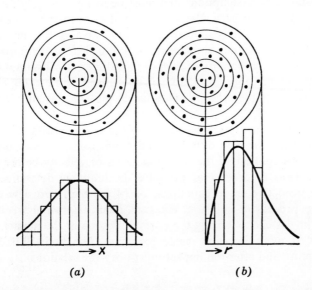

Fig. 6-3. Illustration of Maxwell's distribution law by means of a gun target. Left: the probability of lateral deviations, $f(x) = \text{const} \times e^{-\frac{1}{2}x^2}$. Right: the probability of radial deviations, $W(r) = \text{const} \times re^{-\frac{1}{2}r^2}$. The number of "hits" in each interval is drawn, for comparison, in a horizontal and a radial projection.

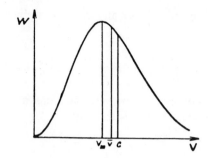

Fig. 6-4. Maxwell's velocity distribution law $W(v) = \text{const} \times v^2 e^{-\frac{1}{2}v^2}$. Compare (41.4). The most frequently occurring velocity; is v_m; $\bar{v}$ is the mean velocity; c is the velocity corresponding to the mean energy, $c^2 = \bar{v}^2$.

total, and not only the kinetic, energy of a molecule with velocity **v** and position x, y, z.

A More Rigorous Deduction of Maxwell's Distribution Law

We shall now give a more rigorous deduction of Maxwell's velocity distribution law, where the mechanical nature of the collision process is more clearly exhibited. In this way, we also obtain a better idea of the minimum number of statistical assumptions that must be made.

We shall discuss two particular kinds of collision; in the former, the velocities of the two molecules are to be $\mathbf{v}_1$ and $\mathbf{v}_2$ before the impact, and $\mathbf{u}_1$ and $\mathbf{u}_2$ after; in the latter, conversely, $\mathbf{u}_1$ and $\mathbf{u}_2$ before the impact and $\mathbf{v}_1$ and $\mathbf{v}_2$ after. If the probability of each of these kinds of collision is the same, the mean velocity distribution will remain unchanged; while, if the one probability is greater than the other, a continual change of this distribution will result.

We assume that the masses m_1 and m_2 of the two colliding molecules are different. Utilizing the laws of energy and momentum, we then obtain four scalar equations connecting the six components of velocity before and after the impact. Thus in order that the latter be uniquely determined by the former, we need two more equations, or (if we like) two new quantities defining the process of collision. As such quantities we may choose the directional angles characterizing the normal vector **s** at the impact.

It follows from our preceding results [compare (39.8) and (39.10)], that the velocities before and after the impact must obey the relations

$$\mathbf{u}_1 = \mathbf{v}_1 - \frac{2m_2}{m_1 + m_2} [(\mathbf{v}_1 - \mathbf{v}_2)\mathbf{s}]\mathbf{s},$$

$$\mathbf{u}_2 = \mathbf{v}_2 + \frac{2m_1}{m_1 + m_2} [(\mathbf{v}_1 - \mathbf{v}_2)\mathbf{s}]\mathbf{s}. \quad (40.15)$$

By subtraction, it follows that

$$\mathbf{u}_1 = \mathbf{u}_2 = \mathbf{v}_1 - \mathbf{v}_2 - 2[(\mathbf{v}_1 - \mathbf{v}_2)\mathbf{s}]\mathbf{s} \qquad (40.16)$$

or

$$(\mathbf{u}_1 - \mathbf{u}_2)\mathbf{s} = -(\mathbf{v}_1 - \mathbf{v}_2)\mathbf{s}. \qquad (40.17)$$

The latter equation says that the relative velocity component along the impact normal changes sign on impact. The other components remain unchanged. Hence it clearly follows that for the two particular kinds of collision here envisaged, that is, the transition from preimpact velocities $\mathbf{v}_1, \mathbf{v}_2$ to postimpact velocities $\mathbf{u}_1, \mathbf{u}_2$, and vice versa, the direction of the impact normal vector must be the same. However, since (40.17) remains invariant when the sign of $\mathbf{s}$ is changed, such a change of sign corresponds to taking the vector $\mathbf{s}$ to point from the molecule m_1 to the molecule m_2.

Let us now, for a moment, return to (39.8) and calculate the functional determinant

$$\frac{\partial(u_1, u_2)}{\partial(v_1, v_2)} = \frac{1}{(m_1 + m_2)^2} \begin{vmatrix} m_1 - m_2 & 2m_2 \\ 2m_1 & m_2 - m_1 \end{vmatrix} = -1. \qquad (40.18)$$

Since the velocity components perpendicular to the impact normal do not change, we therefore find the following important connection between the velocity differentials before and after impact

$$du_{1x}\,du_{1y}\,du_{1z}\,du_{2x}\,du_{2y}\,du_{2z} = dv_{1x}\,dv_{1y}\,dv_{1z}\,dv_{2x}\,dv_{2y}\,dv_{2z} \qquad (40.19)$$

by transformation to an arbitrary fixed coordinate system. This represents a special case of a very general theorem of statistical mechanics: that the "phase cells" (in coordinate-momentum space) has an invariant volume. Equation 40.19 thus presupposes that the molecules have, statistically speaking, a uniform spatial distribution, that the probability of finding a molecule in any infinitesimal element of volume is proportional to the volume $dx\,dy\,dz$ of the element.

Since, by (40.17), the normal component of the relative velocity of the two molecules is the same for both kinds of collision discussed, the spatial distribution will not affect our statistical considerations. We assume that the velocity distributions of the two kinds of molecules are given as functions $w_1(\mathbf{v}_1)$ and $w_2(\mathbf{v}_2)$. The condition for statistical equilibrium is then

$$w_1(\mathbf{v}_1)w_2(\mathbf{v}_2)\,d\mathbf{v}_1\,d\mathbf{v}_2 = w_1(\mathbf{u}_1)w_2(\mathbf{u}_2)\,d\mathbf{u}_1\,d\mathbf{u}_2, \qquad (40.20)$$

where $d\mathbf{u}_1$, and so forth, represent an abbreviated notation for the elements of volume (40.19); from the latter, it then follows that

$$w_1(\mathbf{v}_1)w_2(\mathbf{v}_2) = w_1(\mathbf{u}_1)w_2(\mathbf{u}_2). \qquad (40.21)$$

This must be valid for any collision, that is, for all velocity changes satisfying the laws of energy and momentum

$$U = \tfrac{1}{2}m_1\mathbf{v}_1^2 + \tfrac{1}{2}m_2\mathbf{v}_2^2 = \tfrac{1}{2}m_1\mathbf{u}_1^2 + \tfrac{1}{2}m_2\mathbf{u}_2^2,$$
$$\mathbf{p} = m_1\mathbf{v}_1 + m_2\mathbf{v}_2 = m_1\mathbf{u}_1 + m_2\mathbf{u}_2. \quad (40.22)$$

Hence $w_1(\mathbf{v}_1)w_2(\mathbf{v}_2)$ must be a function of the four scalar quantities U, p_x, p_y, p_z; this may be conveniently expressed as

$$\log w_1(\mathbf{v}_1) + \log w_2(\mathbf{v}_2) = F(U, p_x, p_y, p_z). \quad (40.23)$$

Since $w_1(\mathbf{v}_1)$ depends only on $\mathbf{v}_1$, and $w_2(\mathbf{v}_2)$ only on $\mathbf{v}_2$, the function $F(U, p_x, p_y, p_z)$ must clearly be linear in U and $\mathbf{p}$, so that we may write

$$\log w_1(\mathbf{v}_1) + \log w_2(\mathbf{v}_2) = -\beta(U - \mathbf{v}_0\mathbf{p} + U_0), \quad (40.24)$$

where U_0 is a constant.

A suitable choice of U_0 yields

$$w_1(\mathbf{v}_1) = \left(\frac{\beta m_1}{2\pi}\right)^{3/2} \exp\left[-\frac{\beta m_1}{2}(\mathbf{v}_1 - \mathbf{v}_0)^2\right],$$
$$w_2(\mathbf{v}_2) = \left(\frac{\beta m_2}{2\pi}\right)^{3/2} \exp\left[-\frac{\beta m_2}{2}(\mathbf{v}_2 - \mathbf{v}_0)^2\right]. \quad (40.25)$$

These laws of velocity distribution are slightly more general than the previously established Maxwell's law, since they describe a disordered molecular motion centered around a common translational velocity $\mathbf{v}_0$. Hence they

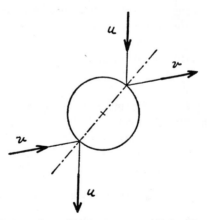

Fig. 6-5. To illustrate opposite collision processes which maintain, on the average, a stationary velocity distribution, provided they are both equally probable. Only the relative velocities, i.e. those of one molecule in a coordinate system where the other remains permanently at rest, are drawn.

correspond to the velocity distribution of a gas in statistical equilibrium, situated in a container that moves with a uniform macroscopic velocity $\mathbf{v}_0$. However, we may refer the motion to any Galilean system by the mechanical principle of relativity. Choosing the container itself as a coordinate system, we arrive back at the previously formulated Maxwell's law. (See Fig. 6-5.)

The parameter β is easily found, as earlier, to have the value

$$\beta = \frac{1}{kT}. \tag{40.26}$$

This is done by calculating the mean kinetic energy of a molecule, and subtracting from this the mean translational energy $\frac{1}{2}m_1\mathbf{v}_0^2$ or $\frac{1}{2}m_2\mathbf{v}_0^2$. The same result may be obtained by calculating directly the average heat energy $\frac{1}{2}m_1(\mathbf{v}_1 - \mathbf{v}_0)^2$ or $\frac{1}{2}m_2(\mathbf{v}_2 - \mathbf{v}_0)^2$.

41. MEAN VALUES FOR COLLISION PROCESSES

Mean Speeds and Mean Square Speeds

We now utilize Maxwell's velocity distribution law to calculate several average quantities which are of importance for the application of kinetic gas theory to physical observations; it is then convenient to write this law in the simplified form

$$w(\mathbf{v}) = \left(\frac{\beta m}{2\pi}\right)^{3/2} \exp\left(-\frac{\beta m}{2}\mathbf{v}^2\right), \qquad \beta = \frac{1}{kT}. \tag{41.1}$$

The mean square speed is found from the previously discussed equation

$$\frac{1}{2}\overline{m\mathbf{v}^2} = \frac{3}{2}kT = \frac{3}{2}\frac{1}{\beta},$$

that is,

$$\overline{\mathbf{v}^2} = \frac{3}{m\beta} = c^2. \tag{41.2}$$

The mean speed $\bar{v} = |\bar{\mathbf{v}}|$ is most easily arrived at in quite a different manner. Instead of the Cartesian velocity components v_x, v_y, v_z, we introduce into "velocity space" spherical polar coordinates v, Θ, Φ, completely analogous to the ordinary spherical polar spatial coordinates. In order to find the mean speed

$$v = (v_x^2 + v_y^2 + v_z^2)^{1/2} \tag{41.3}$$

or powers of v, we may integrate over the coordinates Θ and Φ (thus obtaining a factor 4π) and deduce another function

$$W(v) = \left(\frac{2\beta m}{\pi}\right)^{1/2} \exp\left(-\frac{\beta m}{2}v^2\right)\beta m v^2, \tag{41.4}$$

describing the mean speed distribution.* This means that of a total number of N molecules, a number

$$dN = NW(v)\,dv \qquad (41.5)$$

have mean speeds between v and $v + dv$.

Since

$$\int_0^\infty \exp\left(-\frac{\beta m}{2} v^2\right) v^3 \, dv = \frac{2}{(\beta m)^2},$$

it now follows from (41.4) and (41.5) that

$$v = \overline{|\mathbf{v}|} = 2\left(\frac{2}{\pi\beta m}\right)^{1/2} = \left(\frac{8}{3\pi}\right)^{1/2} c. \qquad (41.6)$$

Hence the mean speed is less than the speed c which, by (41.2), yields the correct value for the average energy. This is easily understood: we may, for instance, take three molecules with speeds 1, 2, and 3 (in suitable units), which yields

$$\bar{v} = 2 \quad \text{and} \quad c = \left(\frac{14}{3}\right)^{1/2} \approx 2.16.$$

Mean Relative Speeds

The mean values of relative speeds will be somewhat larger than those of molecular mean speeds. First, we simplify matters by assuming that the molecules have the same speed c and an isotropic velocity distribution. For molecules traveling in the same direction the relative speed will then be zero, for oppositely directed molecules $2c$, and for molecules with mutually perpendicular trajectories $c\sqrt{2}$. Let the velocity directions of two molecules be drawn in an auxiliary coordinate system (in velocity space) and denote the angle between these two directions by ϑ; the frequency of occurrence of this angle is then proportional to $\sin\vartheta \, d\vartheta$. In this way we easily find, since the relative speed is $2c \sin\vartheta/2$, that

$$\bar{V} = \overline{|\mathbf{V}|} = \frac{1}{2}\int_0^\pi 2c \sin\frac{\vartheta}{2} \sin\vartheta \, d\vartheta = \frac{4}{3}c,$$
$$C^2 = \overline{\mathbf{V}^2} = \frac{1}{2}\int_0^\pi 4c^2 \sin^2\frac{\vartheta}{2} \sin\vartheta \, d\vartheta = 2c^2. \qquad (41.7)$$

Thus we see, once more, that $\bar{v}^2 < \overline{\mathbf{v}^2}$, since $(\frac{4}{3})^2 = \frac{16}{9} < 2$.

The correct mean relative speeds are now found by utilizing the product function

$$w(\mathbf{v}_1)w(\mathbf{v}_2) = \left(\frac{\beta m}{2\pi}\right)^3 \exp\left[-\frac{\beta m}{2}(\mathbf{v}_1^2 + \mathbf{v}_2^2)\right]; \qquad (41.8)$$

* Compare Fig. 6-4.

we then introduce the new variables

$$\mathbf{V} = \mathbf{v}_1 - \mathbf{v}_2, \qquad \mathbf{V}' = \tfrac{1}{2}(\mathbf{v}_1 + \mathbf{v}_2) \tag{41.9}$$

to obtain

$$w(\mathbf{v}_1)w(\mathbf{v}_2) = \left(\frac{\beta m}{2\pi}\right)^3 \exp\left[-\frac{\beta m}{2}\left(\frac{1}{2}\mathbf{V}^2 + 2\mathbf{V}'^2\right)\right]. \tag{41.10}$$

Hence it follows, without any more detailed calculations, that the mean relative speed $|\mathbf{V}|$ is $\sqrt{2}$ times the corresponding mean speed of the separate molecules. The mean values of higher powers are obtained by multiplying the corresponding mean value for a single molecule by the same power of $\sqrt{2}$.

The most important result of the preceding considerations is that

$$\bar{V} = \bar{v}\sqrt{2} = \left(\frac{16}{2\pi}\right)^{1/2} c. \tag{41.11}$$

For two kinds of molecules, with the relative mass m, we shall generally have

$$\sqrt{M}\,\bar{V}' = \sqrt{m}\,\bar{V} = \sqrt{m_1}\,\bar{v}_1 = \sqrt{m_2}\,\bar{v}_2,$$
$$M = m_1 + m_2, \qquad m = m_1 m_2/(m_1 + m_2), \tag{41.12}$$

where V' and V are the center of mass and relative speeds: M and m the total and relative masses of the two molecules.

Collision Number and Mean Free Path of Molecules in a Gas

We are now able to calculate exactly how frequently a molecule of a given size collides with other molecules in a gas where Maxwell's velocity distribution law holds. Let us consider only similar molecules, represented by spheres of diameter d, as previously; let the number of molecules per unit volume be N. In a time interval dt, the probability that a molecule will collide with another will be equal to the probability of finding a molecule within a cylinder of radius d and height $\bar{V}\,dt$. The mean number of impacts per unit time (the collision number) is therefore

$$n = N\pi\,d^2\bar{V} = N\pi\,d^2\bar{v}\sqrt{2}, \tag{41.13}$$

whence it follows that the mean free path is

$$\lambda = \frac{\bar{v}}{n} = \frac{1}{N\pi\,d^2}\sqrt{2}. \tag{41.14}$$

We note that the mean free path is independent of the temperature, and inversely proportional to the density.

Just as the real speed of a molecule will deviate from the mean speed, the real paths traversed between collisions deviate from the mean free path. Let us denote by $g(x_1)$, the probability that a molecule will traverse a path length x_1 without colliding. The probability that it will further traverse another path length x_2 without colliding is then $g(x_2)$, and the probability that it will correspondingly traverse the whole path length $x_1 + x_2$ is

$$g(x_1)g(x_2) = g(x_1 + x_2).$$

This may be written

$$\log g(x_1) + \log g(x_2) = \log g(x_1 + x_2), \tag{41.15}$$

from which it follows that $\log g(x)$ is a linear function

$$\log g(x) = -\frac{x}{\lambda} + \text{const}, \tag{41.16}$$

$$g(x) = \text{const } e^{(-x/\lambda)};$$

the exponential sign must be negative, in order that the probability decrease with distance. The constant of integration in (41.16) must be set equal to unity, since all molecules will certainly traverse an infinitesimal path length between collisions. The parameter λ is, as we shall now demonstrate, equal to the mean free path.

By the preceding considerations we may write

$$g(x) = e^{-(x/\lambda)}. \tag{41.17}$$

The probability that the real path length traversed by a molecule will lie between x and $x + dx$ is then

$$g(x) - g(x + dx) = -dg(x) = e^{-(x/\lambda)} \cdot \frac{dx}{\lambda} \tag{41.18}$$

whence the mean free path is

$$\int_0^\infty e^{-(x/\lambda)} \cdot \frac{x\,dx}{\lambda} = \lambda. \tag{41.19}$$

(See Fig. 6-6.)

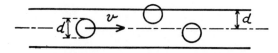

Fig. 6-6. To illustrate the probability of collision of a molecule with velocity v, relative to a group of other molecules.

42. TRANSPORT OF MASS, ENERGY, AND MOMENTUM IN A GAS

Smoothing-Out of the Properties of an Inhomogeneous Gas

Up to now we have dealt exclusively with gases in which the molecules are in statistical equilibrium. Disregarding external fields of force, this means that the gases must be homogeneous, both with regard to composition and physical properties. For gas mixtures each kind of molecules must be uniformly distributed in space and the temperature must be constant; otherwise the gas will not be in equilibrium and a continual smoothing-out will result.

For our simplest model of molecules, representing them as mass points, such a smoothing-out will happen almost instantaneously, since there are no appreciable obstacles to the molecular motions. This will not be the case if the molecules are assumed to have a certain spatial extension, and thus to be continually subject to collisions with neighboring molecules. Having traversed a path length of the order of the mean free path λ, the molecules will then (after the impact) start traversing another path which will be in an opposite direction just as often as parallel to the first path.

The net result of this is that molecules on the whole stay inside a limited spatial volume for a considerable length of time. Thus if we subdivide the spatial region available to the gas into a large number of small elements of volume, we may assume that the molecules inside any one of these elements are in approximate statistical equilibrium.

In order to see how the slow smoothing-out of inhomogeneities takes place, we now assume, for simplicity, that the inhomogeneity only appears in one spatial direction, along the x-axis, say. For instance, the concentration n of a certain kind of molecule may increase or decrease in this direction, so that we have a gradient of concentration $\partial n/\partial x$; or there may be a corresponding variation of temperature, resulting in a temperature gradient $\partial T/\partial x$, and so on.

A feature common to all of these problems is that we may introduce a very drastic (and convenient) simplification by assuming that the thermal motions of the molecules are distributed along the three axes of spatial direction. Thus only a third of the molecules move along the x-axis, traversing between collisions a path length on the average equal to the mean free path λ; and only half of this third, one sixth, move in the positive x-direction, while the other sixth move in the negative x-direction.

That such a drastic simplification actually works may be explained as follows: A molecule whose velocity direction creates an angle ϑ with the x-axis moves, on the average, only on a mean free path $\lambda \cos \vartheta$ in the x-direction. Let E be a property of the gas with a constant gradient in the x-direction. The molecules may then, on the average, be assumed to have the same property. Between collisions they then carry this property with them from the point x, where it belongs, to the point $x + \lambda \cos \vartheta$, where the same property is normally given by

$$E(x + \lambda \cos \vartheta) = E(x) + \lambda \cos \vartheta \, \frac{\partial E}{\partial x}. \qquad (42.1)$$

Thus the difference in E at the two points is proportional to $\cos \vartheta$, and since the direction of motion must also be taken into account for this transport, quantities proportional to $\cos^2 \vartheta$ appear. But $\overline{\cos^2 \vartheta} = \tfrac{1}{3}$ for an isotropic velocity distribution; hence it is this proportionality that makes possible the simplification mentioned above.

Diffusion

In spite of the obstacle to free motion represented by the very short mean free path, a molecule will, nevertheless, in the long run drift from one region of the gas to another; this phenomenon is called *diffusion*. As long as the gas consists of only one kind of molecule, the diffusion is not observable macroscopically. For gas mixtures of variable concentrations it appears as a slow blending, ultimately resulting in complete homogeneity.

Let us assume that we have two gases in a mixture, the molecular concentrations being $n_1(x)$ and $n_2(x)$, with gradients $n_1'(x) = \partial n_1/\partial x$, $n_2'(x) = \partial n_2/\partial x$. Denoting by G_1 and G_2 the number of molecules of each kind crossing per unit time through a unit area perpendicular to the x-axis, we have by definition

$$G_1 = -D_{12} \frac{\partial n_1}{\partial x},$$
$$G_2 = -D_{21} \frac{\partial n_2}{\partial x}, \qquad (42.2)$$

where D_{12} and D_{21} are coefficients of diffusion which will presently be shown to be equal,

$$D_{12} = D_{21} = D. \qquad (42.3)$$

We now calculate G_1 and G_2. Each of these is the difference between two quantities, namely, the numbers of molecules crossing from the left and from the right. Since both may be assumed to have traversed a mean path λ_1 since the last collision, we have to consider the concentrations in $x - \lambda_1$ and $x + \lambda_1$, respectively. If $\bar{v}_1$ is the mean speed, we then obtain (disregarding any translational motion of the gas as a whole)

$$G_1 = \frac{1}{6} \bar{v}_1 [n_1(x - \lambda_1) - n_1(x + \lambda_1)] = -\frac{1}{3} \bar{v}_1 \lambda_1 \frac{\partial n_1}{\partial x}. \qquad (42.4)$$

For G_2 we find a corresponding equation.

However, it must be noted that for a state of statistical equilibrium $n_1 + n_2$ is independent of both position and time. In other words,

$$\frac{\partial n_1}{\partial x} + \frac{\partial n_2}{\partial x} = 0, \qquad (42.5)$$

while the time independence implies that

$$G_1 + G_2 = 0. \qquad (42.6)$$

This, however, is not possible if $\bar{v}_1 \lambda_1 \neq \bar{v}_2 \lambda_2$. Hence if the diffusions of the two gases in opposite directions are different, the resulting change of $n_1 + n_2$ must be compensated by a small displacement of the gas as a whole.

Denoting the velocity of this displacement by V, the true number of molecules crossing the unit area at x then becomes

$$G_1 = n_1 V - \frac{1}{3} \bar{v}_1 \lambda_1 \frac{\partial n_1}{\partial x},$$
$$G_2 = n_2 V - \frac{1}{3} \bar{v}_2 \lambda_2 \frac{\partial n_2}{\partial x}. \qquad (42.7)$$

Eliminating V from these equations and writing

$$G_1 = -G_2 = G, \quad \frac{\partial n_1}{\partial x} = -\frac{\partial n_2}{\partial x} = \frac{\partial n}{\partial x}, \qquad (42.8)$$

we obtain

$$G = -D \frac{\partial n}{\partial x},$$
$$D = \frac{n_2 \bar{v}_1 \lambda_1 + n_1 \bar{v}_2 \lambda_2}{3(n_1 + n_2)}. \qquad (42.9)$$

By (42.2) and (42.8) it follows that (42.3) is satisfied: the two diffusion coefficients D_{12} and D_{21} originally introduced are identical and equal to D.

A weak point of this discussion would seem to be the assumption of mean free paths λ_1 and λ_2 for the two kinds of molecule. These path lengths will, in reality, depend on both concentrations n_1 and n_2. For pure gases, λ is determined chiefly by the molecular size. In the case of a gas mixture, the heavier molecules will, on the whole, have a shorter mean free path than the lighter ones—irrespective of the diameters d_1 and d_2—because of their less rapid thermal motions.

We may also speak of molecular diffusion in a homogeneous gas, even if this phenomenon would be difficult to observe in practice, since the composition of the gas remains unchanged. In this case the coefficient of diffusion takes the form

$$D = \frac{1}{3}\bar{v}\lambda = \frac{\eta}{\rho},\qquad(42.10)$$

where, as we shall see later, η is the viscosity.

The Diffusion Law

In order to find how the concentration of a component of a gas mixture varies with time and position, we must deduce a differential equation, the so-called diffusion law.

Let us consider a rectangular prism with a unit base area and the height dx along the x-axis. The increase in the number of molecules of the kind discussed, $n = n_1$, is then on the one side $dx \cdot \partial n/\partial t$, and on the other side

$$G(x) - G(x + dx) = -\frac{\partial G}{\partial x}dx.$$

By (42.9) this yields the diffusion law

$$\frac{\partial n}{\partial t} = D\frac{\partial^2 n}{\partial x^2}.\qquad(42.11)$$

For more complicated problems in which the concentration gradient is no longer linear, this equation may easily be generalized, to take the form

$$\frac{\partial n}{\partial t} = D\,\mathbf{\nabla}^2 n.\qquad(42.12)$$

Disordered Migration of Molecules

We shall now consider the problem of molecular migration from a somewhat different angle. Assuming that a molecule on the average traverses between collisions a path length equal to the mean free path λ, we find that the projection of all these path lengths on (say) the x-axis must be equal to $\lambda/2$ on the average if projections along positive and negative x-directions are considered separately.

After a total number of N collisions let the numbers of positively and negatively directed projections be $\tfrac{1}{2}N + s$ and $\tfrac{1}{2}N - s$, respectively. The molecule has then drifted a distance $2s \cdot \lambda/2 = s\lambda$ along the x-axis. We shall now determine the probability $P(s\lambda)$ that the x-coordinate of the molecule has this value.

The total number of possibilities for collisions of any kind is

$$2^N = \sum_s \binom{N}{\tfrac{1}{2}N + s}.\qquad(42.13)$$

The number of possibilities for $q = \frac{1}{2}N + s$ "positive" collisions is

$$\frac{N(N-1)\cdots(N-q+1)}{1\cdot 2\cdots q} = \binom{N}{q} = \frac{N!}{(\frac{1}{2}N+s)!\,(\frac{1}{2}N-s)!}. \quad (42.14)$$

Consequently we have

$$P(s\lambda) = \frac{N!\,2^{-N}}{(\frac{1}{2}N+s)!\,(\frac{1}{2}N-s)!}. \quad (42.15)$$

If N, q, and $p = N - q$ are large numbers, we may use Stirling's formula

$$n! = n^n e^{-n}\sqrt{2\pi n} \quad (42.16)$$

for the factorials and approximately put $s/N \approx 0$; this then yields

$$P(s\lambda) = \left(\frac{2}{\pi N}\right)^{1/2}\left(1 - \frac{4s^2}{N^2}\right)^{N/2} = \left(\frac{2}{\pi N}\right)^{1/2}\exp\left(-\frac{2s^2}{N}\right). \quad (42.17)$$

We have here found the probability for the molecule to be situated in the distance range $2s\lambda$, $(2s + 2)\lambda$ from its starting point. The probability (per unit length) that it be at the distance $x = 2s\lambda$ is then found by dividing by 2λ. Moreover, writing

$$N\lambda = \bar{v}t, \qquad \lambda\bar{v} = 2D', \quad (42.18)$$

we obtain the probability distribution

$$n(x) = \frac{1}{\sqrt{2\pi\lambda\bar{v}t}}\exp\left(-\frac{x^2}{2\lambda\bar{v}t}\right) = \frac{1}{\sqrt{4\pi D't}}\exp\left(-\frac{x^2}{4D't}\right), \quad (42.19)$$

where, in accordance with the usual normalizing condition for probability amplitudes, we have

$$\int_{-\infty}^{\infty} n(x)\,dx = 1. \quad (42.20)$$

It is easily checked that (42.19) is a solution of the diffusion equation (42.11), if we replace D by D' in the latter. However, by (42.12) and (42.18), these two coefficients are not equal; rather we have

$$D' = \frac{3}{2}D = \frac{3}{2}\frac{\eta}{\rho}. \quad (42.21)$$

A more detailed theoretical treatment shows that the ratio between the coefficient of diffusion and η/ρ is not unity, as in (42.10), but larger. The exact ratio is difficult to calculate, but certain theoretical considerations lead to a value close to 1.5.

Heat Conduction

In our treatment of heat conduction we once more assume that the gas is homogeneous—apart from inhomogeneity residing in the temperature variation. Hence we need only calculate with one mean free path λ and one mean speed $\bar{v}$. The coefficient of heat conduction, or thermal conductivity, is then defined by

$$Q = -K \frac{\partial T}{\partial x}. \tag{42.22}$$

Here $\partial T/\partial x$ is the temperature gradient in the x-direction and Q the amount of heat flowing through a unit perpendicular to the x-axis per unit time, in the direction of the temperature gradient. If C is the specific heat capacity per unit volume of the gas, the equation of heat conduction will then read

$$C \frac{\partial T}{\partial t} = K \frac{\partial^2 T}{\partial x^2} \quad \text{or} \quad C \frac{\partial T}{\partial t} = K \nabla^2 T. \tag{42.23}$$

The argument now runs quite similarly to that of the preceding section. Let $u(x)$ be the mean heat energy of the molecules (at the same temperature) in a plane x, and let $n(x)$ be the molecular concentration. There will then flow (per unit time) a number of molecules equal to $\frac{1}{6}n\bar{v}$ from each side through a unit area of the plane x, but their heat energies will differ.

Letting C_v be the specific heat capacity per unit mass of the gas, we find that the energy of each molecule then becomes

$$u = mC_v T. \tag{42.24}$$

The molecules arriving from the left must, however, be assumed to have an average heat energy corresponding to the place of their last collision, that is, to the temperature $T - \lambda(\partial T/\partial x)$; similarly the molecules arriving from the right have a heat energy corresponding to the temperature $T + \lambda(\partial T/\partial x)$. This leads to a transport of heat

$$Q = -\frac{1}{3} \lambda \bar{v} n m C_v \frac{\partial T}{\partial x} \tag{42.25}$$

and a thermal conductivity

$$K = \tfrac{1}{3}\lambda \bar{v} n m C_v. \tag{42.26}$$

The latter will presently be compared with the viscosity η.

Viscosity

When fluid layers slide on one another a resistance to the motion results; that is usually referred to as internal friction or viscosity. We assume that the layers lie in the yz-plane, and have a velocity c in the y-direction which increases or decreases with the x-coordinate, that is, perpendicularly to the layers. (See Fig. 6-7.)

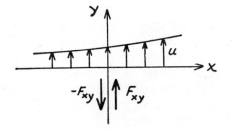

Fig. 6-7. To illustrate the internal friction. The layers on the right drag on the layers to the left with a force $F_{xy} = \eta(\partial c/\partial x)$ per unit area, while the layers on the left drag those on the right in the opposite direction.

Two layers will then act on each other with a force F_{xy} (per unit area) along the y-axis against the motion. The frictional coefficient or viscosity η is defined by

$$F_{xy} = \eta \frac{\partial c}{\partial x}. \tag{42.27}$$

Then $\frac{1}{6}n\bar{v}$ molecules cross the boundary from each side surface; each of them transporting a momentum mc directed long the y-axis. However, the molecules arriving from the right may be assumed to have the mean speed $c + \lambda(\partial c/\partial x)$ and those from the left the mean speed $c - \lambda(\partial c/\partial x)$. The difference between the momenta is thus $2m\lambda\, \partial c/\partial x$, and the net amount of momentum is transported through the surface from right to left.

Further,

$$\frac{1}{3} nm\bar{v}\lambda \frac{\partial c}{\partial x}$$

is the force exerted by the layers to the right of the plane x on the layers to the left, per unit area, that is, it is the force F_{xy} of (42.27). Thus we obtain the following expression for the viscosity

$$\eta = \tfrac{1}{3}nm v \lambda. \tag{42.28}$$

A comparison of (42.26) and (42.28) yields

$$K = \eta C_v. \tag{42.29}$$

This, however, is not in accordance with observations; the correct equation should read

$$K = a\eta C_v, \tag{42.30}$$

where a is a constant characteristic of the gas. For monatomic molecules it turns out to be close to 2.5; for polyatomic molecules, generally less.

424 / Mechanics and Statistics

This discrepancy results from an incorrect theoretical deduction of the thermal conductivity K. An attempt has been made to explain the fact that this quantity actually decreases with an increasing number of atoms in the molecules by using the hypothesis that there is not complete correlation between the translational energy of the molecules and their rotational and vibrational energies. Thus the fastest molecules, which pass most easily across the boundary surface x, transport more heat in the form of internal energy than in the form of translational energy.

43. INTERMOLECULAR FORCES AND THE EQUATION OF STATE

The Virial Theorem of Clausius

Let us consider a gas under the influence of external and internal forces; we assume that the individual molecules are labeled by numbers, that they have masses m_i, and that the net force acting on each molecule is $\mathbf{F}_i$. The equations of motion are then

$$m_i \dot{\mathbf{v}}_i = \mathbf{F}_i. \tag{43.1}$$

We now multiply this equation scalarly by the radius vector from an arbitrarily chosen origin; the resulting equation may be written

$$\frac{d}{dt}(m_i \mathbf{r}_i \mathbf{v}_i) - m_i \mathbf{v}_i^2 = \mathbf{r}_i \mathbf{F}_i, \tag{43.2}$$

or, summing over all the molecules,

$$\frac{d}{dt}\left(\frac{1}{2}\sum_i m_i \mathbf{r}_i \mathbf{v}_i\right) - \sum_i \frac{1}{2} m_i \mathbf{v}_i^2 = \frac{1}{2}\sum_i \mathbf{r}_i \mathbf{F}_i. \tag{43.3}$$

Let us now take the average

$$\frac{1}{\tau}\int_0^\tau d\tau$$

of these quantities over a fairly long period of time τ. If the space available to the gas has a finite extension, as for a gas contained in a closed box, the integral of the first term in (43.3) is finite; hence, when divided by τ, it will vanish when $\tau \to \infty$. As a result we obtain the equation

$$\sum_i \overline{\frac{1}{2} m_i \mathbf{v}_i^2} = -\frac{1}{2}\sum_i \overline{\mathbf{r}_i \mathbf{F}_i}, \tag{43.4}$$

connecting the averages of the other terms.

This is the so-called *virial theorem;* and the quantity on the right-hand side of (43.4) is called (since Clausius) the *virial* of the gas.

Gas Pressure and Virial

All the forces acting on the molecules contribute to the virial. We now subdivide them into the counterpressure exerted by the walls of the container which we denote by P, and the remaining intermolecular forces, which we shall call $\mathbf{K}_i$.

As far as the counterpressure is concerned it is not necessary to study the contribution to the virial from each molecule. We may immediately sum over all molecules adjacent to the walls and write this part of the virial as an integral

$$\tfrac{1}{2}P\int \mathbf{r}\, d\mathbf{S} = \tfrac{1}{2}P\int r\cos\vartheta\, dS = \tfrac{3}{2}PV. \tag{43.5}$$

Drawing the radius vector to each point of the circumference of dS, we obtain a pyramid with base area dS and height $r\cos\vartheta$, whence the integral (43.5) becomes three times the volume of the gas container. (See Fig. 6-8.)

If, moreover, the gas is ideal, exhibiting no other intermolecular forces than those arising during collisions, it may be shown that the remaining part of the virial vanishes, provided that the molecules are assumed to be infinitesimally extended. In the collision of two molecules i and k, there will appear equal and oppositely directed forces $\mathbf{K}_k = -\mathbf{K}_i$, so that the contribution to the virial is $-\tfrac{1}{2}\mathbf{K}_i(\mathbf{r}_i - \mathbf{r}_k)$, which vanishes if the centers of the molecules are regarded as practically coincident. Hence, writing

$$\sum_i \tfrac{1}{2}\overline{m_i v_i^2} = \tfrac{3}{2}NkT \tag{43.6}$$

and dividing by $\tfrac{3}{2}$, we obtain

$$PV = NkT, \tag{43.7}$$

that is, the equation of state for an ideal gas. If the gas molecules are assumed to have finite extension, and intermolecular forces are present, we obtain

$$PV = NkT + \tfrac{1}{3}\sum_i \overline{\mathbf{r}_i \mathbf{K}_i}. \tag{43.8}$$

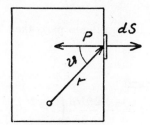

Fig. 6-8. Calculation of the gas pressure by means of the virial.

Form of the Virial for Central Forces

If we assume that the intermolecular forces are purely central, the last term of the virial (43.8) may be considerably simplified. Taking all the molecules to be similar, for simplicity, we may write

$$\mathbf{K}_i = \sum_k \mathbf{K}_{ik}, \quad \mathbf{K}_{ik} = -\nabla_i \Phi(r_{ik}). \tag{43.9}$$

[For dissimilar molecules we should have to introduce several different potential functions $\Phi_{ik}(r)$.] Substitution in (43.8) leads to

$$PV = NkT - \frac{1}{3} \sum_{ik} \overline{r_{ik} \Phi'(r_{ik})}, \tag{43.10}$$

where the summation includes each pair of different indices only once. Since there are altogether $\frac{1}{2}N(N-1) \approx \frac{1}{2}N^2$ such pairs, and since r_{ik} must be assumed to run through the same values in the long run, irrespective of which pairs of molecules is considered, we may write

$$PV = NkT - \tfrac{1}{6}N^2 \overline{r\Phi'(r)}, \tag{43.11}$$

where the last term is averaged over all distances.

This average will now be expressed as an integral; to this end we subdivide the volume V into small cells $d\tau$. We keep one molecule fixed in one cell and place the other successively in all the other cells; in this way, we obtain simply a volume integral of $r(d\Phi/dr)$, divided by V.

However, the other cells are not all equally probable. There is a greater probability that the other molecule will be found in a cell of low potential than in one of high potential. As will be shown later, in the chapter on statistical mechanics, the probability that the molecule will be in a cell of potential $\Phi(r)$ is found by multiplying the probability at $\Phi = 0$ by a factor

$$f(r) = e^{-\beta \Phi(r)}, \quad \beta = \frac{1}{kT}. \tag{43.12}$$

Since the volume of a spherical shell with a radius r is $4\pi r^2\, dr$, we thus obtain (43.11) in the form

$$PV = NkT - \frac{2\pi N^2}{3V} \int_0^\infty r^3 \frac{d\Phi}{dr} e^{-\beta \Phi(r)}\, dr. \tag{43.13}$$

Integrating in parts then leads to the reformulated equations

$$PV = NkT \left(1 + \frac{b}{V}\right) \tag{43.14}$$

and

$$b = -\frac{2\pi N \beta}{3} \int_0^\infty r^3 \frac{1}{\beta} d[1 - e^{-\beta \Phi(r)}]\, dr = 2\pi N \int_0^\infty [1 - e^{-\beta \Phi(r)}] r^2\, dr. \tag{43.15}$$

The Equation of State for Molecules of Finite Size

By representing the molecules as elastic spheres of diameter d, the quantity b may easily be determined both from (43.13) and (43.15). In the former we know that $d\Phi/dr = 0$ for $r > d$, and that the contribution to the integral comes from an infinitesimal interval around $r = d$, where the potential rapidly increases from 0 to ∞. Hence we may put $r = d$, obtaining

$$b = -\tfrac{2}{3}\pi d^3 N \beta \int_\infty^0 e^{-\beta\Phi}\, d\Phi = \tfrac{2}{3}\pi d^3 N. \qquad (43.16)$$

In the latter, on the other hand, we have $1 - e^{-\beta\Phi(r)} = 0$ for $\Phi = 0$, for $r > d$; while $e^{-\beta\Phi} = 0$ for $\Phi = \infty$, for $r < d$. This yields exactly the same result as in (43.16):

$$b = 2\pi N \int_0^d r^2\, dr = \tfrac{2}{3}\pi d^3 N. \qquad (43.17)$$

The last term in (43.14) is small; hence the error introduced by moving it over to the left-hand side and replacing NkT by PV is negligible compared with the correction term already found. In this way we obtain

$$P(V - b) = NkT, \qquad (43.18)$$

that is, the correction for molecular volume in the van der Waals equation.

In elementary texts this phenomenon is explained in roughly the following way: The volume available to the centers of the molecules is not V, but $V - 2b$, since each sphere occupies the volume $4\pi/3\, d^3$. At the walls of the container, where the pressure P is measured, this effect only occurs in one direction, from the molecules inward into the space enclosed. Consequently the volume occupied by the spheres is halved, resulting in an effective available volume

$$V - b.$$

By the virial theorem, however, it should also be possible to calculate this volume effect directly, that is, without utilizing the factor $e^{-\beta\Phi}$ in (43.12) to (43.16), which originates in statistical mechanics. In order to avoid having to deal with the large, short-lived forces arising in collision processes, it seems justifiable to represent these by a uniform hydrostatic pressure P. We must then assume that each separate molecule is subjected, in some way, to this pressure; the problem is to decide where, for the sake of visualization, the pressure is to be applied.

The virial theorem would appear to furnish an acceptable solution to this problem, since it refers to the resultant forces acting at the molecular centers. When any given molecule is hit by its neighbors, the centers of the latter lie

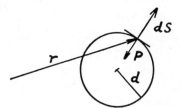

Fig. 6-9. Calculation of the virial for molecular collisions.

on a sphere of radius d around this molecule. As a next step we assume that the molecular impacts on the central molecule are equivalent to a uniform hydrostatic pressure on its surface; the resultant force on the central molecule may then be taken to vanish.

By the action-reaction law, however, a corresponding counterpressure acting on this spherical surface from the inside must be included in our consideration; this counterpressure then represents the resultant impact forces on the centers of the molecules colliding with the central molecule. In this way we may calculate the contribution to the virial from a single molecule; see Fig. 6-9. We thus obtain from (43.8), where the factor $\tfrac{2}{3}$ appears,

$$\frac{1}{3} P \int \mathbf{r} \, d\mathbf{S} = P \cdot \frac{4\pi}{3} d^3, \qquad (43.19)$$

analogously to (43.5).

This quantity is not, however, to be multiplied right away by the molecular number N; this would mean that we counted the contribution from each molecule twice (taking it first as being central, and then as being peripheral). Instead we must multiply by $\tfrac{1}{2} N$, which leads to the term Pb on the right-hand side of (43.18).

The Equation of State for Cohesive Forces

If relatively weak cohesive forces occur, we may also calculate their contribution to the virial without using the statistical factor $e^{-\beta \Phi}$.

Let us consider (43.15). Here $\Phi(\infty) = 0$. For (large) decreasing values of r, $\Phi(r)$ will be initially negative, decreasing down to a minimum, and then increase to, say, $\Phi(r_0) = 0$. Subsequently, for decreasing $r < r_0$, $\Phi(r)$ will as a rule increase rapidly to (positive) infinity. It is mainly the integral of the latter part that provides the volume effect already mentioned. The effect of cohesion, on the other hand, arises in the range $r > r_0$. By neglecting the factor $e^{-\beta \Phi}$, that is, regarding all distances as equally probable and integrating (43.13) in parts, we obtain

$$PV = NkT - \frac{a}{V}$$

or

$$\left(P + \frac{a}{V^2}\right)V = NkT,$$

$$a = -2\pi N^2 \int_0^\infty \Phi(r) r^2 \, dr,$$

(43.20)

where a is a constant which does not depend on P, V, or T.

The combination of this effect and the previously mentioned volume effect then yields (as is easily seen) an equation that is identical, to a first approximation, with the van der Waals equation.

CHAPTER 7

Statistical Mechanics

44. PHASE SPACE AND PHASE CELLS

Phase Space

The motion of an isolated mechanical system with n degrees of freedom is completely determined if $2n$ variables are known at a given time; these variables may be the n generalized coordinates q_i and the n corresponding time derivatives or velocities $\dot{q}_i$. However, we may introduce along with the coordinates q_i, the corresponding n generalized momenta $p_i = \partial L/\partial \dot{q}_i$, which were discussed at the transition from Lagrange's to Hamilton's equations. If all q_i and p_i are known at a given time, their values at any later time is, in principle, uniquely determined by Hamilton's equations

$$\dot{q}_i = \frac{\partial H}{\partial p_i}, \qquad \dot{p}_i = -\frac{\partial H}{\partial q_i}.$$

$$H = H(q, p) = H(q_1, p_1, \cdots, q_n, p_n). \tag{44.1}$$

The state of the system may be described, at any time, as a point or vector in an abstract space of $2n$ dimensions, the so called *coordinate-momentum space* or *phase space;* this space possesses special properties that make it far more convenient to use than the corresponding *coordinate-velocity space*, as will be shown presently. The point representing the state of the system will, during the motion of the latter, describe a curve in phase space; this curve will sometimes, for brevity, be referred to as the trajectory of the system.

Invariance of Volume under Canonical Transformations

When solving the equations of motion for a mechanical system we actually always—consciously or not—perform a so-called canonical transformation; such transformations may be defined by means of certain specified *transformation functions*. The so-called action function S appearing in the Hamilton-Jacobi differential equation is such a function. The most common form of general canonical transformations is

$$p_i = \frac{\partial V}{\partial q_i}, \qquad Q_i = \frac{\partial V}{\partial P_i}, \qquad (44.2)$$

where Q_i and P_i are new canonical coordinates and momenta, and

$$V = V(q, P, t) = V(q_1, P_1, \cdots, q_n, P_n, t) \qquad (44.3)$$

is a transformation function, which depends on the given $2n$ variables and (possibly) explicitly on time.

We now consider the question of how the $2n$-dimensional element of "volume"

$$d\tau = dq_1\, dp_1 \cdots dq_n\, dp_n \qquad (44.4)$$

varies with the transformation (44.2). (See Fig. 7-1.) As is well-known, we may write

$$d\tau = D(q, p, Q, P)\, dQ_1\, dP_1 \cdots dQ_n\, dP_n, \qquad (44.5)$$

where D is the so-called functional determinant

$$D(q, p, Q, P) = \frac{\partial(q_1, p_1, \cdots, q_n, p_n)}{\partial(Q_1, P_1, \cdots, Q_n, P_n)} = \begin{vmatrix} \dfrac{\partial q_1}{\partial Q_1} & \dfrac{\partial p_1}{\partial Q_1} & \cdots \\ \dfrac{\partial q_1}{\partial P_1} & \dfrac{\partial p_1}{\partial P_1} & \cdots \\ \cdots & & \end{vmatrix} \qquad (44.6)$$

This quantity satisfies the relations

$$D(q, p, Q, P) = D(q, p, u, v),\, D(u, v, Q, P),$$
$$D(u, v, Q, P) = \frac{1}{D(Q, P, u, v)}, \qquad (44.7)$$

where u, v is a set of intermediate auxiliary variables.

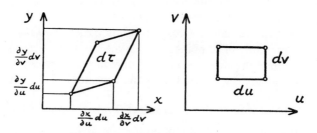

Fig. 7-1. An example of the change of infinitesimal volumes (areas) in coordinate transformations.

$$dr = (dx\, dy) = \begin{vmatrix} \partial x/\partial u & \partial y/\partial u \\ \partial x/\partial v & \partial y/\partial v \end{vmatrix} du\, dv.$$

For simplicity, we shall first examine a system with one degree of freedom, and put $u = q$, $v = P$; we then have $\partial q/\partial u = 1$, $\partial q/\partial v = 0$, $\partial P/\partial u = 0$, $\partial P/\partial v = 1$, so that

$$D(q, p, u, v) = \begin{vmatrix} 1 & 0 \\ \dfrac{\partial p}{\partial u} & \dfrac{\partial p}{\partial v} \end{vmatrix} = \left(\frac{\partial p}{\partial P}\right)_q = \frac{\partial^2 V}{\partial P \, \partial q},$$

$$D(Q, P, u, v) = \begin{vmatrix} \dfrac{\partial Q}{\partial u} & \dfrac{\partial Q}{\partial v} \\ 0 & 1 \end{vmatrix} = \left(\frac{\partial Q}{\partial q}\right)_P = \frac{\partial^2 V}{\partial q \, \partial P}. \quad (44.8)$$

Passing over to a system with n degrees of freedom, we similarly find (putting $u_i = q_i$, $v_k = P_k$) that the two determinants of rank $2n$ reduce to determinants of rank n. By means of the transformations (44.2) these determinants may be written

$$D(Q, p, Q, P) = \left| \frac{\partial^2 V}{\partial P_k \, \partial q_i} \right|,$$

$$D(Q, P, q, P) = \left| \frac{\partial^2 V}{\partial q_i \, \partial P_k} \right|. \quad (44.9)$$

Hence it follows by (44.6) and (44.7) that

$$dq_1 \, dp_1 \cdots dq_n \, dp_n = dQ_1 \, dP_1 \cdots dQ_n \, dP_n \quad (44.10)$$

for any canonical transformation; in other words, the *phase cell volume is invariant*.

Liouville's Theorem

Let us consider a phase cell, determined by the $4n$ coordinate surfaces

$$q_i, q_i + dq_i, p_i, p_i + dp_i, \quad i = 1, 2, \cdots n.$$

We imagine that the points occupying the corners of this cell represent the motions of identical mechanical systems; the phase cell will then move through phase space and change its shape. On the other hand, *it will not change its volume*.

We denote the same variables after a certain (not necessarily infinitesimal) time Δt by Q_i and P_i, thus obtaining a transformation

$$q_i = q_i(Q_k, P_k),$$
$$p_i = p_i(Q_k, P_k), \quad (44.11)$$

which must, from general principles, be canonical (since q, p and Q, P obey sets of canonical equations). Consequently the volume of the cell must have remained constant during its motion in phase space. It is, however, not easy to

434 / Mechanics and Statistics

determine explicitly the transformation function connecting the two sets of variables; and we shall, therefore, prove our assertion in another and (from the point of view of visualization) simpler way.

We imagine the phase space filled with image points, flowing through this space in a fashion somewhat analogous to a fluid in three-dimensional space. If this flow is taken to be *divergence-free*, similar to an incompressible fluid flow, the phase cells will maintain their volumes unchanged during the motion, that is, the density of image points will be uniform.

The equation of continuity in hydrodynamics,

$$\frac{\partial \rho}{\partial t} + \text{div } \rho \mathbf{v} = 0, \qquad (44.12)$$

since

$$\frac{d\rho}{dt} = \frac{\partial \rho}{\partial t} + \mathbf{v} \nabla \rho,$$

$$\text{div } \rho \mathbf{v} = \mathbf{v} \nabla \rho + \rho \text{ div } \mathbf{v},$$

may be reformulated as

$$\frac{d\rho}{dt} + \rho \text{ div } \mathbf{v} = 0. \qquad (44.13)$$

Hence the equations $d\rho/dt = 0$ and div $\mathbf{v} = 0$ are equivalent, both expressing the intuitively simple notion that the fluid is incompressible.

We now define the "velocity vector"

$$\mathbf{v} = (\dot{q}_1, \dot{p}_1, \cdots, \dot{q}_n, \dot{p}_n) \qquad (44.14)$$

in the $2n$-dimensional space and obtain

$$\text{div } \mathbf{v} = \frac{\partial \dot{q}_1}{\partial q_1} + \frac{\partial \dot{p}_1}{\partial p_1} + \cdots + \frac{\partial \dot{q}_n}{\partial q_n} + \frac{\partial \dot{p}_n}{\partial p_n}. \qquad (44.15)$$

It now follows immediately from Hamilton's equations that

$$\text{div } \mathbf{v} = 0. \qquad (44.16)$$

This is *Liouville's theorem*.

However, the vanishing of div $\mathbf{v}$ is not only valid for the total $2n$-dimensional space. Each pair of the $2n$ terms on the right-hand side of (44.15) also vanishes; hence the flow is divergence-free in each pair of corresponding canonical variables q_i and p_i.

45. THERMODYNAMIC PROBABILITY AND ENTROPY

Statistical Description of Complex Systems

We now imagine a large number N of identical noninteracting mechanical systems, each having n degrees of freedom, for example, the molecules of a gas. If the $2n$-dimensional phase space is subdivided into sufficiently small cells, we may describe the state of the system completely by stating in which phase cell each subsystem (say, molecule) is to be found. In this way we define what is generally called the *microstate* of the system.

In thermodynamics we usually have to deal with such a large number of subsystems that a determination of microstates lies outside the bounds of possibility. The quantities that may be observed are the pressure, temperature, energy, mass density, and so forth, of the system. These quantities are the same for several different microstates: those that may be transformed into each other by a permutation of identical subsystems or molecules. We therefore also speak of *macrostates* of a system which may be defined by specifying how many identical subsystems (molecules) are to be found in each phase cell.

The *a priori* Probability of the Phase Cells

We mentioned that the state of a system is completely determined from the point of view of classical mechanics by its microstate, provided that the phase cells may be chosen sufficiently small. Quantum mechanics teaches us that it is meaningless to speak of phase cells with extensions smaller than $\Delta q \, \Delta p = h$ for each pair of variables. By subdivision of phase space into cells of exactly this size, that is, h^n cells for systems with n degrees of freedom, one attains—quantum-mechanically speaking—the complete definition of the state of a system. The consequences of this notion, that the states of systems are *quantized*, are more properly treated in a discussion of quantum mechanics and quantum statistics; we see, however, that the basic ideas of quantum mechanics have a marked statistical character.

In classical statistical mechanics the phase cells are often written by means of differentials. By this we do not mean that they are necessarily infinitesimal, or less than h. On the contrary we often regard them as fairly large, relative to h, so that they may contain a large number of image points (representing the subsystem states). However, they should generally be very small, in the usual (laboratory) sense; for instance, the spatial elements of volume should be small, relative to the gas whose thermodynamical properties we wish to study.

The crucial point of statistical mechanics is that the phase cells all have the same volume. From what has already been said about the invariance of phase-cell volumes, we may then assume that the probabilities of finding, say, a molecule in any two phase cells are equal. Certain restrictions on this general assertion may arise if the system exhibits constants of the motion, such as the total energy E; clearly phase cells with energies higher than E need not be considered in this case, and the number of possibilities will decrease sharply with an increasing number of occupied cells. Apart from such restrictions, however, we assume as a general rule that *the a priori probabilities of any two phase cells with the same volume are equal.*

Thermodynamic Probability and Thermodynamic Equilibrium

The thermodynamic probability of a given macrostate is generally defined as the number of different microstates that may compose this macrostate. Let us consider a gas of N molecules, and number the corresponding phase cells in any well-defined way, by $i = 1, 2, \cdots$. A macrostate is then defined by specifying the number of molecules N_i in each cell. The number of different possibilities are found by permutation of the molecules, which may be done in altogether $N!$ ways. However, permutations of the molecules inside any one phase cell do not lead to any (even in principle) observable change of the system; hence $N_1! \, N_2! \cdots N_i! \cdots$ of the $N!$ permutations represent the same microstate. Thus the thermodynamic probability of a given macrostate is

$$W = \frac{N!}{N_1! \, N_2! \cdots N_i! \cdots}. \tag{45.1}$$

Let us now consider two isolated thermodynamic systems, each composed of a definite number of molecules or subsystems, and let the macrostates of both systems, with the corresponding thermodynamic probabilities W_1 and W_2, be given.

If we now regard these two systems as composing one system (their phase cells are not identical, since the molecules occupy different elements of spatial volume), the latter will also have a well-defined macrostate; the thermodynamic probability for this supersystem is then

$$W = W_1 W_2. \tag{45.2}$$

The problem is now to determine the conditions for a system to be in thermodynamic equilibrium. It seems reasonable to suppose that the probability of finding a certain macrostate actually realized in nature increases with the number of different microstates that may compose this macrostate. Thus the macrostate with the highest probability of occurrence is that of the maximum thermodynamic probability, which is the reason for the name of the latter. The larger W is, the more disordered is the motion of the different parts of the system. We must, therefore, expect that the thermodynamic probability will show a tendency to increase, and that thermodynamic equilibrium for a system is only reached when this probability has attained its maximum value.

Entropy

In thermodynamics we find another quantity which provides a measure for the degree of disorder of the molecular motions; this is the entropy which also has a tendency to increase toward a maximum value, at which the system is in thermodynamic equilibrium. This indicates that the entropy and thermodynamic probability are measures of the same thing, so that there must exist a unique functional connection between the two, which we write as

$$S = f(W). \tag{45.3}$$

Now, entropy is an additive property; if the entropies of two subsystems are S_1 and S_2, the total entropy of the complete system is

$$S = S_1 + S_2. \tag{45.4}$$

On the other hand, it follows from (45.2) that log W is an additive quantity,

$$\log W = \log W_1 + \log W_2. \tag{45.5}$$

It is therefore seen almost immediately (and may, indeed, easily be proved rigorously) that the most general form of the relation (45.3) is

$$f(W) = k \log W,$$

that is,

$$S = k \log W, \tag{45.6}$$

where k is a *universal* constant, identical with the previously discussed *Boltzmann's constant*.

46. THE BOLTZMANN DISTRIBUTION. CONNECTION WITH THERMODYNAMICS

Stirling's Formula

It would, in practice, be extremely inconvenient to work with discontinuous functions like the factorials of (45.1); in particular, the subsequent deduction of general laws would be seriously impeded. Actually the molecular numbers which we usually have to deal with are so large that any single change may be regarded as differential, and all functions correspondingly as continuous. Fortunately a formula exists that expresses the factorial function to a very good approximation by means of more tractable functions, namely, Stirling's formula

$$n! = n^n e^{-n} \sqrt{2\pi n}. \tag{46.1}$$

This is an asymptotic relation, that is, its degree of accuracy increases with n; however, it yields fairly close values for $n!$ even for $n = 1, 2, 3, \cdots$. We shall here utilize it in the form

$$\log n! = n \log n - n, \tag{46.2}$$

438 / Mechanics and Statistics

neglecting the term $\frac{1}{2}\log(2\pi n)$ (and higher corrective terms). This formula, which is sufficiently accurate for our purposes, may be established by replacing $\log 2 + \log 3 + \cdots + \log n$ by

$$\log n! \approx \int_1^n \log x \, dx. \tag{46.3}$$

Condition for Maximum Entropy. The Boltzmann Law

In addition to the occupation numbers N_i we now introduce the relative numbers of occupation $w_i = N_i/N$. Utilizing (46.2), we then find

$$S = k\left\{N \log N - \sum_i N_i \log N_i\right\} = -kN \sum_i w_i \log w_i. \tag{46.4}$$

We shall attempt to determine the relative occupation numbers in such a way that the entropy attains its maximum value, subject to the secondary conditions

$$\sum_i N_i = N \sum_i w_i = N,$$
$$\sum_i N_i U_i = N \sum_i w_i U_i = U, \tag{46.5}$$

where U is the total energy and U_i the energy of the molecules in the corresponding phase cell.

Hence by variation of the occupation number w_i of each cell, we obtain

$$\sum_i (\log w_i + 1) \, \delta w_i = 0, \quad \sum_i \delta w_i = 0, \quad \sum_i U_i \, \delta w_i = 0. \tag{46.6}$$

The use of Lagrange's multiplier method then yields

$$\sum_i \{\log w_i + \alpha + \beta U_i\} \, \delta w_i = 0, \tag{46.6a}$$

$$\log w_i + \alpha + \beta U_i = 0. \tag{46.6b}$$

$$w_i = \frac{1}{Z} e^{-\beta U_i}, \quad Z = \sum e^{-\beta U_i}. \tag{46.7}$$

The term Z is a factor of normalization which is called the sum of state, because of the last expression in (46.7); later it will play an important role in our considerations. Equation (46.7) gives us the relative occupation numbers of the system at thermodynamic equilibrium; it is called Boltzmann's distribution law, and is usually written

$$w_i \sim e^{-E/kT}, \tag{46.8}$$

since, as we shall show, the constant β is equal to $1/kT$.

In our calculations of various thermodynamic quantities, we shall frequently make use of the sum of state Z as a convenient auxiliary function. For instance, since

$$\frac{\partial}{\partial \beta} e^{-\beta U_i} = -U_i e^{-\beta U_i}, \tag{46.9}$$

it is easily found from (46.5) and (46.7) that

$$U = N \cdot \frac{1}{Z} \frac{\partial Z}{\partial(-\beta)} = N \frac{\partial \log Z}{\partial(-\beta)}. \tag{46.10}$$

Similarly we find, using (46.4),

$$S = k\beta U + kN \log Z. \tag{46.11}$$

Temperature, Pressure, and Free Energy

We now assume that the system is supplied with heat energy, in such a way that the chosen subdivision of phase space does not change; this will, for instance, be the case for a gas kept at a constant volume, so that the spatial configuration of cells is unchanged. The energy U_i of each cell is then also constant, and the sum of state changes only through the parameter β.

By (46.10) we must then have

$$dS = k\beta \, dU + kU \, d\beta - kU \, d\beta = k\beta \, dU. \tag{46.12}$$

Comparing with the equation

$$T \, dS = \delta Q = dU + P \, dV \tag{46.13}$$

at a constant volume ($dV = 0$), we thus find $Tk\beta = 1$ or

$$\beta = \frac{1}{kT}, \tag{46.14}$$

as already stated.

For the complete differential of S, which depends on the subdivision into phase cells and hence also on the volume through the sum of state, we find

$$T \, dS = \delta Q = dU + NkT \frac{\partial \log Z}{\partial V} \, dV \tag{46.15}$$

by (46.11). Comparison with (46.13) then yields the gas pressure

$$P = NkT \frac{\partial \log Z}{\partial V}. \tag{46.16}$$

Writing (46.11) in the form

$$U - TS = -NkT \log Z, \tag{46.17}$$

we observe that the quantity most directly determined by the sum of state Z is the free energy

$$F = -NkT \log Z. \tag{46.18}$$

Together with (46.16), this leads to

$$P = -\left(\frac{\partial F}{\partial V}\right)_T \tag{46.19}$$

in accordance with the corresponding thermodynamic equation

$$dF = d(U - TS) = -S\,dT - P\,dV. \tag{46.20}$$

Equation (46.18) may also be solved with respect to Z; we then obtain

$$Z = e^{-F/NkT} \tag{46.21}$$

which demonstrates directly the physical significance of the sum of state.

Deduction of Stirling's Formula

Because of the frequent application of Stirling's formula throughout, we shall here demonstrate its validity and the restrictions on its range of applicability. It is an asymptotic formula that only applies rigorously in the limit $n \to \infty$; or formally,

$$\lim_{n \to \infty} \frac{n!}{n^n e^{-n} \sqrt{2\pi n}} = 1. \tag{46.22}$$

We shall use a method similar to that which, in the theory of complex integration, is called the *saddle point method;* the details of the latter will not be discussed.

We shall utilize the well-known Euler integral

$$n! = \int_0^\infty e^{-x} x^n \, dx, \tag{46.23}$$

noting the fact that the integrand for very high values of n has a very large maximum at $x = n$. The value of the integrand at this point is $n^n e^{-n}$, while the second derivative is equal to $-n^{n-1} e^{-n}$. Close to this maximum we shall have, to a very good approximation,

$$e^{-x} x^n \approx e^{-n} n^n e^{-[(x-n)^2/2n]}, \tag{46.24}$$

as illustrated in Fig. 7-2.

Introducing $(x - n)/\sqrt{2\pi n} = y$ as a new variable, we thus obtain

$$n! \approx e^{-n} n^n \sqrt{2\pi n} \cdot \int_{-\infty}^{\infty} e^{-\pi y^2} \, dy = e^{-n} n^n \sqrt{2\pi n}. \tag{46.25}$$

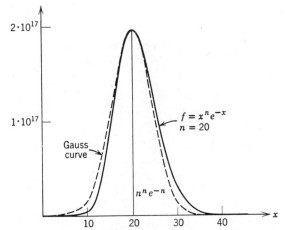

Fig. 7-2. Comparison of $x^n e^{-x}$ with $n^n e^{-n} e^{-(x-n)^2/2n}$.

By computing the third, fourth, and so on, derivatives of the integrand in (46.23), we may easily determine higher-order terms in the asymptotic formula.

47. APPLICATIONS OF STATISTICAL MECHANICS TO GAS KINETICS

Mean Kinetic Energy and the Equipartition Theorem

We shall now demonstrate, by means of the results just obtained, how we may easily arrive back at the equipartition theorem previously discussed; to this end we consider a rigid diatomic molecule with the kinetic energy

$$U = U_x + U_y + U_z + U_\vartheta + U_\varphi,$$

$$U_x = \frac{1}{2m} p_x^2, \text{ etc.}, \quad U_\vartheta = \frac{1}{2I} p_\vartheta^2, \quad U_\varphi = \frac{1}{2I} \frac{p_\varphi^2}{\sin^2 \vartheta}, \quad (47.1)$$

where m and I are the molecular mass and moment of inertia.

The element of volume in phase space is

$$d\tau = dx\, dy\, dz\, d\vartheta\, d\varphi\, dp_x\, dp_y\, dp_z\, dp_\vartheta\, dp_\varphi. \quad (47.2)$$

To find the average of one of the three translational degrees of freedom, we may integrate over the remaining nine variables, thus obtaining, for instance,

$$\bar{U}_x = \frac{1}{Z} \int U_x e^{-\beta U}\, d\tau$$

$$= \frac{1}{Z'} \int_{-\infty}^{+\infty} \frac{1}{2m} p_x^2 \exp\left(-\beta \frac{1}{2m} p_x^2\right) dp_x = -\frac{\partial \log Z'}{\partial \beta}, \quad (47.3)$$

where Z is the total sum of state

$$Z = \int e^{-\beta U}\, d\tau \tag{47.4}$$

and Z' is a reduced sum of state

$$Z' = \int_{-\infty}^{+\infty} \exp\left(-\frac{\beta}{2m} p_x^2\right) dp_x = \left(\frac{2\pi m}{\beta}\right)^{\!\!1/2}. \tag{47.5}$$

From (47.3) and (47.5) it follows that

$$\bar{U}_x = \frac{1}{2}\frac{d \log \beta}{d\beta} = \frac{1}{2\beta} = \frac{1}{2}kT. \tag{47.6}$$

Then $\bar{U}_y$, $\bar{U}_z$, and $\bar{U}_\vartheta$ may be found in exactly the same manner, all turning out to be equal to $\tfrac{1}{2}kT$.

However, $\bar{U}_\varphi$ must be calculated in a somewhat different way, since it depends on ϑ. We may choose to perform the integration over dp_φ before integrating over $d\vartheta$, that is, integrate over $\sin\vartheta\, d\vartheta\, d\varphi$ instead of only $d\vartheta\, d\varphi$, thus obtaining a factor 4π which is the surface of the unit sphere (or, equivalently, the total solid angle subtended by any sphere).

It is simpler, however, to put

$$p_\varphi = \sin\vartheta P_\varphi, \qquad dp_\varphi = \sin\vartheta\, dP_\varphi, \qquad U_\varphi = \frac{1}{2I} P_\varphi^{\,2} \tag{47.7}$$

and integrate over the other variables before the integration over dP_φ. Writing

$$Z'' = \int_{-\infty}^{\infty} \exp\left(-\frac{\beta}{2I} P_\varphi^{\,2}\right) dP_\varphi = \left(\frac{2\pi I}{\beta}\right)^{\!\!1/2}, \tag{47.8}$$

we then find, as expected, that

$$\bar{U}_\varphi = -\frac{\partial}{\partial\beta}\log Z'' = \frac{1}{2}\frac{d}{d\beta}\log \beta = \frac{1}{2}kT. \tag{47.9}$$

Equation of State for an Ideal Gas

To find the equation of state for an ideal monatomic or diatomic gas we calculate the total sum of state Z in (47.4); let us, for the sake of definiteness, still consider a rigid diatomic molecule with the kinetic energy (47.1). Utilizing the results obtained above, we easily find that

$$Z = 4\pi V \left(\frac{2\pi m}{\beta}\right)^{\!\!3/2}\!\left(\frac{2\pi I}{\beta}\right)^{\!\!3/2}. \tag{47.10}$$

However, it then follows from (46.16) that

$$P = NkT \frac{\partial \log Z}{\partial V} = NkT \frac{d \log V}{dV} = \frac{NkT}{V},$$

that is,
$$PV = NkT. \tag{47.11}$$

In other words, the number of degrees of freedom does not affect the form of the equation of state, but only the calculation of the internal energy, which here is identical with the mean free kinetic energy

$$\bar{U} = -\frac{\partial \log Z}{\partial \beta} = \frac{5}{2}\frac{d \log \beta}{d\beta} = \frac{5}{2}\frac{1}{\beta} = \frac{5}{2}kT. \tag{47.12}$$

Mean Free Energy of an Oscillator

In a deformable molecule oscillation will occur; these will then, for small amplitudes, be approximately harmonic. Hence the calculation of the mean thermal energy of a harmonic oscillator is of primary interest in gas kinetics. Such an oscillator has both a kinetic and a potential energy, given by the two terms in the Hamiltonian

$$H = \frac{1}{2m}p^2 + \frac{m}{2}\omega^2 q^2. \tag{47.13}$$

We find the sum of state

$$Z = \int\!\!\int_{-\infty}^{\infty} e^{-\beta H}\, dq\, dp = \left(\frac{2\pi m}{\beta}\right)^{1/2}\left(\frac{2\pi}{m\beta}\right)^{1/2} \cdot \frac{1}{\omega} \tag{47.14}$$

which yields

$$\bar{H} = -\frac{\partial \log Z}{\partial \beta} = kT. \tag{47.15}$$

In order to explain why no perceptible energy of oscillation is observed for approximately rigid diatomic molecules such as N_2 or O_2, we shall calculate the mean energy of a harmonic oscillator by another method, using the principles of quantum mechanics. Actually, what we do first is to perform a canonical transformation from the variables q, p to E, t. The image point in the qp-plane then describes an ellipse, corresponding to a given energy E. Since now $dq\, dp = dE\, dt$, we may integrate over a period in the "coordinate" t. This yields a constant factor $2\pi/\omega = 1/\nu$, which is of no interest for our purposes and may be omitted. The sum of state is

$$Z = \sum_n e^{-\beta E_n} e^{-\beta E_n}. \tag{47.16}$$

By classical mechanics, we must replace this sum by an integral

$$Z = \int_0^\infty e^{-\beta E} \, dE = \frac{1}{\beta}, \tag{47.17}$$

thus obtaining

$$\bar{U} = \frac{d \log \beta}{d\beta} = \frac{1}{\beta} = kT. \tag{47.18}$$

By quantum theory, however, the phase cells have a finite size $\Delta q \, \Delta p = \Delta E \, \Delta t = h$; and since (as already mentioned) $\Delta t = 1/\nu$, where ν is the proper frequency of the oscillator, the extension of the phase cells in the energy direction must be

$$\Delta E = h\nu. \tag{47.19}$$

This results in states where the energy—apart from an additive quantity $\frac{1}{2}h\nu$ which will be omitted here—may be written

$$E_n = nh\nu, \quad n = 0, 1, 2, \cdots. \tag{47.20}$$

The sum of state then becomes

$$Z = \sum_{n=0}^\infty e^{-n\beta h\nu} = \frac{1}{1 - e^{-\beta h\nu}} \tag{47.21}$$

whence we finally obtain

$$\bar{U} = \frac{d}{d\beta} \log(1 - e^{-\beta h\nu}) = \frac{h\nu}{e^{h\nu/kT} - 1}. \tag{47.22}$$

If the frequency of the oscillator is relatively small and the temperature relatively high (more precisely, if $h\nu \ll kT$), the denominator is approximately equal to $h\nu/kT$, which yields the classical energy of the oscillator $\bar{U} = kT$. If $h\nu \gg kT$, on the other hand, we should then obtain

$$\bar{U} \approx h\nu e^{-h\nu/kT}, \tag{47.23}$$

which approaches zero much faster than the classical energy kT, when $T \to 0$. This explains why the molecular energy of oscillation is so small at ordinary temperatures: the order of magnitude of the frequencies of oscillation is known from the band spectra, and these oscillations are too rapid to be able to absorb any perceptible energy from the thermal motion.

Similar results may also, in principle, be deduced for the mean rotational energy of the molecules, since this energy—like that of oscillation, but in contrast to that of translation—appears in separate quantized energy levels

$$E_l = \frac{l(l+1)h^2}{8\pi I}, \quad l = 0, 1, 2, \cdots. \tag{47.24}$$

However, these levels are much more closely spaced than is the case for molecular oscillations; for the band spectra they correspond to the separations between the spectral lines of a band, while the levels of oscillation energy correspond to the separations between different bands. Thus the classical condition $h\nu \ll kT$ is much better realized, at ordinary temperatures, for the rotational energy; so that the classical value obtained for the mean energy of rotation is usually very close to that which is actually observed. However, when we approach very low temperatures, the quantized molecular rotations will also "drop out" gradually, leaving only the continuous energy of translation.

48. GIBBS' FORMULATION OF STATISTICAL MECHANICS

Macrosystems and Microsystems

It is often very difficult to obtain a clear understanding of both the difference and similarity between Boltzmann's and Gibbs's formulation of statistical mechanics. Several texts strongly emphasize the differences, and more or less overlook the close connection between the two. Moreover they are often presented and used simultaneously in such a mixed-up way that the reader is hard put to know which formulation is actually used at any one step of the discussion.

We shall here try to reduce the difference to a minimum; thus we shall formulate our argument in such a way that all equations are transferred directly from the Boltzmann statistics.

In preceding discussions we have tacitly assumed that the identical mechanical systems considered are what we may call *microsystems*, relatively simple mechanical models with not too many degrees of freedom. The molecules of a gas are such *microsystems*: a monatomic molecule has three degrees of freedom, a rigid diatomic molecule five, a deformable diatomic molecule six, and so on.

If we may neglect any interaction forces between molecules not arising from elastic collisions, the statistical treatment disussed above would be quite satisfactory. However, if other interactions are to be investigated, we must treat the whole gas—or at least macroscopic parts of it—as a separate mechanical system; such systems will then be referred to as *macrosystems*.

The apparently very great difference between the statistics of Gibbs and Boltzmann lies primarily in the fact that the former is a statistics for macrosystems while the latter is one for microsystems.

Statistical Ensembles and Canonical Ensembles

An ensemble is, by definition, a large number of identical mechanical systems in different states to be specified in more detail. Such ensembles may be real, in the physical sense, or abstract entities; for instance, identical gas molecules that do not interact (except by collision) constitute a real ensemble.

If we wish to consider the whole gas as one single mechanical system, to be treated statistically, the need for the concept of an ensemble will soon become apparent. We may then construct abstract ensembles in several different ways. In the course of time, the state (i.e. the coordinates and momenta) of the system will change, and the different states of one and the same system may be regarded as constituting an ensemble. We may also imagine that several such identical systems exist side by side in different states; in this way, too, they will constitute an ensemble.

Each of these systems would, if it were insulated from its surroundings, change its state in accordance with the laws of mechanics. Thus if the microstate of the system is known in complete detail at one time, the further development of the system is uniquely determined, and no statistics is necessary.

However, the situation will be different if we allow the system to interact with its surroundings or with other identical systems. We might, for instance, imagine that these systems were all connected with a heat reservoir of a given temperature; the energy of each system would then—even after a thermodynamic equilibrium has been reached—no longer be constant, but undergo small fluctuations.

We may also imagine that the different systems are directly interconnected, so that they together constitute a larger system; while the total energy of this complete system will then be constant, the component subsystems will exchange energy. Let us, for instance, consider as such a complete system a large amount of gas which is thermally insulated from its surroundings and thus has a constant total energy. If the gas is subdivided into elements of volume that are identical and small, though still of macroscopic size, all these elements would then together constitute a real ensemble of identical mechanical systems. In order that these subsystems keep their physical identity, we must assume them to be mutually insulated with respect to the transport of mass (i.e. exchanges of molecules), but not with respect to the transport of energy (through heat conduction).

Such an ensemble of identical systems, which are in thermodynamic equilibrium with each other with respect to exchanges of energy (directly, or by means of a heat reservoir), was called a *canonical ensemble* by Gibbs. Thus the separate subsystems do not have a given energy, but they have the *same temperature*. An important property of canonical ensembles is that the subsystems may be regarded as quasiclosed. This is because the mean energy and other quantities of interest are *volume effects*, that is, proportional to the volume or mass; while the interaction with the other systems of the ensemble is a *surface effect*, that is, proportional to the $\frac{2}{3}$ power of the volume or mass.

Hence the interactional exchange of, say, energy is negligible in comparison with the energy content of the subsystems.

Thermodynamic Quantities in the Gibbs Statistics

We shall now find that the results already deduced in our treatment of the Boltzmann statistics may be utilized to determine all the basic equations of the Gibbs statistics.

Instead of microsystems we now consider an ensemble of N identical macrosystems with a total energy U; the quantities N and U then correspond formally to the N and U of the previous discussions. The possible states of the subsystems are denoted by i, with the corresponding energy U_i which is then equivalent to the energy of the above considered microsystems.

All we have to do now is to calculate the mean values of the thermodynamic quantities characterizing the subsystems, instead of (as before) the corresponding values for the complete system. This means that we only have to substitute $N = 1$ in our previous results or (equivalently) divide by N. Formerly this would have yielded the mean thermodynamic quantities for the individual molecules or microsystems; now, on the other hand, we obtain the corresponding mean values for a macrosystem (i.e. a macroscopic aggregate of molecules).

Let us begin with the distribution law and the sum of state which remain formally unchanged:

$$w_i = \frac{e^{-\beta U_i}}{Z}, \qquad Z = \sum_i e^{-\beta U_i}. \tag{48.1}$$

It should, however, be noted that the number of possible states is now enormously greater than before. If we have N molecules with n degrees of freedom, the dimension of phase space is then $2nN$ instead of $2n$; thus, for, say, non-interacting molecules, we would have $Z = Z_n^N$, where Z_n is the previous Boltzmann sum of state.

The energy of the gas is obtained from (46.10) by putting $N = 1$:

$$U = -\frac{\partial}{\partial \beta} \log Z, \tag{48.2}$$

while the entropy becomes

$$S = k\beta U + k \log Z. \tag{48.3}$$

Furthermore (46.16), (46.18), and (46.21) similarly lead to

$$P = kT \frac{\partial \log Z}{\partial V}, \tag{48.4}$$

$$F = -kT \log Z, \tag{48.5}$$

and
$$F = e^{-F/kT}. \tag{48.6}$$

Apart from this, all thermodynamic relations are identical with those of Section 46.

Finally we remark that the expression (46.4) for the entropy of course changes to
$$S = -k \sum_i w_i \log w_i. \tag{48.7}$$

Because of the change in the definition of Z, and also of w_i, in (48.1), the actual values of S as given by (46.4) and (48.7) are the same.

Equation of State for a Nonideal Gas

We shall consider only one application of Gibbs' statistics, namely, to nonideal monatomic gases, and thus justify a certain assumption made in the previous chapter [compare (43.12)].

First, we must calculate the sum of state Z in (48.1), which will here be regarded as an integral in the $6N$-dimensional phase space of the N molecules. The volume of the phase cells is denoted by $d\pi\, d\tau$, $d\pi = dp_{1x} \cdots dp_{Nz}$, $d\tau = dx_1 \cdots dz_N$. The energy of this cell is

$$U = \varepsilon + \varphi,$$
$$\varepsilon = \sum_{n=1}^{N} \frac{1}{2M} \mathbf{p}_n^2, \qquad \varphi = \sum_{n,m=1}^{N} \Phi(r_{nm}). \tag{48.8}$$

The state sum Z appears as a product of two integrals, the one equal to

$$\left(\frac{2\pi M}{\beta}\right)^{3/2 N},$$

and the other approximately equal to V^N, the value corresponding to $\varphi = 0$. This leads to
$$\log Z_{id} = \tfrac{3}{2} N \log (2\pi MkT) + N \log V \tag{48.9}$$

which, together with (48.4), leads to the equation for an ideal gas.

When $\varphi \neq 0$, we have
$$\log Z = \log Z_{id} + \log \left[1 + V^{-N} \int (e^{-\beta \varphi} - 1)\, d\tau \right] \tag{48.10}$$

or, with sufficient accuracy,
$$\log Z = \log Z_{id} + V^{-N} \int (e^{-\beta \varphi} - 1)\, d\tau. \tag{48.11}$$

We now assume that the range of the interaction forces is very small, of the order of a few molecular diameters. Thus, if the intermolecular separations are much larger, on the average, we shall have $\Phi(r_{mn}) = 0$, except when two or more molecules come very close to each other. The integrand of (48.11) will therefore be sensibly different from zero only if two pairs of coordinates coincide approximately, which happens altogether $\frac{1}{2}N(N+1)$ times.

Let this pair, to start with, be $\mathbf{r}_1$ and $\mathbf{r}_2$. We then put $\mathbf{r} = \mathbf{r}_1 - \mathbf{r}_2$ and consider only a small range of integration over $\mathbf{r}$. If we now integrate over all coordinates $\mathbf{r}_2$ to $\mathbf{r}_N$, and avoid the (small) regions where the other r_{mn} are small, we obtain—apart from negligible corrective terms—the factor V^{N-1}.

Consequently we have the result

$$\log Z = \log Z_{id} + \frac{N(N-1)}{2V} \int (e^{-\beta\Phi(r)} - 1)\,dx\,dy\,dz \qquad (48.12)$$

or, taking $N(N-1) \approx N^2$ and utilizing (48.4) and (48.9),

$$P = \frac{NkT}{V} - \frac{A}{V^2} \qquad (48.13)$$

where

$$A = \tfrac{1}{2}N^2 kT \int (e^{-\beta\Phi(r)} - 1)\,dx\,dy\,dz. \qquad (48.14)$$

Introducing spherical polar coordinates, we finally arrive at

$$A = \frac{2\pi N^2}{\beta} \int_0^\infty (e^{-\beta\Phi(r)} - 1) r^2\,dr = \frac{2\pi N^2}{3} \int_0^\infty e^{-\beta\Phi(r)} \frac{d\Phi}{dr} r^3\,dr, \qquad (48.15)$$

in accordance with (43.13) to (43.15). If we partition the integral at the point $r = r_0$, where $\Phi(r_0) = 0$, and assume that the potential increases sharply for decreasing $r < r_0$ but does not reach any large values for $r > r_0$, we may write

$$A = a - bNkT,$$

where a and b are approximately constant, corresponding to the constants appearing in the van der Waals equation.

49. QUANTUM STATISTICS

Elementary Particles and Individuality

The most crucial difference between the classical and quantum statistics of atomic particles is in the recognition, brought about by quantum theory, of the fact that it is impossible to ascribe individuality to elementary particles. This situation is very well illustrated by the case of atomic electrons. The solutions of a wave-mechanical equation may—because of the symmetry of the Hamiltonian—be subdivided into sets with different symmetry properties

450 / Mechanics and Statistics

with respect to the electrons; the corresponding states then furnish, as is well-known, an exhaustive description of the series of spectral lines which may be observed directly. Furthermore so-called exchange integrals appear during the calculation of mean values of dynamic variables; and the only possible interpretation of these integrals is that the electrons exchange their identities continually. Hence it is impossible to ascribe a certain set of quantum numbers to any one individually discernible electron. We can say that this-or-that quantum state is occupied by an electron; but there are no means of finding out whether this electron is a particular one of the two or more electrons in the atom. On the contrary, we believe that they exchange their positions continually, in accordance with the fact that no one has yet succeeded in determining an atomic system (with two or more electrons) where the probability that the electrons exchange positions vanishes, at least not when this probability is calculated by the rules of quantum mechanics.

That which has been said here about electrons also applies to all other elementary atomic particles, such as protons, neutrons, and so forth. Moreover, it applies even to certain more complex systems; a well-known example is the stable helium nucleus which resembles the elementary particles in many respects. Even for identical atoms and molecules we have that individuality, in the classical sense (i.e. that the behavior of individual components in principle be predictable and controllable throughout an unlimited future time), does not exist. A large number of mutually interacting atoms or molecules thus obeys a statistics subject to the laws of quantum mechanics.

A New Formulation of the Boltzmann Statistics

Since we must abandon the notion of individuality for atomic systems, the basic assumptions on which the Boltzmann statistics was constructed no longer hold. True enough, we determined the macrostates by stating how many molecules or subsystems were to be found in each phase cell; but the number of microstates composing such a macrostate was found by regarding the molecules as individually discernible, so that a permutation of two molecules (from different phase cells) would necessarily lead to a new microstate.

The new quantum statistics may be introduced in many (apparently) different ways, none of which is particularly easy to follow. Indeed, it would seem that confusion and misunderstandings are rather unavoidable if one does not, first of all, try to obtain a clear idea of just how the previous argument leading to the Boltzmann law of distribution is to be modified by quantum considerations.

Let us state, once more, the classical thermodynamic probability

$$W = \frac{N!}{\prod_i N_i!}. \tag{49.1}$$

In the following discussion we shall take the occupation number N_i for a cell to be large, in accordance with the tacit assumption that the size of phase cells may be chosen arbitrarily, subject only to the demands of convenience. However, the extension of elementary cells is not arbitrary, being equal to h per degree of freedom. For such a cell N_i will (in the cases where Boltzmann statistics applies) as a rule be zero, rarely one, and very rarely two or higher. In order that the usual deduction of the distribution law still be valid, we must therefore consider larger cells, containing a fairly large number G of elementary cells, that is, $G \gg 1$.

We may then use Stirling's formula $n! \approx n^n e^{-n}$ to rewrite (49.1) as

$$W = \frac{N^N}{\prod N_i^{N_i}} = \prod_i \left(\frac{N_i}{N}\right)^{-N_i/N} \tag{49.2}$$

or

$$W = \prod_i W_i, \qquad W_i = \left(\frac{N_i}{N}\right)^{-N_i/N} \tag{49.3}$$

The interpretation of this is that every cell with a given occupation number N_i has the thermodynamic probability W_i, and the total thermodynamic probability is then the product of all of these.

Let us now, utilizing the above considerations, try to deduce the thermodynamic probability for each cell, defined as the number of possible different microstates for N_i particles in G cells. In placing the first molecule we have G cells to choose from, that is, G possibilities. Let us assume (somewhat arbitrarily, perhaps) that the *a priori* probability of each cell has not been changed by the placing of the first object, so that we also have G possibilities for the second object, and so forth. Thus there are altogether G^{N_i} combinatorial possibilities; however, since we cannot distinguish between individual particles, the $N_i!$ states resulting from all possible permutations of the particles will represent *the same microstate*. Hence the thermodynamic probability is

$$W_i = \frac{G^{N_i}}{N_i!} \tag{49.4}$$

and the total thermodynamic probability

$$W = \prod_i W_i = \frac{G^N}{\prod_i N_i!}. \tag{49.5}$$

This differs from (49.1) only by a numerical factor which results in an additive constant in the entropy $S = k \log W$.

We now use Stirling's formula to rewrite (49.4) as

$$W_i = \left(\frac{N_i}{G}\right)^{-N_i} e^{N_i} = (n_i^{-n_i} e^{n_i})^G, \qquad n_i = \frac{N_i}{G}. \tag{49.6}$$

Since we have G cells, this may be interpreted as stating that each elementary cell has the thermodynamic probability

$$n_i^{-n_i} e^{n_i}.$$

The mean occupation number n_i per cell is then no longer a very large number; on the contrary, it is very small, since $N_i \ll G$ in general.

We may therefore write the thermodynamic probability as

$$W = \prod_j n_j^{-n_j} e^{n_j}, \qquad (49.7)$$

taking the index j to range over all elementary cells. The entropy becomes

$$S = -k \sum_j (n_j \log n_j - n_j). \qquad (49.8)$$

The condition for thermodynamic equilibrium will now, in order to facilitate the transition to quantum statistics, be formulated somewhat differently. We take the temperature, and not the energy, of the system to be constant; this, as we have seen, is more along the lines of Gibbs's than Boltzmann's statistics. The subdivision of phase space into cells of given energies will, however, be retained, that is, we regard the volume as constant.

Hence the thermodynamic quantity that is most conveniently considered in this situation is not S, which would have been suitable in the event of a *constant energy* and *volume*, since

$$dS = \frac{dU}{T} + \frac{P}{T} dV, \qquad (49.9)$$

but the free energy $F = U - TS$, which is suitable for a *constant temperature* and *volume* due to the relation

$$dF = -S\, dT - P\, dV. \qquad (49.10)$$

Equation (49.8) then leads to

$$F = \sum_j n_j [U_j + kT(\log n_j - 1)]. \qquad (49.11)$$

The condition for thermodynamic equilibrium is now that F has a minimum value, that is, that a variation of the occupation numbers n_j results in

$$\delta F = \sum_j (U_j + kT \log n_j)\, \delta n_j = 0. \qquad (49.12)$$

In addition, the auxiliary conditions

$$\sum_j n_j = N, \qquad \sum_j \delta n_j = 0, \qquad (49.13)$$

must be satisfied; Lagrange's multiplier method then yields the Boltzmann distribution law

$$n_j = e^{(U_0 - U_j)/kT} \tag{49.14}$$

or

$$n_j = Z_0^{-1} e^{-U_j/kT}, \quad Z_0 = e^{-U_0/kT} = \sum_j e^{-U_j/kT}. \tag{49.15}$$

The term U_0 is here constant, that is, independent of the occupation numbers n_j; it will, however, depend on T and V and has the dimension of energy.

The Fermi-Dirac Statistics

It follows from the Pauli principle for electrons—or other particles having half-integral spins—that a quantum state (i.e. phase cell) may be occupied by at most one such particle; this corresponds to the requirement that the quantum-mechanical function of state (spatial wave function plus spin function) must be antisymmetrical in the spatial and spin coordinates of any two particles.

For the statistics this has the consequence that no elementary phase cell may be occupied by more than one particle. A larger cell, composed of G elementary cells, may now be assumed to have the capacity G to start with, since the first particle may be placed in G different ways. However, if the place of this particle has once been determined, then one cell is fully occupied; the capacity will thus have dropped to $G - 1$. Hence we obtain altogether

$$G(G-1)\cdots(G-N_i+1) = G!/(G-N_i)! \tag{49.16}$$

ways of locating N_i particles; on the other hand, $N_i!$ of these are equivalent, since the particles are not to be distinguished from each other. The thermodynamic probability per cell thus becomes

$$W_i = \frac{G!}{N_i!\,(G - N_i)!} \tag{49.17}$$

or, using Stirling's formula with $n_i = N_i/G$,

$$W_i = \frac{G^G}{N_i^{N_i}(G-N_i)^{G-N_i}} = n_i^{-N_i}(1-n_i)^{-G+N_i} = [n_i^{-n_i}(1-n_i)^{-1+n_i}]^G. \tag{49.18}$$

The expression in the square bracket may be regarded as the thermodynamic probability of each of the G elementary cells, for which we use the index j, as previously. This leads to

$$W = \prod_j n_j^{-n_j}(1 - n_j)^{-1+n_j}, \tag{49.19}$$

and

$$S = k \log W = -k \sum_j [n_j \log n_j + (1 - n_j) \log(1 - n_j)]. \tag{49.20}$$

From

$$\delta S = k\,\delta \log W = -k \sum_j \log \frac{n_j}{1-n_j}\,\delta n_j \qquad (49.21)$$

we now obtain by the same procedure as that employed above:

$$\frac{n_j}{1-n_j} = e^{(U_0 - U_j)/kT} = \frac{1}{\alpha}\, e^{-U_j/kT}, \qquad (49.22)$$

$$n_j = \frac{1}{\alpha e^{U_j/kT} + 1}, \qquad \alpha = e^{-U_0/kT}. \qquad (49.23)$$

This is the Fermi-Dirac distribution function.

The Bose-Einstein Statistics

In the Bose-Einstein statistics each elementary phase cell may admit any number of particles; this corresponds to the fact that the quantum-mechanical function of state is symmetrical in the coordinates of all particles.

We shall, as previously, assume that an aggregate of phase cells has the initial capacity G, that the first particle may be placed in any one of these cells in G different ways. One might now, after a superficial consideration of the problem, conclude that the capacity of the aggregate would be for the second and all subsequent particles as well. However, a more detailed analysis shows that the capacity for the second particle is $G + 1$, for the third $G + 2$, and so forth; in other words, *the capacity of the phase cells increases by unity for each new particle admitted.*

This is, of course, no *physical* law but a *combinatorial* one; it is easily seen how it comes about.

The first occupation of a phase cell immediately results in G new microstates. An occupation of the same cell by a new particle leads to G new states; on the other hand, an occupation of an empty phase cell may be done in $G - 1$ different ways, and should hence result in $G(G - 1)$ microstates. However, each of these states is then actually counted twice: If the cells a and b are both occupied, it is immaterial which of them was occupied first. Thus the number of microstates has risen to

$$\frac{G(G-1)}{2} + G = \frac{G(G+1)}{2}. \qquad (49.24)$$

A corresponding analysis of the next step shows that the number of microstates has now risen to

$$\frac{G(G-1)(G-2)}{1\cdot 2\cdot 3} + G(G-1) + G = \frac{G(G+1)(G+2)}{1\cdot 2\cdot 3}. \qquad (49.25)$$

Hence the capacity of the cell complex really increases by unity for each new occupation of a cell and the total thermodynamic probability for N_i particles located in the complex becomes

$$W_i = \frac{(N_i + G - 1)!}{N_i!\,(G-1)!} \approx \frac{(N_i + G)!}{N_i!\,G!}. \tag{49.26}$$

Putting $n_i = N_i/G$, we now utilize Stirling's formula to rewrite this as

$$W_i = \frac{(G + N_i)^{G+N}}{N_i^{N_i} G^G} = \left[\frac{(1 + n_i)^{1+n_i}}{n_i^{n_i}}\right]^G. \tag{49.27}$$

The expression in square brackets may, once again, be interpreted as the thermodynamic probability for a mean occupation number n_i of each phase cell; denoting these by the index j, as previously, we obtain the total thermodynamic probability

$$W = \prod_j \frac{(1 + n_j)^{1+n_j}}{n_j^{n_j}} \tag{49.28}$$

and

$$\log W = -\sum_j [n_j \log n_j - (1 + n_j) \log(1 + n_j)], \tag{49.29}$$

Hence

$$F = U - TS = \sum_j \{U_j n_j + kT[n_j \log n_j - (1 + n_j)\log(1 + n_j)]\}, \tag{49.30}$$

$$\delta F = \sum_j \left(U_j + kT \log \frac{n_j}{1 + n_j}\right) \delta n_j. \tag{49.31}$$

The form of the distribution function now depends on whether the particle number N is taken to be constant or not. For photons, which obey Bose-Einstein statistics, no such definite number exists, since they are continually absorbed and reemitted. Consequently (49.31) yields, for a photon gas,

$$\frac{n_j}{1 + n_j} = e^{-U_j/kT}, \qquad n_j = \frac{1}{e^{U_j/kT} - 1}. \tag{49.32}$$

For a gas of material Bose-Einstein particle (say, helium atoms), on the other hand, we find by similar methods as for Fermi-Dirac statistics

$$n_j = \frac{1}{\alpha e^{U_j/kT} - 1}, \qquad \alpha = e^{-U_0/kT}, \tag{49.33}$$

where U_0 or α is a constant, to be determined by the relation

$$\sum_j n_j = N. \tag{49.34}$$

(See Fig. 7-3.)

456 / Mechanics and Statistics

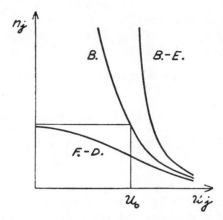

Fig. 7-3. Comparison of statistical distribution functions: B = Boltzmann statistics; F.-D. = Ferm-Dirac statistics; B.-E. = Bose-Einstein statistics.

Entropy and Nernst's Theorem

It should be clearly understood that the Boltzmann statistics is not an independent statistical theory, to be considered beside the two quantum statistical theories. To be sure, one may formally (i.e. by assuming the capacity of an aggregate of G elementary cells to be constantly equal to G) construct such a statistics; similarly, by introducing different assumptions, it is possible to construct several other statistical theories. In physics, however, the Boltzmann statistics appear as a limiting case of both the Fermi-Dirac and Bose-Einstein statistics, namely, in the limit of very small mean occupation numbers for the phase cells; this is easily seen to follow from (49.8), (49.20), and (49.29), since

$$(1 - n_j) \log (1 - n_j) \to -n_j, \qquad (1 + n_j) \log (1 + n_j) \to n_j \quad \text{for} \quad n_j \to 0.$$

Physically this may be explained as follows: When the density of the particles in space is sufficiently low or if their energy is spread over a large range (due to high temperatures), the mean occupation numbers of the cells will be very small. The actual occupation number will then be mostly $n_j = 0$, rarely $n_j = 1$, and the case of n_j being two or higher will practically never occur. Hence it is immaterial which of the two quantum statistical theories we employ.

Since the Boltzmann statistics is not an independent theory, but only a simplified form of the other two, the results presently to be deduced will have general validity.

At the absolute zero point of the temperature T the particles will aggregate into the phase cells having the smallest possible energy. In the Fermi-Dirac

Statistical Mechanics / 457

statistics, therefore, the mean occupation number will initially all be $n_j = 1$; for phase cells above a certain energy limit $U_j = U_0$, however, they will suddenly drop to $n_j = 0$. In both cases we have

$$n_j \log n_j + (1 - n_j) \log (1 - n_j) = 0. \tag{49.35}$$

In the Bose-Einstein statistics, on the other hand, all particles will crowd into the same phase cell: that of smallest energy. We shall then have, say, $n_1 = N$, $n_2 = n_3 = \cdots = 0$. Substitution of the latter in (49.29) yields zero, while we obtain

$$(N + 1) \log (N + 1) - N \log N \tag{49.36}$$

for the first phase cell. This is not zero, even though it is very small compared to the particle number N. However, the deduction of (49.28) was based on the notion of an aggregate of G phase cells, where G was taken to be a large number. Since all the particles are now crowded into one cell, this assumption is no longer valid and consequently we may not use Stirling's formula. Hence we must revert to the exact equation (49.26); substituting $G = 1$ in this, we obtain

$$W_j = 1, \quad W = 1, \quad \log W = 0, \tag{49.37}$$

in accordance with Nernst's theorem (compare the introduction to chapter 5).

50. APPLICATIONS OF THE NEW STATISTICS TO GAS KINETICS

The Photon Gas and the Law of Radiation

As a first example, we shall study the energy distribution in a photon gas, since (49.32) represents a special case; we assume that the photons or radiation are contained in a box of volume V. The radiation is assumed to be isotropic, so that we need only consider a spherical shell in momentum space, having the volume $4\pi p^2 \, dp$; taken in conjunction with V, this "momentum-volume" then represents a number of $4\pi p^2 \, dp \cdot V/h^3$ elementary phase cells. However, a photon is not only characterized by its momentum, but also by its spin, which is integral and given by the quantum numbers ± 1. This doubles the number of possibilities; and we obtain a number of photons, having an energy corresponding to the momentum range considered which is equal to

$$dN = V \frac{8\pi p^2 \, dp}{h^3} \frac{1}{e^{U/kT} - 1}. \tag{50.1}$$

By the light quantum hypothesis, $U = h\nu$ and $p = h\nu/c$, where ν is the photon frequency. Multiplying by $h\nu$ and dividing by $V\,d\nu$, we thus obtain the energy density

$$u_\nu = \frac{U_\nu}{V} = \frac{8\pi\nu^2}{c^3} \frac{h\nu}{e^{h\nu/kT} - 1}. \tag{50.2}$$

The Virial Theorem in Quantum Statistics

In the kinetic theory of gases we found, for a gas composed of mass points, that the pressure was $\tfrac{2}{3}$ of the kinetic energy per unit volume. This result could also have been deduced from the virial theorem.

We must assume that the same law holds in quantum statistics. If this is so, a considerable amount of work may be saved; we may, for instance, omit any direct calculations of the pressure P (by means, say, of the relation $P = -(\partial F/\partial V)_T$), and content ourselves with finding U.

First of all, we check whether our assumption is compatible with Boltzmann's statistics. We have by (46.10) and (46.16):

$$U = -N\frac{\partial \log Z}{\partial \beta}, \quad P = NkT\frac{\partial \log Z}{\partial V}, \quad Z = \sum_i e^{-\beta U_i}. \tag{50.3}$$

Here we must, evidently, determine how the energy U_i of the quantized energy levels depends on the volume V. For a given total momentum p the volume of all phase cells with a smaller energy will be

$$\frac{4\pi}{3} p^3 V,$$

and this volume will then contain a certain number of phase cells. A change of the volume V must result in a corresponding change of the total momentum p, if this number is to remain constant; thus we must have

$$p^3 V = \text{const}, \tag{50.4}$$

or, writing

$$U_i = \frac{1}{2m} p^2, \tag{50.5}$$

$$U_i^{3/2} V = \text{const}, \quad U_i = \text{const } V^{-2/3}. \tag{50.6}$$

It follows from this that

$$V\frac{\partial U_i}{\partial V} = -\frac{2}{3} U_i \tag{50.7}$$

whence, by (50.3),

$$PV = NkTV\frac{\partial \log Z}{\partial V} = NkT\frac{2}{3}\beta\frac{\partial \log Z}{\partial(-\beta)} = \frac{2}{3} U. \tag{50.8}$$

We shall now show that the same result also applies for Fermi-Dirac and Bose-Einstein statistics. Let us consider the entropy

$$S = -k \sum_j [n_j \log n_j - (n_j \mp 1) \log (1 \mp n_j)] \quad (50.9)$$

and the distribution function

$$n_j = \frac{1}{\alpha e^{U_j/kT} \pm 1}, \quad (50.10)$$

where the upper and lower sign apply throughout to the Fermi-Dirac and Bose-Einstein statistics, respectively. First of all, we calculate the free energy $F = U - TS$,

$$\begin{aligned} F &= U + kT \sum_j \left[n_j \log \frac{n_j}{1 \mp n_j} \mp \log (1 \mp n_j) \right] \\ &= kTN \log \alpha \mp kT \sum_j (1 \mp \alpha^{-1} e^{-U_j/kT}) \end{aligned} \quad (50.11)$$

by means of

$$kT \log \frac{n_j}{1 \mp n_j} = -kT \log \alpha - U_i \quad \text{and} \quad \sum_i n_j = N, \sum_j n_j U_j = U. \quad (50.12)$$

We then find the pressure, utilizing the relation $P = -(\partial F/\partial V)_T$, to be

$$P = N \frac{\partial \log \alpha}{\partial V} - \sum_j \frac{(\partial \log \alpha/\partial V) + (\partial U_j/\partial V)}{\alpha e^{U_j/kT} + 1} = -\sum_j n_j \frac{\partial U_j}{V}. \quad (50.13)$$

Finally multiplication by (50.7) leads to

$$PV = \frac{2}{3} \sum_j n_j U_j = \frac{2}{3} U. \quad (50.14)$$

Thus we have proved that the virial theorem is valid also in quantum statistics.

For a photon gas we find $PV = \frac{1}{3}U$ or $P = \frac{1}{3}u$; this is because the relation between the energy and momentum of a photon, $p = U/c$, differs from the corresponding relation $p = 2U/v$ valid for material nonrelativistic particles.

Sums of State in Quantum Statistics

Finally we shall determine the quantities that are to take the place of the previously discussed sum of state Z in the Fermi-Dirac and Bose-Einstein statistics. From (50.11) it follows that

$$e^{-F/kT} = \begin{cases} \alpha^N \prod_j \left(1 + \frac{1}{\alpha} e^{-U_j/kT}\right) = Z_{FD}, \\ \alpha^N \prod_j \dfrac{1}{1 - \dfrac{1}{\alpha} e^{-U_j/kT}} = Z_{BE} \end{cases} \quad (50.15)$$

for the Fermi-Dirac and Bose-Einstein statistics, respectively.

If we denote the functions on the right-hand side by Z, it follows from the preceding considerations that

$$P = kT \left(\frac{\partial \log Z}{\partial V}\right)_T, \quad (50.16)$$

analogously to the classical statistics of Gibbs. Similarly, the corresponding equation for the internal energy

$$U = \left[\frac{\partial \log Z}{\partial(-1/kT)}\right]_V \quad (50.17)$$

is easily verified.

Ideal Gases with Weak Degeneration

From the distribution function (50.10) we see that the classical statistics will be approximately valid if

$$\alpha e^{U_j/kT} = e^{(U_j - U_0)/kT} \gg 1. \quad (50.18)$$

For an ideal gas, where U_j may be as small as $U_j = 0$, this means that α must be fairly large. Hence we obtain a fairly rapidly converging series

$$n_j = \alpha^{-1} e^{-U_j/kT} \mp \alpha^{-2} e^{-2U_j/kT} + \alpha^{-3} e^{-3U_j/kT} \mp \cdots. \quad (50.19)$$

Putting the sum of this series equal to N, we may determine $a = e^{-U_0/kT}$; however, an expression in closed form for this quantity can unfortunately not be obtained. Multiplying by U_j and summing, we find U, and thus once more $PV = \frac{2}{3}U$.

For a gas we must replace the sum over the index j by the integral

$$\frac{4\pi V}{h^3} \int_0^\infty \cdots p^2 \, dp \quad (50.20)$$

and write

$$U_i = \frac{1}{2m} p^2. \quad (50.21)$$

Defining
$$K = \frac{Nh^3}{V(2\pi mkT)^{3/2}}, \tag{50.22}$$
we thus obtain
$$K = \alpha^{-1} \mp 2^{-3/2}\alpha^{-2} + 3^{-3/2}\alpha^{-3} \mp \cdots, \tag{50.23}$$
and
$$\frac{\tfrac{2}{3}E}{NkT} K = \frac{PV}{NkT} K = \alpha^{-1} \mp 2^{-5/2}\alpha^{-2} + 3^{-5/2}\alpha^{-3} \mp \cdots. \tag{50.24}$$

The division of (50.24) by (50.23) then leads to the equation of state
$$\frac{PV}{NkT} = \frac{1 \mp 2^{-5/2}\alpha^{-1} + 3^{-5/2}\alpha^{-2} \mp \cdots}{1 \mp 2^{-3/2}\alpha^{-1} + 3^{-3/2}\alpha^{-2} \mp \cdots}. \tag{50.25}$$

Taking $\alpha^{-1} = K$ as a first approximation, we obtain an equation that is correct to the first order in K. Since higher-order corrections require a more accurate determination of α, we shall here give only the equation of state in the first-order approximation
$$PV = NkT(1 \pm 2^{-5/2}K). \tag{50.26}$$

Hence it follows that a weakly degenerate Fermi-Dirac gas is less compressible, and a corresponding Bose-Einstein gas more compressible, than a classical ideal gas; in other words, the Fermi and Bose statistics lead to apparent repulsive and cohesive forces, respectively. For gases exhibiting somewhat stronger intermolecular forces, however, these purely statistical effects are completely insignificant.

Strong Degeneration of a Bose-Einstein Gas

We have seen that the degeneration of a gas obeying the Bose-Einstein statistics gives rise to apparent cohesive forces. Initially the degeneration is approximately proportional to the quantity K in (50.22), so that it increases with increasing concentration and decreasing temperature. It is then natural to ask: Will an increasing degree of degeneration ultimately result in an actual condensation of the gas, even in the absence of ordinary intermolecular cohesive forces?

In order to obtain some insight into this mechanism for a strong degeneration, we calculate the number of particles in the gas by means of the expression
$$N = \frac{4\pi V}{h^3} \int_0^\infty \frac{p^2\, dp}{\alpha e^{U_j/kT} - 1}, \quad U_j = \frac{1}{2m} p^2. \tag{50.27}$$

If this is to have any meaning, we must have $\alpha \geq 1$. Thus the number of particles contained in the volume V will increase for decreasing α, up to the limit $\alpha = 1$; the value of N in this case is found by (50.22) and (50.23) to be

$$\frac{N}{V} = \frac{(2\pi mkT)^{3/2}}{h^3} \sum_{n=1}^{\infty} n^{-3/2} = \frac{2.612}{h^3}(2\pi mkT)^{3/2}. \tag{50.28}$$

This, then, is the maximum density attainable by a Bose-Einstein gas in thermodynamic equilibrium. At greater densities the excess part must disappear from the gas phase and then presumably pass over into the liquid phase, that is, condense.

Similarly we deduce from (50.24) that

$$P = \frac{NkT}{KV} \cdot \sum_{n=1}^{\infty} n^{-5/2} = 1.341(2\pi mkT)^{3/2}kT. \tag{50.29}$$

Thus the pressure depends on the temperature alone, and not on the volume. This is just the situation obtaining at the ordinary condensation of a gas, where the isothermals in a PV-diagram are horizontal, that is, parallel to the V-axis.

In order to assess roughly at what temperatures such a condensation would be expected to take place, we put $K = 1$, or

$$T = \frac{h^2}{2\pi mk}\left(\frac{N}{V}\right)^{2/3}. \tag{50.30}$$

Taking

$$h \approx 10^{-27}, \quad 2\pi \approx 10, \quad m \approx 10^{-24}, \quad k \approx 10^{-16}, \quad \frac{N}{V} \approx 10^{24},$$

we then find

$$T \approx 10. \tag{50.31}$$

Hence the condensation only occurs at temperatures close to absolute zero; consequently this effect would primarily be of importance for low melting point gases such as helium.

Zero-Point Energy of a Fermi Gas

In the case of strongly degenerate Fermi gases new phenomena occur; in particular, we mention the appearance of a zero-point pressure and energy for the gas at the absolute temperature $T = 0$. The physical explanation of these effects is, of course, that the molecules cannot all crowd into the lowest quantum state, due to the Pauli principle; the majority maintain a considerable kinetic energy and thus contribute to a correspondingly high gas pressure, even at $T = 0$. The free electrons of a metal constitute such a strongly degenerate Fermi gas; and the Fermi-Dirac statistics therefore play an important role in the exact theory of metallic conductivity. We shall not

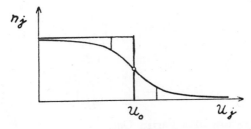

Fig. 7-4. The distribution function in Fermi-Dirac statistics.

consider this problem in detail but shall here only deduce the zero-point energy and zero-point pressure of a Fermi gas.

In order to examine the energy distribution in a state of strong degeneration, we once more write the mean occupation number for each phase cell as

$$n_j = \frac{1}{e^{(U_j - U_0)/kT} + 1}, \qquad e^{-U_0/kT} = \alpha. \tag{50.32}$$

A strong degeneration then means that α is very small, and U_0 or U_0/kT fairly large.

Let us assume a fixed value for U_0; when $T \to 0$, n_j will then take the values 1 or 0, according as U_j is less or greater than U_0. Thus n_j is a discontinuous step function with the intermediate value $n = \frac{1}{2}$, taken for a cell with the energy $U_j = U_0$; for $T > 0$, the function is continuous in this energy region. The distribution is plotted, as a function of the energy, in Fig. 7-4.

We note that this function has the curious property that $n_j - \frac{1}{2}$ is antisymmetric in the argument $U_j - U_0$; thus if we rotate the curve 180° around the point with the ordinate $n_j = \frac{1}{2}$, it will coincide with the original (unrotated) curve.

In the limit $T \to 0$ it is very easy to calculate N and U as functions of U_0, and hence also U as a function of N and V. We obtain as before, writing $p^2\, dp = \frac{1}{2}(2m)^{3/2} U_j^{1/2}\, dU_j$:

$$N = \frac{2\pi V}{h^3}(2m)^{3/2} \int_0^{U_0} U_j^{1/2}\, dU_j = \frac{4\pi}{3}\frac{V}{h^3}(2m)^{3/2} U_0^{3/2}, \tag{50.33}$$

$$U = \frac{2\pi V}{h^3}(2m)^{3/2} \int_0^{U_0} U_j^{3/2}\, dU_j = \frac{4\pi}{5}\frac{V}{h^3}(2m)^{3/2} U_0^{5/2}, \tag{50.34}$$

so that

$$U = \tfrac{3}{5} N U_0. \tag{50.35}$$

Thus U_0 is actually, apart from the factor $\frac{5}{3}$, the zero-point energy per molecule of the gas. Its value for a given volume V and a given number of particles

N is by (50.33):

$$U_0 = \left(\frac{3N}{4\pi V}\right)^{2/3} \frac{h^2}{2m}. \tag{50.36}$$

The zero-point energy $U = E_0$ of the gas at $T = 0$ is thus

$$U = E_0 = \frac{3}{5}\frac{h^2}{2m}\left(\frac{3N}{4\pi V}\right)^{2/3} N. \tag{50.37}$$

Strong Degeneration of a Fermi Gas

In order to determine just how a Fermi gas behaves at a temperature close to absolute zero, we must consider the integrals for N and U in more detail; we shall then utilize Fig. 7-4 and exploit the properties already found for the distribution function.

First of all, we clearly commit a fairly small error (of the order of $e^{-U_0/kT}$) by integrating only up to $U_j = 2U_0$. In this region the functions $U_j^{1/2}$ and $U_j^{3/2}$ may be expanded in convergent power series of the argument $U_j - U_0$,

$$U_j^{1/2} = U_0^{1/2} + \tfrac{1}{2}U_0^{-1/2}(U_j - U_0) + \cdots, \tag{50.38}$$
$$U_j^{3/2} = U_0^{3/2} + \tfrac{3}{2}U_0^{1/2}(U_j - U_0) + \cdots.$$

In the region $0 \leq U_j \leq U_0$, we now subtract the integrals (50.33) and (50.34), so that we have to integrate over the functions $-(1 - n_j)$. As already mentioned, however, we have

$$1 - n_j[U_0 + (U_j - U_0)] = n_j[U_0 + (U_0 - U_j)]. \tag{50.39}$$

Thus, putting

$$U_0 - U_j = \xi, \quad U_j < U_0,$$
$$U_j - U_0 = \xi, \quad U_j > U_0, \tag{50.39a}$$

we find

$$N \doteq (N)_{T=0} + \frac{2\pi V}{h^3}(2m)^{3/2}U_0^{-1/2}\int_0^{U_0} \frac{\xi\, d\xi}{e^{\xi/kT} + 1}, \tag{50.40}$$

and correspondingly for U. For small kT and large U_0 the error introduced by integrating only up to U_0 may once again be neglected; we then obtain, after some calculation,

$$N = \frac{4\pi}{3}\frac{V}{h^3} U_0^{3/2}(2m)^{3/2}\left[1 + \frac{3}{2}\frac{(kT)^2}{U_0^2} c_2\right], \tag{50.41}$$

and

$$U = \frac{4\pi}{5}\frac{V}{h^3} U_0^{5/2}(2m)^{3/2}\left[1 + \frac{15}{2}\frac{(kT)^2}{U_0^2} c_2\right], \tag{50.42}$$

where

$$c_2 = \int_0^\infty \frac{x\, dx}{e^x + 1} = \sum_{n=1}^\infty \frac{(-1)^{n-1}}{n^2} = \frac{\pi^2}{12}. \tag{50.43}$$

Division of (50.42) by (50.41) and substitution of the approximate value (50.36) of U_0 in the corrective term now leads to

$$\frac{U}{N} = \frac{3}{5} U_0 \left[1 + \frac{\pi^2}{2} \left(\frac{4\pi V}{3N} \right)^{4/3} \left(\frac{2m}{h^2} \right)^2 (kT)^2 \right]. \qquad (50.44)$$

However, a correction of (50.36) by means of (50.41) yields

$$U_0 = \left(\frac{3N}{4\pi V} \right)^{2/3} \frac{h^2}{2m} \left[1 - \frac{\pi^2}{12} \left(\frac{4\pi V}{3N} \right)^{4/3} \left(\frac{2m}{h^2} \right)^2 (kT)^2 \right]; \qquad (50.45)$$

hence we finally obtain, by eliminating U_0, (50.44) in the form

$$\frac{U}{N} = \frac{3}{5} \frac{h^2}{2m} \left(\frac{3N}{4\pi V} \right)^{2/3} \left[1 + \frac{5\pi^2}{12} \left(\frac{4\pi V}{3N} \right)^{4/3} \left(\frac{2m}{h^2} \right)^2 (kT)^2 \right]. \qquad (50.46)$$

From this the equation of state for the degenerate Fermi gas is easily found by using the equation

$$PV = \tfrac{2}{3} U. \qquad (50.47)$$

Thus we see that the gas is characterized by a large internal zero-point energy and a large zero-point pressure. In addition, a term proportional to T^2 appears in the expressions for both these quantities, for low temperatures. Hence the specific heat capacity at constant volume will be proportional to T and will approach zero when $T \to 0$,

$$C_V = \left(\frac{\partial U}{\partial T} \right)_V = \text{const } T. \qquad (50.48)$$

Further details concerning the behavior of strongly degenerate Fermi gases will not be considered here. This topic is of particular importance in the theory of metallic electrons; however, a discussion of this theory would transcend by far the scope of this book.

Author Index

d'Alembert, J. le Rond, 213, 263, 267
Archimedes, 213, 263
Avogadro, A., 404

Bernoulli, D., 341, 398
Bernoulli, Jacob, 263
Bernoulli, Johann, 263
Bjerknes, V., 344
Boltzmann, L., 214, 398, 437, 450
Bose, S. N., 454
Boyle, R., 371
Brown, R., 397

Carnot, S., 369, 375
Cauchy, A. L., 324
le Chatelier, 388
Clapeyron, B. P. E., 388
Clausius, R., 214, 386, 398, 424
Coriolis, G., 255

Dalton, J., 214, 400
Davy, H., 214, 369
Dirac, P. A. M., 453

Einstein, A., 213, 454
Euler, L., 213, 270, 301, 338

Fahrenheit, G. D., 370
Faraday, M., 214
Fermat, P. de, 269
Fermi, E., 453
Foucault, J. B. L., 259
Fresnel, A., 214

Galilei, Galileo, 213, 215, 219

Gibbs, J. W., 214, 388, 398, 445

Hamilton, W. R., 213, 244, 261, 269
Helmholtz, H., 214, 346, 375
Hertz, H., 214, 217
Hooke, R., 321
Huygens, Chr., 213

Jacobi, K. G. J., 213, 290
Joule, J. P., 214, 375, 389

Kant, I., 216
Kapitza, P., 390
Kelvin, Lord, 344, 370, 389
Kepler, J., 213
Kirchhoff, G. R., 217

Lagrange, J. L., 213, 268, 270, 338
Laplace, S. P., 213, 330
Leibniz, G. W., 270
Liouville, J., 433
Lorentz, H. A., 214

Mariotte, E., 371
Maupertuis, P. L. M. de, 213, 270, 271
Mayer, R. J., 214, 375, 376
Maxwell, J. C., 214, 389, 398, 407

Nernst, W., 369, 456
Newton, I., 213, 215-219

Pauli, W., 453
Poisson, S. D., 327

Réaumur, R. A. F., 370

Rumford, Earl, 214, 369, 375

Stevin, S., 263
Stirling, J., 440

Torricelli, E., 341

van der Waals, J. D., 393

Young, T., 214

Subject Index

Abbreviate cosine z, 138
Absolute conditional convergence, 140
  derivative, 332, 333
    maximum, 92
    minimum, 92
    temperature, 371, 378, 391
      scale, 371, 465
    units, 218
    value, 6, 123
      of a product, 171
    zero, 371, 377, 392, 462
Absolutely convergent hypergeometric series, 181
  convergent series, 140
Acceleration, angular, 255
  centripetal, 255
  Coriolis, 255
  hydrodynamical, 332
  law of, 216
  relative, 255
Accuracy, of approximate expression, 135
Action, 217, 270, 271, 281
  function, 244, 272, 290, 293, 431
  integral, 244
  least, 213, 270
  variable, 290
Action-reaction principle, 217, 249
Actual wave lengths, 93
Additional condition, normalization as, 104
Addition of tensors, 44
Additivity law, 65, 66
Adiabatic processes, 379
Adiabatics, 374, 385

Affine coordinate systems, 50
Aggregation, 372
Air resistance, 223
d'Alembert's principle, 261, 267, 268, 273
Alternative definition of gamma function, 156
Amplitude, 225, 228, 232
Analytical expression of vector products, 12
Analytic continuation, 146
  continuation of power series, 145
  function, 130
    different power series of, 145
  functions, 125, 127, 139
    of the complex variable z, 125
    power theory for, 134
    geometry, 11
Angular acceleration, 255
  frequency, 80
  velocity, 255, 345
Anharmonic oscillator, 239
Anticommutative law, 11
Antisymmetric functions with zero value, 188
  part of the relative displacement, 50
  tensor, 44
    and tensor products, 46
Aperiodic oscillations, 229
Approximate eigenvalues, 108, 109
Approximation, $(n$-$1)$, 110
Arbitrarily chosen fundamental region, 84
  oriented Cartesian system, 25
Arbitrary exponents, 137
Arc length, 17

470 / Subject Index

Area of a surface of rotation and the brachystochrone, 95
Areas, law of, 250, 251, 281
Argand's diagram, 121
Argument, 126
Arguments, 123
Arrangement, of differential equations according to their single points, 176
Artificially nonanalytic, 130
Associate Laguerre functions, 195
   Legendre function of the second kind, 191
Asymptotic, 192
  behavior of eigenfunctions, 117
  behavior of eigenvalues, 114, 119
  form, 175
  solution of Bessel's equation, 176
  solutions, 174
  value and asymptotic expansions, 204
  value of a Bessel function, 203
Atomic orbits, quantum theory of, 153
  spectra, 196
Auxiliary equation, 135
Avogadro's number, 404
Axial vector, 46

Base powers, 83
Berninoulli, 95
  theorem, 341
Bessel coefficients, 207
  differential equation, 201
  equation, 174, 175, 178-9, 180
    asymptotic solution of, 176
  function with imaginary arguments, 206
  functions, 201 *et seq.*
    Hankel function, 203-204
    Neumann function $N(z)$, 203
    of the second kind, 203
    of the third kind, 203-204
Binding, non-polar, 239
  polar, 239
Bjerknes' circulation theorem, 344
Bohr atomic theory, 196
Boltzman constant, 400, 404, 437
  distribution law, 409, 437, 450
  statistics, 445, 450-452, 455
Bose-Einstein distribution law, 455
  gas, 461
  statistics, 454, 456, 457, 459, 461
Boundary conditions, 81, 336, 349, 356

of the domain of integration, 107
for Euler differential equation, 100
Boyle-Mariotte's law, 371, 377, 398, 399
Brachystochrone, the area of the surface of rotation, 95
Branches, of a function, 149
Branching points, 151
Brownian motion, 397

Calculation of D^2, 55
Calculus, of variational, 116
  of variations, elements of, 91, 92
Caloric, 369
Canonical ensemble, 445
  equations, 213, 286, 287
  transformations, 247, 287, 288, 292-294
Carnot cycle, 375, 383
Cartesian components, 8
  coordinate system, 7, 56
Cauchy integral theorem, 130, 131
  relations, 324-327
  -Riemann equation, 125, 127, 129
  theorem, 129, 130, 184, 186
Cauchy's theorem, 194
  and complex integrals, 128
Central field, 288, 296
Central forces, 249, 426
Centre of mass, 217, 220, 247-251, 299, 303
  law, 248, 299
Centrifugal force, 256
Centripetal acceleration, 255
Cgs system, 5
Chain, end points of, 96
  line, 96
le Chaletier's principle, 388
Change in unit volume, 71-72
Characteristics of vector transformation, 44
Choice of Cartesian coordinates, 126-127
Circle, of convergence, 143, 144, 146, 147, 171, 181
  of convergents, diagram of, 136
  for first inequality, 171
  of a series, 147
Circular apertures, 58
  plates, 58
Circulation, 344, 346
  of the vector, 23
  law, Bjerknes', 345
    Kelvins', 344
Classification of differential equations, 167

Clausius-Clapeyron formula, 388
Closed region, Laplacian vector field in, 132
Coefficients of the tensor, 43
Cohesion, 393, 428, 461
Coincidence of eigenvalues or degeneration, 106
Collision number, 415
Collisions, intermolecular, 400
Combination of eigenvalue equations, 107
Combined central and linear field, 59
Commutation law for complex multiplication, 133
Commutative and distributive law, 9
Comparison, series, 141
  with simpler eigenvalue equations, 112
Completeness, of a functional system, 89
  of hydrogen eigenvalue function, 198
  of Legendre polynomial systems, 189
  properties for solutions of eigenvalue equations, 115
  of a system in an infinite region, 116
    proof, 115
  of systems of functions, and the delta functions, 161
  relation, 89, 91, 119
Complete system of orthogonal functions, 117
Complex analytic function, definition of, 126
  conjugate numbers, 121
  exponents, 137
  integral, 128
  integrals and Cauchy's theorem, 128
  integration, and contour integrals, 150
    evaluation of definite integrals by, 150
  numbers, 121
    as vectors, 121
    in polar coordinates, 123
  number theory, 132
  plane, 121, 124, 125
  variable z, 128
Components of a function on the axis of an orthonormal system by least square deviation, 88
Composite transformations, multiplication of matrices, 70
Compressibility, 329
Compressible fluid, 334, 358
Compression, modulus of, 329, 339
Concept of the delta function, 161

Condensation, 370-372, 374, 389, 462
Conditionally convergent, 140, 147
  series, 143
    interchange of terms in, 141
Confluence of singular points, 177
Confluent hypergeometric equations, 180
  functions, 192 *et seq.*
Confocal rotational ellipsoids, 57
  hyperboloids, 57
Conjugate complex quantities, 73
Connection of analytic complex functions with vector analysis, 132
Conservative forces, 24, 266-290, 337
Constant density or tension function, 113
Constraint, forces of, 261-265
Constraints, 235
  holonomic, 261
  nonholonomic, 262
Contact forces, 217
  transformations, 291
Continuity, equation of, 334
Continuous eigenvalue spectrum, 198
  functions, 34
  parameter, orthogonality of function with, 163
  unit matrix, 162
Contour curve, 156
  integral, 130, 134, 155, 207
  integral encircling one or both singular points, 158
  integrals, 151, 153
  and complex integration, 150
Contraction, 315
Contravariant components, 50
Convergence, absolute and conditional, 140
  criterion of hypergeometric series, 143
  of infinite series, 140
Convergent contour integral, 182
  product integral, 85
Coordinate function, 83
  -momentum, 291, 431
  surfaces, 53
  systems, Galilean, 218, 253
  vectors, 83
  -velocity space, 431
Coordinates, curvilinear, 53
  generalized, 213, 265-269, 280-287, 431
  identification, 331
  instantaneous, 331, 332
  positional, 261, 266, 288

prolate and oblate rotational elliptic, 57
relative, 256
three-dimensional general, 53
Coplanar vector, 68
zero volume for, 65
Coriolis acceleration, 255
force, 256, 257
Corresponding true orthogonal functions, 85, 86
Covariant components, 50, 51
Criterion of absolute convergence, 142
Critical point, 395
pressure, 374, 394
temperature, 374, 394
volume, 374, 394
Crystal symmetry, 325
Curl, **A**, 23, 24
components, 24
as a true vector, 25
v=j, 133
Curvature and torsion of a curve, 17
Curvilinear coordinates, 53, 104
"Cut," 148, 151, 152
Cuts, between singular points, 126, 190
in the z plane, 149
Cutting of a series, 175
Cyclic permutation, 13
variables, 287, 291, 294

Dalton's law, 400
Damped oscillations, 228
Damping, 228
constant, 229
force, 228
Decomposition of symmetric and antisymmetric tensors, 44
Decrement, logarithmic, 229
Definition, of an analytic function, 125
of complex integrals, 128
of the gamma function, 154
of unit vector, 63
of unity of volume, 65
Deflection, 257-260
Deformation, 50
of paths of integration, 129, 152, 153
Degeneration of gases, 460-462
Delta function, 89, 162, 163, 188
and the completeness of systems of functions, 161
of fourier series, 164

of gegenbauer polynomials, 188
Delta-like function, 189
Density, and tension functions, positive, 85
of zeros, 114
Density function, 86, 90
Dependent variables, 112
Derivative, absolute, 332, 333
individual, 332, 333, 336
of function theorem integral, 131
local, 332
partial, 332
of tensor fields, 48
total, 332, 333
Determinant, 13, 111
and linear equation, 64
of a matrix, 64
minor, 67
multiplication, 71
of the tensor, 64
Development of determinants in terms of minors, reciprocal vectors, 67
Deviation, 257-260
Difference equation, 80
square integral, 88-89
Different expressions for the derivatives, 125
power series of analytic function, 145
Differential dz, 128
equation for the Laguerre polynomial, 197
equation of the second order, 176
equations, classification of, 167
with not more than two singular points, 176
ordinary, 167
partial, 167
and particular functions, 167
with three or more singular points, 177
variational problems leading to linear, 103
formula, 193
operative, vectorial, 25
operators, 200
principle, 273
solid angle, 37
Differentiation along the normal unit vector, 102
of vectors, 16
Diffraction law, 272
Diffusion, 418

Subject Index / 473

coefficient, 418, 421
   equation, 420
Dilation, 303
Dimension, 5
Dipole field, 37
Dirac definition, 162
   improper function, 162
Direct complex integration, 152
   derivation of the integrand, 131
   methods of solution, the Ritz method, 107-109
Discontinuities of solenoidal fields, 38
Discontinuous fields, 35
   series, 144
Discrete set of negative eigenvalues, 118
Dispersion, 329
Displacement, 263, 311
   and deformation, 49
   of the kth mass point, 79
   virtual, 263, 267
Distance force, 217
Distribution law, Boltzmann, 437
   Bose-Einstein, 455
   Fermi-Dirac, 454, 462
   Maxwell, 398, 407-415
Distribution of approximate eigenvalues by the Ritz method, 109
Distributive law, 9-10
   for vector products, 11
Div $\mathbf{v} = \rho$, 133
Divergence, and the Laplacian in total curvilinear coordinates, 55
   of a tensor, 48
   of a vector, 21
Duplication, formula, 160, 204
   and multiplication formulas, 159
Dyadic products, 9, 43
   of vectors, 42
Dyadics, 42

Efficiency, 381-385
Eguatorial angle, 18
Eigenfrequencies, 79, 80
Eigenfunctions, 85, 90, 107, 115, 119
   of more complete eigenvalue equations, 117
   of a Sturm-Liouville problem, 117
   of a vibrating homogeneous string, 86
Eigenvalue, 74, 75, 78, 85, 107
   equation, 106
   combinations of, 107

Sturm-Liouville, 90
   parameter, 90
   problem, Sturm-Liouville, 85
   problems, 100, 106
   variational problems with additional conditions, 103
   spectra, 196
   spectrum, 196
Eigenvalues, 90, 106, 110, 196
   asymptotic behavior of, 119
   and eigenfunctions, 106
   condensing at a finite value, 118
   number and distribution of, 112
   of a quadratic form, 73
Eigenvectors, 74, 75, 78
Eigenvibrations, 79, 80, 81, 82
Eikonal, 271
Elastic forces, 225, 313, 321, 322
   media, 316, 319
   oscillations, 225
Elasticity, coefficients of, 319
Electric oscillatory circuit, 224, 228, 233
Electromagnetic field, 277, 283, 284
Elementray curve and surface integrals, 26
   particles, 449
   permutations, 66
   theory of maxima and minima, generalization of, 91
Elements, of calculus of variations, 91
   of a matrix, 70
End points, a and b, 128, 129
   of chains, 96
Energy, equation, 221, 243
   exchange between molecules, 402, 407
   free, 388, 439, 443
   function, 281-283
   internal, 371, 376, 388, 399, 460
   kinetic, 241, 251, 268, 273, 280, 282-285, 299-302
   mean, 399
   mechanical, 241
   of motion, 241
   potential, 241, 269, 320
   principle, 95
   total, 243, 282, 284, 288
   zero-point, 462-465
Ensembles, 445
Enthalpy, 388, 390
Entropy, 369, 374, 378, 384-389, 398, 436, 447, 452

474 / *Subject Index*

Equation, inhomogeneous linear with vanishing determinants, 68
Equation of state, 373, 374, 425
  for degenerate Fermi gas, 465
  for ideal gas, 371, 373, 378, 425, 442, 461
  for molecular gas, 399
  for molecules of finite size, 427
  for non-ideal gas, 392, 448
  for van der Waals gas, 373, 393-396, 427, 449
  with cohesive forces, 428
Equilibrium equations connecting displacements and elastic forces, 41
Equipartition theorem, 404
Equipotential surfaces, 20
Equivalence of Gauss and Green theorems in a plane, 29
Essential singular point, 174
Euler constant, 158
  derviative, 97
  differential equation, 100
    boundary conditions for, 100
    of a variational integral with boundary conditions, 98
  equation, 95, 97, 98, 104, 105
    of a variational problem, 103
  hydrodynamical equations, 338
  integral, of the first kind, 158, 181
    of the second kind, 155
  partial differential equation, 102
  variational derivative, 100
Eulerian angles, 301
Evaluation of definite integrals by complex integration, 150
Evaporation, 386, 387
Exact eigenvalues, 109
  expansion, 188
Examples of power series, 136
Expansion, of a function, 188
  of functions by orthonormal systems, 88
  of gases, coefficient of, 371, 376, 390
  of hydrogen function, 197-198
  in power series, 147
Explicit power series, 146
Exponential function, 123
External forces, 215, 216, 225, 249, 250
  work performed by, 372
Extremum, 77
  value, 74

Factorial function, 155
  or gamma function, 154
Factor matrices, 71
Fahrenheit scale, 370
Falling motion, 220, 222, 257
Fermat's principle, 92, 95, 269
Fermi-Dirac distribution function, 454
  gas, 462
  statistics, 453, 460
Fields, scalar and vector, 19
First integral, 243, 281
First law of thermodynamics, 369-377
First solution of the Bessel equation, 201
Flow, 19
  curl-free, 340, 349
  around a cylinder, 351
  potential, 349
  around a sphere, 352
  stationary, 335, 340
  vortex, 343
Fluid, compressible, 334, 358
  frictionless, 336, 338, 340, 343
  homogeneous, 337, 340, 354, 355
  ideal, 340, 343, 346, 349, 352, 354, 358
  incompressible, 334, 336, 339-341, 345, 349
  inhomogeneous, 334, 344
  viscous, 336, 338
Flux, 19
  of curl, 30
  mean, 22
  total, 23
"Forbidden" use of n values, 110
Force, action of, 216-218
  central, 249, 426
  centrifugal, 256
  cohesive, 393, 428, 461
  conservative, 241, 274-290, 337
  contact, 217
  Coriolis, 256, 257
  damping, 228
  distance, 217
  elastic, 225, 313, 321-322
  electromagnetic, 277, 278, 283-285
  external, 215, 216, 225, 249, 250, 372
  generalized, 266, 269, 274, 279
  inertial, 267-269, 274, 279
  intermolecular, 425
  internal, 249, 264, 337
  position-dependent, 221, 225, 242

reaction, 256, 273
static, 218, 312
surface, 319, 321, 322
time-dependent, 221
unit of, 218
velocity-dependent, 221, 278
volume, 321, 327, 337
Forced oscillations, 78
Forces of constraint, 261, 264, 265
Formal definition of tensor, 41-42
  solution of Bessel equation, 206
Formalism, 127
Foucault pendulum, 259
Fourier integrals, 91
  series, 188
  transform, 164
Fourth and third approximation of an eigenvalue problem, 111
Freedom, degrees of, 220, 261, 405, 431
Free energy, 388, 439, 452, 459
  oscillations, 222, 225, 226, 229
  path, 415, 417, 419
Frequencies, 81
Frequency, 225, 228
  proper, 225, 230, 232
  resonance, 232
Friction, 216, 223, 229, 243, 338, 381, 422
  coefficient of internal, 338, 340, 423
Functional determinant, 54
  system, completeness of, 89
Function of functions, 92
  on the axis of an orthonormal system by
    least square deviation, 88
  space, 82; *see also* Hilbert space
Functions, analytic, 125
  isoparametric, 180 *et seq.*
  many valued, 148
  of Laguerre, integral representations and
    generating, 194
Fundamental equation in the calculus of
  variations, 100
  operations, 122
  vectors, 43

Gamma function, (factorial function), 154
  alternative definition of the, 156
  expressed by infinite products, 157
  functions, 181
Gas, Bose-Einstein, 461
  constant, 371, 378, 399, 404
Fermi-Dirac, 461, 462, 464
  ideal, 377, 399, 425, 442-460
  inhomogeneous, 417
  non-ideal, 390, 392, 448
  permanent, 370, 373
  photon, 457
  thermometer, 370-373, 377, 392
  van der Waals, 394
Gases, kinetic theory of, 214, 397
  thermal expansion coefficient of, 371, 390
Gauss, and Argand, 121
  curve, 162
  gamma function, 201
  and Green theorems, 27
  equivalence in a plane, 29
  Green and Stokes integral theorems, 26
  plane, 121
  and Stokes theorem, 134
  theorem, 22, 27
Gegenbauer functions, 189
  polynomials, 88, 186-188, 189
General features of variational problems, 103
  homogeneous boundary conditions, 107
  rules of convergence, 140
  theorems for analytic functions, 138
Generalization of the elementary theory of
  maxima and minima, 91
Generalized Cartesian coordinate systems, 61
  coordinates, 213, 265, 280-287, 431
  forces, 266, 267, 269, 274, 279
  Legendre functions, 85
    polynomial, 85
  momenta, 280-287, 431
  operator, 55
  vector space, 61
  velocities, 269, 431
  Wronskian, 202-203
Generating function (x,t), 185
  function for Legendre polynomials, 185
Geometric series, 143, 144-145
  theory of functions, 145
Geometrical addition of complex numbers, 140
  interpretation, 40
German literature, 6, 9
Gibbs' notation (original and modified), 9
  potential, 388

statistics, 398, 445, 452
Giorgi system, 5; *see also* Mks system; Mksa system
(Grad u)^2 of the integrand, 102
Gradient, of a scalar field, 20
 of a vector, 49
Gravity waves, 354
Green, method, 32
 theorem, 27, 106
Gyrocompass, 310
Gyroscope, 303

Half-curve, 93
Hamiltonian, 284-288, 293
Hamilton-Jacobi differential equation, 245, 290, 294, 431
Hamilton's equation, 213, 284-288, 431
 principle, 98, 213, 269, 275, 286, 292, 316
Hankel functions, (Bessel functions of the third kind), 203-204
Harmonic oscillations, 225, 235
 oscillator, 198, 225, 238
Heat, 214, 369, 375, 397
 amount of, 370-377
 engine, 381-386
 function, 398, 390
 mechanical equivalent of, 369, 375, 376
 specific, 375, 378, 422, 465
Helmholtz' theorem, 346
Hermite equation, 178
 polynomial, 180
 see also Hermitian
Hermite's polynomials, 199
 and functions, 198
Hermitian, 75
 forms, 73
Higher derivatives, 136
 eigenvalues, 196
Hilbert space, 62, 82, 83
 vectors in, 88
Holomorphic (analytic) functions, 150
Holonomic constraints, 261
Homogeneous boundary conditions, 90, 102
 equation in a regular point, solution by power series, 170
 equations, 78
 Euler equation, 102
 fluid, 337, 340, 354
 linear equations, 68

Hooke's law, 321
Hyperbolic functions, 95
Hypergeometric equation, Legendre equation, 179
 functions, 180 *et seq.*
 series, 179, 180
 integral representation of, 181
 surface, 101
 type polynomials, 85

Ideal fluid, 340, 343, 346, 349, 352, 354, 358
 gas, 371, 372, 377, 381, 399, 425, 442, 460
Identity of fields from a double layer and a bordering vortex line, 39
 transformation, 64, 72
Imaginary arguments, Bessel functions with, 206
 axis, 121, 125
 components, 121
 unit, 122
Imponderabilia, 369
Incomplete eigenvalue system, 118
Independent proof of the completeness relation, 118
 solutions of the Bessel equation, 202
 variable, 112
Individuality, 449
Inertia, 215, 219, 299
 law of, 215, 216, 219
 moment of, 247, 248, 299
 tensor, 299, 300
Inertial force, 256, 267-269, 281
Infinite number of eigenvalues, 114
 power series, 132
 products, gamma function expressed by, 157
 series, convergence of, 140
 of complex terms, 140
Infinity, 124
Inflexion point, 92
Inhomogeneous boundary conditions, 103
 linear equations, 168
 solution of, 67
 with vanishing determinants, 68
"Inner properties," 146
Inside point, 147
Integral exponent n, 126
 principles, 213, 273-275

representation, 207
  of hypergeometric series, 181
  of $P_n(z)$, 184
  representations and generating functions of Laguerre, 194
  theorem, Cauchy, 131
  variation principle, 101
Integrand, 103, 104
  (grad u)² of, 102
Integration in the complex s-plane, 204
Intensity of a point source, 35
Interchange, of factors and transformation of tensors, 43
  of terms in conditionally convergent series, 141
Intermixing positive and negative terms, 141
Intermolecular forces, 424
Invariants, of a quadratic form, 77
  of the tensor, 47
Inverse orthogonal transformation, 110
  transformations, 72
  and reciprocal matrices, 72
Inversion, temperature of, 393, 394
Irrational exponents, 126
Irregularity as a pole of the fourth order, 174
Irregular points, 172
  singular points, 138, 173
  singularity, 177
Irreversible process, 379-381, 384, 386
Irrotational field with vanishing curl, 29
Isentropic process, 379
Isentropics, 374
Isoparametric problems, 93
Isothermals, 374, 385, 395
Isotropic bodies, 326, 327

Joule-Kelvin experiment, 389

Kelvin, circulation theorem of, 344
  temperature scale, 371, 373, 391
Kinematics and dynamics of elastic bodies, 49
Kinetic energy, 61, 98, 241, 251, 268, 273, 280-285, 299-302, 399, 405
  theory of gases, 214, 397
Kronecker symbol, 161

Lagrange's equation, 268, 269, 275, 277, 281, 285, 301, 431

hydrodynamic equations, 338
method of indefinite multipliers, 105
multiplier, 97
Lagrangian, 269, 275, 277-281, 283, 286
Laguerre differential equations, 192
  function, 117, 180
  polynomials, 86, 116, 180, 193
  and orthogonal functions, 192
Lamellar and Solenoidal vector fields, 29
  field, 30
Lame's equation, 178
Laplace's equation, 32, 125, 126, 185, 336, 349
  in polar coordinates, 127
  integral, 184
  operator in spherical coordinates, 185
Laplacian in curvilinear orthogonal coordinates, 56
  fields and uniqueness of solution, 33
  operator, 59
  of a scalar, 55
  vector, 134
  vector field in a closed region, 132
Lattice symmetry, 325
Laurent series, 207
Least action, 213, 270
Legendre differential equation, 182
  equation, 178, 179, 190
  solutions of, 182
  functions, 184
  of the first kind, 182
  of the second kind, 100, 190
  polynomials, 83-84, 86, 88, 100, 179, 185, 186, 190
Length coordinate, 80
Lever, 263
Light path, 93
Limiting case of vibrating mass points, vibration of a continuum as, 79
Line integral, 28
  and normal elements, 129
  vortices, 38
Linear equations, 167
  determinants and, 64
  Euler equation, 102
  homogeneous boundary conditions, homogeneous and linear Euler equations and, 102
  homogeneous differential equations of the second order, 168

478 / Subject Index

transformation and matrices, 69
transformations, 70
 applied in succession, 70
 extremum and variational problems, 61
 of vectors, 41
 vector functions, 41
Liouville's theorem, 131-132, 157, 433
Log (-1), 148
Log z, 148
Logarithmic character of singularities, 190
 derivative of the gamma function, 159, 160
 functions, 138
 second solution, 173
 solutions, 172
Lorentz force, 277

Macrostate, 435, 450
Macrosystem, 445
Magnitude of eigenvalues, 114
Main axis of a symmetric tensor, 46, 47
Many-valued functions, 126, 148
Mass, 215-219, 249, 279
 point, 81
 transport of, 417
Mathieu's equation, 178
Matrices, linear transformation and, 69
Matrix, defined by n basic vectors, 64
 element of, 70
 product, 70
Matrix algebra, 69
 elements of the harmonic oscillator, 201
 multiplication, 71
 notation, 71
Maupertuis' principle (principle of least action), 97, 269-271
Maximum or minimum, absolute, 92
Maxwell's thermodynamic relations, 389
Mean square error, 115
Mechanical energy, 241
 equivalent of heat, 369
Melting, 386
Membrane, vibrating, 314, 318
Meromorphic functions, 150
Microstate, 435, 450, 454
Microsystem, 445
Minimum properties of eigenvalues, 111
Minor, determinants, 67
 development of determinants in terms of, 67

Mks system, 5
Mksa system, 5
Modified Laguerre polynomials, 198
Modulus, 123, 126, 140
 of a sum, 144
 $|z|=1$, 143
Molecular binding, 405
 oscillation, 239
 potential, 239, 240
Momentum, 216, 243, 249, 270, 280-287, 431
 angular, 247, 281, 300
 law, 216, 247, 250
 transport of, 417
Monotonously and from the upper sides, 109
Motion, constants of, 287, 291, 294
 energy of, 241
 quantity of, 215, 216
 relative, 219, 251, 253
Multiconstant theory, 324-327
Multiple k-value, 142
 products of vectors, 13
Multiplication in division by a vector, 133
 of vectors by scalars, 8
 theorem, 160
Mutually orthogonal eigenfunctions, 106
 orthogonal polynomials, 87
Mutual orthogonality, 83, 84
 orthogonalization, 107

$(n-1)^{\text{th}}$ Approximation, 110
Natural boundary conditions, 102
 path of a particle, 97-98
n-Dimensional space, volume in, 64
 vectors, 79
Nerst's theorem, 369, 456
Net flux, 22
Neumann function $N_m(z)$, 203
 Bessel function of the second kind, 203
Newton's binomial coefficients, 136, 137
 first law, 215, 219
 second law, 41, 216, 217, 219, 249, 253, 279
 third law, 217
Nominators, 110
Noncomplete systems of eigenfunctions, 117
Nondivergent, 31
Nonlinear equations, 94, 167

Subject Index / 479

Nonorthogonal functions, 109
Non-polar binding, 239
Nonuniform convergence, 144
Normal element, 132
 unit vector, differentiation along, 102
Normalization, 87
 as an additional condition, 104
 integral, 84, 184
Normalized expression, 88
 function, 105, 115
 solution, 47
North and south pole (N and S), 124
Notation, 6
 Re and Im, 121
n-Particle system, 79-80
nth-Order determinant equation, 109
 secular equation, 109; *see also* Secular equation
Number and distribution of eigenvalues, 112
 of roots, 111
Numerical evaluation of determinants, 69
Numbers, complex, 121
 zero and infinity, the unit circle and the number sphere, 124
Nutation, 304
 of the Earth, 255, 307

Oblate elliptic coordinates, 58
Occupation number, 438, 451-454
Operations with complex numbers, 122
Operators, 70
Operator $\nabla \mathbf{x}$, 48
Order of magnitude of an integral, 132
 of simultaneous equation, 167
Ordinary differential equations, 167
Oriented closed curve, 27
Orthogonal approximate eigenfunctions, 112
 functions, 63
 polynomials, 83, 84
 properties of eigenfunctions, 91
 system of polynomials, 79, 91
 transformation, 73, 77, 109
 unit vectors, 10
 vectors, 8, 18, 102
Orthogonalities, sense of, 85
Orthogonality of functions with a continuous parameter, 163
 properties of aromic spectra, 197
Orthonormality, 90, 108

 of the orthonormal Laguerre functions, 194
 relations, 119
Orthonormal polynomials, 90
 relation, 90
 system, 63, 188
 expansion of functions by, 88
 with a density function, 108
 vectors, 47, 51, 52, 56, 75
 set of, 64
Oscillations, aperiodic, 229
 damped, 222, 228-233
 elastic, 225
 forced, 222, 225, 226, 230-232
 free, 222, 225, 226, 229
 harmonic, 225, 235
 molecular, 239
 monochromatic, 225
 periodic, 229
 proper, 225, 226
 undamped, 222, 230
Oscillator, anharmonic, 239
 harmonic, 225, 443
Oscillatory circuit, electric, 224, 227, 232
Osculating plane, 18
Outward vector flux, 55

Parabolic coordinates, 58
 transformation from Cartesian, 59
Parallelepiped volume, 50, 54
Parameters of position, 261, 266
Partial differential equations, 167
 in several independent variables, 101
 integration, 108
 pressures, 400
Particular functions, differential equations and, 167
Passage from real to complex numbers, 122
Path of integration, 128
 deformation of, 129
Pauli principle, 453, 462
Pendulum, cycloid, 237
 Foucault, 259
 mathematical, 234
 physical, 235
Periodic peaks of area, 165
Perpetual motion machine of the first kind, 380
 of the second kind, 380, 381
Phase, 225

cells, 441, 433, 435, 450-457
  integrals, 289
  shift, 232
  space, 291, 431, 452
  in thermodynamics, 386
  velocity, 359
Photon gas, 457
Pillbox-shaped surface, 36
Planck's constant, 435, 444
  radiation law, 458
Planetary motion, 213, 239-241, 251, 252
Point and surface double sources, 36
  doublet, 36
Poiseuille's law, 364
Poisson equation, solution of, 32
Polar coordinates, 126, 136
Pole of the first order, 172
  of the first or higher orders, 151
  of the fourth order, irregularity as, 174
Polynomials, 34
  Gegenbauer, 88
  of the hypergeometric series, 85
  of the nth degree, 193
  orthogonal, 83
  Tschebyscheff, 88
Position-dependent force, 221, 225, 242
Positive definite integrals, 104
  definite quadratic forms, 76
  density and tension functions, 85
  density function, 85
  eigenvalues, 77, 118
  or negative power of z^n, 126
  normal of a surface area, positive direction on the limiting curve, 11
Postfactors, 42
Potential, 242, 269, 274, 337, 349, 426
  energy, 242, 269, 318, 405
  flow, 349
  Gibbs', 388
  of an irrational field, 38
  molecular, 240
  theory, 84
  thermodynamic, 373, 388, 389
  velocity, 336
Power, 270
  series, analytic continuation of, 145
    at singular points, 172
    solution of differential equations by, 167
    theories for analytic functions, 134

Precession, 303
  of the Earth, 255, 307
Prefactor, 15, 42
Pressure, 217, 370-374, 388, 425, 435, 439, 461
  critical, 394
  hydrostatic, 337, 398, 427
  waves, 358
  zero-point, 462, 465
Principal axis, transformation on, 75, 76
  normal, 18
Principle of least action (Maupertuis' principle), 97
Probability, thermodynamic, 398, 436-456
Product determinants, 71
  matrix, 70
  operator, 49
  sum, 71
Projection of the gradient vector, 20
Prolate and oblate rotational elliptic coordinates, 57
Proof of completeness of a system, 115
  of convergence of nonuniform series, 145
  of reflection formula, 161
Proper frequency, 225, 226, 230-232
  oscillations, 225, 226
Pseudoscalar, 13
Pure number, 5

Quadratic equation, 100
  forms, 73, 104
  and their stationary values, 73
Quadratures, 169
Quantitative variation and derivation, 101
Quantum statistics, 398, 435, 449-457
  theory of atomic orbits, 153

Radiation law, 457
Radicand, 154
Radius of convergence, 135, 139, 143, 146
Rank of tensor fields, 48
Rariconstant theory, 324-327
Rate of strain, 338
Rational exponents, 126
Ratio of two consecutive terms, 142, 143
Ray of light, 92
Reaction forces, 256, 273
Real axis, 121, 125, 149
  components, 121
  irrational number, 149

Re and Im notation, 121, 147, 155
Reciprocal basic vectors, 51, 52
  function of determinant, 54
  matrices, 72
  vector systems, 50, 52
Recurrence formula, 192, 193
Reflection formula, 157
  for symmetric factors, 160
Refraction index, 93
Refrigeration machines, 390, 394
Region, 83
  arbitrarily chosen fundamental, 84
  of convergence, 171
Regular point, 135, 170, 172
  singular point, 174, 178
Relative acceleration, 255
  coordinates, 256
  displacement, 49
  motion, 219, 253, 259
  velocity, 254, 402, 414
Relativity, theory of, 213-215, 279, 283, 284
Remaining sum after nth term, 145
Resonance, 227, 231, 232
  frequency, 232
Residue, 151, 153
Residum, 151, 153
Resonance, 78
Restriction of choice of trial functions, 104
Reversed vector, 7
Reversible process, 379-381, 384
Right-hand system, 11
Rigid bodies, 300
Rigidity, 329, 337
  modulus of, 329
Ritz method, 109
  direct methods of solution, 107-109
  distribution of approximate eigenvalues by, 109
Rodriques formula, 184, 191
Rotation, 299, 301-303, 313, 338, 345, 406
  of coordinate systems, 61
Rotational velocity, 14
Rows and columns in determinants, 71

Scalar, 5
  multiplication, 13
  product, 9, 42, 90
  products, triple, 14
  surface element, 27

Scattering theory of light and particles, 58
Schafli's integral, 184, 186, 190
Schwartz's inequality, 62
Scientific system of units, 218
Secondary normal, 18
Second law of thermodynamics, 369, 378-386
  order, differential equation of, 105, 176, 177
  equations, 100
  orthogonal transformation, 77
Secular equation, 110; see also nth-Order secular equation
Self-adjoint form, 105, 112, 199
Semiconvergence, 175
Separation constant, 127
  into an irrotational and solenoidal field, 32
  of variables, 294
Series, hypergeometric, 180-181
Set of orthonormal vectors, 64
  of scalars, 61
Severl dependent variables, simultaneous Euler equations, 101
  independent variables, partial differential equations, 101
Shear, 45
Simpler eigenvalue functions, comparison with, 112
Simplest example of a continuous function, 125
Simultaneous differential equation, 101, 167
  Euler equation, several dependent variables, 101
  transformations, 76
Sine function, 81
  z (define), 138
Single discontinuity, 35
  k-value, 142
  value functions, 126
Singularities, 90, 100
  at infinity, 132
  irregular, 177
  limit boundary curve displacement, 135
Singular point, 90, 148, 152
  surpass, 130
Singular points, 130, 135, 137, 138, 139
  of an analytic function, 126
  arrangement of differential equations ac-

cording to, 176
differential equations with not more than two, 176
differential equations with three or more, 177
irregular, 173
power series at, 172
Six-, nine-, and twelve- dimensional coordinate systems, 61
Smaller moduli, 141
Snell's law, 93
Solenoid, 30
Solid angle, 35
Solution of differential equations by power series, 167
of eigenvalue equations, completeness properties for, 115
of a given variational problem, 98
of inhomogeneous eigenvalue equation, 78
of inhomogeneous linear equation, 67
of Legendre equations, 182
logarithmic, 172
by power series, the homogeneous equation in a regular point, 170
Sound waves, 330, 358
Sources and sinks, 35
of a field, point and surface divergence, 35
of positive and negative sign, 35
Source strength, 37
Space vectors, 16, 79
Specified functions of general coordinates, 53
Spherical convergence, 135
coordinates, 56
harmonics, 34, 85
"Spur," 48, 50, 78
Square-differential line element, 59
Standard hypergeometric series, 181
Stark effect, 59
Starting point in the theory of Bessel functions, 207
State, sum of, 438, 442, 459
State diagrams, 374, 385
quantities, 372, 377, 379
variables, 372
Statement of problems, function of functions, 91
States, 372
Static force, 218, 312
Stationary atomic energy states, 197

of the integral, 94
values, 92, 108
Statistical ensemble, 445
Statistics, Boltzmann, 445, 450, 456
Bose-Einstein, 454-462
Fermi-Dirac, 453-463
Gibbs, 445-449
Stirling's formula, 437, 440
Stokes theorem, 24, 28, 30, 129
Strain, 311, 314, 321
components of, 319-321, 338
rate of, 338
tensor, 50, 312
Stresses, 319, 322, 337, 338
Stress tensor, 322, 338
Strictly identical poles, 156
String, vibrating, 314-318
Strongest boundary conditions, 103
Sturm-Liouville eigenvalue equation, 90, 91
eigenvalue problem, 85
equation, 115, 168
Surface forces, 319, 321-323
integral, 29
tension, 357
vortices, 38
Surpass, a singular point, 130
Symmetric functions, 188
matrix, 73
tensor, 43, 44, 45, 73
Symmetry and boundary, 188
Systems in an infinite region, completeness of, 116

Tangential component, 24
Temperature, 370-374, 377, 399, 439, 446
absolute, 378, 390-392, 456
critical, 374, 394
of inversion, 393, 394
scale, 370, 378, 381
absolute, 371, 372, 376
centigrade, 370, 371
individual, 371, 377, 378, 381
Kelvin, 371, 373, 391
thermodynamic, 371, 373, 377, 399
Tension, as product of two symmetric shears, 46
Tensors, 63
addition and subtraction of, 44
coefficients of, 43
field, 41, 48

fields and their derivatives, 48
general definition of, 43
interchange of factors and transformation of, 43
invariants of, 47
main axis of symmetric, 46
second rank, 6
and shears, 44
symmetric, 44
and tensor products, antisymmetric, 46
transposition of the, 43
Theorem of residues, 150
Theorems, variational, 118
Theory of determinants, 66
of eigenvalues, 103
of Fourier series, 144
of functions, 121
of the gamma function, 157
of residues, 194
Thermal conduction, 370, 378, 381
coefficient of, 422
equation of, 422
Thermodynamic equilibrium, 379-400, 407, 436, 446, 452, 462
potentials, 373, 388, 389
probability, 398, 436, 450-455
quantities, 372, 447
relations, 389
Thermodynamics, 369, 397
Thermometer, gas, 370-372, 377, 392
hydrogen, 370, 371, 373
liquid, 370
mercury, 370
Third law of thermodynamics, 364, 457
Three dimensional general coordinates, 53
Three or more eigenfunctions, 107
Tidal motion, 256
Time-dependent deformation of a body, 49
force, 221
Torque, 250, 300
Torques, law of, 247-251, 299-300
Torricelli's law, 341
Torsion, 18
Total number of eigenvalues, 114
solid angle, 38
Trajectory, 245, 294
Transformation, canonical, 247, 287, 288, 291-294, 431
from Cartesian to parabolic coordinates, 59

contact, 291
coordinate, 247, 266-268, 293
function, canonical, 292-294, 431
Galilean, 219, 220
identity, 293
Lorentz, 214, 219, 220
matrix, 76
orthogonal, 73
on the principal axis, 75
relativistic, 214, 219, 220
unitary, 73
Transformations, composite and multiplication of matrices, 70
Transposed matrix, 64
Transposition of the tensor, 43
Transversal forces, 79
Trial function, 98
Triple scalar product, of vectors, 13
True eigenfunctions, 112
orthogonal density function, 85
Tschebyscheff polynomials, 87, 88
Two-dimensional Gauss theory, 129
Two-term recurrence formula, 199

Undamped oscillations, 222, 231
Uniform convergence, 144
Unique and continuous derivative, 125
Unit, 5
fundamental, 5
matrix, 72
quadratic forms, 73
vector, 12, 17, 63
Unitary matrix, 63-64
transformation, 73
Units of measurement, CGS, 218
absolute, 218
technical, 218
Unity of volume, definition of, 65
Upper limit of integrals, 94
side, monotonously and from the, 109

Value of $F(a,b,c,1)$, 182
of the integral, 150
Variable action, 289
cyclic, 287, 291, 294
state, 372
upper limit of integrals, 94
Variational integrals, 101, 103, 105, 108
with boundary conditions, Euler differential equation of, 98

problems, 94, 116
  with additional conditions, eigenvalue problems, 103
  leading to linear differential equations, 103
  theorems, 118
Variation of a function, 98
  of derivatives, 99
Variations, 264
Vector, 6
  addition and decomposition, 7
  addition and subtraction, 7
  analysis, connection with complex analytic functions, 132
  components, nonorthogonal coordinate systems, 50
  curve element, 132
  fields, determination from divergence and curl, 31
    lamellar and solenoidal, 29
  flux, outward, 55
  functions, linear, 41
  generalized space, 63
  multiplication, 13
  notation, usage of, 121
  potential and curl, 30, 31
  product, 10, 13
  transformation, characteristics of, 44
  triple products, 14
Vectors, axial, 46
  coplanar, 68
  curl of, 23
  as dependent variables, 16
  differentiation of, 16-17; see also Differentiation of vectors
  divergence of, 21
  dyadic products of, 42
  from fixed to variable point, 133
  gradient of, 49
  in Hilbert space, 83, 88
  linear transformations of, 41
  normal, 18
  in n-space, 61
  parallelepiped, 13
  polar and axial, 14
  reciprocal, 67
  in a space of infinite dimensions, orthogonal system of polynomials and functions, 79
Velocities, mean, 399, 413, 423

directional distribution of, 400
Velocity, angular, 254, 345
  and distance traversed by a ray of light, 93
  distribution law, 398, 400, 407, 410, 413, 415
  generalized, 269, 431
  phase, 359
  potential, 336
  relative, 254, 402, 423
Velocity-dependent field, 331, 347
  force, 221, 278
Vena contracta, 342
Very simple extremum problem, 109
Vibrating homogeneous string, eigenfunction of, 86
  membrane, 314, 318
  string, 79, 314-317
Vibrational problems, 103
Vibration of a continuum as a limiting case of vibrating mass points, 79
Virial, 424-426
  theorem, 424, 427, 458
Virtual displacements, 263, 267
  work, 261, 263, 267
Viscosity, 338, 422
Viscous fluids, 336, 338
Volume, critical, 374, 394
  element, elementary volume, 21
  force, 321, 327, 337
  in n-space, 64
  invariance of, 411, 431, 435
  by vector components, 14
Vortex, 345-349
  filament, 345
  homogeneous, 347
  lines, 345
  rectilinear, 347
  strength, 346
Vortices, line, 38
  surface, 38
Vorticity, 339, 340, 349

van der Waals gas, 393
  equation of state for, 373, 393-396, 429, 449
Wave front, 271
Wave mechanics, 192
Waves, curl-free, 314, 329
  dipole, 359
  divergence-free, 314, 329

gravity, 354
in isotropic media, 327
longitudinal, 314, 329, 359
pressure, 358
sound, 330, 358
spherical, 359
transverse, 314, 329
Weakest boundary condition, 103
Weierstrass, 146, 158
product, 159
Weight, 217, 218, 222
Wessel, C., 121
Work, 217, 218, 242, 251, 267, 268, 299

virtual, 261, 263, 265
performed by gases, 375, 376, 382
Wronskian, 206
determinant, 191
of second-order equation, 168
of a solution, 169

Zero fields, 33
point,
absolute, 371, 372, 392, 462
energy, 462, 465
pressure, 462, 465
volume for coplanar vector, 65